AF334104

Alberta Oil Sands: Energy, Industry and the Environment

Developments in Environmental Science

Alberta Oil Sands:
Energy, Industry and the Environment

Kevin E. Percy
Wood Buffalo Environmental Association (WBEA),
Fort McMurray, Alberta, Canada

AMSTERDAM · BOSTON · HEIDELBERG · LONDON · NEW YORK · OXFORD
PARIS · SAN DIEGO · SAN FRANCISCO · SYDNEY · TOKYO

Elsevier
The Boulevard, Langford Lane, Kidlington, Oxford, OX5 1 GB, UK
Radarweg 29, PO Box 211, 1000 AE Amsterdam, The Netherlands

First edition:
© 2013 Elsevier Ltd. All rights reserved.

British Library Cataloguing in Publication Data
A catalogue record for this book is available from the British Library

Library of Congress Cataloging-in-Publication Data
A catalog record for this book is available from the Library of Congress

For information on all **Elsevier** publications
visit our web site at store.elsevier.com

Printed in China
13 14 15 16   12 11 10 9 8 7 6 5 4 3 2 1

ISBN: 978-0-08-099443-7

Working together to grow
libraries in developing countries
www.elsevier.com | www.bookaid.org | www.sabre.org

ELSEVIER   BOOK AID International   Sabre Foundation

Contents

13. WBEA Receptor Modeling Study in the Athabasca Oil Sands: An Introduction   311

S. Krupa

14. Method for Extraction and Multielement Analysis of *Hypogymnia physodes* samples from the Athabasca Oil Sands Region   315

E.S. Edgerton, J.M. Fort, K. Baumann, J.R. Graney,
M.S. Landis, S. Berryman, and S. Krupa

15. Coupling Lead Isotopes and Element Concentrations in Epiphytic Lichens to Track Sources of Air Emissions in the Athabasca Oil Sands Region   343

J.R. Graney, M.S. Landis, and S. Krupa

16. Mercury Concentration and Isotopic Composition of Epiphytic Tree Lichens in the Athabasca Oil Sands Region   373

*J.D. Blum, M.W. Johnson, J.D. Gleason, J.D. Demers,
M.S. Landis, and S. Krupa*

Contents

Contributors

Numbers in Parentheses indicate the pages on which the author's contributions begin.

K. Baumann (315), Atmospheric Research & Analysis, Inc., Cary, North Carolina, USA

S. Berryman (315), Integral Ecology Group Ltd., P.O. Box 23012, Cook St. RPO, Victoria, British Columbia, Canada

J.D. Blum (373), Earth and Environmental Sciences, University of Michigan, Ann Arbor, Michigan, USA

L.-W.A. Chen (145, 171), Division of Atmospheric Sciences, Desert Research Institute, Reno, Nevada, USA

J.C. Chow (145, 171), Division of Atmospheric Sciences, Desert Research Institute, Reno, Nevada, USA

T. Dann (47), Ottawa, Ontario, Canada

M.J.E. Davies (267), Stantec Consulting, Calgary, Alberta, Canada

J.D. Demers (373), Earth and Environmental Sciences, University of Michigan, Ann Arbor, Michigan, USA

E.S. Edgerton (315), Atmospheric Research & Analysis, Inc., Cary, North Carolina, USA

V. Etyemezian (145), Division of Atmospheric Sciences, Desert Research Institute, Reno, Nevada, USA

J.M. Fort (315), Atmospheric Research & Analysis, Inc., Cary, North Carolina, USA

J.D. Gleason (373), Earth and Environmental Sciences, University of Michigan, Ann Arbor, Michigan, USA

J.R. Graney (315, 343, 427), Geological Sciences and Environmental Studies, Binghamton University, Binghamton, New York, USA

M.C. Hansen (47, 93), Wood Buffalo Environmental Association, Fort McMurray, Alberta, Canada

D.R. Jaques (219), Ecosat Geobotanical Surveys Inc., North Vancouver, British Columbia, Canada

R.K.M. Jayanty (391), RTI International, Post Office Box 12194, Research Triangle Park, Durham, North Carolina, USA

M.W. Johnson (373), Earth and Environmental Sciences, University of Michigan, Ann Arbor, Michigan, USA

S.D. Kohl (145, 171), Division of Atmospheric Sciences, Desert Research Institute, Reno, Nevada, USA

S. Krupa (311, 315, 343, 373, 391, 427, 469), Plant Pathology, University of Minnesota-Twin Cities, St. Paul, Minnesota, USA

M.S. Landis (315, 343, 373, 427), US EPA, Office of Research and Development, Research Triangle Park, Durham, North Carolina, USA

A.H. Legge (93, 113, 171, 193, 219), Biosphere Solutions, Calgary, Alberta, Canada

M. Lowey (35), Institute for Sustainable Energy, Environment and Economy (ISEEE), University of Calgary, Calgary, Alberta, Canada

B. Mayer (243), Department of Geoscience, University of Calgary, Calgary, Alberta, Canada

D.G. Maynard (193), Canadian Forest Service, Pacific Forestry Centre, West Victoria, British Columbia, Canada

E.M. Nosal (93), Wood Buffalo Environmental Association, Fort McMurray, Alberta, Canada

M. Nosal (93), Department of Mathematics and Statistics, University of Calgary, Calgary, Alberta, Canada

R.J. O'Brien (113), VOC Technologies & Portland State University, Damascus/Portland, Oregon, USA

R.L. Orbach (1), The Energy Institute, The University of Texas at Austin, Austin, Texas, USA

J.P. Pancras (427), Allion Science and Technology, Research Triangle Park, North Carolina, USA

K.E. Percy (47, 113, 171, 193, 427, 469), Wood Buffalo Environmental Association, Fort McMurray, Alberta, Canada

B.C. Proemse (243), Department of Geoscience, University of Calgary, Calgary, Alberta, Canada

J.H. Raymer (391), RTI International, Post Office Box 12194, Research Triangle Park, Durham, North Carolina, USA

D.A. Sodeman (171), County of San Diego Air Pollution Control District, San Diego, California, USA

R.K. Stevens (427), Cary, North Carolina, USA - Formerly with U.S. EPA Office of Research and Development, Research Triangle Park, North Carolina, USA

G. Stringham (19), Canadian Association of Petroleum Producers, Calgary, Alberta, Canada

W.B. Studabaker (391), RTI International, Post Office Box 12194, Research Triangle Park, Durham, North Carolina, USA

X.L. Wang (145, 171), Division of Atmospheric Sciences, Desert Research Institute, Reno, Nevada, USA

J.G. Watson (145, 171), Division of Atmospheric Sciences, Desert Research Institute, Reno, Nevada, USA

Acknowledgments

The Wood Buffalo Environmental Association (WBEA) is an independent, not-for-profit organization that operates through consensus among its membership. Sincere appreciation is extended to the WBEA nongovernmental, governmental, aboriginal, and industry members who from 2008 have supported this work and continue to do so. Financial support for the May 23 International Symposium "Alberta Oil Sands: Energy, Industry and the Environment" and the associated 43rd International Air Pollution Workshop was provided by the following WBEA industry members: Suncor Energy Inc., Syncrude Canada Ltd., Shell Albian Sands, Canadian Natural Resources Ltd., Nexen Inc., Imperial Oil, Total E&P Canada, Devon Canada, Husky Energy, MEG Energy, Conoco Phillips Canada, and Williams Energy, with additional assistance from Finning Canada.

The key role of my WBEA science advisor colleagues Drs. Allan Legge (Biosphere Solutions, Calgary, AB) and Douglas Maynard (NRCan-Canadian Forest Service, Victoria, BC) in monitoring program design, as well as the science advisory contributions from our university/agency colleagues Drs. Sagar Krupa (University of Minnesota, USA), Robert Stevens (retired U.S.EPA, USA), Dale Johnson (University of Nevada, USA), Tom Nash III (University of Wisconsin, USA), Sandy McLaughlin (retired Oak Ridge National Laboratory, USA), Mike Miller (Argonne National Laboratory, USA), Ted Hogg (NRCan-Canadian Forest Service, Victoria, BC), Ken van Rees (University of Saskatchewan, SK), Suzanne Visser (University of Calgary), and Neil Cape (Centre for Ecology and Hydrology, UK) is gratefully acknowledged.

The WBEA member committees have played an important role in providing regional context, scoping, and oversight of the work contained within this book. The WBEA Terrestrial Environment Effects Committee and its Science Subcommittee are especially recognized here, as they have enabled most of the scientific content presented. The role of the former WBEA Executive Director Carna MacEachern for her early vision in seeing the need for the new science represented in this book cannot be underestimated.

Appreciation is extended to all the peer reviewers whose incisive and constructive comments enhanced the final quality of the chapters. The editor duly acknowledges the dedicated, expert, and diligent editorial assistance provided by Dr. Sagar Krupa, which made the publication of this book a reality in 2012.

Kevin E. Percy

Since the Industrial Revolution began in the 1750s, humans have increasingly depended upon the burning of fossil fuels for essential activities such as heating, cooking, manufacturing, electricity, and transportation. The Organization of Economic Cooperation and Development reported that, since 1800, the world's per capita income has increased 10-fold, while the world's population has increased over sixfold. According to the Nobel Laureate Robert E. Lucas, Jr., "... the living standards of the masses of ordinary people have begun to undergo sustained growth ... Nothing remotely like this economic behavior has happened before." To fuel this growth, the world initially turned to coal and, post WWII, oil. In 1900, global crude oil (petroleum) production was minimal, rose steadily to approximately 10 Mb/d (barrels per day) in 1950, and has risen exponentially ever since. Between 2003 and 2011, annual global oil and condensate production has ranged between 72 and 74 Mb/d, despite concerted efforts made on efficiencies in combustion technology. Against this backdrop, global energy demand is expected to grow by 39% by 2030, or 1.6% annually. By 2030, BP (http://www.BP.com) predicts that coal, oil, and gas together will contribute to meeting most of this demand.

The Canadian Oil Sands deposits are located in the Peace River, Cold Lake, and Athabasca deposits in Alberta and Saskatchewan (Figure 1). Together, they comprise the third largest oil reserve in the world, with 170 Bb recoverable using today's technology. Eighty percent of the oil can be recovered by *in situ* drilling and 20% by mining where deposits are sufficiently close to the surface. Canada's oil sands currently contribute 1.5 Mb/d toward meeting this global demand. By 2030, this is expected to have increased by 2 Mb/d. The oil sands are predicted to contribute 12% of the global production increase expected to have occurred by 2030!

With any large-scale industrial development activity comes the challenge of managing it in a sustainable manner, in order to reduce its impact on the environment. Alberta's oil sands lay under 142,220 km^2 of Boreal Forest. Until the 1960s, this area was largely undisturbed. In recent years, the development in the Athabasca Oil Sands Region (AOSR), in particular, has come under intense public scrutiny. During 2010 and 2011, the federal and provincial governments convened expert panels to examine current monitoring activities and to recommend enhancements. These reports were made public and a *Joint Canada/Alberta Implementation Plan for Oil Sands Monitoring* (http://www.ec.gc.ca; http://www.environment.alberta.ca) has emerged.

FIGURE 1   A map of Alberta Oil Sands Region (AOSR). *Courtesy of: Wikipedia.*

It is noteworthy that one monitoring organization in the AOSR embarked on a scientific enhancement of its activities several years prior to the panels being convened. The Fort McMurray-based Wood Buffalo Environmental Association (WBEA; http://www.wbea.org) traces its origins back to 1985, when the Air Quality Task Force was established to address environmental concerns raised by the Fort McKay First Nation (aboriginal community).

In 1997, WBEA was officially incorporated under the Alberta Societies Act. Today, WBEA operates as a consensus-driven, independent, community-based not-for-profit association with 29 members representing nongovernmental, governmental, aboriginal, and industry sectors. Its membership along with a professional staff and science advisors plan, execute, and oversee large air-quality and terrestrial ecosystem monitoring programs, along with human exposure monitoring currently focused on odors.

Recent federal and provincial expert panels have positively recognized WBEA for its scientific underpinnings and capacity in environmental monitoring, evaluation, and reporting. In mid-2007, WBEA members supported a review of the terrestrial monitoring program. This external scientific review led to a proposal from science advisors to members in the fall of 2007 for a major scientific enhancement in both the extent and the intensity of monitoring activities. With a significant funding increase on January 1, 2008, which has been sustained since, WBEA has built an international, multidisciplinary scientific team of over 35 senior scientists who are measuring and monitoring the environment at key points along the air pollutant pathway.

This volume presents the results from a number of the WBEA projects conducted as part of the strategic science enhancement begun in 2008. We believe that this *practical science* will contribute needed new knowledge and information, and inform stakeholders, the public, and decision makers engaged in air-quality regulation and environmental management.

Kevin E. Percy
Book Editor

The Oxford Dictionary (http://oxforddictionaries.com) defines the adjective *practical* as "... concerned with the actual doing or use of something rather than with theory and ideas; or, ... likely to succeed or be effective in real circumstances; feasible." In January 2008, the Wood Buffalo Environmental Association (WBEA) embarked on a multiyear enhancement of its environmental monitoring activities. Projects enabled by a threefold, step-change increase in funding from WBEA members were at the outset designed to undertake *practical*, science-based measurement and monitoring. Between 2008 and 2011, contracted principal investigators from Canadian, U.S., and European universities, research labs, and governmental agencies carried out a number of projects in the Athabasca Oil Sands Region of northeastern Alberta, Canada. Critical infrastructure and staff were put in place by WBEA in support of air-quality and forest-health monitoring, and the determination of potential effects of air emissions from oil sands development on the boreal ecosystem.

On May 23, 2011, in Fort McMurray, Alberta, Canada, WBEA hosted the International Symposium "Alberta Oil Sands: Energy, Industry and the Environment" in conjunction with the 43rd International Air Pollution Workshop (May 24–26). This book comprises chapters with content presented at the symposium, as well as others representing selected WBEA projects initiated under the 2008 science enhancement. Bound together, these chapters provide original scientific data on emissions, transport, deposition, and source contributions to terrestrial receptors. This information will be used by decision makers on air-shed management, contribute new knowledge to support environmental impact assessments, inform stakeholders and the public on air quality, provide guidance on what has been learned since 2008 and what is still needed to achieve a more holistic evaluation of the role of oil sands development on the air and terrestrial environments.

This book is organized into six sequential groupings of chapters followed by a summary chapter. We begin with three chapters that review energy production and the history and place of Canada's oil sands in that production and provide a summary of environmental challenges being addressed by the oil sands producers. Next, we move to the three chapters that report on air quality, new approaches to processing air-quality data, and new technology being used to co-measure odor-causing compounds. Two chapters follow that report on "real-world" characterization of emissions from fixed (stacks) and mobile (mine heavy haulers) sources operating in the region. Then, the reader will find three chapters that describe a now-deployed deposition/ecosystem-based approach to tracing the fate of emissions and assessing the health of boreal forests that receive those emissions. Chapter 12 that follows summarizes some 30+ years of regional source plume dispersion modeling carried out within the AOSR, reports on a recent model run, and relates model predictions to actual measured WBEA receptor data.

Chapters 13–18 comprise the WBEA Receptor Modeling (Source Apportionment) Project that uses original receptor and source data to link the air-quality and terrestrial systems. This section begins with the introductory Chapter 13 authored by the project leader and providing the rationale, theory, and practice for the five interlinked studies. Chapter 14 evaluates the utility of naturally occurring epiphytic tree lichens in modeling approaches and presents new methodologies developed to measure a wide range of heavy metals (e.g., As) in the AOSR. Chapters 15 and 16 report on novel studies in which stable lead and mercury isotopes analyzed in the lichens across the region have been used to attribute receptor concentrations and to discriminate between natural and anthropogenic sources. Chapter 17 reports novel data on PAH concentrations in lichens and their potential use in source attribution. Chapter 18 concludes this grouping and brings individual study results into the wider context. For the first time in the Athabasca Oil Sands Region, concentrations of sulfur, nitrogen, and trace elements measured in lichens at 359 sites extending out over 150 km from oil sands plants and mines have been apportioned back to source type and contributions to deposition estimated in a scientifically defensible manner.

In 2005, Dr. Raymond L. Orbach (later undersecretary for science, U.S. Department of Energy), delivered an address to the annual meeting of the American Association for the Advancement of Science (AAAS) in which he stressed the need for "scientific literacy." This address eloquently presented the case for what I have termed above *practical science*. It is logical then that Dr. Orbach was invited to be the first speaker at the May 23 Symposium and to be the author of the first chapter in this book.

The concluding chapter entitled "Concluding Remarks" attempts to put the book's content into context. While WBEA members can be justly proud of the contents of this book, they also recognize that this represents one contribution to improved understanding of the environmental effects from oil sands development. The book, therefore, concludes with some insights by several chapter authors and discussion gleaned from the closing symposium panel discussion to summarize key findings in light of what has been learned, the gaps, challenges, and work remaining.

Kevin E. Percy
Book Editor

Environmental pollution has played a critical role in human lives since the early history of the nomadic tribes. During the past millennium, the Industrial Revolution, increased population growth, and urbanization have been the major determinants in shaping our environmental quality.

Initially, primary air pollutants such as sulfur dioxide and particulate matter were of concern. For example, the killer fog of London in 1952 resulted in significant numbers of human fatalities leading to the advent of major new air pollution control measures. During the 1950s, scientists also began to understand the cause and atmospheric mechanisms for the formation of the Los Angeles photochemical smog. We now know that surface-level ozone and photochemical smog are a worldwide problem at regional, continental, and intercontinental scales. As economic development, urbanization, and the combustion of fossil fuels continues worldwide, large geographic areas of agriculture, forestry, and natural resources, including their biological diversity, are increasingly at risk. As scientific advances increase our understanding of atmospheric photochemical processes, air pollutant transport, their transformation and removal mechanisms, so too is the effort increasing to control the emissions of primary pollutants (sulfur dioxide, oxides of nitrogen, hydrocarbons, carbon dioxide, and carbon monoxide).

During the mid-1970s, environmental concerns regarding the occurrence of "acidic precipitation" began to emerge to the forefront. Since then, our knowledge of the adverse effects of air pollutants on human health and welfare (visibility, terrestrial and aquatic ecosystems and materials) has begun to rise substantially. Similarly, studies have been directed to improve our understanding of the accumulation of persistent inorganic (heavy metals) and organic (polyaromatic hydrocarbons, polychlorinated biphenyls) chemicals in the environment and their impacts on sensitive receptors, including human beings. Use of fertilizers (excess nutrient loading) and herbicides and pesticides in both agriculture and forestry and the related aspects of their atmospheric transport, fate, and deposition and their direct runoff through the soil and impacts on ground and surface water quality and environmental toxicology have become issues of much concern.

In recent times, environmental literacy has become an increasingly important factor in our lives, particularly in the so-called developed nations. Currently, the scientific, public, and political communities are much concerned with the increasing global-scale air pollution and the consequent global climate change. There are efforts being made to totally ban the use of chlorofluorocarbon and organobromine compounds at the global scale. However, during this millennium, many developing nations will become major forces governing environmental health, as their populations and industrialization grow at a rapid pace. There is an ongoing international debate regarding policies and the mitigation strategies to be adopted to address the critical issue of increasing energy demand being met by combustion of fossil fuels and climate change. Human health and environmental impacts and risk assessment and the associated cost–benefit analyses, including global economy through cross-border resource management, are germane to this controversy.

An approach to understanding environmental issues in general, and in most cases, mitigation of the related problems, requires a systems analysis and a multi- and interdisciplinary philosophy. There is an increasing scientific awareness to integrate environmental processes and their products in evaluating the overall impacts on various receptors. As momentum is gained, this approach constitutes a challenging future direction for our scientific and technical efforts.

The objective of *Developments in Environmental Science* is to facilitate the publication of scholarly works that address any of the described topics. In addition to edited or single- and multiauthored books, the series also considers conference proceedings for publication. The emphasis of the series is on the importance of the subject topic, the scientific and technical quality of the content, and timeliness of the work.

Sagar Krupa
Chief Editor, Book Series

Chapter 1

Energy Production: A Global Perspective

R.L. Orbach[1]

The Energy Institute, The University of Texas at Austin, Austin, Texas, USA
[1]Corresponding author: e-mail: orbach@energy.utexas.edu

ABSTRACT

Canada has the world's third largest oil reserves, with 97% of these (170 Bb, billion barrels) in the oil sands. Of these, 20% are recoverable with mining, while most (80%) are recoverable only by drilling (*in situ*). Production from the oil sands has been rapidly increasing (millions of barrels of oil per day): 0.1 (1980), 1.5 (2010), and 3.5 (2030) expected. In 2010, there were 91 active oil sand projects, and of these, four were mining with the remainder using various *in situ* recovery methods. It is in this context that future world demand for energy resources will be analyzed. In the next two decades, as the world appetite for energy continues to increase, the oil sands will produce *one-eighth of the total increase in global oil-based liquids*. The major presence of oil sand production in the world's energy markets will mean that many of the same constraints that face major producers elsewhere will be felt in Alberta. This includes CO_2 production associated with global climate change. Current methods for CO_2 capture and storage are not cost-effective, and have been slow (if not absent) to introduce at scale. This chapter describes research into some potentially economically feasible approaches: cost-effective capture and storage of CO_2 through energy production from methane-saturated saline aquifers, fuels from sunlight without CO_2 production, and large-scale electrical energy storage for intermittent (and even constant) electricity generating sources.

1.1 THE SITUATION

The world's commercial energy usage will continue to increase. The BP Energy Outlook (2011) displays an inexorable global increase in energy

Disclaimer: The content and opinions expressed by the author in this chapter do not necessarily reflect the views of the Wood Buffalo Environmental Association (WBEA) or of the WBEA membership.

Developments in Environmental Science, Vol. 11. http://dx.doi.org/10.1016/B978-0-08-099443-7.00001-6

appetite by some 46% over the next 20 years (Figure 1.1). The analysis of this chapter will focus on this period, but there are some who are not convinced that this increase can go on indefinitely. An example is the warning by Patzek (2012) that the original predictions of Hubbert are with us now. Figure 1.2

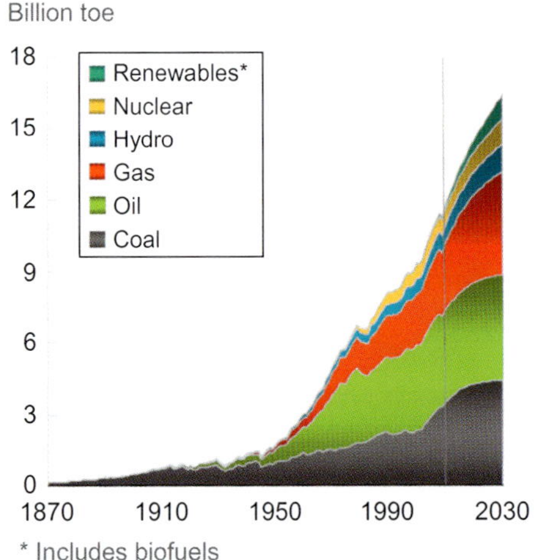

FIGURE 1.1   Global energy consumption by type from 1870 through 2030. TOE, tons of oil equivalent. *Modified from BP Energy Outlook (2011).*

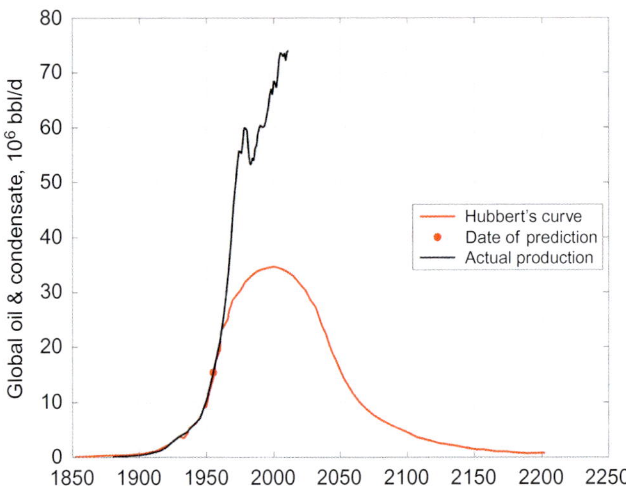

FIGURE 1.2   Hubbert's original prediction of global oil production. Superposed is the actual history of the world's production of crude oil (petroleum) and the associated lease condensate liquids in millions of barrels of oil per day, 10^6 bbl/d. *Modified from Patzek (2012).*

displays Hubbert's curve for Global oil and condensate production, "superimposed with the actual history of world production of crude oil (petroleum) and the associated lease condensate liquids. . .Over the last 8 years, this rate oscillated around 72-74 (millions of barrels of oil per day) despite of the whole world trying to do its best to meet the ever-growing demand for petroleum." From Patzek's perspective, it is remarkable that Hubbert was only off by a factor of two in his predictions. Patzek contends that the "supergiant oilfields have now been discovered and it is increasingly unlikely that more will be discovered in the future." This perspective suggests that there is not much more time to develop alternatives to combustion of oil for transportation. Hubbert's curve falls precipitously by 2050, so that there is a mere 40 years to find alternatives. And, as discussed below, these alternatives must be essentially CO_2 free. This is a tall order.

The expectation is that liquids production will increase to meet growth in consumption. Figure 1.3 suggests this may be true for the 20-year period covered by the BP Energy Outlook, but it would be prudent to express concern beyond. Within these 20 years, the mix of liquids that satisfy the growth in consumption will change on a global scale. Figure 1.3 projects the liquids supply by type.

From Figure 1.3, the global liquids supply will rise by about 16.5 Mb/day from 2010 to 2030. But the sources of growth in liquids production will

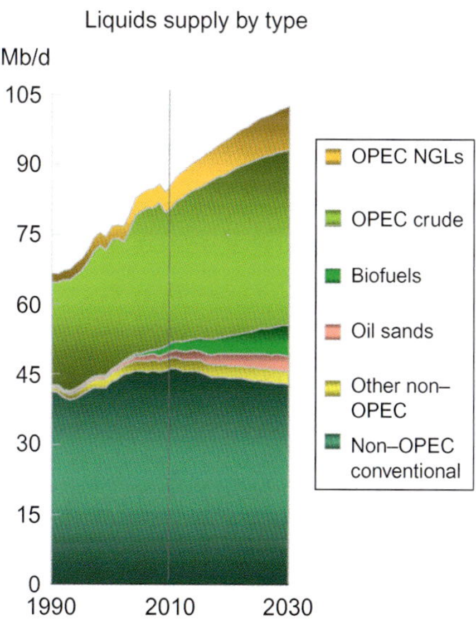

FIGURE 1.3   Total liquids growth. This is based on the assumption that liquids production meets growth in consumption. NGL, natural gas liquids; OPEC, Organization of Petroleum Exporting Countries. *Modified from BP Energy Outlook (2011).*

change the global balance. The projection for biofuels production (largely ethanol) is in excess of 6.5 Mb/day by 2030, from 1.8 Mb/day in 2010. First-generation biofuels are expected to account for most of the growth.

Figure 1.3 provides evidence of the significant importance of oil sands to the global liquid supply. In particular, BP projects an increase of 2 Mb/day from 2010 to 2030, amounting to 12% of the increase in global liquids. Local projections are much higher, projecting an increase of 2.5 Mb/day by 2020.

The importance of oil sands to global liquids growth is made even more evident by examining liquids growth from Non-OECD (Organization of Economic Cooperation and Development) countries (BP Energy Outlook 2030) in Figure 1.4. It is a major component of the Non-OPEC (Organization of Petroleum Exporting Countries) growth over the 20 years from 2010 to 2030.

The assumption of hydrocarbons to meet demand takes on a more striking consequence when one considers the consequence of hydrocarbon combustion, and its concomitant contribution to atmospheric concentrations of CO_2. As seen from Figure 1.5, hydrocarbon sources (especially coal) will remain major contributors to world electricity production, and are a major source of this greenhouse gas. What is worse, the projection for electricity production from coal and natural gas continues to increase over the next two decades, further exacerbating anthropogenic contributions to atmospheric warming.

Breaking down the Non-OECD and OECD contributions to world CO_2 production (Figure 1.6), it becomes clear that the former will be the major

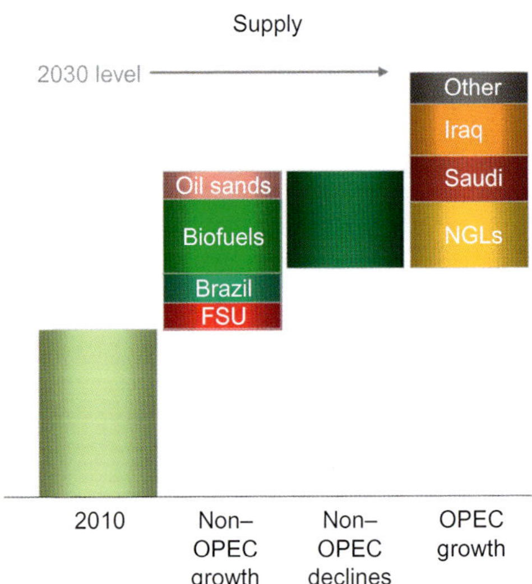

FIGURE 1.4 Liquids growth from Non-OECD Countries. OPEC, Organization of Petroleum Exporting Countries; OECD, Organization of Economic Cooperation and Development. *Modified from BP Energy Outlook (2011).*

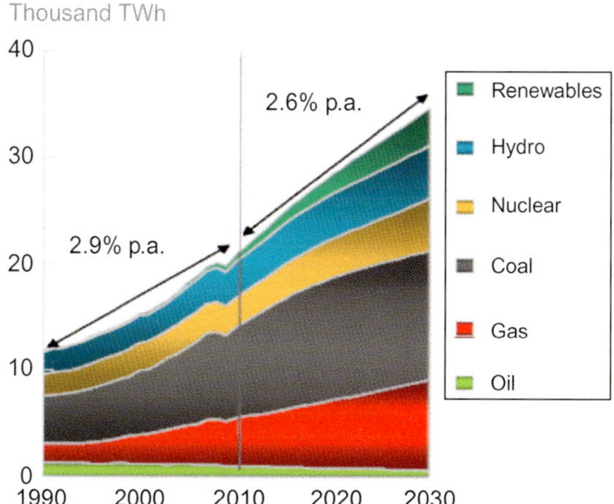

FIGURE 1.5   World electric power generation by source type and year. TWh, terawatt hours. *Modified from BP Energy Outlook (2011).*

contributor over the next two decades, even while the OECD countries begin to restrain their CO_2 production. By 2030, the Non-OECD output of CO_2 is projected to be more than *twice* that of OECD countries. Further, hydrocarbon production of CO_2 continues to increase, with coal being an ever-increasing "bad actor."

This alarming prediction is often softened by the exhortation to conserve. Put in the simplest term, if we use less electricity, we burn less coal, and therefore we emit less CO_2. The problem is that, historically, that does not

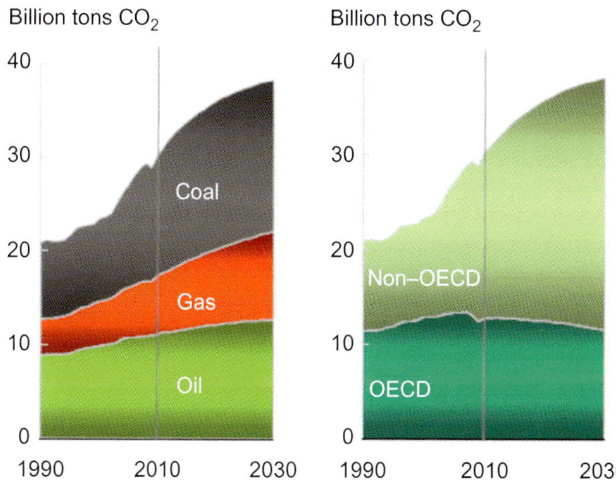

FIGURE 1.6   Projected world production of CO_2 by source type and nation clusters. *Modified from BP Energy Outlook (2011).*

happen. This is often referred to as "Jevons' Paradox" after the nineteenth century economist William Stanley Jevons. Put directly: "Improving economic efficiency enables the creation of more new energy uses than energy savings. The net effect is to increase the rate of resource depletion" (Jevons, 1866). P.F. Henshaw put it most clearly in Figure 1.7: CO_2 is being produced at the same increasing rate as total energy use. New "clean energy" sources are not replacing any fossil fuel use.

The consequences are stark. Keeping CO_2 production at the same increasing rate as total energy use will lead to an explosive atmospheric concentration. Indeed, even keeping CO_2 emissions constant will lead to increasing atmospheric concentrations over time. As Figure 1.8 shows that only through reductions of CO_2 emissions by 80% can one stabilize atmospheric CO_2 concentrations, the 2050 target of 80% reduction of CO_2 is required, as difficult as it is, if the CO_2 atmospheric concentration is to be kept constant.

Atmospheric CO_2 concentrations have been increasing since 1880, when the world energy use of coal surpassed that of wood, rising from a historical value of 280 ppmv (parts per million by volume) to today's concentration in excess of 390 ppmv, and continuing to increase. Is there evidence that atmospheric temperatures have increased, as one would assume from an increase in greenhouse gasses?

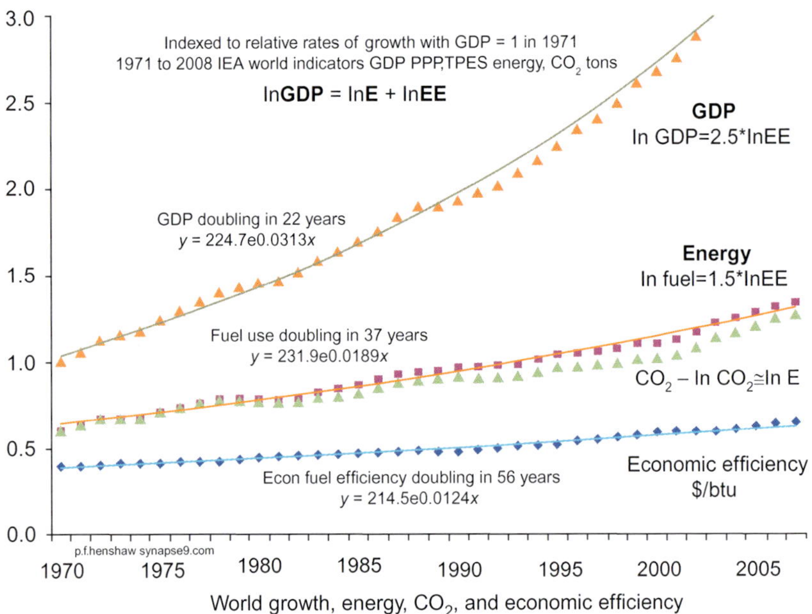

FIGURE 1.7   World growth, energy, CO_2, and energy efficiency, indexed to relative rates of growth with GDP=1 in 1971. *Modified from P. F. Henshaw.*

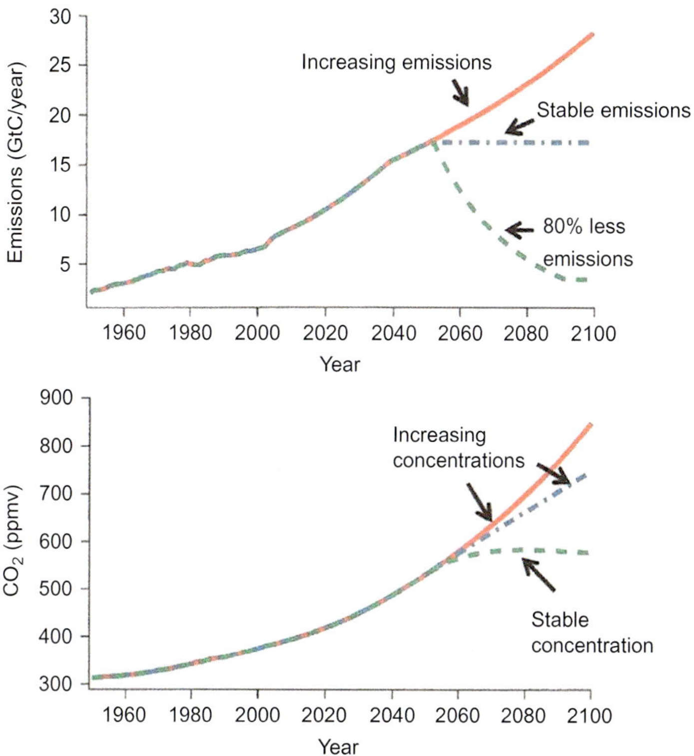

FIGURE 1.8   Two plots displaying the consequence of increasing, stable, 80% less emissions rate of CO_2 in comparison to atmospheric concentrations of CO_2, respectively. GtC, Global fossil CO_2 emissions (GtC/year); GtC = 1/3.67. *Modified from National Academies Press (2010).*

Figure 1.9 displays the evidence for warming of the lower atmosphere from 1980 through 2010 (Santer et al., 2011). On the one hand, the upper curve displays clear evidence for a monotonic increase in atmospheric temperature when a moving average is taken over sufficiently many years (20 in this instance). On the other hand, for a moving average of only 10 years, there appear to be periods where atmospheric temperature changes are relatively flat in time. The difficulty arises from the noise in atmospheric temperature measurements. Santer et al., 2011 show that when averaging over a 10-year period, the fluctuations are comparable to the temperature change (i.e., the "signal to noise" is roughly unity). However, when averaging over periods longer than 17 years, the temperature change is four times the fluctuations (i.e., the signal to noise is four). Hence, results presented in Figure 1.9 are evidence of consistent warming over a period of 30 years.

It is interesting to look at a longer period for atmospheric temperatures, specifically from 1880 when global energy production switched from wood to coal. Figure 1.10 shows that there were periods of modest increases and

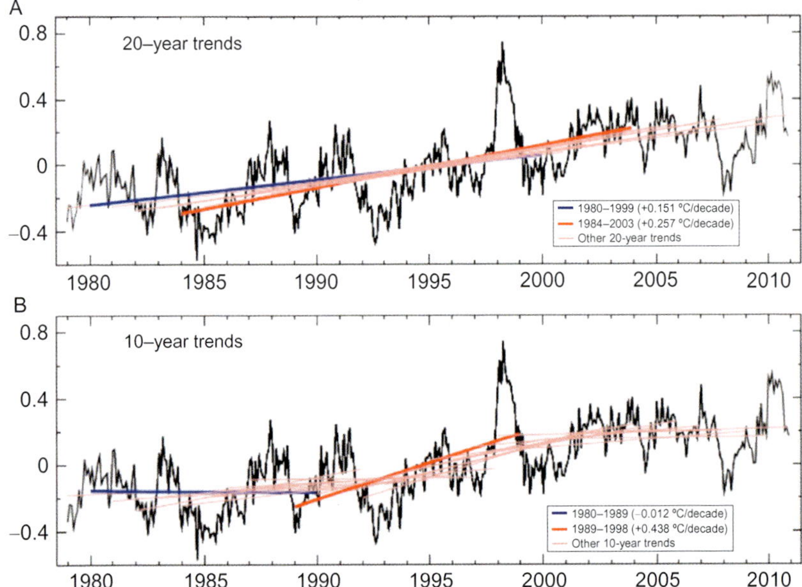

FIGURE 1.9   Differing time average intervals for atmospheric temperature changes. Note that a 20-year moving average displays a monotonic increase of temperature over the full interval period, while a 10-year moving average displays periods of relatively constant temperature. *Modified from Santer et al. (2011).*

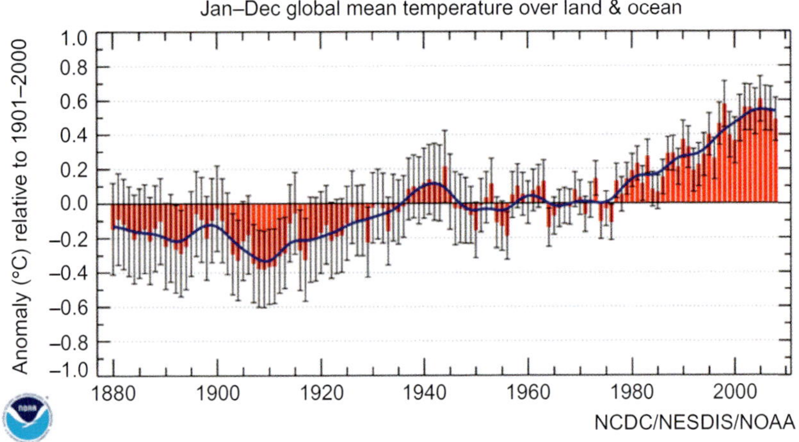

FIGURE 1.10   Temperature changes since the "crossover" between biomass and coal global energy production in 1880, displayed in Figure 1.11. *Modified from NOAA/National Climatic Data Center (2011).*

decreases, but the "inexorable" trend for temperature increase began in earnest in 1980. What happened in 1980? Nothing. This chapter contends that the increase in 1980 was a consequence of the net anthropogenic increase in CO_2 as a consequence of switching from energy generation from biomass fuels to coal in 1880.

Figure 1.11A displays the primary global energy supply from 1700 through the beginning of the twentieth century. There is a crossover in

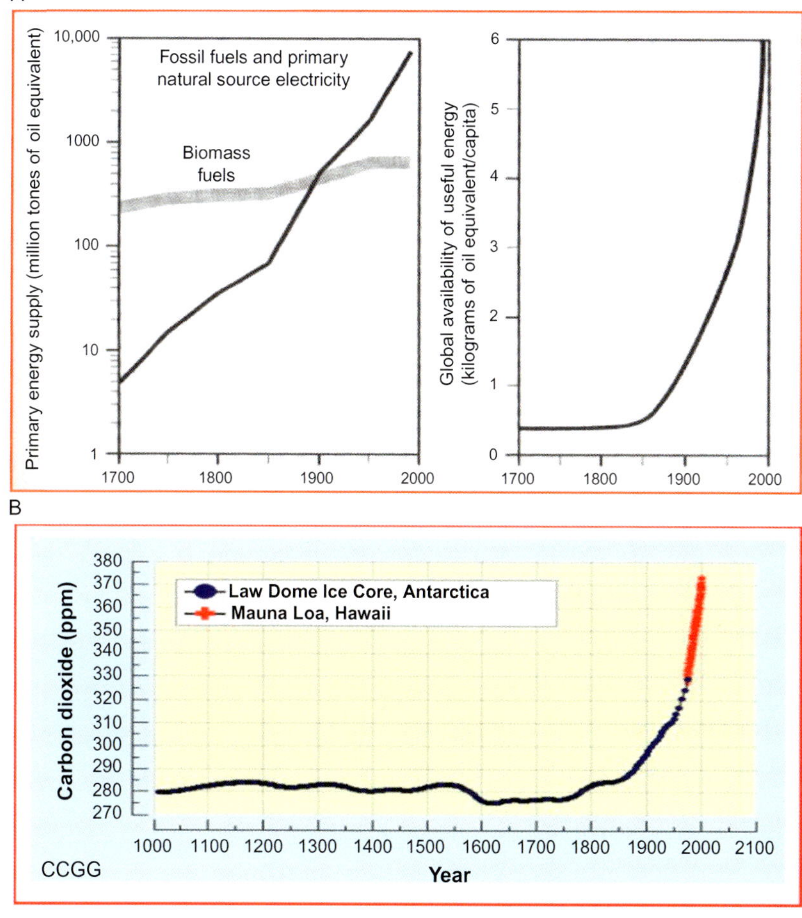

FIGURE 1.11   (A) Modified: Primary (natural source) global energy supply. Though the industrial revolution began much earlier in Great Britain, in 1880 the export of technology from Great Britain to the European continent and to South America began a global increase in energy supply. Modified from Smil (1994). (B) Atmospheric concentrations of CO_2 measured from ice core data, and directly from Mauna Loa, Hawaii. CCGG, CCGG Carbon Cycle Greenhouse Gasses Group, Earth System Research Laboratory, National Oceanic & Atmospheric Administration. *Modified from Smil (1994).*

1880, with a sharp increase in global availability of useful energy at that time. Figure 1.11B displays the atmospheric concentration of CO_2 over the same time interval. The increase from a rather stable 280 ppm takes place around 1880, and rises precipitously in proportion to the global availability of useful energy.

Why the delay in atmospheric temperature increases due to the increase in CO_2? The thermal capacity of the upper layers of the ocean is much larger than that of the atmosphere. As a consequence, the upper surface layers of the ocean take roughly a century to come into thermal equilibrium with the lower atmosphere. Hence, what is seen in 1980 was a consequence of the increase in thermal energy of the ocean-atmosphere system. This suggests that temperatures for 100 years beyond 1980 will continue to increase, *even if no further CO_2 were generated.* That is, the industrial revolution that spread globally in 1880, and continues to this day, generates increasing temperatures roughly 100 years from when the original levels of anthropogenic CO_2 was produced. Put bluntly, the CO_2 added to the atmosphere today will result in increased atmospheric temperatures five generations hence. From Figure 1.6, global CO_2 generation will continue to increase for at least another 20 years beyond today, increasing atmospheric temperatures at least to 120 years from today.

1.2 SOME REMEDIES

There are few if any easy choices for significant reductions of CO_2 emissions, given the global energy appetite exhibited in Figures 1.1 and 1.3. A global response is required, given what appears to be an inexorable increase in atmospheric temperatures, and the need to achieve an 80% reduction in CO_2 emissions simply to stabilize atmospheric CO_2 concentrations and, presumably, stabilize atmospheric temperatures 100 years hence.

The first target should be coal. Is there an economical process to capture and sequester CO_2 economically from coal-fired electric power plants? Next, is there a process to use solar energy to produce fuels for transportation? And, finally, is there a way to store electrical energy at base-load levels so that intermittent sources of electricity such as wind and solar power can respond to demand and stabilize the electricity grid. There are other options available for research and development, but the three options listed seem achievable on time scales of interest.

1.2.1 Cost-Effective Capture and Storage of CO_2 Through Energy Production from Saline Aquifers

The current methods for capture and sequestration of CO_2 from coal-fired power plants are pure cost, amounting to roughly one-third of the power plants energy. In monetary terms, the cost of capture and sequestration is *at*

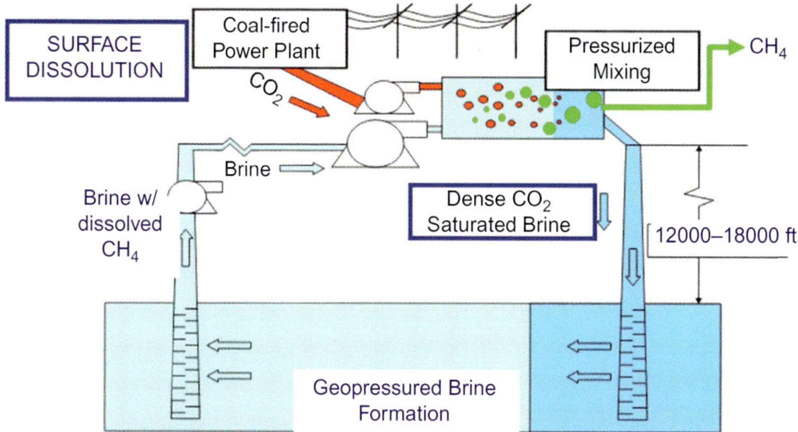

FIGURE 1.12   A schematic for cost-effective capture and storage of CO_2 through energy production from saline aquifers. *Modified from Pope (2011).*

least \$50 per ton of CO_2 and can be as high as \$75 per ton, using amine liquids for capture (Rochelle, 2009). This cost is prohibitive in competitive markets, and is probably not supportable in terms of a price on carbon. Recently, as shown in Figure 1.12, a proposal has been put forward that reduces the net cost of capture and sequestration, and adds to the efficiency of capture (Burton and Bryant, 2009).

The production of energy from geothermal aquifers has evolved as a separate, independent technology from the sequestration of CO_2 and other greenhouse gases in deep saline aquifers. A game-changing new idea combines these two technologies and adds another:

- Dissolution of CO_2 into extracted brine which is then reinjected;
- Production of methane from the extracted brine;
- Production of energy from the extracted brine offsets the cost of capture, pressurization, and the subsequent injection of brine containing CO_2 back into the aquifer;
- Methane production plus thermal energy offsets the cost of carbon capture and sequestration to a point that it can survive in a competitive market environment without subsidies or a price on carbon.

As noted in Figure 1.12, instead of direct injection of CO_2 into an aquifer, saline water is pumped to the surface, and the CO_2 captured from the flue gas is injected under modest pressure ($\sim$1000 psi) into the water. This immediately reduces the cost of CO_2 pressurization. Further, when CO_2 contacts water with dissolved methane in it, the methane is expelled from solution resulting in a wave front of methane that can be captured, and then either sold commercially or used to generate the lost electrical energy through CO_2

capture. The saline water comes to the surface from original reservoir temperatures of the order of 300 °F. This heat can be used to assist the energy required for CO_2 capture, with preliminary estimates suggesting offsets comparable to the value of the released methane. Pressurization is required to return the saline water with injected CO_2 into the aquifer, but injection is aided by gravity and less costly energetically than pumping the same amount of CO_2 directly into the aquifer. Finally, the saline solution is taken from a different portion of the aquifer than the returned saline water. This provides a much more robust permanence for CO_2 storage.

Issues that need further consideration to optimize and predict the potential for production of dissolved methane and geothermal energy by CO_2 injection are the following: (1) the locations of suitable aquifers; (2) the volume and concentration of methane in the brine; (3) the most favorable aquifer conditions; (4) the fraction of dissolved methane that can be produced; (5) the best strategies for injecting CO_2 and producing methane and geothermal energy; and (6) the best strategy for well types, locations, and the operating conditions. Pursuit of these issues may well provide a vehicle for capture and storage of CO_2 from coal-fired power plants that would be attractive economically as well as environmentally.

1.2.2   Solar Energy to Produce Transportation Fuels

Photosynthesis combines CO_2, sunlight, and water in plants to produce ATP and NADPH, the "fuels" or energy that enable them to grow and reproduce (with the production of O_2, responsible for our atmosphere). "Artificial photosynthesis" has been the elusive target for solar energy researchers, but the stability of CO_2 has frustrated ready success. At present, an achievable goal is to use sunlight to split water into H_2 and O_2 through solar-powered photoelectrochemical (PEC) reactions and reactors, as sketched in Figure 1.13.

The production of H_2 without CO_2 would contribute to reduction of CO_2 emissions. Currently, H_2 is produced by reforming natural gas, leading to one CO_2 molecule produced for every four molecules of H_2. A typical petroleum refinery uses roughly a billion cubic feet of H_2 a day. If it could obtain H_2 without CO_2 emissions, it would reduce its "carbon footprint" by 30% to 40%. So, solar PEC production of H_2 is a further example of a carbon "remedy."

The core of any practical PEC system involves:

1. *Photocatalysts*: This material is responsible for the transduction of solar energy into electron–hole pairs. The key needed characteristics are high efficiency of light capture and electron–hole pair formation, high mobility of these carriers, low recombination rates of the carriers, and high stability under irradiation. These should be composed of earth-abundant materials of low

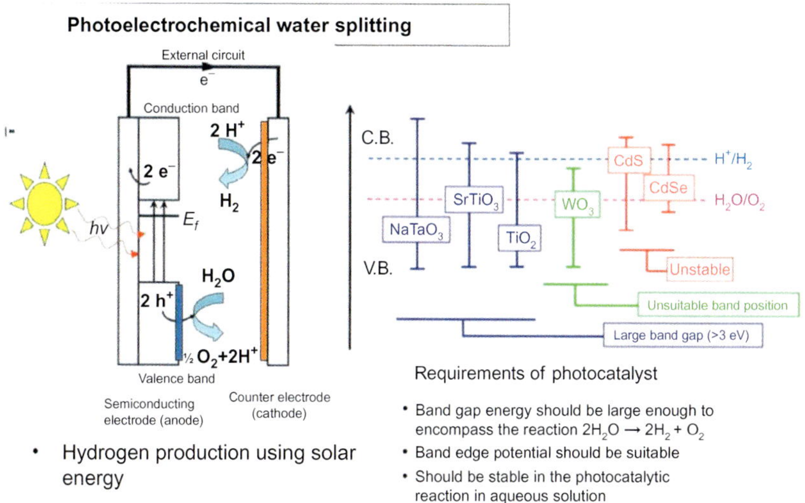

FIGURE 1.13   A schematic of the photoelectrochemical splitting of water into H_2 and O_2. *Modified from Bard (1995).*

cost. It is probable that to attain practical efficiencies, multijunction systems that can capture light across the whole solar spectrum will be needed.

2. *Electrocatalysts*: These promote the capture of holes at the photocatalyst surface to oxidize water in the oxygen evolution reaction (OER) and capture electrons to reduce water in the hydrogen evolution reaction (HER). The efficiency of these materials contributes to the overall efficiency of the system. They must also be efficient, stable, and inexpensive.

3. *Electrolyte*: This should be highly conductive, noncorrosive to photocatalysts and electrocatalysts, environmentally safe, and conducive to both the anodic (OER) and cathodic (HER) reactions.

4. *PEC*: In integration of the components in steps 1–3, cell design should minimize ion flow resistance; prevent mixing of the evolved hydrogen and oxygen; and design electrodes and flow path to minimize gas stagnation. Preferably, the cells will not require an expensive conductive membrane to separate the anodic and cathodic sites.

A promising approach is through the use of semiconducting metal oxides. New tools (Liu et al., 2010) allow for a combinatorial approach for making a multitude of different complex compositions of metal oxides, all unique, and testing them rapidly for their promise as photoelectrocatalysts. The method is based on scanning electrochemical microscopy. Broad arrays of potential active materials and structures may be examined using this rapid new technique as shown for Cd–In–Bi oxides in Figure 1.14.

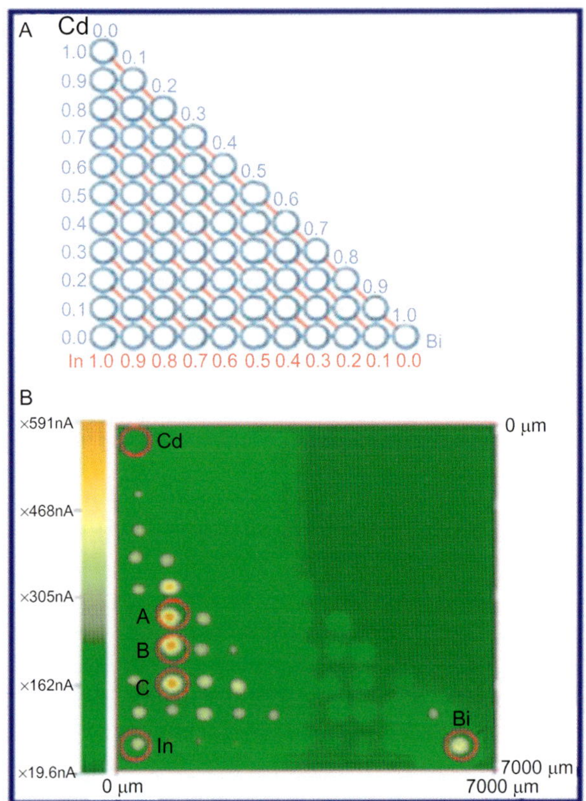

FIGURE 1.14   A display of photoelectrochemical activity for an array of different compounds of Cd–In–Bi oxide. The bright spots, labeled A, B, and C, represent Cd–In–Bi concentration ratios of 40:50:10, 30:60:10, and 20:70:10 (atom), respectively. *Modified from Wen Liu (2012)*.

The "scale-up" to laboratory prototypes requires geometries that allow the photo-induced electrons and holes to flow to the surface in order to produce H_2 and O_2, respectively. A geometry that can optimize solar radiation absorption and promote PEC activity has been developed by Flaherty et al. (2007), and is exhibited in Figure 1.15.

1.2.3   Electrical Energy Storage at Base-Load Levels

Electricity production from wind and solar sources seems a helpful alternative to CO_2 emitting coal-fired power plants. However, their electricity production is intermittent. In west Texas, for example, the wind is strongest at night when the demand is least. And wind gusts and other fluctuations imply intermittent wind-based electricity production. Though solar intensities peak at roughly the same time as demand, there are still significant fluctuations in

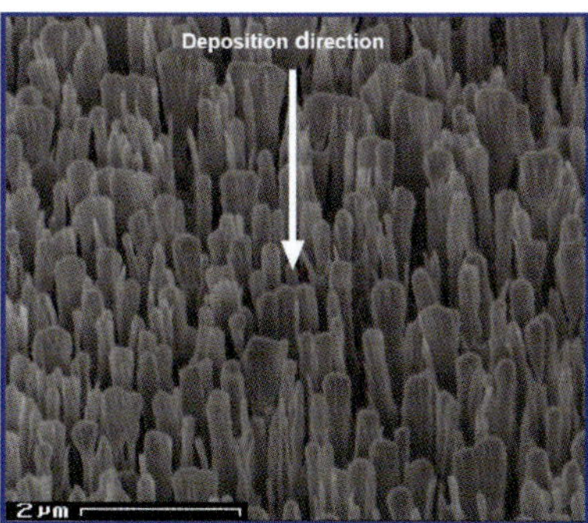

FIGURE 1.15   Structural morphology of films that optimize the geometrical properties for photoelectrochemical activity. The upper figure is a scanning electron microscope image of a film grown at 70°, 100 K; the lower at 85°, 100 K. Solar radiation is absorbed along the (long) axis of the columns, and photoelectrochemical production at the surface with transverse (short) electron–hole migration distances. *Modified from Flaherty et al. (2007), reprinted with permission, Copyright 2007, American Chemical Society.*

photovoltaic electricity production even on cloudless days. These fluctuations can have a destabilizing effect on the grid. Even worse, heavy reliance on intermittent sources requires "back-up" power, which for the most part remains mostly idle.

If there were some way to store electricity as base-load levels, the electrical energy generated at night could be made available during peak demand periods during the day, and the fluctuations inherent in wind and solar sources could be smoothed. The difficulty is that no efficient, inexpensive, and large capacity electrical energy storage system exists.

"Normal" batteries are ill suited to base-load storage requirements. However, "flow" batteries do offer base-load storage capacity. A flow battery replaces one of the solid electrodes found in usual batteries with a liquid that can be stored in relatively unlimited quantities. Because the energy of a charged battery is stored in the cathode, flow batteries have a liquid cathode. This requires a solid electrolyte electronic insulator (that separates the cathode from the anode) with sufficient ionic conductivity for rapid charge and discharge, and having the appropriate band gap relative to the cathode and anode

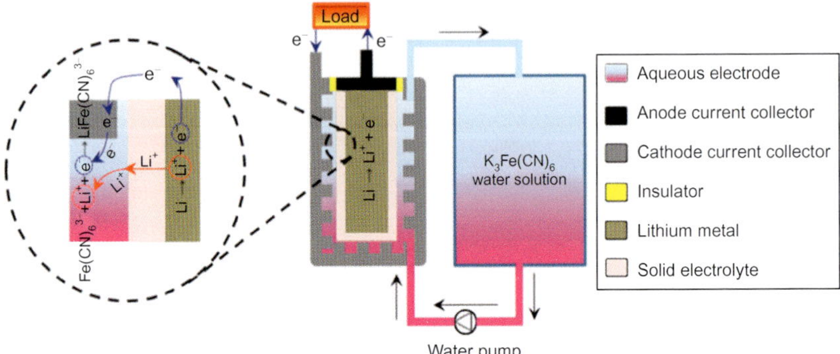

FIGURE 1.16 A schematic diagram of a lithium-water rechargeable battery. *Adapted from Goodenough (2012).*

materials. An aqueous flow battery, in addition to these requirements, must also possess aqueous electrode properties (Goodenough, 2012): (1) high specific energy density, (2) ambient temperature operation, (3) proper redox potentials, (4) no side-reactions, (5) good stability in water, (6) good reversibility, (7) reliable safety, and (8) low cost.

A schematic example is displayed in Figure 1.16. It utilizes Li ions in conjunction with a Fe^{2+}–Fe^{3+} redox couple. The difficulty is the solid electrolyte. Not only must it possess sufficient transport for Li^+ ions, but it must also not be permeable to protons in order to keep water separated from Li metal. More research is required to find such a material that satisfies all of the requirements for a solid electrolyte. Nevertheless, this class of device has a nearly unlimited electrical storage capacity. When charging, the liquid cathode can pass through the battery to be stored externally. This conversion of electrical to chemical energy is reversible. When electricity delivery is required, the liquid cathode can pass through the battery and be discharged to produce the required electrical power. This class of device, operating at base-load levels, would make intermittent wind and solar electricity practicable at large scales, again reducing the need for CO_2-producing coal-fired power plants.

1.3 SUMMARY

The many figures presented in the chapter make it clear the need for fossil fuels as a global energy source for the foreseeable future. Canada has the world's third largest oil reserves, with 97% of such fuels (170 Bb) in the oil sands. In the next two decades, as the world's appetite for energy continues to grow, oil sands will produce one-eighth of the total increase in global liquids. The major presence of oil sand production in the world's energy market means that many of the same constraints that face other major producers will be felt in Alberta.

New "clean energy" sources have not replaced any fossil fuel use. And CO_2 production will increase at the same rate as total energy use. Yet, CO_2 emissions must be reduced by 80% in order to stabilize the CO_2 concentration in the atmosphere. New technological developments are required to reduce CO_2 emissions in the face of increasing use of fossil fuels. Examples are:

- Cost-effective capture and storage of CO_2 through energy production from saline aquifers
- Solar energy to produce transportation fuels
- Electrical energy storage at base-load levels.

This is hardly an exhaustive list, but it is illustrative of the opportunities to stabilize atmospheric CO_2 concentrations, and ultimately, to stabilize global temperatures. Failure to do so will create an unsustainable increase in global temperatures, affecting the very future of the planet.

REFERENCES

Bard, A.J., Fox, M.A., 1995. Artificial photosynthesis: solar splitting of water to hydrogen and oxygen. Acc. Chem. Res. 28, 141–145.

BP Energy Outlook, 2011. www.BP.com. The figures 1.1, 1.3, 1.4, 1.5, and 1.6 were taken from BP Energy Outlook, 2011. A link to Energy Outlook 2030 now generates BP Energy Outlook, 2012. The Respective figures are only slightly altered from those in the text.

Burton, M., Bryant, S.L., 2009. Eliminating buoyant migration of sequestered CO_2 through surface dissolution: implementation costs and technical challenges. SPE Res. Eval. Eng. 12 (3), 399–407. http://dx.doi.org/10.2118/110650-PA.

Committee on Stabilization Targets for Atmospheric Greenhouse Gas Concentrations, Board on Atmospheric Sciences and Climate, National Research Council, The National Academy of Sciences, 2010. Climate Stabilization Targets: Emissions, Concentrations, and Impacts over Decades to Millennia. The National Academies Press, Washington, DC. http://dels.nas.edu/basc/Reports-Academies-Findings; a Report-in-Brief can be found through that site, or directly through http://www.scribd.com/doc/44922181/Climate-Stabilization-Targets-Report-in-Brief.

Flaherty, D.W., Dohnalek, Z., Dohnalkova, A., Arey, B.W., McCready, D.E., Ponnusamy, N., Mullins, C.B., Kay, B.D., Reactive ballistic deposition of porous TiO_2 films: growth and characterization. J. Phys. Chem. 111 (12), 4765–4773. Reprinted with permission, Copyright 2007. American Chemical Society, http://dx.doi.org/10.1021/jp067641m.

Goodenough, J.B., 2012. Rechargeable batteries: challenges old and new. J. Solid State Electrochem. 16, 2019–2029. Figure 11. With kind permission from Springer Science and Business Media, http://dx.doi.org/10.1007/s10008-012-1751-2.

Henshaw P.F., 2009. Inside Efficiency – Why Efficiency Multiplies Consumption. Biophysical Economics 09, Syracuse NY Oct 16–17; Notes & Slides. http://synapse9.com/pub/EffMultiplies.htm.

Jevons, W.S., 1866. The Coal Question, second ed. Macmillan and Co., London.

Liu, W., Bard, A.J., 2012. Screening of metal doped In_2O_3 photocatalysts by scanning electromechanical microscopy, unpublished communication.

Liu, W., Ye, H., Bard, A.J., 2010. Screening of novel metal oxide photocatalysts by scanning electrochemical microscopy and research of their photoelectrochemical properties. J. Phys. Chem. 485 (1–3), 231–234.

National Climatic Data Center, Global Surface Temperature Anomalies: National Oceanic and Atmospheric Administration. Effective November 2011. http://www.ncdc.noaa.gov/cmb-faq/anomalies.php (accessed 4 September 2012).

Patzek, T.W., 2012. http://patzek-lifeitself.blogspot.com/2012/04/world-is-finite-isnt-it.html. (accessed 24 April 2012).

Pope, G., Bryant, S.L., Ganjdanesh, R., Orbach, R.L., Rochelle, G.T., and Sepehrnoori, K., 2011. Offsetting the costs of CO_2 capture and storage by energy production from saline aquifers, unpublished.

Rochelle, G.T., 2009. Amine scrubbing for CO_2 capture. Science 325, 1652–1654. http://dx.doi.org/10.1126/scienc1176731.

Santer, B.D., Mears, C.A., Doutriaux, C., Caldwell, P.M., Gleckler, P.J., Wigley, T.M.L., Solomon, S., Gillett, N., Ivanova, D.P., Karl, T.R., Lanzante, J.R., Meehl, G.A., Stott, P.A., Taylor, K.E., Thorne, P., Wehner, M.F., Wentz, F.J., 2011. Separating signal and noise in atmospheric temperature changes: the importance of time scale. J. Geophys. Res. 116 (D22), 105–124. http://dx.doi.org/10.1029/2011JD016263.

Smil, V., 1994. Energy in World History. Reproduced with permission of WESTVIEW PRESS in the format republish in a book via Copyright Clearance Center, 186.

Chapter 2

Energy Developments in Canada's Oil Sands

G. Stringham[1]

Canadian Association of Petroleum Producers, Calgary, Alberta, Canada

[1]*Corresponding author: e-mail: greg.stringham@capp.ca*

ABSTRACT

Canada has the world's third largest proved reserves of crude oil. Ninety-seven percent of these, or 169 billion barrels, are located in the oil sands. The oil sands cover an area of about 142,000 km^2. Surface mining will impact 3% of the total land that overlies the oil sands. The remainder is recoverable by *in situ* drilling techniques.

Oil sands are a mixture of sand, water, clay, and bitumen. Bitumen is heavy oil that is too thick to flow or to be pumped without being diluted or heated. The resource has been known since early fur traders arrived and found Aboriginal people using bitumen to seal their canoes. Technologies used to produce oil from the oil sands have evolved considerably since first commercial development in 1967. Industry continues to find ways to make the process more efficient and improve environmental performance.

Current production from oil sands is 1.6 million bbl/day (barrels per day) and is expected to increase to 3.7 million bbl/day by 2025. Oil sands meet energy needs in Canada and make up a large share of Canadian crude exports to the United States. Canada is the United States' largest energy supplier. Global demand for secure and stable energy supplies continues to grow with the IEA predicting demand growth of 47% by 2035 (IEA, 2011), which drives increasing global interest, most recently from Asia.

The challenge faced by oil sands producers is to continually improve environmental and economic performance, while providing a secure source of energy for the world. A few specific examples of recent progress include reduced carbon dioxide emissions by 26% per barrels since 1990, tailings management technology reducing time it takes to reclaim tailings facilities from decades to months, increased water recycling and use of saline water.

Disclaimer: The content and opinions expressed by the author in this chapter do not necessarily reflect the views of the Wood Buffalo Environmental Association (WBEA) or of the WBEA membership.

Developments in Environmental Science, Vol. 11. http://dx.doi.org/10.1016/B978-0-08-099443-7.00002-8

2.1 INTRODUCTION

No nation can long be secure in this atomic age unless it be amply supplied with petroleum … It is the considered opinion of our group that if the North American continent is to produce the oil to meet its requirements in the years ahead, oil from the Athabasca area must of necessity play an important role.

> J. Howard Pew (at the opening of Canada's first Commercial oil sands plant near Fort McMurray, Alberta in 1967)

Truer words were never spoken. And from humble beginnings in 1967 when the first oil sands company produced about 12,000 barrels per day (bbl/day), the region now produces about 1.6 million bbl/day, and CAPP's 2011 forecast projects production will double by 2025 to about 3.7 million bbl/day (CAPP, 2011). At 169 billion barrels, Canada's oil sands are the third largest source of developable oil in the world (Figure 2.1).

These reserves offer energy security for Canada, North America—and, indeed, the world—while providing unprecedented opportunities for economic prosperity, jobs, and government revenue generation through taxes and royalties. With opportunity, of course, comes challenge. To achieve and maintain these benefits, oil sands must continue to be developed responsibly by industry, keeping a tight focus on reducing any adverse environmental impacts, appropriate fiscal terms need to remain in place, and companies must be able to get the products to markets.

New oil sands development is expected to contribute more than $2.1 trillion (2010 dollars) to the Canadian economy over the next 25 years—about $84 billion per year. The industry will pay an estimated $766 billion in

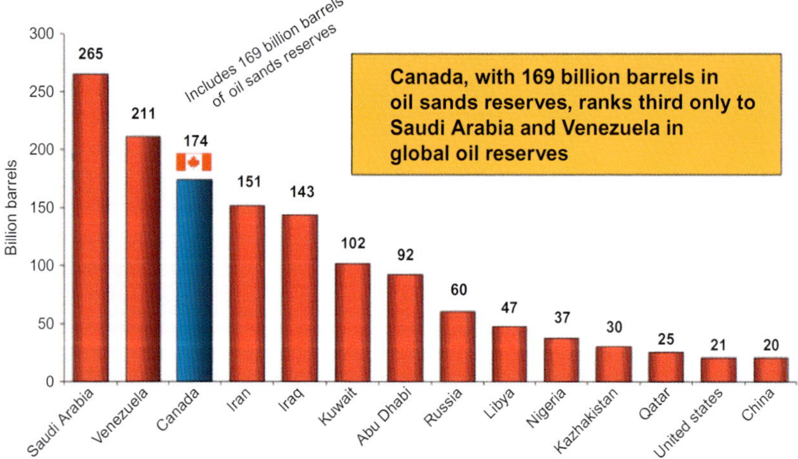

FIGURE 2.1 Crude oil reserves by country. Source: *Oil and Gas Journal, Dec 2011.*

provincial taxes ($105 billion), federal taxes ($311 billion), and royalties ($350 billion). Employment in Canada is expected to grow from 75,000 jobs in 2010 to 905,000 jobs in 2035, including 126,000 jobs sourced in provinces outside Alberta. U.S. employment is expected to grow from 21,000 jobs in 2010 to 465,000 jobs in 2035 (CERI, 2011).

Goods, materials, and services used to construct and operate *in situ* oil sands projects; mines and upgraders come from throughout North America to across the globe. Many of the components—tires, trucks, gages, valves, and pumps—are produced in central and eastern Canada. Every dollar invested creates about $75 in total economic impact over 25 years, and much of the impact is generated outside Alberta in the rest of Canada, the United States, and around the world (CERI, 2011).

The local benefits of oil sands development can be seen clearly in the Wood Buffalo region in Alberta. Fort McMurray is one of the fastest growing communities in North America, and more than 21,000 residents are directly employed in oil sands operations jobs. Solid relationships with Aboriginal communities have created employment and business opportunities. In 2010, oil sands companies contracted over $1.3 billion for goods and services from Aboriginal-owned businesses, and more than 1700 Aboriginals were employed in permanent operations jobs. Also in 2010, oil sands companies provided $5.5 million to support Aboriginal community programs.

2.2 EARLY DAYS

Canada's oil sands, a natural mix of sand, water, clay, and bitumen, are found in three deposits—the Athabasca, Peace River, and Cold Lake deposits in Alberta and Saskatchewan (Figure 2.2). Bitumen is oil that is too heavy or thick to flow or be pumped without being diluted or heated. Some bitumen is found within 200 ft of the surface; the majority is found deeper underground.

Oil sands are recovered through one of two primary methods—mining and drilling. About 20% of the reserves are close to the surface and can be mined using large shovels and trucks (Figure 2.3). The remaining 80% are found much deeper. They can be recovered "in place," or *in situ*, by drilling wells and injecting heat to warm the bitumen, so it can be pumped to the surface through recovery wells.

The first recorded mention of Canada's bitumen deposits was June 12, 1719. According to an entry in the *York Factory* journal, on that day, a Cree Indian brought a sample of oil sands to Henry Kelsey of the Hudson's Bay Company. When fur trader Peter Pond traveled down the Clearwater River to Athabasca in 1778, he saw the deposits and wrote of "springs of bitumen that flow along the ground." A decade later, Alexander Mackenzie saw Chipewyan Indians using bitumen and dirt to caulk their canoes. Oil sands have been both a marvel and a technical and environmental puzzle for almost a century. Early explorers did not know quite what to make of the bituminous

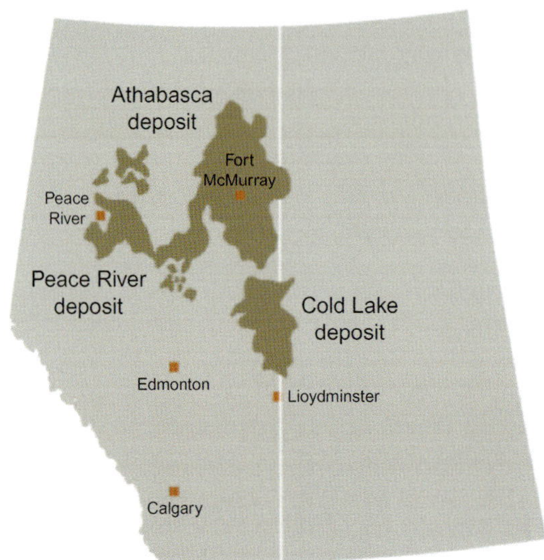

FIGURE 2.2   Oil sands deposits with operating and constructing projects.

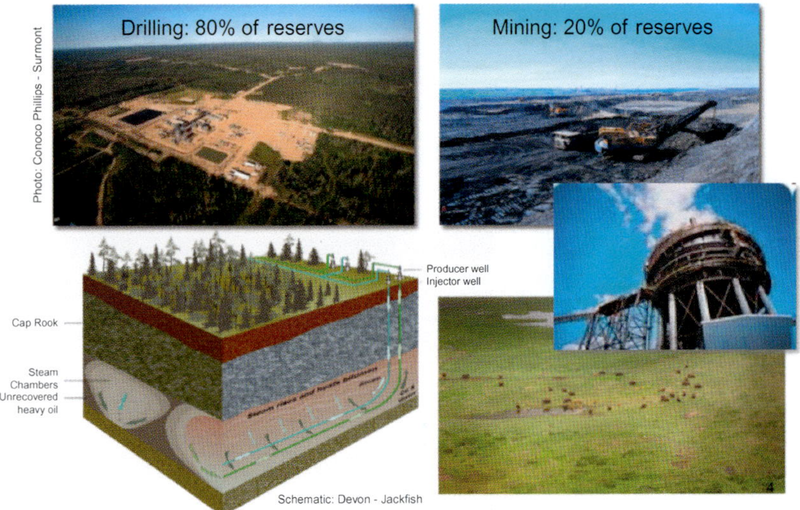

FIGURE 2.3   Oil sands recovery by drilling and mining.

sands. And contrary to the popular "enterprising oil man," stereotype neither did early entrepreneurs.

In fact, until about 40 years ago, private enterprise was barely involved in oil sands technical and commercial development. Under intense postwar demand for energy and economic diversification, it was the federal and provincial governments of the day that pushed innovation as both a national and provincial priority. As then University of Alberta President Henry Marshall Tory put it, State-funded scientific research would "lay the foundations of accurate knowledge upon which we can build our industries with security for the future." And it largely did, although it took almost 50 years for technology, economics and policy to evolve and support the first foray into commercial development.

The Alberta government created the first oil sands policy in 1962. The next year, the Great Canadian Oil Sands (GCOS) venture was formed. It was followed in 1964 by Syncrude, which was then a research consortium. The first commercial oil sands developments followed with GCOS in 1967 and Syncrude in 1978. But in the 1980s, volatile commodity prices and the federal government's National Energy Program changed the game—a few years of economic recession and bad policy moved 50 years of proved technology and expertise to the back burner. And by the early 1990s, the oil sands outlook was becoming increasingly negative.

The recession and the oil price slump were unrelenting. Alberta leaders knew something had to be done to support the economy and ensure oil sands development had a future. The National Oil Sands Task Force was formed in 1995 to examine the options, and in 1996, it produced "A Declaration of Opportunity," signed by all parties including then Prime Minister Jean Chretien. That declaration established a goal of 1.2 million bbl/day of oil sands production by 2020—at the time, an ambitious target.

The declaration paved the way for the necessary fiscal terms required for industry to manage the risks associated with large, long-term capital investments. Most importantly, the declaration ushered in a new era of cooperation among the federal government, the provincial government and industry aimed at advancing the oil sands industry for the benefit of all Canadians. By 2004, economic conditions were much improved. Oil prices were up—first to $50, then $100 in January 2008, and finally, $147 per barrel in July 2008. Thanks to the sudden economic uplift, the declaration's ambitious goal of production for 2020 was exceeded in 2007.

But Alberta's attractive fiscal policy, record-high oil prices and low interest rates created an unintentional perfect economic storm, the eye of which was the small community of Fort McMurray. Costs, both local and global, increased quickly. Booms create short-term imbalance, economically as well as socially. And infrastructure, housing, services, and other pressures became acute in Fort McMurray. In contrast to the 1990s, the pendulum now had swung the other way, with environmental and social issues taking on increasing prominence, in part due to local and regional issues in the Wood Buffalo region and in part due to changes in broader societal expectations.

Questions emerged about whether Alberta's and Canada's environmental policies were prepared for this pace and level of resource development, especially in light of increasing global focus on climate change. Then, as it started to grow, the oil sands industry took a downward turn, a victim of local pressures and a global economic recession. Although given the multiyear timescale for regulatory approvals and construction, some projects under construction continued and a major project began its construction phase.

Today, 3 years after the most recent downturn began, economic stability is returning and the oil sands industry is on the upswing. A number of projects are back on the books, and some new projects have been announced, even though we continue to be in an uncertain period in terms of policy and the economy; economic returns from oil sands projects remain challenging.

At the same time, the industry continues to implement new technologies with lower environmental impacts. However, some environmentalists are advancing an aggressive anti-oil sands campaign. They often focus on environmental performance, ignoring the economic, and energy security aspects of this resource—all three are fundamental. They claim the energy industry can be made "greener" in very short periods of time through a global transformation to renewable forms of energy—a laudable goal over the longer term but unrealistic for the near to medium term. In many respects, the issue is not oil sands. Rather, it's broader climate policy and the off-hydrocarbons agenda that underlies these campaigns, and the objective is to accelerate the energy transformation process.

2.3 OPPORTUNITIES AND CHALLENGES

The world will need all forms of energy, developed responsibly, and would be well served to stop the "my energy is better than your energy" debate and focus on working together to improve the full portfolio of energy supply while improving environmental performance. Canada has 174 billion barrels of oil that can be recovered economically with today's technology, including the 169 billion barrels located in the oil sands (ERCB, 2011; Oil and Gas Journal, 2011). And while some 79% of the world's oil reserves are owned or controlled by national governments, more than half of the remaining 21% accessible for private sector investment are found in Canada's oil sands.

According to the International Energy Agency's (IEA, 2011) report, worldwide primary energy demand will increase by about 40%—under its "new policies" scenario, and hydrocarbons remain the dominant energy source (Figure 2.4). With conventional oil supply declining, the need for unconventional resources like oil sands will increase. Consequently, there is opportunity for oil sands to play a pivotal role in meeting Canada's energy needs and contributing to global demand for several decades.

Canada is uniquely positioned to provide an abundance of safe, secure energy. The oil sands industry is a secure supplier with a growing need for labor, producing jobs, economic prosperity, and government revenue. The challenges? We need to continue developing these resources responsibly. We need to remain focused on reducing environmental impact. We need to

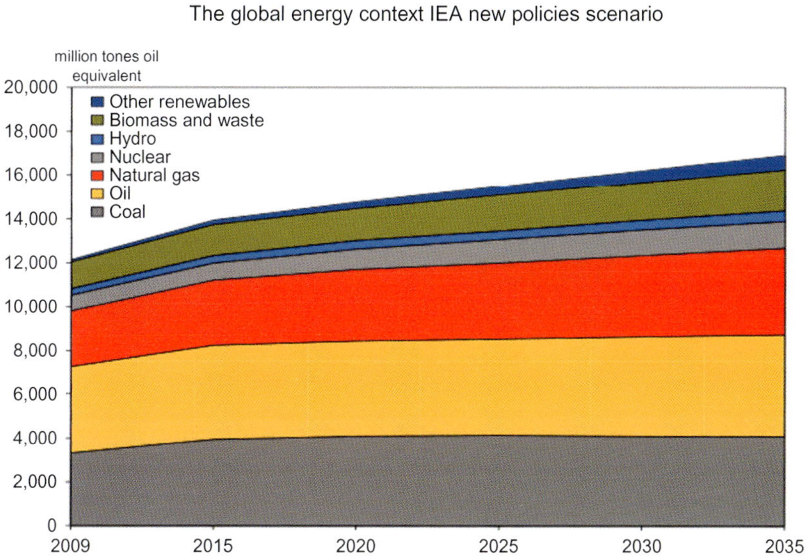

The global energy context IEA new policies scenario

FIGURE 2.4   Global primary energy demand. Source: *International Energy Agency. World Energy Outlook (2011).*

remain vigilant and ensure that the fiscal regime supports investment and development of the capital-intensive projects. And we need access to markets for our products.

The key to realizing the opportunities and addressing the challenges is new technology and innovation. The oil sands industry invests billions of dollars to discover develop and implement new and better technology. These initiatives are ongoing on a major scale because it is good business—they lower capital and operating costs and reduce environmental impact.

2.3.1   Greenhouse Gases and Air Quality

Canada produces about 2% of the global greenhouse gas (GHG) emissions. Oil sands account for 6.9% of Canada's GHG emissions and just over 0.1% (1/1000th) of global GHG emissions (Figure 2.5). Despite these low levels in the global context, industry continues to reduce GHG intensity. Since 1990, GHG emissions associated with every barrel of oil sands crude produced have been reduced by 26%. New and emerging technological innovations will continue to improve environmental performance. Examples include processes that combine hydrocarbons, like propane or butane (which are recovered with the bitumen and recycled), with steam to improve oil recovery rates and reduce GHG emissions per barrel; technologies to process bitumen and remove the heaviest hydrocarbon components on-site and non-aqueous extraction techniques to reduce the need for water and settling ponds.

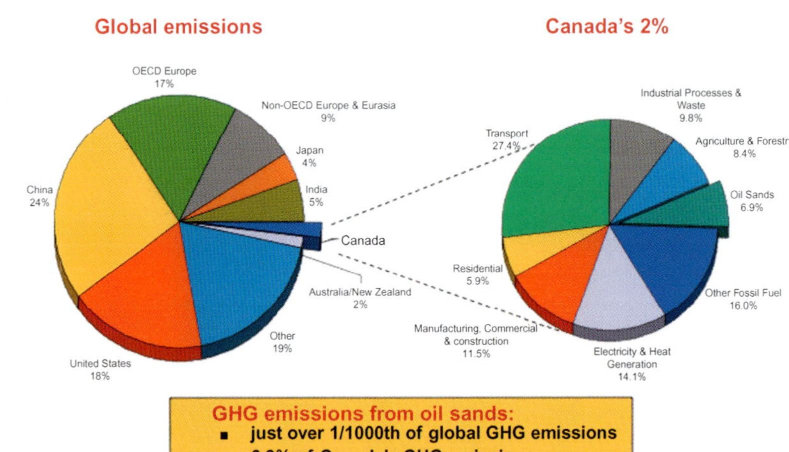

FIGURE 2.5   Oil sands contribution to global anthropogenic CO_2 emissions.

Further, climate policy initiatives are already in place in some jurisdictions in Canada, including Alberta, where industry must reduce its GHG emissions by 12% or pay a price on carbon of $15 per ton under Alberta legislation—comparable to the price paid under Europe's current carbon trading regime. The key feature of this legislation is that revenue accumulates in a technology fund earmarked by the Alberta government for reinvestment in GHG reduction-related solutions such as carbon capture and storage and efficiency advancements.

Producing and burning fossil fuels for electricity generation, industrial uses, and transportation and for heating in homes and buildings emit carbon dioxide, one of the most commonly cited GHGs. About 80% of oil-related carbon dioxide comes from combustion, not production. The industry measures carbon dioxide emissions from the start of production (wells) through to combustion (wheels)—a process called wells-to-wheels, life-cycle intensity analysis. Oil sands crude has similar carbon dioxide emissions (Figure 2.6) to other heavy oils and is 6% more intensive than the average U.S. crude oil supply on a wells-to-wheels basis (IHS CERA, 2010).

The Wood Buffalo Environmental Association (WBEA) monitors the air in the oil sands region in and around Fort McMurray—the center of oil sands production—24 h a day, 365 days a year. Monitoring is science based, transparent, and credible. WBEA's air quality monitoring network is one of the most extensive in North America, and the organization provides monitoring information in real time on its Web site (http://www.wbea.org/).

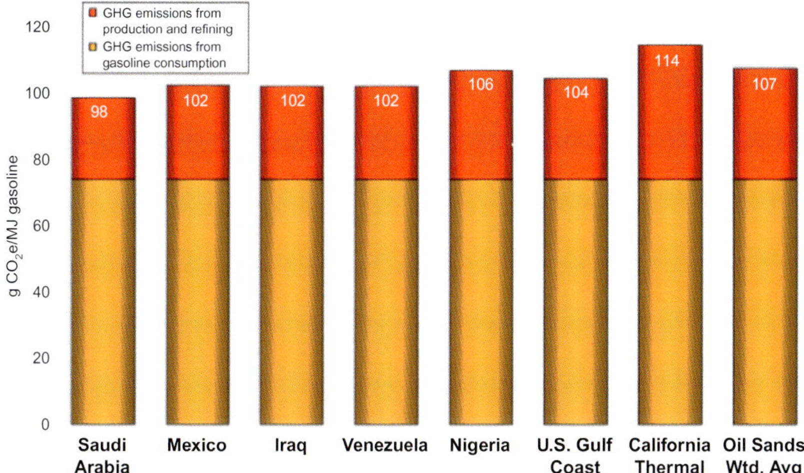

FIGURE 2.6   Full cycle GHG emissions. Source: *Jacobs Consultancy, Life Cycle Assessment Comparison for North America and Imported Crudes, June 2009.*

According to data provided by the Alberta Clean Air Strategic Alliance (http://www.casadata.org/comparison/index.asp), air quality in Fort McMurray is better than many North American cities, including Toronto, Edmonton, and Seattle. Based on analysis of average concentrations of common air pollutants, air quality generally has not significantly deteriorated in the Wood Buffalo region, despite an increase in emissions associated with population growth and oil sands development (Kindzierski, 2010).

2.3.2   Water Use, Tailings Ponds, and Quality

It takes water to produce most things that define our quality of life—food, electricity, oil and gas, the green spaces we all enjoy, a cup of coffee, even a pair of jeans. The oil and gas industry believes responsible water use; economic growth and energy security are necessary, achievable and not mutually exclusive.

The Alberta government closely regulates the use of water and industry must apply for permits to divert fresh water from its original source. The amount of water allocated under the permits is based on sustaining Alberta's groundwater and surface water. In 2009, for example, irrigation and agriculture represented 44% of total Alberta water allocations to industry and the oil sands industry portion represented 7%. Even then, the oil and gas industry typically uses less than one-third of its total annual water allocation.

Oil sands producers recycle between 80% and 95% of the water used to produce oil and continue to look for ways to reduce fresh water use. At present, *in situ* production requires an average half-barrel and mining requires

- Accelerating reclamation
- Minimizing future tailings

- Industry tailings collaborative
  - Accelerate new technology
    implementation

Suncor Pond 1

Research on dry and consolidated tailings

FIGURE 2.7 Tailing pond before and after reclamation.

two-to-four barrels of water for each barrel of oil. The Athabasca River is the main source of water for mining projects, about 80% of which is recycled. In 2009, the industry withdrew 107 million m^3 of water, 0.5% of average total river flows and about 3.4% of the lowest weekly winter flow. *In situ* projects recycle about 95% of the water they use and draw no water from the Athabasca River. *In situ* operators are shifting to the use of saline water from subsurface aquifers and most new projects are using 100% saline water for steam.

Settling or "tailings" ponds are found at oil sands mining sites. These large, engineered dam and dyke systems are designed to contain and settle the water, sand, fine clays, silts, and residual bitumen (by-products of the oil sands mining and extraction process), allowing the water to be recycled.

All oil sands operators are required under law to have plans in place to convert fine tailings to reclaimable landscapes. After separation, the middle layer has the consistency of yogurt. This combination of water and clay can take up to 30 years to separate and dry (Figure 2.7). New, game-changing technology has recently been introduced that accelerates drying time to months rather than decades. And late past year, major oil sands operators announced a collaborative agreement to advance tailings pond management. The announcement by Canadian Natural Resources, Imperial Oil, Shell Canada, Suncor Energy, Syncrude Canada Ltd., Teck Resources, and Total E&P Canada reflects the companies' commitments to socially and environmentally responsible operations.

Each company pledged to share its existing tailings research and technology and to remove barriers to collaborating on future tailings research and development. Bringing all of the companies' scientific expertise together creates a strong foundation of resources that will lead to improvements in tailings management. The companies have agreed to these core principles to guide the actions of the research collaboration:

- Make tailings technical information more broadly available to industry members, academia, regulators, and others interested in collaborating on tailings solutions;
- Collaborate on tailings-related research and development and technology among companies as well as with research agencies;
- Eliminate monetary and intellectual property barriers to the use of knowledge and methods related to tailings technology and research and development; and
- Work to develop an appropriate framework, so tailings information is organized, verified through peer review and kept current.

In addition to developing new technology to reduce how much water it uses, the industry focuses on ensuring water quality is not compromised, and existing tailings are reclaimed faster. In 2010, the Royal Society of Canada commissioned an Expert Panel of Canadian Scientists to review and assess evidence relating to several perceived environmental impacts of oil sands, including impact on regional water supply. In its report (RSC, 2010), the panel stated "Current evidence on water quality impacts on the Athabasca River system suggests that oil sands development activities are not a current threat to aquatic ecosystem viability."

2.3.3   Land—Impact and Reclamation

Canada's oil sands industry is committed to reducing its footprint, reclaiming all land affected by operations and maintaining biodiversity—and its work to date is successful. A recent report from the Alberta Biodiversity Monitoring Institute—an independent, third-party, multistakeholder group—says the Lower Athabasca region's living resources are 94% intact (ABMI, 2009).

Alberta's oil sands lay under some 142,000 km^2 of land. Only about 3%, or 4802 km^2, could ever be impacted by mining because the remaining reserves that underlie 97% of the surface area are too deep to mine and only recoverable by *in situ* methods, which require little surface land disturbance (Figure 2.8). Since operations began 44 years ago, about 715 km^2 of land (slightly larger than the size of the City of Edmonton, Alberta, population of $\sim$820,000), including 0.02% (2/1000th) of Canada's boreal forest, has been disturbed by oil sands mining operations.

Some claim the oil sands are destroying an area the size of England (about 130,000 km^2). In fact, the total mining footprint covers an area about 0.5% the size of England, and 10% of that land has been or is being reclaimed. The total area that could be impacted by mining is about 4% the size of England.

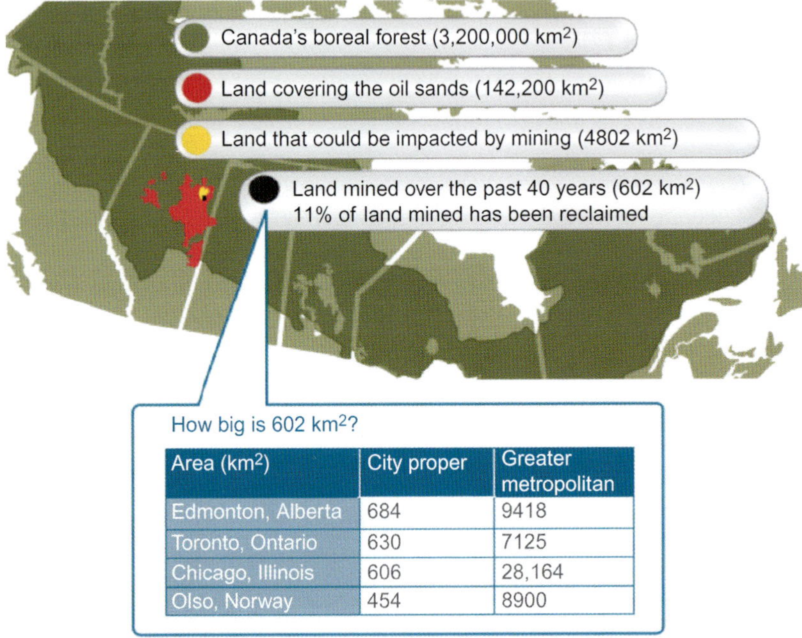

Area (km²)	City proper	Greater metropolitan
Edmonton, Alberta	684	9418
Toronto, Ontario	630	7125
Chicago, Illinois	606	28,164
Olso, Norway	454	8900

FIGURE 2.8   Land use and reclamation in the boreal forest.

Alberta law requires reclamation of all lands disturbed by oil sands operations. All companies must develop a reclamation plan that spans the life of any project. The reclamation process involves monitoring, seeding, fertilizing, tree planting, seed collecting, topsoil salvaging and replacing, landform creation and contouring. Companies apply for reclamation certification when vegetation is mature, the landscape is self-sustaining as a thriving ecosystem, and the land can be returned to the Crown for public use.

2.3.4 Market Access

Market diversification is critically important to the industry because companies making these large, long-term investments to develop Canada's oil sands need to ensure they will have access to markets for their product. Our business relies on export markets, in the same way Canada's forestry, agriculture, high-technology, and automotive industries rely on exports. Canada is a trading nation and a global resource powerhouse, but we need to improve access to oil and gas markets to strengthen and continue to build Canada's position as a safe, secure, and reliable global energy provider.

Canada produces more crude oil than it can consume domestically. In 2010, available crude supply from western Canada was 2.7 million bbl/day.

This supply included about 1.6 million bbl/day from the oil sands, 60% of which is upgraded in Alberta. After meeting the domestic demand for western Canadian crude oil of 828,000 bbl/day, the remaining supply—over 1.9 billion bbl/day—was exported, virtually all of it to the United States.

Canada is already the largest supplier of crude oil to the United States. In 2010, U.S. imports from Canada were 2 million bbl/day (22% of total imports) including just over 1 million bbl/day from the oil sands. We will provide more oil in the future, much of it from growing oil sands production. This will create jobs in both countries and replace imports into the United States from less reliable sources.

A recent poll (CAPP, 2011) shows Americans strongly prefer oil from Canada rather than other supply sources such as countries in the Middle East. The poll indicates a vast majority of Americans believe U.S. government policies should support the use of oil from Canada's oil sands. Moreover, a major opportunity exists to sell Canadian oil to growing markets in Asia. In 2009, China became the world's top energy consumer. Japan and South Korea have expressed interest in developing long-term access to Canadian oil and gas. Asia's growing energy demands can decrease Canada's dependence on exports to the United States and raise revenue for provincial governments. Canada has the resources. United States and Asia markets need and want the products. Greater access will help to expand transportation networks that facilitate global supply chains between Canada and the markets. New pipelines also will strengthen Canada's economy by providing new sources of revenue and jobs for Canadians.

It's important to recognize local and regional perspectives on proposals for greater access and balance these with the perspective that responsible development and operation of new pipelines are feasible, necessary, and in Canada's national interest. Effects on the environment, including wildlife, fisheries, proximity to shipping routes for fishing and marine areas, air quality, coastal marine life and communities, environmental standards, and increased access to sensitive areas are examples of the issues to be addressed through the extensive, open, and thorough Canadian and U.S. regulatory processes.

2.4   THE PATH FORWARD

Go-forward actions are required of governments and industry to realize the full economic opportunities offered by the oil sands while meeting the public's expectations for responsible environmental and social performance. Many of these initiatives are already under way, but we need to improve continuously.

We need to balance energy security, economics, and environmental and social imperatives while developing our resource in the context of a Canadian energy strategy that defines and guides our actions as a responsible provider on the global energy stage.

2.4.1   Governments

From a policy perspective, we need

- Policy that is right for Canada;
- Policy that advances economic interests and energy security, while ensuring responsible environmental and social outcomes—this is not a one-dimensional challenge;
- Policy that maintains open borders to enable trade;
- Policy that encourages investment and use of the technologies and innovation necessary to enable development, reduce costs and improve environmental performance;
- Policy founded on ensuring competitiveness to attract financial and intellectual capital, which are fundamental to maintaining a healthy energy sector; and
- Policy that seeks economy-wide solutions—our economic structure and our environmental performance improvement objectives require leadership throughout the energy system.

Governments need to make the case for the oil sands, not as an advocate for industry, but as advocates for sound policy and regulation and as part of their legitimate role in advancing Alberta's and Canada's interests. And governments need to have clear, straightforward conversations about energy and environmental policy. We cannot resolve these issues if we shy away from transparency about the impacts and implications on consumers of different policy choices. In the end, energy consumers must be willing to bear some impact and to change behaviors if we are to make transformative changes to the energy system.

2.4.2   Industry

Industry must ensure that it maintains the social license to develop and operate in the oil sands region. We have significantly improved performance across a broad spectrum—operations, cost recovery, and environment—but we need to continue to improve. Our reputation depends on our performance and how we communicate with our stakeholders. On performance, technology and innovation will be the key enabler. We need to

- Continue investing aggressively in technology development;
- Collaborate effectively among ourselves and with governments, academia, and other industries, particularly regarding technology that enables environmental performance improvements;
- Seek new approaches to funding, development, and deployment of technology;
- Work with multistakeholder organizations like WBEA, governments and others to ensure we have science-based measurement and monitoring

systems that stand the test of third-party review and are improved as new information becomes available;

- Encourage and participate in the creation of a Canadian energy strategy that defines and guides our actions;
- Contribute constructively to the ongoing policy dialog about our energy future and about climate change policy;
- Reduce GHG emissions—over time, we need to work toward making an oil sands barrel as good or better than competing barrels on a life-cycle basis;
- Encourage pragmatic actions such as energy efficiency and conservation that just make good sense;
- Be as transparent as possible in performance reporting and on our commitment to continuous performance improvement; and
- Ensure our minds are always open to credible new sources of information.

2.4.3 Working Together

Governments and industry need to provide balanced and credible information to inform decisions that will be made by the public. We need to

- Continue addressing the public's concerns about environmental and social impacts of oil sands development;
- Tangibly demonstrate industry's commitment to performance improvement through technology and innovation; and
- Ensure we stay focused on both performance and communication—they are interdependent.

The oil sands are a success story in which we should take great pride as a country.

They represent a tremendous opportunity for Canadians. Realizing this opportunity and addressing the challenges that come with it requires technology and innovation to improve recovery, drive down costs and address environmental concerns. It requires bold, consistent, creative leadership by industry and governments—knowing we are doing the right things—and the passion to improve. Clearly, this is of global interest.

REFERENCES

Alberta Biodiversity Monitoring Institute (ABMI), 2009. The status of birds and vascular plants in Alberta's lower Athabasca planning region 2009: preliminary assessment. Oil & Gas Journal December 5, 2011 Volume 109, Issue 49.

Canadian Association of Petroleum Producers (CAPP), 2011. Oil & Gas Journal December 5, 2011 Volume 109, Issue 49.

Canadian Energy Research Institute, Economic Impact of New Oil Sands Projects in Alberta (2010–2035), Study 124.

Energy Resources Conservation Board ST98-2011, Alberta's Energy Reserves 2010 and Supply/ Demand Outlook 2011–2020.

International Energy Agency (IEA), 2011. Oil & Gas Journal December 5, 2011 Volume 109, Issue 49.

IHS CERA, 2010. Canadian Oil Sands Energy Dialogue. Oil & Gas Journal December 5, 2011 Volume 109, Issue 49.

Kindzierski, W., 2010. Ten-year trends in regional air quality for criteria pollutants in the Athabasca oil Sands Region. Extended abstract 2010-A-1079-AWMA, June 22–25, Calgary, Alberta.

Oil and Gas Journal, 2011. Oil & Gas Journal December 5, 2011 Volume 109, Issue 49.

Royal Society of Canada (RSC), 2010. Environmental and health impacts of Canada's oil sands industry. Oil & Gas Journal December 5, 2011 Volume 109, Issue 49.

Chapter 3

Energy and Environment: Toward Achieving the Balance in Alberta

M. Lowey[1]

Institute for Sustainable Energy, Environment and Economy (ISEEE), University of Calgary, Calgary, Alberta, Canada
[1]*Corresponding author: e-mail: mlowey@ucalgary.ca*

ABSTRACT

Alberta's energy-and-environment story is a lot more than the chapter on "dirty oil" from the oil sands, and even that chapter requires some much-needed context. The province certainly faces some significant environmental challenges, especially in continuing to expand development of its vast oil sands deposits. However, Alberta also has pioneered environmental stewardship programs, although sometimes the province's "deliverables" do not match its good intentions on the environment. This presentation puts into global and North American context the economic benefits of Alberta's energy resources. It looks at air pollution and greenhouse gas (GHG) emissions and how Alberta, through such organizations as the Energy Resources Conservation Board and the Clean Air Strategic Alliance, has led North America—and indeed the world—in: reducing air pollution, requiring large emitters to reduce their GHG emissions intensity, and providing significant support to help implement industrial-scale carbon capture and storage projects. This presentation also looks at Alberta's and the oil sands industry's management of water, land, and waste. It provides context for the improvements that have been made in these areas, as well as some thoughts on what still needs to be done to create a truly world-class monitoring system for oil sands development.

3.1 INTRODUCTION

Alberta's energy-and-environment story is a lot more than the chapter on "dirty oil"—and even that chapter needs to be put into perspective. The western Canadian province certainly faces some significant environmental challenges,

Disclaimer: The content and opinions expressed by the author in this chapter do not necessarily reflect the views of the WBEA or of the WBEA membership.

Developments in Environmental Science, Vol. 11. http://dx.doi.org/10.1016/B978-0-08-099443-7.00003-X

especially in continuing to expend development of its vast oil sands deposits. However, Alberta also has pioneered environmental stewardship programs, although sometimes the province's "deliverables" do not match its good intentions on the environment.

First, let us put Canada and Alberta's energy resource, and the economic benefits from that resource, into a global context. The world consumed about 86 million barrels of oil per day (b/d) in 2008. By 2035, consumption will be approximately 112 million b/d, according to the most recent forecast by the U.S. Energy Industry Administration (EIA, 2011). Demand for natural gas—the world's fastest-growing fossil fuel—is expected to grow from about 110 trillion ft^3 in 2010 to approximately 169 trillion ft^3 in 2035, the EIA says. Fossil fuels will still account for 75% of global primary energy consumption in 2035, down slightly from 81% in 2010, according to the Organization for Economic Cooperation and Development (OECD)'s International Energy Agency.

Canada is the world's sixth largest crude oil producer and the third largest natural gas producer. And Alberta produces 69% of Canada's crude oil and 80% of its natural gas. In 2010, the petroleum industry's net capital spending in Canada amounted to $51 billion (including $38 billion in Alberta), generating $18 billion in taxes and royalties paid to governments, which represents member companies that produce about 90% of Canada's natural gas and crude oil. Capital spending in Alberta's oil sands alone was $17.2 billion in 2010, generating $3.7 billion in royalties for the province (Table 3.1).

For Albertans, natural gas was the largest single source of nonrenewable resource development revenue, accounting for about 63%, or more than $42.6 billion in provincial royalties, from fiscal year 2000/2001 to 2006/2007. In FY 2008/2009 alone, natural gas and by-product royalty revenue amounted to $5.8 billion, according to the Government of Alberta. Conventional crude oil production was the third largest source of this revenue for Albertans during the 2007/2008 fiscal years, amounting to $1.655 billion and increasing to $1.8 billion in FY 2008/2009.

According to the provincial government's energy department, Alberta Energy (www.energy.gov.ab), one out of every working seven Albertans—or 170,000 people—is directly or indirectly employed in the energy industry. The Canadian Association of Petroleum Producers (CAPP) says the oil and gas industry currently supports 550,000 jobs across Canada. Oil, gas, and their by-products account for more than 70% of Alberta's exports (Table 3.1).

With the EIA, OECD and other organizations forecasting an increase in demand for energy in the United States and worldwide, Alberta's strategic position as a secure and reliable energy supplier will become more important. But how has Canada's energy province done in balancing development of its abundant fossil fuel resources with environmental stewardship?

TABLE 3.1 Energy and Environment in Alberta—By the Numbers

Economic impact of Canada's petroleum industry
  (i) $51 billion—net capital spending by the industry in 2010 across Canada
  (ii) $12.2 billion—Alberta oil and gas royalty revenues (fiscal year 2008/2009) that contributed more than 30% of the Government of Alberta's total revenue. Of this amount, $3 billion in royalties came from oil sands projects
  (iii) 550,000—jobs across Canada supported directly or indirectly by the energy industry
  (iv) 170,000—people directly or indirectly employed in Alberta's energy industry
  (v) 44%—percentage of oil sands-related employment outside of Alberta

Future economic impact of Canada's petroleum industry (forecast over the next 25 years)
  (i) $3.56 trillion—economic impact generated to Canada's GDP by a $1.09-trillion investment by the petroleum industry
  (ii) $1.1 trillion—government revenues generated by Canada's petroleum industry, including:
    • $408 billion in federal tax
    • $711 billion in provincial tax and royalties
  (iii) 24.5 million person years (mmp/yr)—employment created by Canada's petroleum industry, including 14.1 mmp/yr in Alberta

Alberta's oil sands resource
  (i) 1.7 trillion barrels (bbl)—estimated oil contained in the oil sands
  (ii) 170 billion bbl—oil in oil sands reserves (deposits that are recoverable with current technologies), the second largest oil reserves in the world behind Saudi Arabia
  (iii) million bbl—oil sands production in 2010
  (iv) 3.5 million bbl—oil sands production expected by 2020
  (v) 73%—percentage of Canada's oil production forecast to come from oil sands by 2020
  (vi) 80%—percentage of Alberta's *in situ* oil sands deposits (too deep to be mined from the surface), recoverable through steam-assisted gravity drainage and other technologies that do not require mine pits or tailings ponds
  (vii) 100—active oil sands projects in Alberta (as of November 2011). Of these, six mining projects have been approved, including five projects currently producing bitumen. The remaining projects use various *in situ* recovery methods

Alberta's fossil energy resources
  (a) Oil
    (i) 503,000 bbl—conventional crude oil production per day in 2008
    (ii) 1.37+ million bbl—amount of crude oil exported daily to U.S. markets in 2008
  (b) Natural gas
    (i) 87 trillion ft^3—estimated recoverable conventional natural gas remaining in Alberta
    (ii) 500 trillion ft^3—estimated coal bed methane (unconventional gas) deposits in Alberta's coal seams; it is not known yet how much of this is economically recoverable

---

Continued

TABLE 3.1 Energy and Environment in Alberta—By the Numbers—Cont'd

    (iii) 5 trillion ft^3—natural gas produced per year in Alberta (enough to heat every home in the province for about 35 years)

    (iv) 80%—percentage of Canada's natural gas production that is from Alberta

    (v) 1.15 billion ft^3—natural gas that Alberta exports daily to the rest of Canada

    (vi) 2.48 billion ft^3—natural gas that Alberta exports daily to the United States

(c) Coal

    (i) 34–37 billion tons—coal reserves remaining (70% of Canada's coal reserves, enough to last for about 400 years at current usage rates)

    (ii) 32.5 Mt—total marketable coal production in 2008 25 Mt—amount of coal that Alberta uses per year to generate electricity

Electricity

    (i) 13,000 MW—current generation capacity

    (ii) 5000 MW+—new generation facilities built since 1998.

    (iii) 521 MW—wind power generation.

    (iv) 21,000 km (13,049 miles)—length of transmission lines

Greenhouse gas emissions in Alberta

    (i) 113.1 Mt—total CO_2 equivalent (CO_2e) emissions in 2009

    (ii) 49.9 Mt—CO_2e emissions from Alberta's utilities sector in 2007 or 44% of the province's total reported GHG emissions

    (iii) 8.9 Mt—total CO_2e emissions from Alberta's *in situ* oil sands projects.

All financial figures are in Canadian dollars.

Sources: Government of Alberta (Alberta Energy and Alberta Environment), Canadian Association of Petroleum Producers, Canadian Energy Research Institute, and Canadian Centre for Energy Information.

3.2 AIR POLLUTION AND GREENHOUSE GAS EMISSIONS

Alberta was the first jurisdiction in Canada to create a governmental environment department, in 1971. The province's oil and gas regulator, the Energy Resources Conservation Board (ERCB) is seen as a leader by jurisdictions and organizations around the world—including the World Bank which adopted the ERCB's petroleum-flaring emissions-reduction model.

Driving much of the effort to clean up air pollution is the Clean Air Strategic Alliance (CASA), a multistakeholder partnership of representatives selected by industry, government, and nongovernmental organizations that recommend strategies to assess and improve air quality in Alberta. Through CASA's work, Alberta's petroleum industry had, as of 2010, reduced flaring emissions (from burning off solution gas that cannot be conserved) by 77%, to 307 million ft^3 compared with a 1996 baseline of 1.34 billion ft^3, according to the Energy Resources Conservation Board's "Upstream Petroleum Industry Flaring and Venting Report, 2010" (ERCB ST60B-2100) (Figure 3.1). However, flaring emissions increased about 17% in 2010 compared with the previous year, mainly due to increase in new oil sands bitumen production during

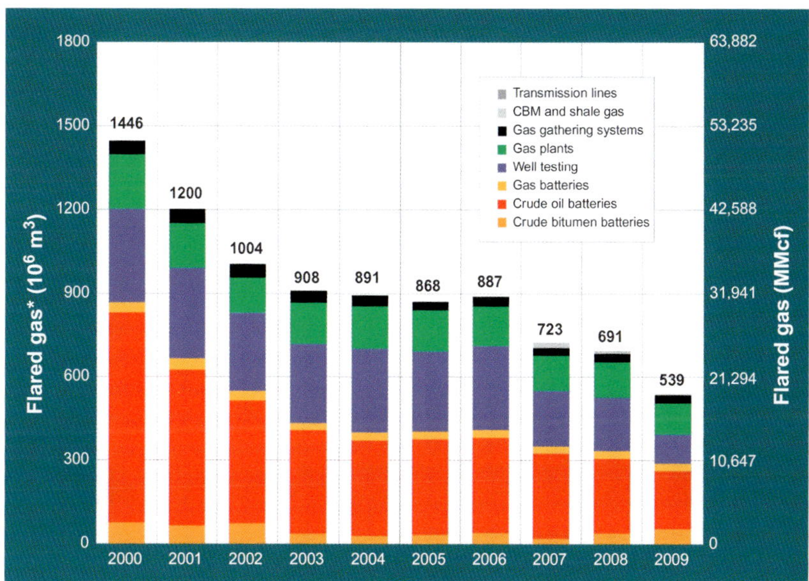

FIGURE 3.1   Flared from all upstream oil and gas sources. *Due to rounding, some totals may not exactly match the sum of the source categories. Source: *Energy Resources Conservation Board, AB.*

the year and low gas prices, which makes the economic viability of conserving solution gas more challenging (Figure 3.2). Venting emissions (from unburned gas releases) have been cut by about 52%, to 335 million ft^3 compared with a 2000 baseline of 704 million ft^3, according to the ERCB. Sulfur emissions from "grandfathered" natural gas plants, processing so-called sour gas (containing hydrogen sulfide) in the province, also have been reduced by 61%, to about 77 ton/day in 2010 compared with a baseline of about 202 ton/day in 2000, according to the ERCB's "Sulphur Recovery and Sulphur Emissions at Alberta Sour Gas Plants" annual report for 2010 (Figure 3.3). Grandfathered plants are those that do not meet the sulfur recovery requirements listed in ERCB Interim Directive 2001–2003. All grandfathering will end by December 31, 2016.

Alberta's air quality monitoring programs include a program, operated by the Wood Buffalo Environmental Association (WBEA), in the Regional Municipality of Wood Buffalo. This region includes Fort McMurray, a city of about 65,000 people in northern Alberta around which most of the province's oil sands industry is concentrated. The WBEA's round-the-clock monitoring shows that according to CASA, even with the oil sands as a next-door neighbor, Fort McMurray's air quality is consistently better for criteria air pollutants than in many large Canadian cities, including Toronto, Montreal, and Vancouver. Odors, however, remain a concern for some communities in the region.

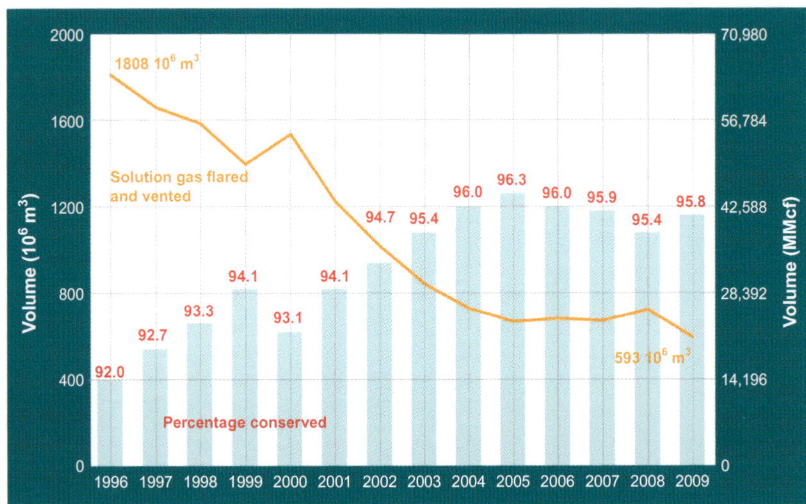

FIGURE 3.2 Solution gas conserved, flared, and vented. Source: *Energy Resources Conservation Board (ERCB), AB.*

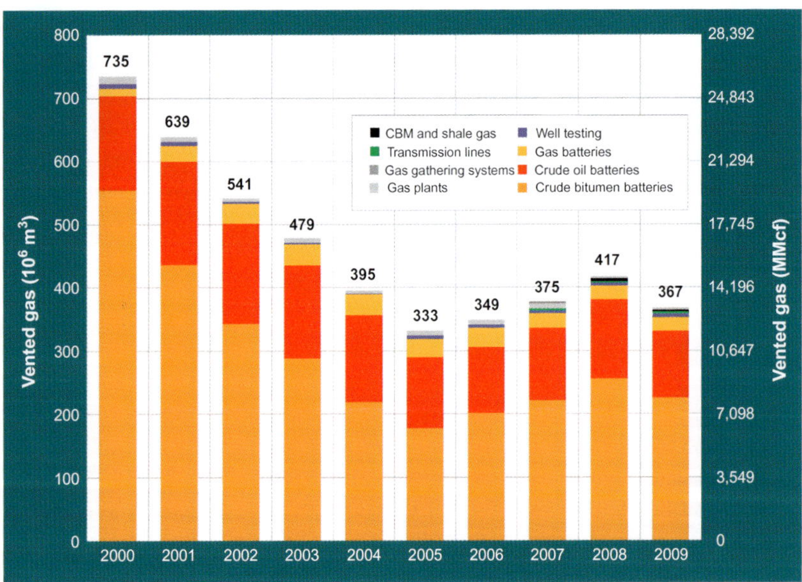

FIGURE 3.3 Vented from all upstream oil and gas sources. Source: *Energy Resources Conservation Board (ERCB), AB.*

When it comes to greenhouse gas (GHG) emissions, Canada contributes 2% of the world's overall GHGs, and Alberta is the largest emitter of GHGs in the country. The largest source of GHGs in the province historically has been coal-fired power plants (as it is in many U.S. states), not the oil sands. However, by 2010, the oil sands sector and the electricity/heat sector each contributed about 21% of Alberta's total 233 Mt of GHGs.

Alberta's oil sands industry contributed 38.4 million tons (Mt) of GHGs or approximately 5% of Canada's GHG emissions in 2007 and 0.1% of overall GHGs in the world (Table 3.1). For comparison, coal-fired power plants in the United States produce close to 2 billion tons of carbon dioxide (CO_2) per year, or 27% of total U.S. emissions, according to the Pew Center on Global Climate Change. The total GHG emissions from Alberta's oil sands, therefore, are equal to about 2% of U.S. power generation GHG emissions. Nevertheless, Alberta's oil sands do represent the fastest-growing source of GHG emissions in Canada, rising to 41.9 Mt or 6.5% of the country's GHGs in 2009.

Although the GHG emissions produced by Alberta's oil sands industry represent a small percentage of Canada's emissions and barely a fraction of global GHGs, this is not to say that these emissions are insignificant or unimportant in the worldwide effort to reduce emissions. However, making comparisons across industrial sectors helps prioritize where money and effort should be spent to reduce GHGs to achieve the greatest global benefit. For example, significantly cutting emissions from Alberta's coal-fired power plants—and especially from coal plants in the United States—would have a greater global net benefit than reducing the same percentage of emissions from oil sands operations. But all countries, provinces (and states), municipalities, and industrial sectors need to do their fair share.

Alberta was the first jurisdiction in North America to enact legislation requiring large facilities to reduce their GHG emissions intensity (i.e., the amount of GHGs per barrel of oil produced or other unit of production). As of July 1, 2007, facilities emitting more than 100,000 tons of GHGs annually are required to reduce emissions intensity by 12%. Through this measure, Alberta is targeting a 200-Mt CO_2-equivalent (CO_2e) reduction annually over business-as-usual projects by 2050.

Companies can meet their reductions in three ways:

- Making operational improvements;
- Buying Alberta-based credits; or
- Contributing, at $15 per ton for their emissions over target, to the province's Climate Change and Emissions Management Fund.

During the first 18 months of this initiative, Alberta's large emitters achieved 6.5 Mt of GHG reductions through a combination of offsets and process improvements. After 2 years of operation, the Climate Change and Emissions Management Corporation (CCEMC, http://ccemc.ca/), which manages the Climate Change and Emissions Fund, has allocated approximately

$167 million to support 32 clean technology projects, including $50 million for renewable energy projects. The CCEMC indicates it expects these 32 projects to reduce GHG emissions by 17 Mt (verifiable by an independent auditor) over 10 years.

In addition, Alberta has committed $2 billion to help implement four commercial-scale carbon capture and storage (CCS) projects in the province, although the industry partners for one of these projects, Project Pioneer (http://www.projectpioneer.ca/), announced in April 2012 that they will not proceed with the project due to insufficient market for carbon sales and the price of emissions reductions. The target of the province's initiative is to see 5 Mt of CO_2e sequestered annually by 2015. This is the equivalent of taking 1 million vehicles—or one third of all registered vehicles in Alberta—off the road. The provincial government has said that in the longer term, the bulk of the province's emissions reductions in its climate strategy will come from CCS.

Oil sands companies have reduced the amount of GHG emissions per barrel of oil produced by an average 29% from 1990 to 2009. Suncor Energy Inc., for example, has voluntarily reduced GHG intensity by 51% at its oil sands-mining operation (NRCAN, 2012). Nevertheless, absolute emissions from Suncor's plant have increased due to significant production growth. When full life-cycle GHG emissions are factored in, Canadian oil sands bitumen is comparable with crude oil imported by the United States from other countries, including Venezuela, Mexico, and Nigeria, according to a 2001 study by McCann and Associates (2007) in Calgary and commissioned by the oil sands industry-funded Oil Sands Developers Group. This study found that full-cycle emissions for synthetic oil made from Canada's oil sands are 568 kg of CO_2e per barrel, compared with 579 kg/CO_2e for Venezuelan crude, 557 for Mexico's product and 541 for Nigeria's. Full cycle includes emissions from production, transportation to markets, refining, refining by-products, and end-use consumption.

However, a study done by the U.S. National Energy Technology Laboratory (2008) and sponsored by NETL's Office of Coal and Power Research & Development, found the life-cycle GHG emissions from Canadian oil sands to be among the highest among crude oil types imported by the United States. Canadian oil sands emissions were second highest after Nigeria's oil, with Angola's product having the third highest life-cycle emissions.

A study published by Charpentier et al. (2009) reviewed 13 studies of GHG emissions associated with oil sands. The authors concluded that there were wide differences among the studies in arriving at GHGs from oil sands (although most studies found life-cycle emissions to be higher for oil sands than those for conventional oil), and that "a consensus on the characterization of life-cycle emissions of the oil sands industry has yet to be reached in the public literature." The U.S. Department of Energy's National Energy Technology Laboratory (2008) reported that producing a barrel of oil from the oil sands produces 3.2–4.5 times more GHGs than a barrel of conventional oil.

3.3   WATER MANAGEMENT

In Alberta, about 90% of the water flows through the northern part of the province. Yet, approximately half the province's population lives in the southern part, accounting for 88% of the water demand in Alberta.

In 2003, the Alberta government released "Water for Life: Alberta's Strategy for Sustainability" (GOA, 2003). This strategy established 10 watershed planning and advisory councils, with the local knowledge and expertise to assess each of the province's major watershed basins and develop plans and activities to address issues. An independent review by eight environmental and community groups, called the "Water for Life" strategy "a positive step forward for water management in Alberta . . . If implemented, it has the potential to greatly improve the ways Albertans use and think about water and poises Alberta as a leader in protecting watersheds, but unbalanced progress in implementing the strategy's actions has limited its effectiveness to date." (Water Matters, 2007).

As of 2006, Alberta had allocated more than 9.2 billion m^3 (325 billion ft^3) of water for a variety of uses across the province. Ninety-eight percent of this allocation is from surface water sources; the remaining 2% comes from groundwater. Most of the water (45.1%) is allocated to agricultural uses, followed by commercial (30.5%) and municipal uses (11.9%).

The oil and gas industry is allotted about 7.2% of the water licensed in Alberta, which includes about 1.9% for water- and steam-injection operations to recover oil and gas. Companies are required to investigate alternatives to fresh water for these enhanced recovery operations. Typically, the industry uses about one-third to one-half of its total water allocation, depending on different project stages and life cycles.

Currently, in oil sands pit-mining operations near Fort McMurray, between 2.2 and 5 barrels of freshwater—depending on the operation—are withdrawn from the Athabasca River to produce each barrel of synthetic oil, according to Alberta Energy. For *in situ* operations using steam-assisted gravity drainage to extract the bitumen, up to half a barrel of freshwater is required to produce each barrel of bitumen, although this can increase to two barrels or more in some pilot projects. Approximately 90% of the water used by *in situ* projects is recycled and reused; for mining operations, the figure is 80%.

Existing oil sands-mining operations are currently using 1% of the Athabasca River's total average flow, and 5% of record low weekly winter flow rates. The maximum projected use by existing operations is 2% of annual flow and 10% of record low weekly winter flow rates. Current allocations in the Athabasca River Basin are sufficient to sustain oil sands mining to 2030.

The Pembina Institute, however, points out that current oil sands-mining operations are licensed to divert 349 million m^3 of water per year from the Athabasca River, or twice the amount of water used by the City of Calgary, which has a population of more than 1 million. At least 90% of this water ends up in oil sands waste tailings ponds, which now cover more than 130 km^2, according to the Pembina Institute.

3.4   LAND USE AND WASTE MANAGEMENT

Alberta's oil sands underlie 140,200 km^2 (54,114 square miles) or approximately 37% of Alberta's boreal forest—an area the size of Florida. The area that can be mined for oil sands is 4802 km^2. The mining area currently under development is 530 km^2, or approximately 0.1% of Alberta's boreal forest.

More than 65 km^2 of disturbed lands are in the process of being reclaimed. In March 2008, the Alberta government issued a reclamation certificate to Syncrude Canada Ltd. for a 104-ha (257-acre) parcel of land near Fort McMurray—the first piece of oil sands land to be reclaimed in the province.

The Alberta government and private industry have each invested more than $1 billion in oil sands research, aimed at continuously improving the efficiency and reducing the environmental footprint of oil sands recovery and upgrading. The provincial government has invested at least $7 million on research to speed up reclamation of large oil sands tailings ponds. Oil sands mine tailings consist of bitumen, clay, and process water, along with unrecovered hydrocarbons and naphthenic acids; the ponds are a source of methane, volatile organic compounds, and hydrogen sulfide emissions.

In February 2009, the Alberta government (GOA, 2009) released a comprehensive oil sands management plan: "Responsible Actions: A Plan for Alberta's Oil Sands." The plan outlined "an integrated approach for all levels of government, for industry and for communities to address the economic, social and environmental challenges and opportunities in the oil sands regions." A progress report was issued in 2010.

Alberta is also implementing a new Land-Use Framework that consists of seven strategies to improve land-use decision-making in the province. A key strategy is using cumulative effects management at the regional level to manage the impacts of development on land, water, and air.

3.5   SUMMARY

Although government and industry have made progress on managing the environmental impacts of oil sands development, recent reports by two expert panels found that there is considerable room for improvement. The Royal Society of Canada (2010) report stated: "The environmental regulatory capacity of the Alberta and Canadian Governments does not appear to have kept pace with the rapid expansion of the oil sands industry over the past decade." This panel also said: "The EIA process relied upon by decision-makers to determine whether proposed projects are in the public interest has serious deficiencies in relation to international best practice. Environmental data access for cumulative impact assessment needs to improve."

The federal minister of Environment's Oil Sands Advisory Panel, in its report, A *Foundation for the Future: Building an Environmental Monitoring System for the Oil Sands* (EC, 2010) said: Until Alberta does have a world-class environmental monitoring system, "there will continue to be uncertainty and public distrust, both of industry's environmental performance and governments' oversight."

REFERENCES

Charpentier, A.D., Bergerson, J.A., MacLean, H.L., 2009. Understanding the Canadian Oil Sands Industry's greenhouse gas emissions. Environ. Res. Lett. 4, 014005. http://dx.doi.org/10.1088/1748-9326/4/1/014005.

Environment Canada, 2010. A foundation for the future: building an environmental monitoring system for the oil sands. http://www.ec.gc.ca/pollution/E9ABC93B-A2F4-4D4B-A06D-BF5E0315C7A8/1359_Oilsands_Advisory_Panel_report_09.pdf. Accessed 23 August 2012.

Government of Alberta (GOA), 2003. Water for life: Alberta's strategy for sustainability. http://www.waterforlife.alberta.ca/. Accessed 23 August 2012.

Government of Alberta (GOA), 2009. Responsible actions: a plan for Alberta's oil sands. http://infrastructure.alberta.ca/Content/docType3700/Production/GOA_ResponsibleActions_web.pdf. Accessed 23 August 2012.

International Energy Agency (IEA), 2011. International Energy Outlook 2011. http://www.eia.gov/forecasts/ieo/pdf/0484(2011).pdf. Accessed 23 August 2012.

McCann and Associates, 2007. Typical heavy crude and bitumen derivative greenhouse gas life cycles in 2007. http://www.oilsandsdevelopers.ca/wp-content/uploads/2009/06/GHG-Life-Cycle-in-2007-Aug.-12-08.pdf. Accessed 23 August 2012.

National Energy Technology Laboratory (NETL), 2008. http://www.netl.doe.gov/energy-analyses/pubs/NETL%20LCA%20Petroleum-based%20Fuels%20Nov%202008.pdf. Accessed 23 August 2012.

Natural Resources Canada (NRCAN), 2012. http://www.nrcan.gc.ca/energy/sites/www.nrcan.gc.ca.energy/files/files/OilSands-GHGEmissions_e.pdf. Accessed 23 August 2012.

Royal Society of Canada (RSC), 2010. Environmental and health impacts of Canada's oil sands industry.http://www.rsc.ca/documents/expert/RSC%20report%20complete%20secured%209Mb.pdf. Accessed 23 August 2012.

US National Energy Technology Laboratory Technical, 2008. Petroleum-based fuels, http://www.netl.doe.gov/energy-analyses/pubs/NETL%20LCA%20Petroleum-based%20Fuels%20Nov%202008.pdf. Accessed 23 August 2012.

Water Matters, 2007. http://www.water-matters.org/docs/water-for-life-renewal-analysis.pdf. Accessed 23 August 2012.

Chapter 4

Air Quality in the Athabasca Oil Sands Region 2011

K.E. Percy[*,1], M.C. Hansen[*] and T. Dann[†]

[*]*Wood Buffalo Environmental Association, Fort McMurray, Alberta, Canada*
[†]*Ottawa, Ontario, Canada*
[1]*Corresponding author: e-mail: kpercy@wbea.org*

ABSTRACT

The Wood Buffalo Environmental Association (WBEA) is a multistakeholder, not-for-profit association located in Fort McMurray, Alberta. WBEA has been monitoring air quality in the Athabasca Oil Sands Region (AOSR) of north-eastern Alberta, Canada since 1997. In 2011, WBEA operated 15 air monitoring stations that used continuous and time-integrated techniques to report on ambient air quality. Ambient air quality continuously measured in 2011 at WBEA compliance, attribution, and community air monitoring stations are presented. Maximum 1-h SO_2 concentrations ranged from 56 to 122 ppb at compliance stations and 12 to 83 ppb at community stations. There were no exceedances of the Alberta Ambient Air Quality Objectives (AAAQOs) for SO_2 in 2011. Maximum 1-h NO_2 concentrations ranged from 52 to 154 ppb at compliance stations and 42 to 66 ppb at community stations. There were no exceedances of the 1-h AAAQO for NO_2. Maximum 1-h O_3 concentrations measured at community stations ranged from 77 to 89 ppb. In 2011, there were 15 exceedances of the 82 ppb AAAQO for O_3, all in the period of intense forest fire activity and smoke episodes. In 2011, ambient fine particulate matter ($PM_{2.5}$) levels were highly influenced by the heavy particulate loading in fire smoke resulting in periods of extremely reduced visibility. Maximum 1-h $PM_{2.5}$ concentrations ranged from 406 to 451 $\mu g\,m^{-3}$. In 2011, there were 97 exceedances of the 24-h AAAQO for $PM_{2.5}$ of 30 $\mu g\,m^{-3}$. Beta attenuation technology being evaluated by WBEA measured short-term concentrations as high as 900 $\mu g\,m^{-3}$. Ninety-four exceedances (97% of total) of these occurred during the fire event. Maximum 1-h H_2S/TRS concentrations ranged from 6 to 98 ppb at compliance/attribution stations, and from 3 to 7 ppb at community stations. Maximum 24-h concentrations ranged from 1 to 23 ppb at compliance/attribution stations and from 1 to 3 ppb at community stations. In 2011, there were 23 exceedances of the 1-h AAAQO.

Disclaimer: The content and opinions expressed by the authors in this chapter do not necessarily reflect the views of the WBEA or of the WBEA membership.

Developments in Environmental Science, Vol. 11. http://dx.doi.org/10.1016/B978-0-08-099443-7.00004-1

This was 74% less than in 2010 and 89% less than in 2009. There were no exceedances in 2011 at community stations. Time-integrated air quality measurements included canister sampling for 60 VOCs and 20 RSCs at nine air monitoring stations. Most frequently, measured VOCs were benzene, toluene, acetone, butane, isopentane, isobutane, m,p-xylene, 2-methylpentane, hexane, pentane, and ethylbenzene. Most frequently, measured RSC were carbonyl sulfide, carbon disulfide, H_2S, and dimethyl disulfide. Twenty-three polycyclic aromatic hydrocarbon (PAH) species were routinely measured in low concentrations at four community stations. Phenanthrene, acenapthene, acenapthylene, fluoranthene, fluorene, and pyrene had highest mean concentrations.

Between October 21, 2010 and May 31, 2011, Environment Canada operated a gaseous mercury analyzer at the Patricia McInnes air monitoring station in Fort McMurray. Excluding data collected during the forest fire period in May 2011, average ambient total gaseous mercury (TGM) concentrations averaged 1.40 ± 0.15 ng m^{-3}. This value is similar to average TGM concentrations measured elsewhere in Canada.

Air quality monitoring observations in 2011 were dominated by a massive forest fire complex north of Fort McMurray that consumed some 700,000 ha and burnt to within a few kilometers of the community of Fort McKay. Data presented on O_3 and $PM_{2.5}$ were particularly influenced and skewed to higher than normal values. Air quality health index (AQHI) values calculated for the four reporting WBEA air monitoring stations indicated that air quality posed a low risk to health between 96% and 99.3% of the time in 2011. AQHI values presenting a high health risk (0.9–1.3% of available 8760 h per station) all occurred during the forest fire smoke episodes. While exceedances of the odor perception-based AAAQO for H_2S/TRS of 10 ppb for 1 h, and 3 ppb for 24 h have decreased significantly since 2009, odors, nevertheless, remain a concern in some communities.

4.1 THE WOOD BUFFALO ENVIRONMENTAL ASSOCIATION AMBIENT AIR QUALITY MONITORING NETWORK

The history of air quality monitoring by the Wood Buffalo Environmental Association (WBEA) has recently been reviewed by Phillips (2010). WBEA was initially established to monitor air quality in the Regional Municipality of Wood Buffalo (RMWB) using continuous and time-integrated filter-based methods. During 1997–1998, WBEA transitioned from industry (Syncrude and Suncor) operation and maintenance of their respective compliance air monitoring stations (AMSs), and Alberta Environment (AENV) operation and maintenance of the community stations located in Fort McMurray and Fort McKay, to WBEA operating and maintaining the ambient air monitoring network. During this time period, station upgrades to the community stations in Fort McKay and Fort McMurray (Athabasca Valley) also occurred and additional community stations were commissioned (Fort McMurray—Patricia McInnis and Fort Chipewyan). The first full year of operation of the independent WBEA air monitoring network was 1998. At that time, there were eight continuous AMSs operating. Since then, seven additional stations have been commissioned (Table 4.1).

In 2011, the WBEA monitoring network comprised 15 monitoring stations located from Fort Chipewyan (AMS 8, 58.70 lat.) in the north to Anzac (AMS 14, 56.50 lat.) in the south (Figure 4.1). The five types of stations are as follows:

TABLE 4.1 History and Purpose of the WBEA Air Monitoring Stations

Station ID	Type	Date installed	Initial purpose
AMS 1 Fort McKay	Community/ health	1983 (original) 1997 (moved)	Originally installed in 1983 as an Alberta Environment station to address concerns from the community.
AMS 2 Mildred Lake	Compliance	1979	Installed by Syncrude as a compliance station close to Syncrude.
AMS 3 Lower Camp	Compliance (original) Meteorological	1975	Installed by Suncor as a compliance station close to Suncor. Later, the purpose of this station was changed to monitor meteorological parameters.
AMS 4 Buffalo View Point	Compliance	1979	Installed by Syncrude as a compliance station close to Syncrude.
AMS 5 Mannix	Compliance Meteorological	1975	Installed by Suncor as a compliance station close to Suncor. Later, this station was also used to monitor meteorological parameters.
AMS 6 Patricia McInnis Fort McMurray	Community/ health	1998	Installed as a community health station following a WBEA strategic review.
AMS 7 Athabasca Valley Fort McMurray	Community/ health	1977	Installed as an Alberta Environment monitoring station. Also considered a federal air quality station.
AMS 8 Fort Chipewyan	Background community/ health	1998	Installed by Syncrude as a regulatory requirement (Aurora Approval) and due to the desire by the community.
AMS 9 Barge Landing	Attribution	2000	Installed by Albian as a regulatory requirement and due to the desire by the community. Located between Albian's tailings pond and Fort McKay.
AMS 10 Albian Mine	Compliance	2000–2009	Installed by Albian as a regulatory requirement. Station relocated in 2009 as a result of mining activity.

Continued

TABLE 4.1 History and Purpose of the WBEA Air Monitoring Stations—Cont'd

Station ID	Type	Date installed	Initial purpose
AMS—11 Lower Camp	Compliance	1975 (original) 2000 (moved)	Installed by Suncor as a compliance station close to Suncor. Station was moved <1 km in 2000.
AMS 12 Suncor Millenium	Compliance	2001	Installed by Suncor as a regulatory requirement. Station located close to active truck and shovel mining.
AMS 13 Syncrude UE-1	Attribution	2002	Installed by Syncrude as a regulatory requirement and due to the desire of the community. Located between Syncrude and Fort McKay.
AMS 14 Anzac	Community/ health	2006	Installed by the Long Lake Project as a regulatory requirement and due to the desire of the community.
AMS 15 CNRL Horizon	Compliance	2007	Installed by CNRL as a compliance station located close to CNRL. Entered WBEA network in January 2008.
AMS 16 Shell Albian Muskeg River	Compliance	2009	AMS 10 (Albian Mine) was relocated during 2009 as a result of mining activity.

1. *Compliance*: located in areas where maximum ambient concentrations from specific projects are predicted by plume dispersion modeling. The purpose of this compliance function is to determine if Alberta Ambient Air Quality Objectives (AAAQOs) are being met.
2. *Community/Health*: These stations are located in communities of varying sizes and source profiles. The purpose is to measure potential outdoor exposure to humans.
3. *Attribution*: located between fixed sources and local communities. The purpose is to determine the specific oil sands operation responsible for an air quality event.
4. *Background*: located in areas largely unaffected by industrial emissions. The purpose is to provide background levels.
5. *Meteorological*: These stations with tall towers are located at WBEA AMS 3 (Lower Camp) and at WBEA AMS 5 (Mannix). Their purpose is to collect a profile of meteorological conditions for use in regional dispersion modeling.

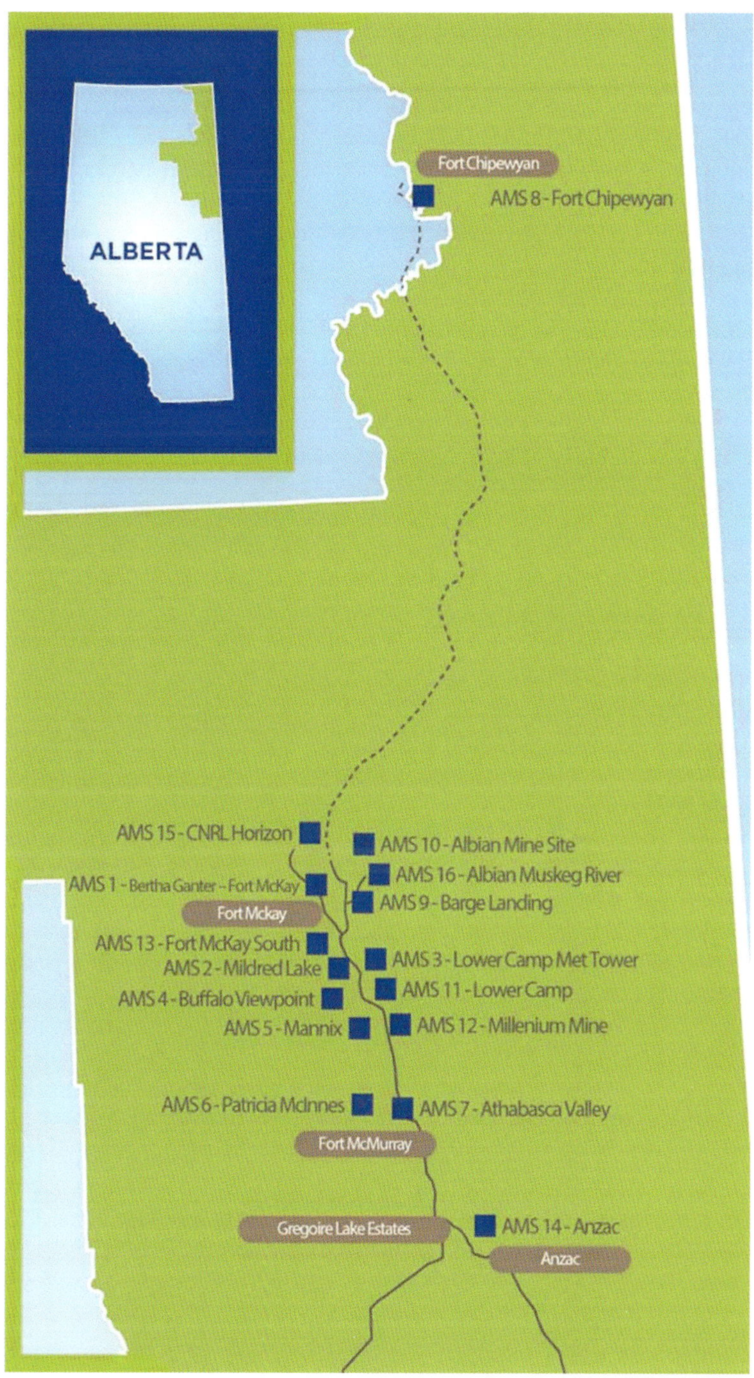

FIGURE 4.1   Map of WBEA air monitoring station locations.

4.2 MAJOR EMISSION SOURCES IN THE ATHABASCA OIL SANDS

There are both combustion and noncombustion emission sources in the AOSR. According to Davies (this volume), combustion sources that produce sulfur and nitrogen emissions to air include the following: (1) stacks (gas-fired heaters and boilers, gas-fired combustion turbine cogeneration, coke-fired power plants, reciprocating engines); (2) mines (combustion exhaust from diesel-powered fleets); and (3) nonindustrial sources (community and on-road sources).

Noncombustion sources in the region include fugitive emissions from the following: (1) plant emissions (small leaks associated with valves, pipe fittings, seals, and vents); (2) mine areas (exposed mine faces yielding hydrocarbons and dust); and (3) tailing ponds (substantial source of hydrocarbons, potentially reduced sulfurs, and resuspended dust). Biogenic volatile organic compound (VOC) emissions from the surrounding boreal forest can be substantial in certain periods of the growing season, and participate in secondary reactions in the atmosphere. Over the period 2003–2008, SO_2 emissions ranged between 274 and 314 tons/day. NO_x emissions have increased from 148 to 200 tons/day over the same period (Chapter 12).

In Alberta, similar to other Canadian provinces and territories, the primary source of emissions inventory data is the National Pollutant Release Inventory (NPRI). The NPRI is a federally legislated, Environment Canada-administered, and publicly accessible inventory of pollutant releases that is comprised of two key components: data reported by facilities on releases, disposals, and recycling of over 300 pollutants and more comprehensive information on all sources of emissions of certain key air pollutants. The NPRI (2012) lists the following emission amounts for the year 2010 for Alberta oil sands (three deposits including the Athabasca) mining, extraction, and processing: 2552 tons of SO_x, 3560 tons of NO_x, 34,557 tones of VOC, and 243 tons of $PM_{2.5}$. Oil sands *in situ* extraction and processing in Alberta produced 10,249 tons of SO_x, 14,288 tons of NO_x, 1389 tons of VOC, and 470 tons of $PM_{2.5}$.

4.3 CONTINUOUSLY MONITORED AIR POLLUTANTS

Air pollutants continuously measured by the WBEA network include sulfur dioxide (SO_2), hydrogen sulfide (H_2S), or total reduced sulfur (TRS), nitrogen oxides (NO, NO_x, and NO_2), total hydrocarbons (THC), fine particulates ($PM_{2.5}$), ozone (O_3), and meteorological parameters (barometric pressure, relative humidity, temperature, wind speed, and direction). Ammonia (NH_3) is measured continuously at two sites and carbon monoxide (CO) at one site. A number of other specialized measurements are made at selected sites including continuous dual detector pneumatically focused GC for VOCs and volatile reduced sulfur compounds (RSCs).

In the following sections, air quality data for each monitored pollutant is presented for the year 2011. In many cases, data are related to North American air quality objectives or standards as listed in Table 4.2. The United States National Ambient Air Quality Standards (NAAQS) for SO_2 and NO_2 are also used for comparison since the Canadian Ambient Air Quality Objectives have not been recently updated.

TABLE 4.2 North American Air Quality Objectives and Standards

Pollutant	Agency	Level	Averaging period	Form
SO_2	AAAQO[a]	172 ppb	1 h	
		48.0 ppb	24 h	
		11.0 ppb	30 days	
		8.0 ppb	Annual	
	NAAQS primary[b]	75 ppb	1 h	99th percentile of 1-h daily maximum concentrations, averaged over 3 years
	NAAQS secondary	500 ppb	3 h	Not to be exceeded more than once per year
NO_2	AAAQO	159 ppb	1 h	
		24 ppb	Annual	
	US EPA primary	100 ppb	1 h	98th percentile, averaged over 3 years
	US EPA primary and secondary	53 ppb	Annual	Annual mean
O_3	AAAQO	82 ppb	1 h	Daily maximum
	NAAQS primary and secondary	75 ppb	8 h	Annual fourth highest daily maximum 8-h concentration, averaged over 3 years
	CWS[c]	65 ppb	8 h	By 2010, achievement based on the fourth highest measurement annually, averaged over three consecutive years

Continued

TABLE 4.2 North American Air Quality Objectives and Standards—Cont'd

Pollutant	Agency	Level	Averaging period	Form
$PM_{2.5}$	AAAQO	$30\ \mu g\ m^{-3}$	24 h	
	NAAQS primary and secondary	$15\ \mu g\ m^{-3}$	Annual	Annual mean, averaged over 3 years
		$35\ \mu g\ m^{-3}$	24 h	98th percentile, averaged over 3 years
	CWS	$30\ ug\ m^{-3}$ (micrograms)	24-h	By 2010, achievement based on the 98th percentile ambient measurement annually, averaged over three consecutive years
H_2S	AAAQO	10 ppb	1 h	Odor perception
		3 ppb	24 h	Odor perception
	AAQC[d]	$7\ \mu g\ m^{-3}$ ($\sim$5.2 ppb)	24 h	Adverse effects on respiratory system
		$13\ \mu g\ m^{-3}$ ($\sim$9.3 ppb)	10 min	Odor effects
		$10\ \mu g\ m^{-3}$ ($\sim$7.2 ppb)	30 min	Odor and health effects
Benzene	AAAQO	9.0 ppbv	1 h	
Acetone	AAAQO	2400 ppb	1 h	
Acetaldehyde	AAAQO	50 ppb	1 h	
Toluene	AAAQO	499 ppb	1 h	
		106 ppb	24 h	
Ethylbenzene	AAAQO	460 ppb	1 h	
Carbon disulfide	AAAQO	10 ppb	1 h	

[a]AAAQO: Alberta Ambient Air Quality Objective (http://environment.gov.ab.ca/info/library/5726.pdf).
[b]U.S. Environmental Protection Agency National Ambient Air Quality Standards (http://www.epa.gov/air/criteria.html).
[c]Canada-Wide Standard Standards (http://www.ccme.ca/assets/pdf/pmozone_standard_e.pdf).
[d]Ontario Ministry of the Environment Ambient Air Quality Criteria (http://www.ontla.on.ca/library/repository/mon/20000/277839.pdf).

4.3.1   Sulfur Dioxide

Sulfur dioxide is measured continuously at 13 sites in the WBEA network using pulsed fluorescence gas analyzers. The current set of analyzers consist of Thermo Scientific Models 43A, 43C, and 43i trace-level analyzers. These analyzers are operated on 0–1000 ppb range. The detection limits observed under field conditions varied from 0.5 to 1 ppb.

The 1-h maximum SO_2 concentrations measured at compliance/attribution stations in 2011 ranged from 56 ppb at CNRL Horizon (AMS 15) to 122 ppb at Mannix (AMS 5) (Figure 4.2). At community stations, maximum 1-h SO_2 concentrations ranged from 12 ppb at Fort Chipewyan (AMS 8) to 83 ppb at Fort McKay (AMS 1). The 99th percentile SO_2 concentrations ranged from

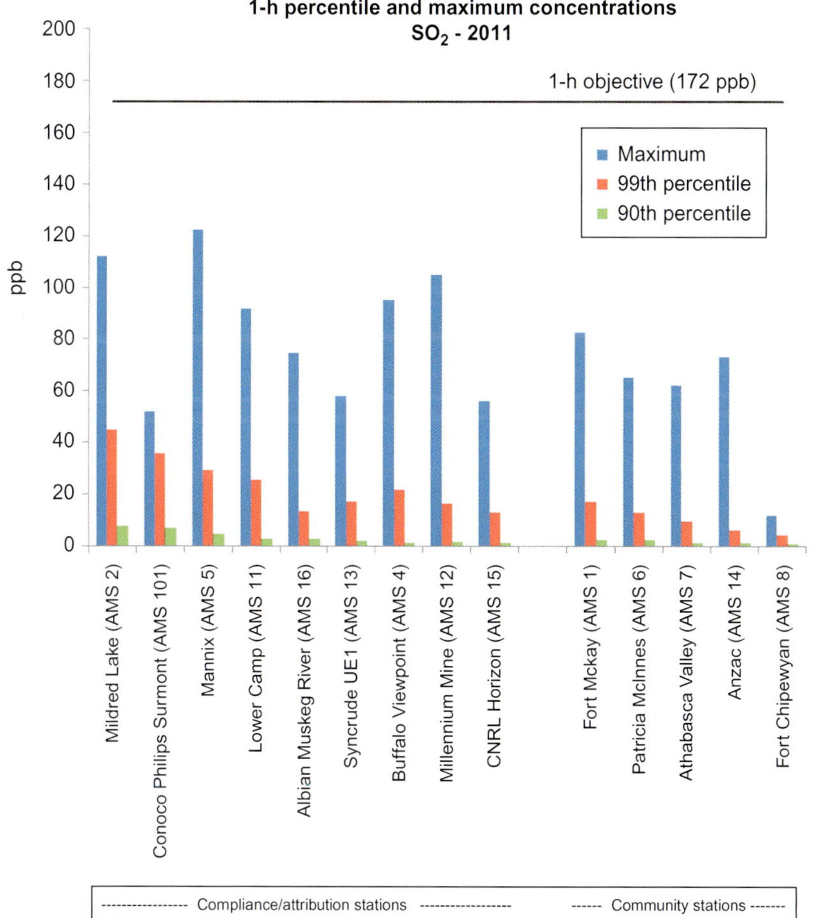

FIGURE 4.2   2011 1-h average ambient maximum, 99th and 90th percentile SO_2 concentrations at WBEA air monitoring stations.

4 ppb (AMS 8) to 45 ppb (AMS 2). There were no exceedances reported in 2011 of the AAAQO of 172 ppb for SO_2 averaged over 1 h (Figure 4.2). There were also no exceedances of the 24-h AAAQO for SO_2 (48 ppb), or the Alberta annual average AAAQO for SO_2 (8 ppb).

4.3.2 Nitrogen Dioxide

Chemiluminescence gas analyzers from Thermo Scientific and Teledyne Advanced Pollution Instrument were utilized for measurements of atmospheric nitrogen oxide, nitrogen dioxide, and oxides of nitrogen at nine stations. Models 17C, 17i, 42C, 42C trace level, 42i from Thermo Scientific, and model 200E from Teledyne are used in the network. These analyzers are operated on 0–500 ppb and 0–1000 ppb ranges. The detection limits observed under field conditions range from 0.5 to 1 ppb.

The maximum 1-h NO_2 concentrations in 2011 at compliance/attribution stations ranged from 52 ppb at CNRL Horizon (AMS 15) to 154 ppb at Millenium Mine (AMS 12) (Figure 4.3). At community stations, maximum 1-h concentrations ranged from 42 ppb at Fort Chipewyan (AMS 8) to 66 ppb at Anzac (AMS 14). The maximum 1-h average concentration of 154 ppb measured at AMS 12 was 97% of the AAAQO of 159 ppb NO_2 (Figure 4.3). There were no exceedances reported of the 1-h AAAQO level.

The 98th percentile NO_2 concentrations at compliance/attribution stations ranged from 28 ppb (AMS 13, AMS 15) to 44 ppb (AMS 12). The 98th percentile concentrations at community stations ranged from 9 ppb at AMS 8 to 35 ppb at AMS 7. Decreasing order of 98th percentile concentrations at community stations was Athabasca Valley (AMS 7)>Fort McKay (AMS 1)>Patricia McInnes (AMS 6)>Anzac (AMS 14)>Fort Chipewyan (AMS 8) (Figure 4.3).

4.3.3 Ozone

Tropospheric ozone (O_3) is a secondary pollutant that results from photochemical reactions of NO_x and VOCs involving generally well-understood reactions (Jenkins and Clemitshaw, 2000). Increases in ambient concentration come from primary anthropogenic pollutant precursor emissions, forest fires, and episodic stratospheric injection in the spring. An UV Photometric analyzer, Thermo Scientific Model 49C, was used for the measurement of O_3 in ambient air at six stations. These analyzers are operated on a range of 0–500 ppb with a detection limit of 1 ppb.

Maximum 1-h O_3 concentrations ranged from 77 ppb (AMS 7) to 91 ppb (AMS 13) (Figure 4.4). At the five community stations where O_3 is measured, maximum 2011 1-h concentrations ranged from 77 ppb at Athabasca Valley (AMS 7) to 89 ppb at Fort McKay (AMS 1). The 1-h AAAQO for O_3 of 82 ppb was exceeded 15 times at WBEA monitoring stations in 2011. All exceedances occurred in the month of May, the month associated with large, widespread, and persistent forest fires in the region. This mid-May period also

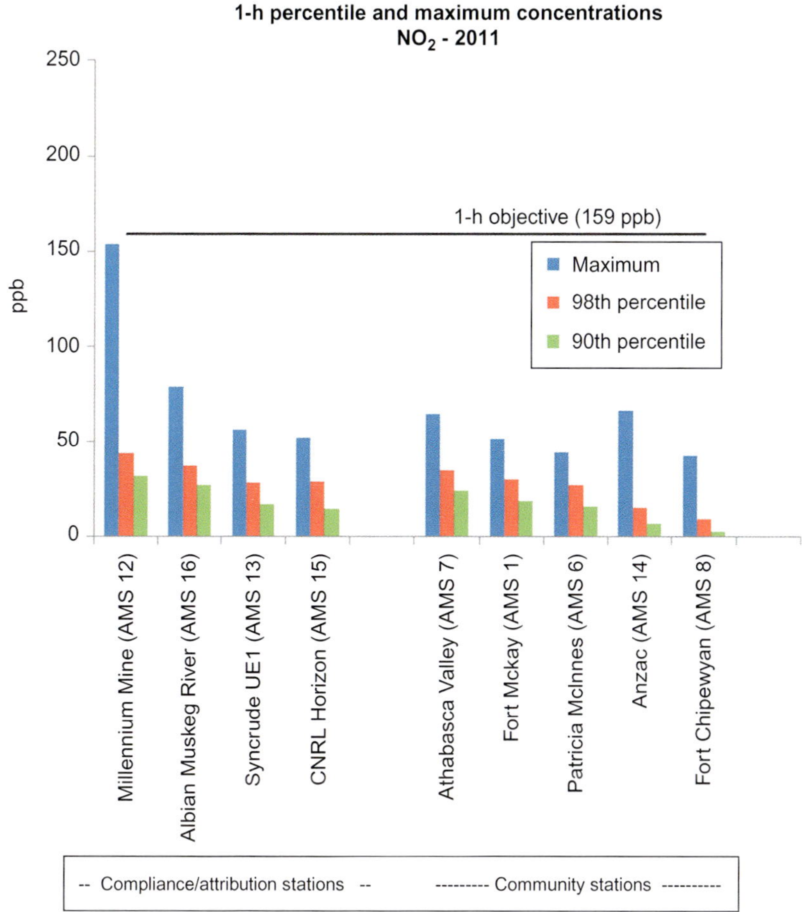

FIGURE 4.3   2011 1-h average ambient maximum, 98th and 90th percentile NO$_2$ concentrations at WBEA air monitoring stations.

coincided with expected stratospheric O$_3$ intrusion into the troposphere (Lefohn et al., 2001; Monks, 2000). The spring O$_3$ maximum is a northern hemisphere phenomenon and is a defining feature of the annual cycle at remote and rural sites (Monks, 2000). There appear to be several mechanisms responsible for the spring maximum, including UV-enhanced photochemistry in the free troposphere (Dibb et al., 2003), stratospheric–tropospheric exchange (Diem et al., 2003; Tarasick et al., 2005), and enhanced hemispheric transport (Jaffe et al., 2008). This observation of spring time O$_3$ maxima is contrary to the notion that O$_3$ peaks during the summer, when local photo-chemistry is usually greatest (Environment Canada, 2012).

Calculated 2011 fourth highest daily maximum 8-h ozone concentrations (the Canada Wide Standard (CWS) metric) exceeded 65 ppb at both AMS 13

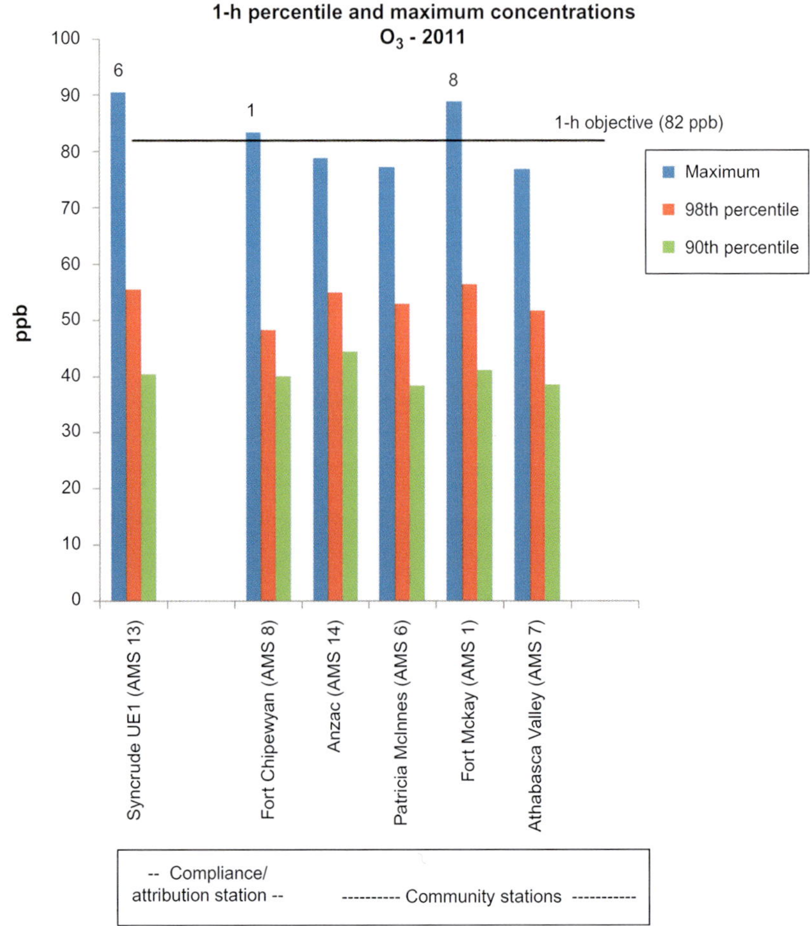

FIGURE 4.4 2011 1-h average ambient maximum, 98th and 90th percentile O_3 concentrations at WBEA air monitoring stations.

(67 ppb) and AMS 1 (74). These high values were associated with the May forest fires and were the highest values recorded in any year of the data record. The fourth highest daily maximum 8-h ozone values at the other sites ranged from 58 to 63 ppb. For the 3-year averaging period (2008–2010) prior to the May–June 2011 fire event, the calculated CWS values ranged from 53 to 55 ppb.

4.3.4 Fine Particulate Matter (PM$_{2.5}$)

During 2011, WBEA was in the process of transitioning from the filter and microbalance-based technology PM$_{2.5}$ to analyzers that employ a combination beta attenuation and light scattering technology. Prior to this transition period in 2011, the technology used in the network was the Tapered Element Oscillating Microbalance (TEOM) analyzers. The TEOM analyzers were operated at -50–$450\ \mu g\ m^{-3}$ with a constant column temperature of 40 °C. The Synchronized Hybrid Ambient Real-time Particulate (SHARP) monitor range is set to 0–$1000\ \mu g\ m^{-3}$. The detection limit on these analyzers is $1\ \mu g\ m^{-3}$.

In 2011, the region experienced a significant air quality event in the form of a 700,000-ha wildfire. At its most southern extent, the fire front was within a few kilometers of both AMS 15 and AMS 1. The TEOM analyzer's operating range's upper limit is $450\ \mu g\ m^{-3}$ and this limit was often exceeded causing the instruments to shut down. WBEA had been evaluating Beta attenuation technology (upper range limit $900\ \mu g\ m^{-3}$) prior to the fire event. With a requirement for WBEA instruments to report continuously during air pollution events, the network is adopting Beta attenuation instrumentation at all nine stations by the end of 2012. The new $PM_{2.5}$ analyzers are Thermo Scientific Model 5030 SHARP monitors.

Maximum 1-h $PM_{2.5}$ concentrations in 2011 ranged between $406\ \mu g\ m^{-3}$ (AMS 8) and $451\ \mu g\ m^{-3}$ (AMS 12), with the exception of one station in the south. Less affected by fire smoke, the maximum 1-h $PM_{2.5}$ concentration at Anzac (AMS 14) was $229\ \mu g\ m^{-3}$ (Figure 4.5). The 99th percentile concentrations at compliance/attribution stations ranged from $110\ \mu g\ m^{-3}$ at Millennium (AMS 12) to $176\ \mu g\ m^{-3}$ at Syncrude UE1 (AMS 13). Concentrations at community stations ranged between $8\ \mu g\ m^{-3}$ at Fort Chipewyan (AMS 8) to $172\ \mu g\ m^{-3}$ at Fort McKay (AMS 1). These concentrations at each station were considerably lower than the maximum 1-h values experienced during the most extreme smoke events. The highest concentrations at each station were greater than the 99th percentile concentrations by factors ranging from 2 to 4 and were greater than the 95th percentile concentrations by factors ranging from 14 to 56. In other words, the highest 5% of concentrations were 14–56 times greater than concentrations observed 95% of the time.

In 2011, the 24-h AAAQO for $PM_{2.5}$ of $30\ \mu g\ m^{-3}$ was exceeded 97 times. Of these, 94 exceedances (97%) occurred in the months of May and June, coincident with the widespread forest fires. In prior years without the forest fire influence, there were many fewer reported (WBEA, 2011, 2010, 2009) exceedances of the $PM_{2.5}$ 24-h AAAQO; 22 in 2010, 7 in 2009, and 4 in 2008.

The AAAQO was exceeded 49 times at community stations, and 48 times at compliance/attribution stations. There were 9 exceedances reported at Syncrude UE1 (AMS 13), 13 at CNRL Horizon (AMS 15), 10 at Muskeg River (AMS 16), and 16 at Millennium Mine (AMS 12). At community stations there were 12 exceedances at Fort McKay (AMS 1), 17 at Athabasca Valley (AMS 7), 3 at Fort Chipewyan (AMS 8), 11 at Patricia McInnes (AMS 6), and 6 at Anzac (AMS 14). The CWS metric is based on the 98th percentile of daily average values. Because of the forest fire influence all seven WBEA stations that met data completeness criteria exceeded the CWS metric of $30\ \mu g\ m^{-3}$ with 98th percentile values ranging from 52 (AMS 16) to $132\ \mu g\ m^{-3}$ (AMS 1).

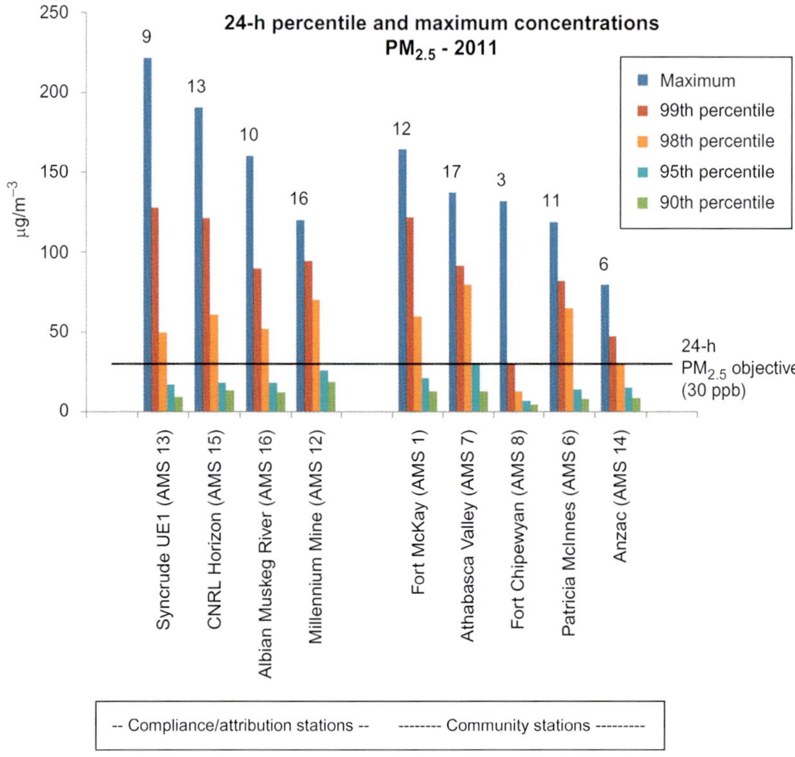

FIGURE 4.5 2011 24-h averaged maximum and percentile PM$_{2.5}$ concentrations at WBEA air monitoring stations.

4.3.5 Ammonia

Ammonia was monitored at two community stations, Fort McKay (AMS 1) and Patricia McInnes (AMS 6). Thermo Environmental 17 series analyzers were used to measure ambient NH$_3$ concentrations. The three active channels were nitric oxide (NO), total oxides of nitrogen (NO$_x$), and total nitrogen (N$_t$). Nitrogen dioxide (NO$_2$) and NH$_3$ were derived internally in the analyzer using the following calculations:

$$NO_2 = NO_x - NO$$

$$NH_3 = N_t - NO_x.$$

This analyzer is cycled between these three active channels approximately every 10 s to measure each active channel on a continuous basis. All three measurements are completed using a single detector in the 17 series analyzers. The range settings were from 0 to 500 ppb for the NO and NO$_x$, and 0 to 5000 ppb for N$_t$ channels (AAAQO for NH$_3$ is 2000 ppb).

Measurements of NH_3 concentrations were below the detection limit of the analyzer most of the time. The 2011 maximum 1-h concentrations were 123 ppb at Patricia McInnes (AMS 6) and 49 ppb at Fort McKay (AMS 1). The 98th percentile concentration at both sites was below detection indicating the very infrequent measurements of elevated NH_3 concentrations in ambient air. There were no exceedances of the 2000 ppb NH_3 1-h AAAQO.

4.3.6   Total Reduced Sulfur/Hydrogen Sulfide

Hydrogen sulfide (H_2S) and TRSs were measured with pulsed fluorescence gas analyzers that detect SO_2 formed by the catalytic conversion of hydrogen sulfide or sulfur compounds. WBEA used Thermo Scientific models 45C and 450i analyzers for measuring H_2S and TRS. These analyzers are operated on a 0–100 ppb range. The detection limits observed under field conditions ranged from 0.1 to 0.2 ppb. Hydrogen sulfide (H_2S) is measured at four stations in the WBEA-monitoring network (Mildred Lake—AMS 2, Buffalo Viewpoint—AMS 4, Mannix—AMS 5, and Lower Camp—AMS 11). TRS is measured at seven other stations. It should be noted that H_2S continuous analyzers at the AMSs may detect other RSCs in addition to detecting H_2S. This is an interference acknowledged by WBEA (www.wbea.org) based upon technical literature provided by the manufacturer, as well as new confirming, on-site comeasurements made by O'Brien et al. (Chapter 6).

Maximum 1-h TRS and H_2S concentrations in 2011 at compliance/attribution stations ranged from 6 ppb at CNRL Horizon (AMS 15) to 98 ppb at Mildred Lake (AMS 2) (Figure 4.6). At community stations, the maximum 1-h concentrations ranged from 3 ppb at Patricia McInnes (AMS 6) to 7 ppb at Anzac (AMS 14). The 99th percentile concentrations ranged from 1 to 13 ppb, and the 95th percentile concentrations ranged from 0.7 to 3 ppb. Twenty-four-hour averaged TRS concentrations in 2011 were 3 ppb at Fort McKay (AMS 1), 2 ppb at Patricia McInnes (AMS 6) and Athabasca Valley (AMS 7), and 1 ppb at Anzac (AMS 14). Concentrations of TRS or H_2S at compliance/attribution stations ranged from 23 ppb at Millenium (AMS 12), to 5 ppb at Lower Camp (AMS 11), to 1 ppb at CNRL Horizon (AMS 15). All 24-h AAAQO exceedances were at compliance stations. There were 23 exceedances at Millennium Mine (AMS 12), 4 at Mildred Lake (AMS 2), 3 at Lower Camp (AMS 11), and 1 at Conoco Phillips Surmont (portable station AMS 101). There were no exceedances of the 24-h objective at community stations. While odors were still apparent in some communities, the 31 exceedances of the 24 h AAAQO in 2011 set for odor perception (Table 4.2) was less than the 118 exceedances in 2010 and the 252 exceedances in 2009.

The 10 ppb 1-h AAAQO for H_2S was exceeded 169 times in 2011. This was 74% less than the 614 exceedances in 2010, and 89% less than the 1625 exceedances in 2009. All exceedances in 2011 were recorded at compliance stations (Figure 4.6). There were 133 exceedances at Millennium Mine

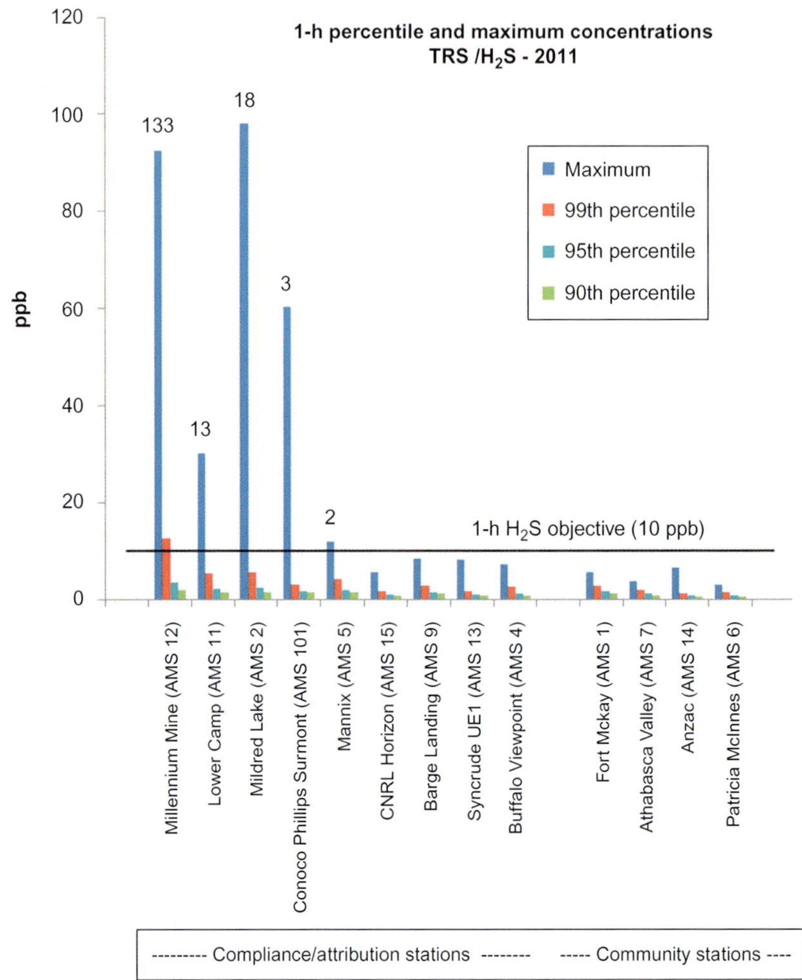

FIGURE 4.6 2011 1-h average ambient maximum and percentile hydrogen sulfide (H₂S) and total reduced sulfur (TRS) concentrations at WBEA air monitoring stations.

(AMS 12) adjacent to an active tailings pond, 13 at Lower Camp (AMS 11), 18 at Mildred Lake (AMS 2), 3 at Conoco Phillips Surmont (AMS 101; October–December), and 2 at Mannix (AMS 5). All exceedances of AAAQO recorded by WBEA are reported automatically and immediately to Alberta Environment and Sustainable Resource Development (AESRD), and all industries are also informed by WBEA of each exceedance. For each event, industry undertakes an internal investigation to determine a cause and follows up with AESRD on findings and mitigative actions. There were no exceedances of the 1-h AAAQO at any community station.

4.3.7   Hydrocarbons

THC were measured using flame ionization detector instruments, Model 51i from Thermo Scientific and model 400A from Rosemount. The analyzers were operated on 0–25 ppm range. The detection limits for these analyzers were 0.1 ppm. The maximum 1-h THC concentration in 2011 at compliance/attribution stations ranged from 6 ppm at Lower Camp (AMS 11) to 9 ppm at CNRL Horizon (AMS 15) (Figure 4.7). At community stations, the maximum 1-h concentrations ranged from 3 ppm at Anzac (AMS 14) to 5 ppm at Fort McKay (AMS 1). The highest 1-h THC concentrations at all stations in 2011 ranged from 3 to 9 ppm. The 99th percentile concentrations ranged from 2 to 5 ppm, and the 95th percentile concentrations ranged from less than 2 to 3 ppm. At each station, the maximum concentrations were one to three times greater than the 99th percentile concentrations. Median THC concentrations were comparable to average concentrations, indicating

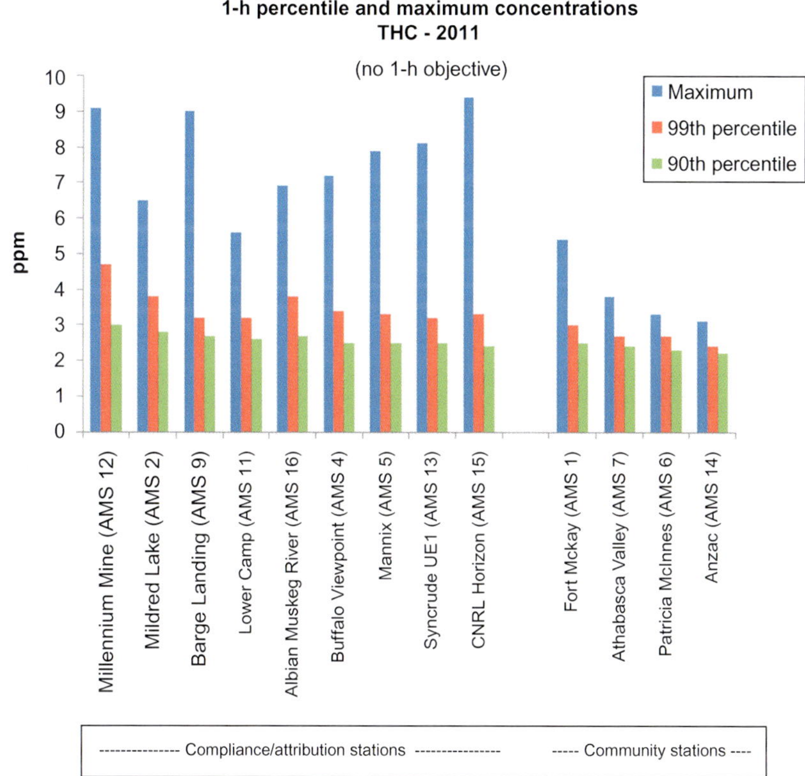

FIGURE 4.7   2011 1-h average ambient maximum, 99th and 90th total hydrocarbon percentile concentrations at WBEA air monitoring stations.

absence of significantly high concentrations, and the presence of a large number of very small concentrations. Global methane (CH_4) background levels are 1.8–1.9 ppm (NOAA, 2008) comprising most of the 90th percentile concentrations measured (Figure 4.7).

4.4 TIME-INTEGRATED MEASUREMENTS

There is also a large time-integrated sampling program carried out at WBEA monitoring stations. The integrated sampling program includes measurements of particulates $\leq 10\,\mu m$ aerodynamic diameter (PM_{10}), particulates $\leq 2.5\,\mu m$ aerodynamic diameter ($PM_{2.5}$), VOC and RSC by canister (VOC), polycyclic aromatic hydrocarbons (PAHs), metals, and passive sampling.

4.4.1 Volatile Organic Compounds

VOCs were sampled using 6.0-liter evacuated canisters. The sampling procedure and analysis are performed as prescribed in United States EPA Method TO-14 (U.S. EPAa,b, 1999). In 2011, the samples were taken every 12 days at four compliance/attribution stations: Barge Landing (AMS 9), Millennium Mine (AMS 12), Syncrude UE1 (AMS 13), and CNRL Horizon (AMS 15). VOCs were also collected at four community stations: Fort McKay (AMS 1), Patricia McInnes (AMS 6), Athabasca Valley (AMS 7), and Anzac (AMS 14). Samples at the WBEA NAPS-designated station (AMS 7) were collected every 6 days. Samples were analyzed for 60 VOCs and 20 target RSCs. Duplicate samples and replicate analyses were introduced at selected sites for selected samples. Beginning in 2012, all VOC/RSC samples are being collected according to complete NAPS protocols.

Descriptive statistics for the most frequently measured canister collected compounds in 2011 are listed in Table 4.4. Mean concentrations of VOCs measured in 50% or more of canister collections averaged across WBEA stations were the C_2–C_{10} alkanes butane (C_4H_{10}) (1.31 ppb), isobutane (C_4H_{10}) (0.68 ppb), pentane (C_5H_{12}) (1.09 ppb), isopentane (C_5H_{12}) (1 ppb), hexane (C_6H_{14}) (0.24 ppb), and 2-methylpentane (C_6H_{14}) (0.34 ppb). Octane (C_8H_{18}) (0.27 ppb) was measured in 79 (26%) of canister samples. Aromatic compounds measured in $\geq 50\%$ of canisters were benzene (C_6H_6) (0.68 ppb), toluene (C_7H_8) (0.44 ppb), and $m+p$-xylene (0.20 ppb). Cyclopentane (C_5H_{10}) was measured in 52 (17%) of samples and had a mean concentration of 0.52 ppb. Ethylbenzene (C_8H_{10}) was measured in 94 (31%) of canister samples. The oxygenate measured in $\geq 50\%$ of canisters was acetone ($CH_3)_2CO$ with a mean concentration of 3.37 ppb. Methyl ethyl ketone (2-butanone) was measured in 87 (29%) (mean concentration 0.52 ppb) canister samples. Methanol (CH_3OH) was not measured in any canister samples in 2011. The two most common monoterpene compounds were alpha(α)-pinene (0.14 ± 0.15 ppb) and beta(β)-pinene (1.03 ± 2.66 ppb), measured in 126 (44%) and 45 (15%) canister samples, respectively.

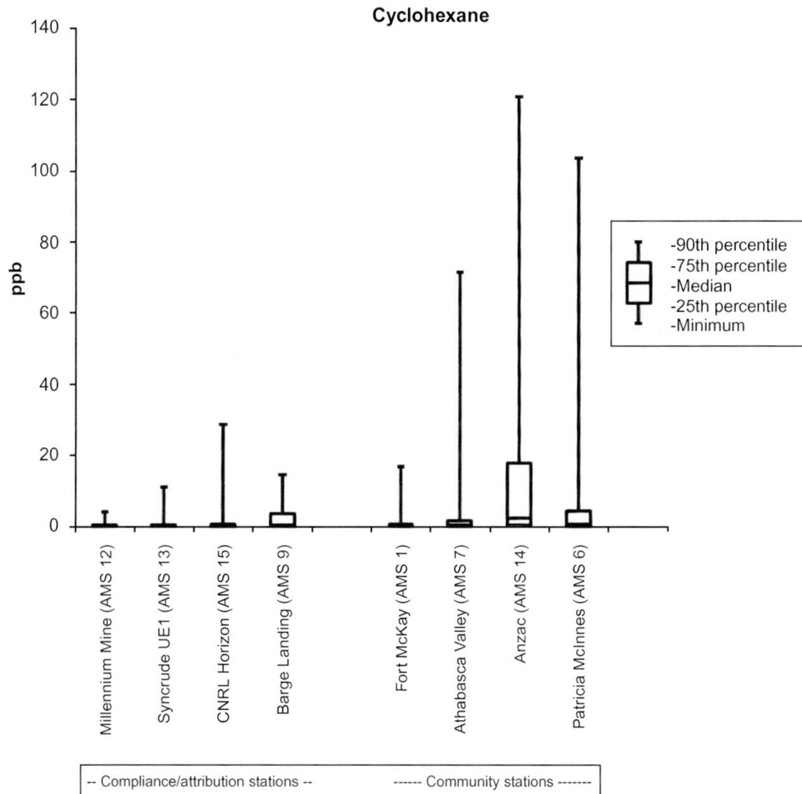

FIGURE 4.8   Cyclohexane concentrations measured in 2011 in 24-h time-integrated sampling at WBEA air monitoring stations. Data are displayed as maximum, median, and 90th, 75th, 25th, and 10th percentile values.

At compliance/attribution stations, the VOCs having the highest measured 24-h concentrations were cyclohexane (C_6H_{12}) (Figure 4.8) with maximum 24-h averaged concentrations of 520 ppb at Barge Landing (AMS 9) and 163 ppb at CNRL Horizon (AMS 15), isopentane with a maximum concentration of 78 ppb at CNRL Horizon (AMS 15) (Figure 4.9), butane with a maximum concentration of 73 ppb at CNRL Horizon (AMS 15) (Figure 4.10), and acetone with a maximum concentration of 43 ppb at Barge Landing (AMS 9) (Figure 4.11). VOCs observed in every sample from at least one compliance/attribution station included acetone, benzene (Figure 4.12), and toluene (Figure 4.13).

At community stations, the VOC with the highest measured 24-h concentrations was cyclohexane, with maximum concentrations of 121, 120, 71, and 59 ppb at Anzac (AMS 14), Patricia McInnes (AMS 6), Athabasca Valley (AMS 7), and Fort McKay (AMS 1), respectively. Other VOCs with highest measured 24-h concentrations were isopentane (Figure 4.9) with a maximum

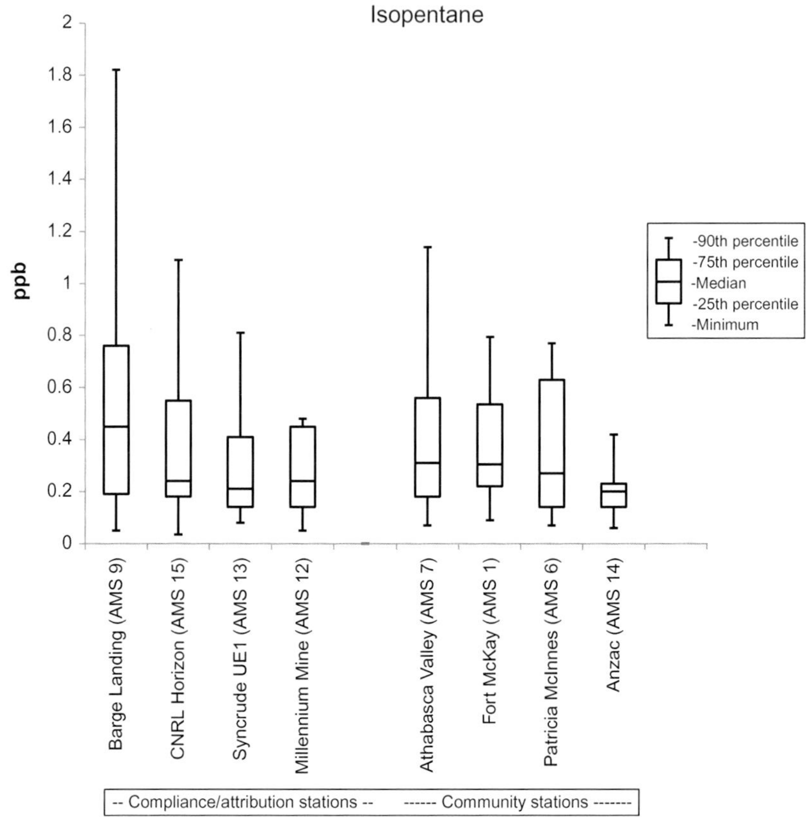

FIGURE 4.9 Isopentane concentrations measured in 2011 in 24-h time-integrated sampling at WBEA air monitoring stations. Data are displayed as maximum, median, and various percentile values.

concentration of 42 ppb and butane (Figure 4.10) with a maximum concentration of 39 ppb at Athabasca Valley (AMS 7). Next highest concentrations were for acetone, with a maximum concentration of 19 ppb at Anzac (AMS 14). The only VOC observed in all samples at any community station was benzene at Anzac (AMS 14). VOCs observed in all but two samples at any station included acetone, benzene, butane, isobutane (Figure 4.14), isopentane, and toluene. Acetaldehyde (CH_3CHO) was only measured in 5 (1%) of canister samples with a mean concentration of 5.9 ± 4.76 ppb. It was measured twice at AMS 9, and once each at AMS 7, AMS 12, AMS 13, and AMS 15.

On July 10, 2008, Simpson et al. (2010) sampled VOCs in the boundary layer (altitudes 720–850 m asl) above the AOSR surface mines as part of the Composition of the NASA Troposphere from Aircraft and Satellites field mission. Ten-day back trajectory analysis indicated that the air masses had arrived from the west. Average ($n = 17$ collections) concentrations for some alkanes were butane (0.202 ppb), pentane (0.116 ppb), hexane (0.044 ppb), octane

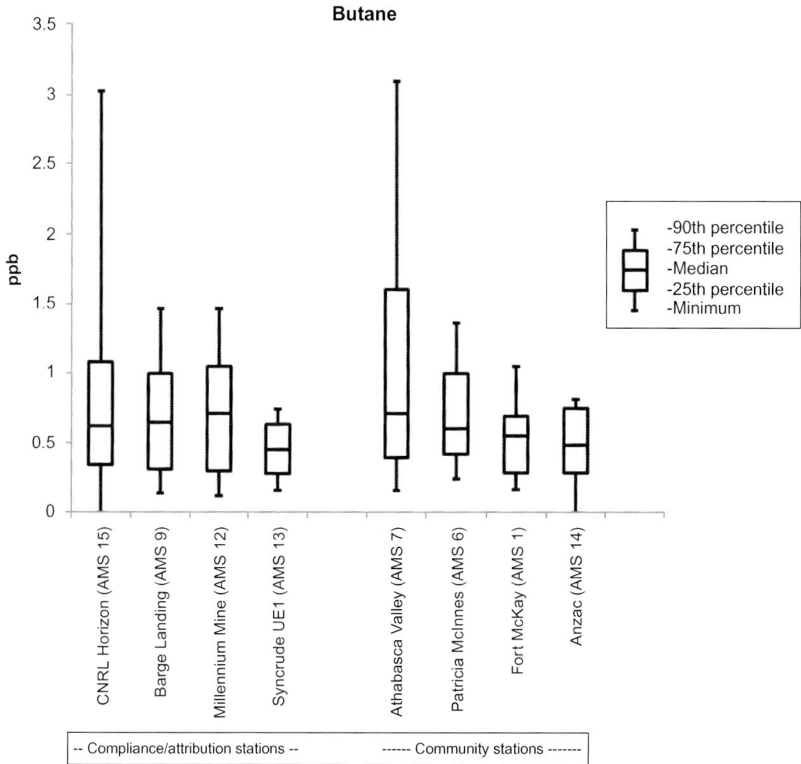

FIGURE 4.10   Butane concentrations measured in 2011 in 24-h time-integrated sampling at WBEA air monitoring stations. Data are displayed as maximum, median, and various percentile values.

(0.04 ppb), and $2+3$ methylpentane (0.057 ppb). Concentrations of some cycloalkanes were cyclopentane (0.009 ppb) and cyclohexane (0.023 ppb). Aromatic compound concentrations were benzene (0.024 ppb), toluene (0.050 ppb), ethylbenzene (0.008 ppb), $m+p$-xylene (0.029 ppb), and o-xylene (0.014 ppb). Monoterpenes were ubiquitous with average concentrations for α-pinene of 0.067 ppb and β-pinene of 0.226 ppb. Average concentrations of oxygenates measured by Simpson et al. (2010) included: methanol (2.515 ppb), acetone (0.644 ppb), and methyl ethyl ketone (2-butanone) (0.065 ppb).

4.4.2   Reduced Sulfur Compounds

Twenty target RSCs were coanalyzed with VOC using 6.0 l evacuated canisters. Descriptive statistics for the most frequently measured RSCs in 2011 are listed in Table 4.3. Carbon disulfide was measured 231 times (77%), carbon disulfide 55 times (18%), hydrogen sulfide 41 times (14%), dimethyl disulfide ($C_2H_6S_2$) 10 times (3%), and thiophene (C_4H_4S) only twice. At compliance/

TABLE 4.3 Descriptive Statistics for Most Frequently Sampled TRS and VOCs in 2011

Compound	N	Mean (ppb)	SD (ppb)	Minimum (ppb)	Median (ppb)	95% (ppb)	99% (ppb)	Maximum (ppb)
Carbonyl sulfide	231	0.8	0.8	0.2	0.6	2.3	3.3	9
Carbon disulfide	55	0.7	2	0.1	0.2	2.7	13.4	13.4
Hydrogen sulfide	41	0.5	0.6	0	0.3	1.6	2.7	2.7
Dimethyl disulfide	10	0.2	0.2	0	0.2	0.5	0.5	0.5
Thiophene	2	5.1	2.4	3.4	5.1	6.8	6.8	6.8
Benzene	241	0.68	1.31	0.06	0.21	3.6	7.71	8.69
Toluene	237	0.44	1.28	0.02	0.18	1.12	5.68	16.25
Acetone	232	3.37	4.03	0	2.17	9.67	16.33	42.46
Butane	216	1.31	5.65	0	0.61	2.63	8.35	73.25
Isopentane	216	1	5.98	0.04	0.26	1.34	5.27	78.21
Isobutane	195	0.68	2.64	0	0.26	2.51	16.65	32.5
"*m,p*-Xylene"	163	0.2	0.44	0.03	0.1	0.5	2.68	4.49
2-Methylpentane	151	0.34	1.27	0	0.13	0.97	6.4	14.29
Hexane	146	0.24	0.62	0	0.12	0.52	3.28	6.65
Pentane	138	1.09	2.92	0	0.57	2.85	14.01	30.01

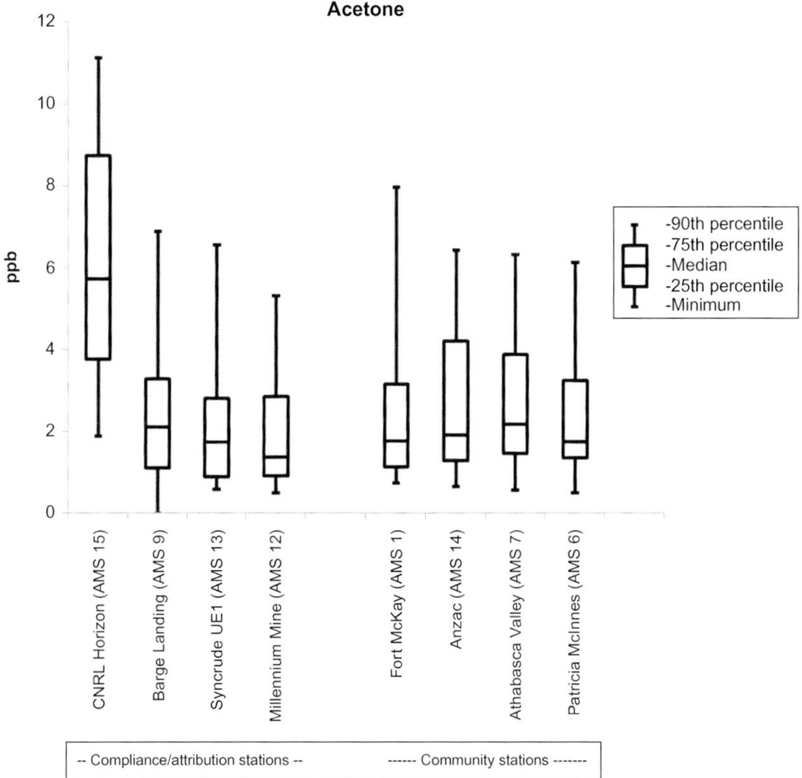

FIGURE 4.11   Acetone concentrations measured in 2011 in 24-h time-integrated sampling at WBEA air monitoring stations. Data are displayed as maximum, median, and various percentile values.

attribution stations, the RSCs having the highest 24-h concentrations were thiophene (6.8 ppb) at Millenium Mine (AMS 12) (Table 4.3), carbonyl sulfide (3.4 ppb) at CNRL Horizon (AMS 15) (Figure 4.15), hydrogen sulfide (2.7 ppb) at CNRL Horizon (AMS 15) (Figure 4.16), and carbon disulfide (2.7 ppb) at CNRL Horizon (AMS 15) (Figure 4.17). Corresponding annual average concentrations at each station were 6.8, 1.8, 1.7, and 0.4 ppb.

At community stations, the RSCs with the highest 24-h concentrations were carbon disulfide 13.4 ppb (Figure 4.17) and carbonyl sulfide 9.0 ppb (Figure 4.15), both measured at Patricia McInnes (AMS 6). Corresponding annual average concentrations were 4.6 and 0.8 ppb. Other compounds measured a community stations included hydrogen sulfide (Figure 4.16) and dimethyl disulfide (Figure 4.18).

The Fort McKay Sustainability Department operates an event-based, canister-sampling program within the First Nations community of Fort McKay. Canister samples were collected on May 11 and on June 1, 2010

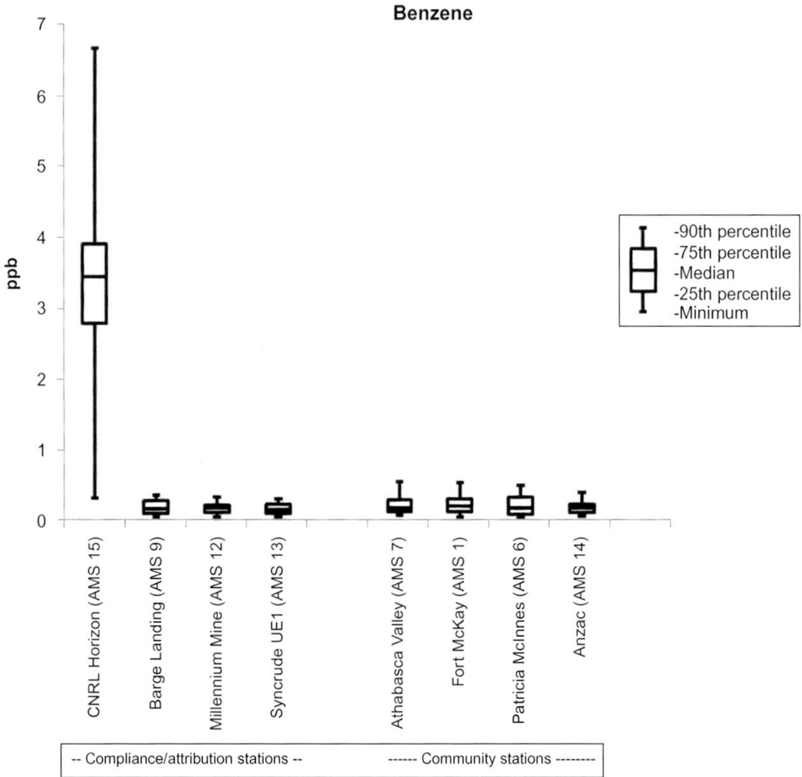

FIGURE 4.12   Benzene concentrations measured in 2011 in 24-h time-integrated sampling at WBEA air monitoring stations. Data are displayed as maximum, median, and various percentile values.

during odor events. Analysis of the canisters by the same laboratory and following the same protocols as used in WBEA canister sampling, indicated there were differences in compounds detected between the two events. DMDS, thiophenes, and allyl sulfide were detected in one event, but not in the other. Acetaldehyde was stated as potential candidate odor compound (Spink and Dennis, 2010).

4.4.3   Polycyclic Aromatic Hydrocarbons

Polycyclic aromatic compounds (PAH) were sampled at four community stations: Fort McKay (AMS 1), Patricia McInnes (AMS 6), Athabasca Valley (AMS 7), and Anzac (AMS 14). During the year, there were a maximum 14 sets of results from each station. PAHs were sampled using a combination of Poly-Urethane Foam cartridges and XAD resin tubes. Samples were collected with three-stage URG (Model 3000AQ) samplers consisting of

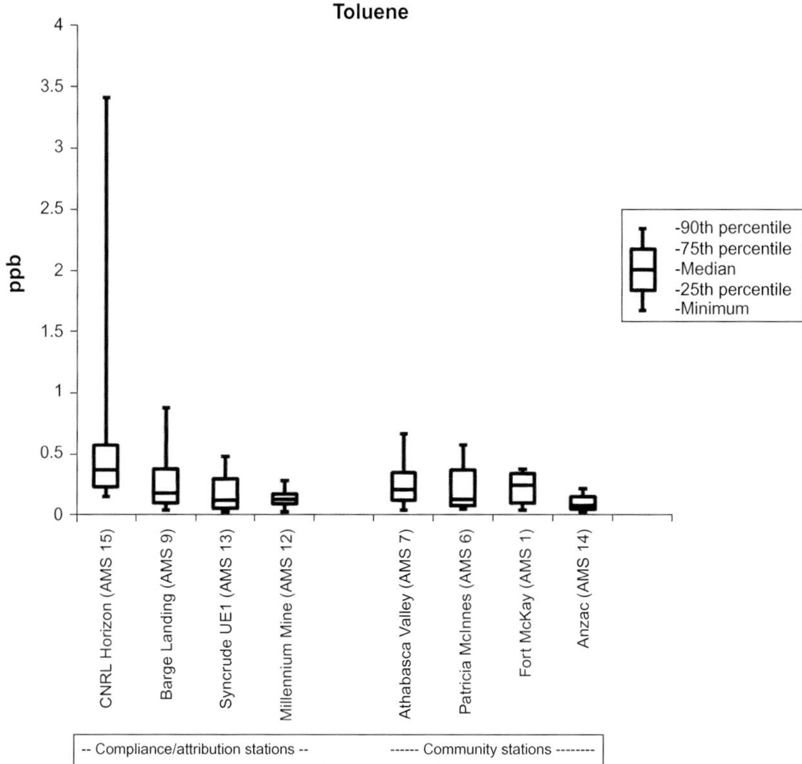

FIGURE 4.13   Toluene concentrations measured in 2011 in 24-h time-integrated sampling at WBEA air monitoring stations. Data are displayed as maximum, median, and various percentile values.

47-mm glass fiber filter elements followed by two sequential 1 "diameter by 3" long PUF media. The filter holder and PUF media holders were constructed of Teflon and glass, respectively. The filter media were prepared by firing in the oven at 300 °C, and the PUF media were extracted in a Soxhlet apparatus with two batches of hexane over 48 h. The PUF media were then dried under vacuum. Media blanks were tested prior to assembly of samplers. The samplers were cleaned and charged with sampling media at the laboratory (Airzone, http://www.airzoneone.com) and then sealed for shipment. At the sites, samplers were deployed over approximately 11-day sampling intervals and collected 100–200 m^3 of air. At the end of the sampling period, samplers were sealed, repackaged, and returned to the laboratory.

At the laboratory, samples were catalogued and checked for inconsistencies and other problems, then extracted with a Dionex, Automated Solvent Extractor with 20 ml of hexane/acetone (70/30) at 100 °C and 1200 PSI (Dionex, 2011). The solvent was spiked with Phenanthrene-d^{10} and Chrysene-d^{12} to assess extraction efficiency for each extraction. Filter and PUF media were extracted together thus providing data on total concentrations of 23 target

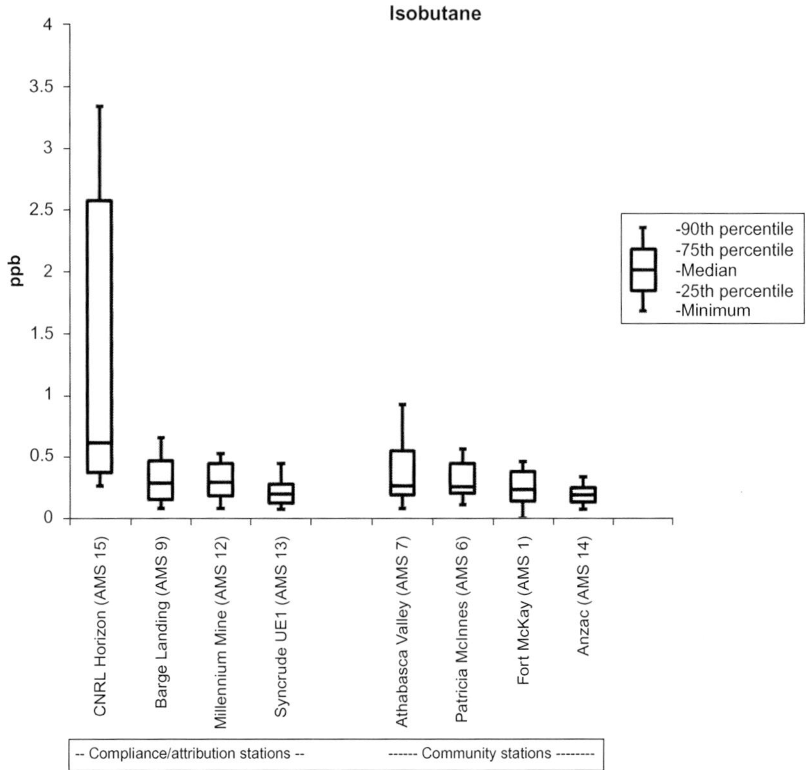

FIGURE 4.14 Isobutane concentrations measured in 2011 in 24-h time-integrated sampling at WBEA air monitoring stations. Data are displayed as maximum, median, and various percentile values.

PAH compounds. Samples were then concentrated under a UHP nitrogen gas stream to 500 μl. Then Pyrene-d^{10} and Fluoranthene-d^{10} were added to each extract to assess integrity during gas chromatograph (GC) analysis. Aliquots (1 μl) of extracts were injected into a gas chromatograph/mass spectrometer (GC/MS), fitted with 12 m DB-5MS (0.25 mm ID coated with 0.25 μm film thickness) chromatographic column. Chromatographic conditions and mass spectrometry parameters were optimized to achieve resolution and quantification of 23 target PAH.

During 2010 and 2011, an extra PUF element was introduced into the measurement system to assess break through. The back-up PUF was extracted and analyzed separately. Although a comprehensive evaluation of the results has not been undertaken, a sampling of results collected in the summer and winter indicates that breakthrough was up to 39% for naphthalene in winter and up to 49% for dibenzo(*a,h*)pyrene in summer. Higher molecular weight compounds, such as benzo(*ghi*)perylene had breakthrough levels of <10%. These results were for concentrations more than 10 times the MDL.

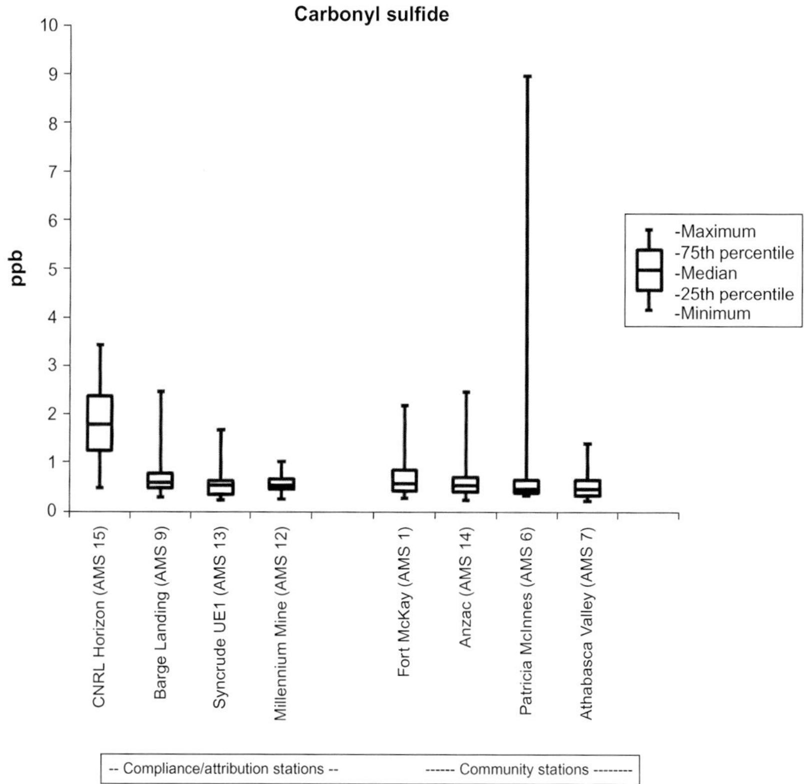

FIGURE 4.15   Carbonyl sulfide concentrations measured in 2011 in 24-h time-integrated sampling at WBEA air monitoring stations. Data are displayed as maximum, median, and various percentile values.

Occasionally, samples exhibited greater complexity and required a cleanup step using silica column chromatography prior to analysis. The method used was as listed in EPA SW-846 METHOD 3630C (U.S. EPA 1999b, 2000). Method detection limits ranged from 0.03 to 0.06 ng m^{-3} for naphthalene to benzo(*ghi*)perylene with this method. Recoveries ranged from 87% for naphthalene to 92% for benzo(*ghi*)perylene. Precision of analysis based on deuterated compounds introduced at the extraction stage ranged up to 18% relative standard deviation.

Table | 4.4 lists the 2011 descriptive statistics for the 23 PAH species averaged across the four stations. Phenanthrene was measured in the highest concentrations and was detected at all four community stations (Figure 4.19). The maximum concentration of 33.3 ng m^{-3} was recorded at Fort McKay (AMS 1). Mean concentrations for the PAH species decreased in the order: phenathrene   (4.35 $\pm$ 5.1 ng m^{-3}) > fluorene   (1.0 ng m^{-3}) > fluoranthene (0.8 ng m^{-3}) > pyrene (0.7 ng m^{-3}) > anthracene (0.4 $\pm$ 0.4 ng m^{-3})

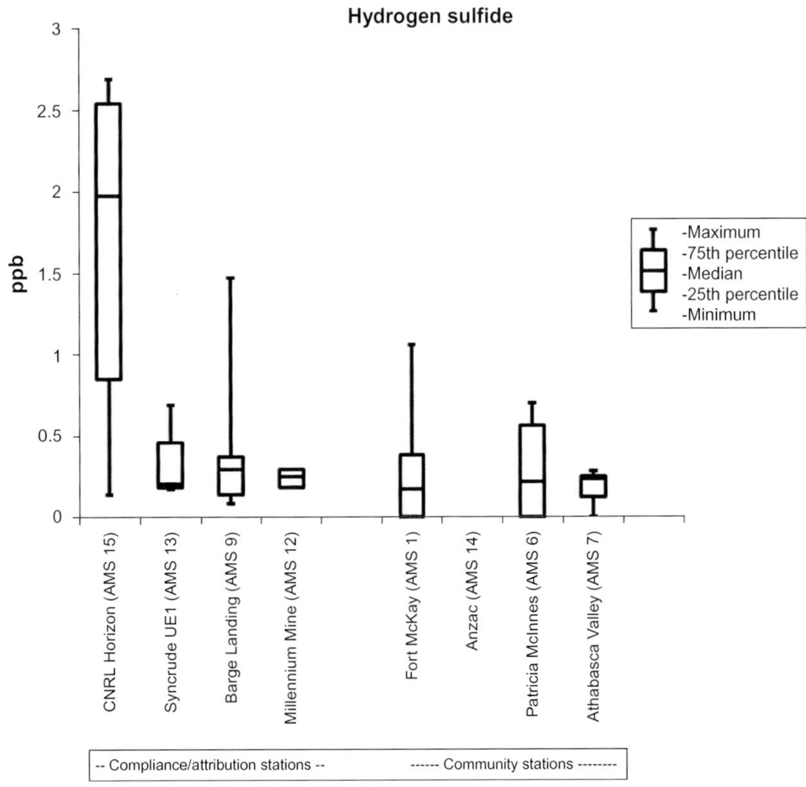

FIGURE 4.16 Hydrogen sulfide concentrations measured in 2011 in 24-h time-integrated sampling at WBEA air monitoring stations. Data are displayed as maximum, median, and various percentile values.

$> 7,12$-dimethylbenz(a)anthracene $(0.3 \pm 0.4$ ng m$^{-3}) >$ acenaphthene $(0.25$ ng m$^{-3}) >$ acridine $(0.23 \pm 0.27$ ng m$^{-3}) >$ naphthalene $(0.18 \pm 0.2$ ng m$^{-3})$ (Table 4.4).

4.4.4 Total Gaseous Mercury Monitoring at AMS 6 Patricia McInnes

Mercury exists in the three forms in the atmosphere. Its predominant form is gaseous elemental mercury (GEM) which makes up around 95–97% of the airborne Hg. The remainder of mercury exists in the atmosphere as inorganic reactive gaseous mercury (RGM) and particulate bound mercury (PHg) (Lin and Pehkonen, 1999; Schroeder and Munthe, 1998). Mercury can be removed from the atmosphere through wet or dry deposition processes; however, mercury is a unique species, as it can be reemitted after being deposited (Schroeder and Munthe, 1998). Industrial mercury emission sources in the

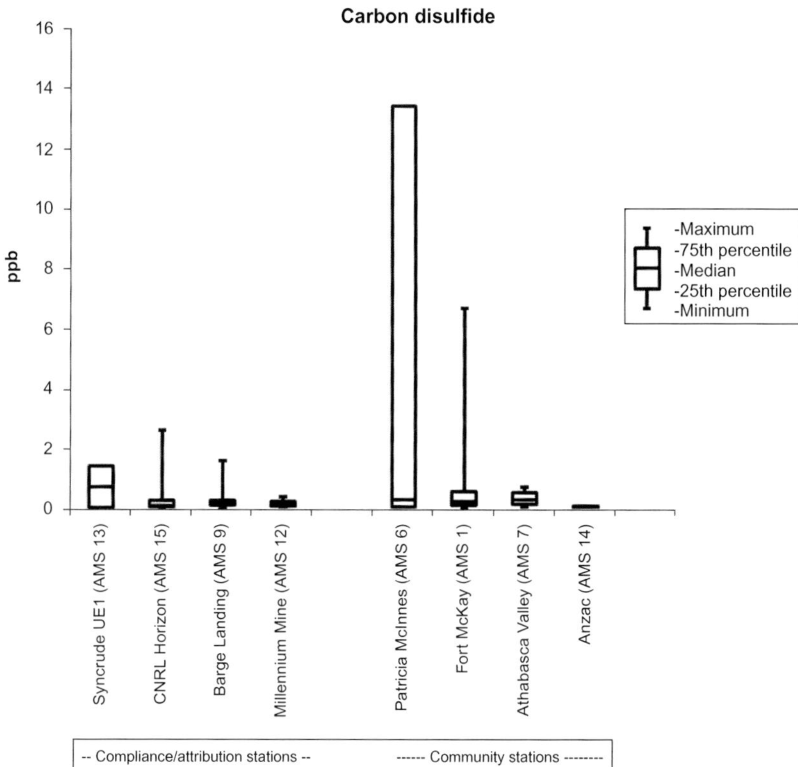

FIGURE 4.17   Carbon disulfide concentrations measured in 2011 in 24-h time-integrated sampling at WBEA air monitoring stations. Data are displayed as maximum, median, and various percentile values.

region include oil sands extraction, bitumen upgrading, and power generation facilities related to the oil industry (NPRI, 2010). Additional influences on this site include urban sources associated with the town site of Fort McMurray (i.e., transportation and residential fuel combustion emissions, landfill, wastewater treatment) (NPRI, 2010) and natural emissions from the surrounding boreal forest and wetlands in the region (i.e., degassing mercury from surface water, soil, and vegetation; wind erosion of soil; and biomass burning) (Richardson et al., 2003). Forest fires are significant in mercury cycling due to revolatilization of mercury deposited on plant surfaces and fire-heated soil (Friedli et al., 2003, Turetsky et al., 2006). Moreover, forest fires are episodic sources that can greatly influence total gaseous mercury (TGM) concentration measurements over short-time spans.

Since October 2010, Environment Canada has measured TGM on a continuous basis from WBEA's Patricia McInnes air monitoring station (AMS 6). This community-based site is located in Fort McMurray, near the western edge of the Timberlea subdivision. The gaseous mercury analyzer (Tekran

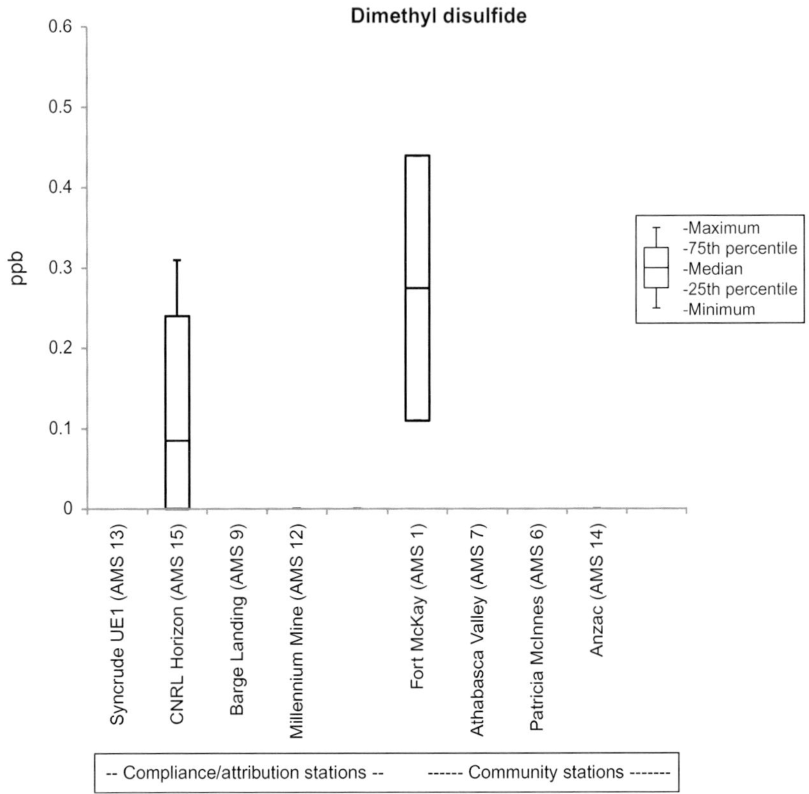

FIGURE 4.18 Dimethyl disulfide concentrations measured in 2011 in 24-h time-integrated sampling at WBEA air monitoring stations. Data are displayed as maximum, median, and various percentile values.

2537) at the Patricia McInnes station captured atmospheric mercury on gold traps, followed by thermal desorption and cold vapor atomic fluorescence spectroscopy detection (Tekran, 2007). The analyzer has a detection limit of $0.1 \, \text{ng m}^{-3}$. Sample air was collected over a 5-min period and parallel gold traps allow for one cartridge to be sampled while the other is analyzed. The instrument was auto calibrated every 23 h to correct for any instrument drift. Raw concentration data were quality controlled using Environment Canada's Research Data Management Quality and subsequent manual quality control procedures as described by Steffen et al. (2012).

The ambient TGM concentration data presented here spans from October 21, 2010 through to May 31, 2011. Table 4.5 lists a statistical summary of the TGM concentration dataset for the full date range and also subsets of the TGM concentration data during forest fire activity. Figure 4.20 presents the TGM time series for the entire dataset. Measured TGM concentrations reflect a combination of background, regional, and local sources. TGM

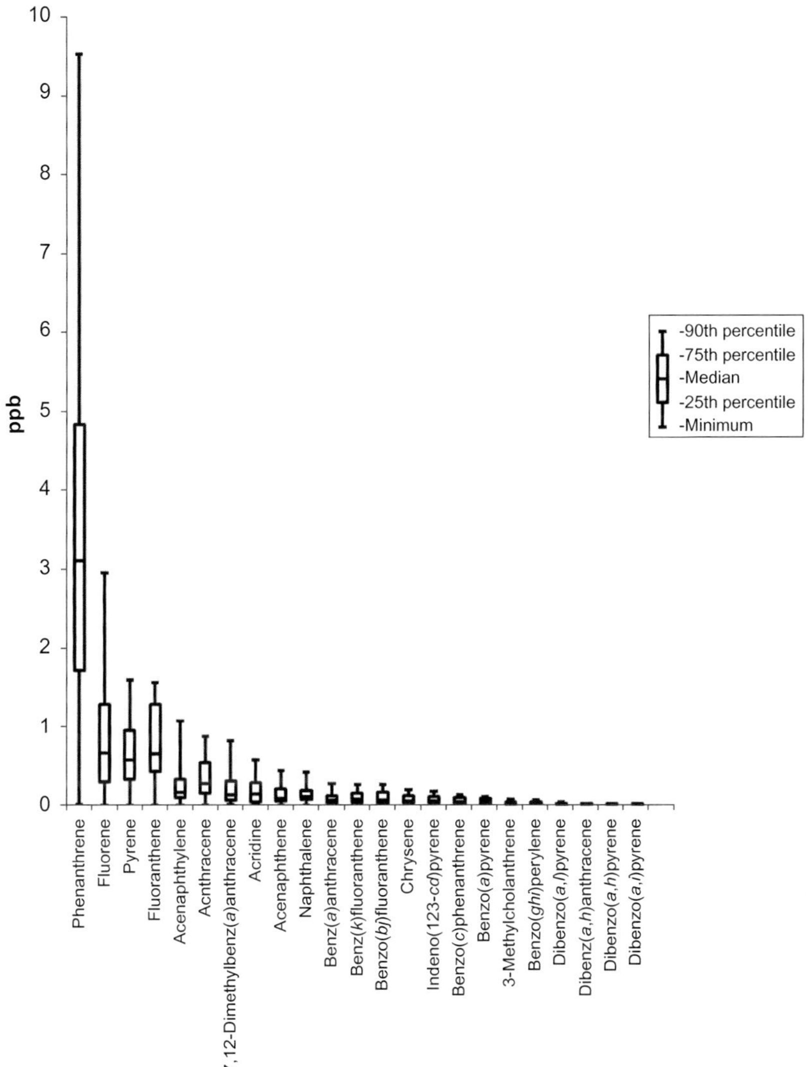

FIGURE 4.19   Ambient concentrations of 23 polycyclic aromatic hydrocarbons measured at four WBEA community air monitoring stations in 2011. Data are displayed as maximum, median, and various percentile values.

concentration measurements varied depending on wind direction, with influence from either oil sands or town site sources from the north-northwest and east-southeast, respectively. The average measured TGM concentration to date, excluding data with forest fires, is similar to average TGM concentrations measured elsewhere in Canada (Kellerhals et al., 2003; Temme et al., 2007); values measured in Alberta have ranged from 1.36 to 1.65 ng m^{-3}

TABLE 4.4 Descriptive Statistics for Concentrations of 23 PAH Species Measured in 2011

Compound	N	Mean $(ng\ m^{-3})$	SD $(ng\ m^{-3})$	Minimum $(ng\ m^{-3})$	Median $(ng\ m^{-3})$	95% $(ng\ m^{-3})$	99% $(ng\ m^{-3})$	Maximum $(ng\ m^{-3})$
Dibenzo(*a,h*) anthracene	54	0.02	0.01	0	0.01	0.04	0.06	0.06
Dibenzo(*a,h*)pyrene	54	0.02	0.02	0	0.01	0.03	0.12	0.12
Dibenzo(*a,i*)pyrene	54	0.01	0.01	0	0.01	0.03	0.06	0.06
Dibenzo(*a,l*)pyrene	54	0.02	0.02	0	0.02	0.06	0.07	0.07
3-Methylcholanthrene	54	0.04	0.03	0.01	0.03	0.13	0.15	0.15
Acenaphthene	54	0.25	0.64	0	0.09	0.79	4.64	4.64
Acenaphthylene	54	0.46	0.87	0.01	0.16	2.67	4.21	4.21
Acridine	54	0.23	0.27	0	0.15	1.01	1.1	1.1
Anthracene	54	0.42	0.46	0.01	0.27	1.17	2.74	2.74
Benz(*a*)anthracene	54	0.11	0.12	0	0.07	0.35	0.63	0.63
Benz(*k*)fluoranthene	54	0.11	0.11	0	0.08	0.32	0.54	0.54
Benzo(*a*)pyrene	54	0.06	0.05	0	0.05	0.15	0.24	0.24
Benzo(*bj*) fluoranthene	54	0.11	0.1	0.01	0.07	0.3	0.5	0.5
Benzo(*c*) phenanthrene	54	0.09	0.17	0	0.05	0.23	1.21	1.21

TABLE 4.4 Descriptive Statistics for Concentrations of 23 PAH Species Measured in 2011—Cont'd

Compound	N	Mean (ng m^{-3})	SD (ng m^{-3})	Minimum (ng m^{-3})	Median (ng m^{-3})	95% (ng m^{-3})	99% (ng m^{-3})	Maximum (ng m^{-3})
Benzo(*ghi*)perylene	54	0.03	0.02	0	0.02	0.09	0.1	0.1
Chrysene	54	0.09	0.1	0	0.06	0.34	0.5	0.5
Fluoranthene	54	0.84	0.54	0	0.65	1.68	2.43	2.43
Fluorene	54	1.04	1.11	0.01	0.67	3.61	4.83	4.83
Indeno(123-*cd*)pyrene	54	0.08	0.06	0.01	0.05	0.2	0.26	0.26
Naphthalene	54	0.18	0.21	0	0.11	0.64	1.01	1.01
Phenanthrene	54	4.35	5.13	0.01	3.1	10.5	33.38	33.38
Pyrene	54	0.74	0.54	0	0.57	1.85	2.39	2.39

TABLE 4.5 Ambient Total Gaseous Mercury (TGM) Statistics for Full Dataset and Subsets of Data With and Without Forest Fire Activity at the Patricia McInnes Site

Descriptor	Date	N	Avg. (ng m^{-3})	Median (ng m^{-3})	Max. (ng m^{-3})	Min. (ng m^{-3})	Std. dev. (ng m^{-3})
All data	10/21/2010 to 5/31/2011	4914	1.42	1.41	3.71	0.64	0.21
Forest fire only	5/18/2011 to 5/31/2011	171	1.99	1.77	3.71	1.20	0.55
Excluding forest fire	10/21/2010 to 5/18/2011	4743	1.40	1.41	2.10	0.64	0.15

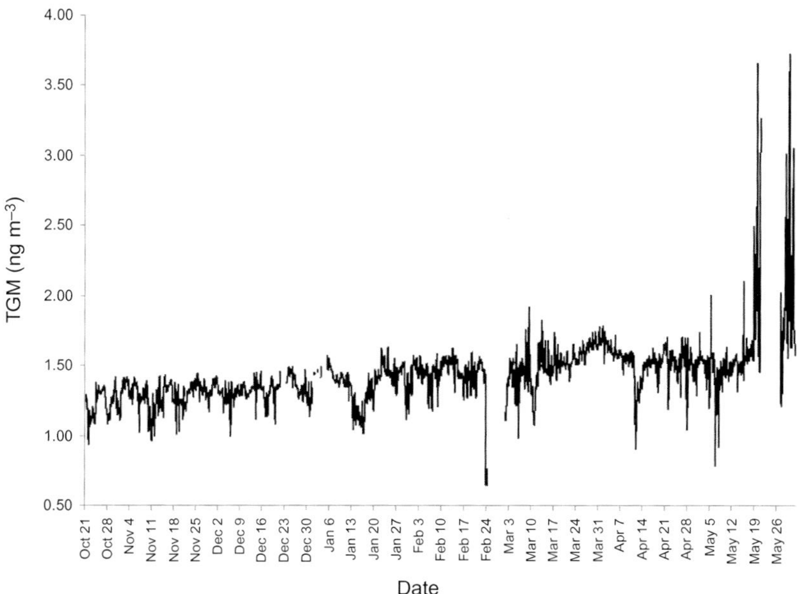

FIGURE 4.20 Ambient total gaseous mercury (TGM) concentrations measured from October 21, 2010 to May 31, 2011 at the Patricia McInnes air monitoring station in Fort McMurray.

(Mazur et al., 2009). Seasonal trends have been observed for TGM concentration measurements with highest concentrations in the winter–spring and lowest concentrations in summer–fall (Temme et al., 2007). Although the Patricia McInnes data set reported here is not sufficiently long to firmly establish such trends, these data suggest a similar seasonal trend is likely with lower values in the fall and higher values in the spring.

From May 18–31, 2011, TGM concentrations were also influenced by forest fire activity north of Fort McMurray, approximately 100 km north of the monitoring site. According to the Fire Information for Resource Management System (FIRMS) Web site using Moderate Resolution Imaging Spectroradiometer (MODIS) data (Justice et al., 2002), appreciable forest fires were detected north of Fort McMurray starting May 14, 2011. Winds were directly from the north starting late May 18, 2011, resulting in elevated TGM concentrations as shown in Figure 4.20. Fine particulate matter ($PM_{2.5}$) is plotted in addition to the TGM levels. Fine particulate matter ($PM_{2.5}$) concentration has been shown to correlate with biomass burning area (Jaffe et al., 2008) and thus $PM_{2.5}$ is plotted in addition to the TGM levels in Figure 4.21. This figure shows that concentration of TGM varies with that of $PM_{2.5}$, suggesting these elevated TGM concentration measurements are likely due to forest fire activity. Heavy smoke (particulate matter) levels from the forest fire interrupted data collection between May 21 and 27. As shown in Section 4.4, the highest TGM concentration and strongest

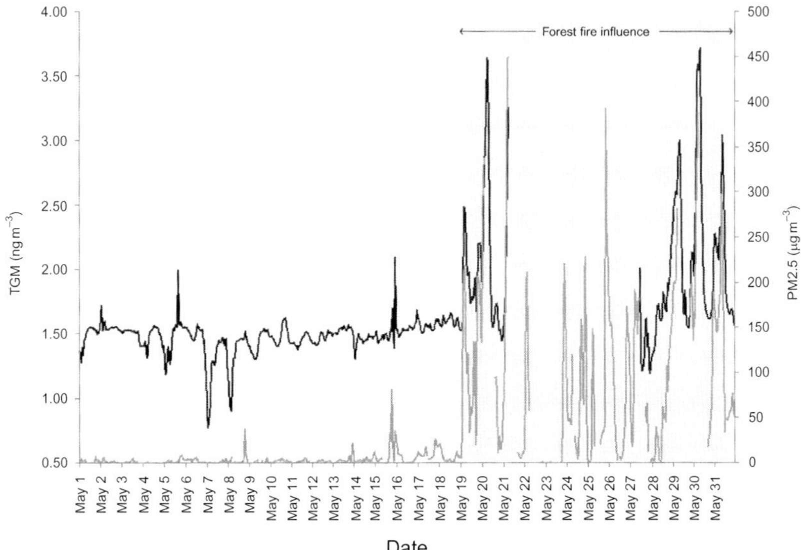

FIGURE 4.21   Ambient total gaseous mercury (TGM) concentration measurements (black line) and particulate matter (PM$_{2.5}$) concentration measurements (gray line) during May 2011 at the Patricia McInnes site. *Note*: the elevated TGM concentration measurements influenced by forest fire activity from late May 18, 2011. The TGM analyzer was not operated from May 21 to 27, 2011 due to heavy fire smoke (particulate) levels.

variation in the data are reported during this forest fire event. Figure 4.21 shows the typical variability of TGM concentration measurements leading up to the forest fire activity, followed by the elevated TGM concentrations during the forest fire events.

4.5   2011 AIR QUALITY HEALTH INDEX VALUES

The Air Quality Health Index (AQHI) is a public information tool that helps Canadians protect their health on a daily basis from the negative effects of air pollution. This tool has been developed by Health Canada and Environment Canada, in collaboration with the provinces, and key health and environment stakeholders (Environment Canada, 2012). The AQHI value is calculated by using a formula which combines the readings of three specific pollutants—fine particulate matter, ozone, and nitrogen dioxide. The AQHI value is calculated in other provinces by using PM$_{2.5}$, O$_3$, and NO$_2$. However, because of Alberta's energy-based economy, other pollutants monitored in the province are included when calculating the AQHI. The AQHI number presented in Alberta considers hourly comparisons of individual pollutant concentrations to Alberta's Ambient Air Quality Objectives (AAQOs). If hourly air pollutant concentrations are higher than Alberta's AAQOs, then the AQHI value is replaced with the appropriate "High" or "Very High" risk value. This adjustment is relevant for particulate matter, ozone, nitrogen dioxide, sulfur dioxide, and carbon monoxide. In addition, the AQHI will be modified if hydrogen sulfide and TRS concentrations exceed a specified health threshold (AESRD, 2012).

In 2011, the AQHI was measured in more than 20 communities across Alberta. These included the three WBEA-monitored community stations of Fort Chipewyan (AMS 8), Fort McKay (AMS 1), Fort McMurray Athabasca Valley (AMS 7), and the attribution station at Fort McKay South (Syncrude UE-1). WBEA raw air quality data are transmitted hourly to AESRD, where the calculation is made and the AQHI value then uploaded to an AESRD Web site, the WBEA Web site, and WBEA building's outdoor digital sign and community display. The 2011 hourly AHQI values were provided by AESRD, and the percent of the values within each of the four risk categories were calculated.

As can be seen, air quality at the four WBEA stations presented a low health risk between 96% (Fort McMurray) and 99.3% (Fort Chipewyan) of the time in 2011 (Figure 4.22). AQHI values presenting a moderate risk to health at WBEA stations occurred between 0.18% (Fort Chipewyan) and 2.1% (Fort McMurray) of the time in 2011. In 2011, AQHI values presenting a high health risk occurred between 0.09% (Fort Chipewyan) and 1.3% (Fort McMurray) of available hours. Almost all hourly AQHI values recorded in the moderate (4–6) to high (7–10) risk to health categories at WBEA stations were reported during the extended May 14 to June 2011 regional wildfire episode.

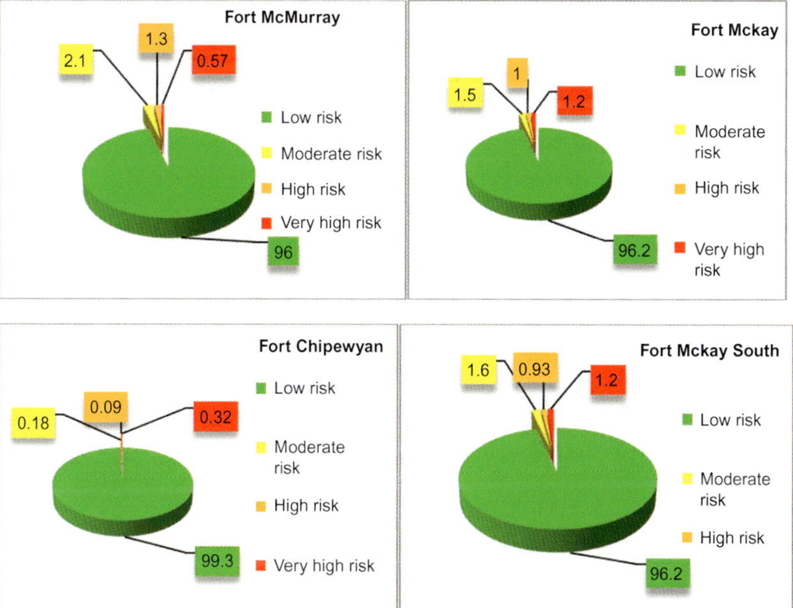

FIGURE 4.22 2011 air quality health index percentages by risk level for four WBEA monitoring.

According to AESRD, when the amount of air pollution is abnormally high, such as air quality associated with wildfire smoke, the AQHI number may exceed 10. AQHI numbers above 10, indicating very high health risk, occurred between 0.32% (Fort Chipewyan) and 1.2% (Fort McKay, Fort McKay South) of the time. All hours with AQHI values above 10 occurred during the most intense portions of the regional fire smoke episode. Extremely high concentrations of fine particulate matter, causing reduced visibility, were measured by WBEA during this event, as well as occasional elevated levels of ground-level ozone.

It is important to note that the pollutants used to calculate the AQHI do not include RSCs and VOCs that contribute to odor events experienced in several communities within the RMWB. Therefore, the AQHI values should not be used to evaluate a potential health risk associated with detection of odors.

4.6 TRENDS AND OTHER REGIONS

4.6.1 Long-Term Trends

In 2009, WBEA commissioned a regression-based trending analysis of air quality over a 10-year period from 1998 to 2007. WBEA hourly average SO_2, NO_2, $PM_{2.5}$, O_3, and H_2S/TRS data from 11 monitoring stations operating continually during that period were used in the analysis by Kindzierski et al. (2009).

At Fort McKay (AMS 1), there was no indication of any change for NO_2 at the lower percentile concentrations more typically experienced on any given day (Kindzierski 2010). However, there was an indication of increases at the higher percentile concentrations experienced, on average, much less frequently. A decreasing trend was observed for SO_2 at the low percentile concentrations and no change noted at high percentile concentrations. A decreasing trend was observed for $PM_{2.5}$ at almost all of the percentile concentrations. No change was observed for O_3 at all of the percentile concentrations. A small, but increasing trend was observed for TRS at all percentile concentrations, although these increases were relatively small (1 ppb/yr).

Decreasing hourly concentrations were observed for $PM_{2.5}$ at the two Fort McMurray stations (AMS 6, AMS 7) and at Fort Chipewyan (AMS 8) between 1998 and 2007. Small and consistently increasing hourly concentrations were observed for NO_2 in Fort McMurray and Fort McKay (AMS 1) over the 10-year period. Air quality at Fort Chipewyan (AMS 8) was unique and separate from that observed at the other community AMSs. Fort Chipewyan appeared to be far enough away from sources and activities influencing the other communities to varying degrees in the airshed. Kindzierski (2010) concluded that, "Despite increasing oil sands development activity in region during the 10-year period, this study found that changes to air quality in the communities were either small or not observed for most of the air pollutants examined."

4.6.2   Other Regions

At the time of writing this chapter, national-quality-assured data for 2011 were not available in the NAPS database. Therefore, NAPS 2010 data have been used to relate WBEA air quality data to Alberta and the other provinces. Table 4.6 provides a provincial comparison of the 99th percentile of daily maximum 1-h SO_2 concentrations (metric used for U.S. NAAQS). In 2010, 16 Canadian sites exceeded the U.S. NAAQS. Three of these sites were in Alberta: Redwater (376 ppb), Mannix-AMS 5 (107 ppb), and Mildred Lake-AMS 2 (91 ppb).

Table 4.7 provides a provincial comparison of the 98th percentile of daily maximum 1-h NO_2 concentrations. In 2010, no Canadian sites exceeded the U.S. NAAQS. The two highest 98th percentile values, however, were measured in Calgary and Redwater, Alberta, and the third highest value was measured at WBEA Millenium-AMS 12.

Table 4.8 provides a provincial comparison of the 98th percentile of daily mean $PM_{2.5}$ concentrations. In 2010, 38 sites in Canada recorded a 98th percentile value $>30\ \mu g\ m^{-3}$, with two WBEA sites Fort Chipewyan (AMS 8) and Anzac (AMS 14) included. There were 13 sites in other parts of Alberta

---

TABLE 4.6 2010 Provincial Comparison of the 99th Percentile of Daily Maximum 1-h SO_2 Concentrations (ppb)

Province/territory/region	No. of sites	Minimum (ppb)	Median (ppb)	Maximum (ppb)
Northwest territories	3	2	2	4
British Columbia	36	1	28	409
Alberta—Oil sands	12	5	60	107
Alberta—other	20	4	15	376
Saskatchewan	3	1	6	12
Manitoba	2	11	–	492
Ontario	16	4	20	103
Quebec	13	4	35	279
New Brunswick	4	46	63	119
Nova Scotia	1	11	11	11
Newfoundland	4	3	4	9

The number of reporting sites in each province, the concentration at the median site and at the highest and lowest site are provided. On June 2, 2010, the U.S. EPA established a new 1-h National Ambient Air Quality Standard (NAAQS) for SO_2 at 75 ppb.

TABLE 4.7 2010 Provincial Comparison of the 98th Percentile of Daily Maximum 1-h NO_2 Concentrations (ppb)

Province/territory/region	No. of sites	Minimum (ppb)	Median (ppb)	Maximum (ppb)
Northwest territories	2	25	25	25
British Columbia	33	17	34	49
Alberta—Oil sands	8	15	36	69
Alberta—other	23	10	47	79
Saskatchewan	4	41	48	55
Manitoba	3	46	47	52
Ontario	35	14	42	66
Quebec	17	28	44	50
New Brunswick	5	28	33	42
Newfoundland	4	23	25	40

The number of reporting sites in each province, the concentration at the median site and at the highest and lowest site are provided. On June 2, 2010, the U.S. EPA established a new 1-h National Ambient Air Quality Standard (NAAQS) for NO_2 at 100 ppb.

TABLE 4.8 2010 Provincial Comparison of the 98th Percentile of Daily mean $PM_{2.5}$ Concentrations ($\mu g\ m^{-3}$)

Province/territory/region	No. of sites	Minimum ($\mu g\ m^{-3}$)	Median ($\mu g\ m^{-3}$)	Maximum ($\mu g\ m^{-3}$)
Northwest territories	2	10	–	15
British Columbia	37	7	16	140
Alberta—Oil sands	7	15	22	41
Alberta—other	24	15	31	76
Saskatchewan	3	16	19	20
Manitoba	3	15	19	33
Ontario	40	12	22	40
Quebec	36	15	25	35
New Brunswick	8	15	16	25
Nova Scotia	1	15	15	15
Newfoundland	6	10	11	15

The number of reporting sites in each province, the concentration at the median site and at the highest and lowest site are provided. The CWS metric is set at 30 $\mu g\ m^{-3}$ (based on a 3-year average).

TABLE 4.9 2010 Provincial Comparison of fourth Highest Daily Maximum 8-h Ozone Concentrations (ppb)

Province/territory/region	No. of sites	Minimum (ppb)	Median (ppb)	Maximum (ppb)
Northwest territories	3	44	48	48
Nunavut	1	48	48	48
British Columbia	33	42	50	61
Alberta—Oil sands	5	52	54	57
Alberta—other	21	51	58	72
Saskatchewan	5	43	54	58
Manitoba	3	56	64	64
Ontario	47	57	70	87
Quebec	47	45	61	68
New Brunswick	13	50	55	62
Nova Scotia	7	45	55	60
Newfoundland	6	44	46	53

The number of reporting sites in each province, the concentration at the median site and at the highest and lowest site are provided. The CWS metric is set at 65 ppb (based on a 3-year average).

(6 in Edmonton), 11 in British Columbia, 9 in Quebec, and 1 in Manitoba that exceeded 30 μg m^{-3}.

Table 4.9 provides a provincial comparison of the fourth highest daily maximum 8-h ozone concentrations. In 2010, 44 sites in Canada exceeded the 65 ppb value, 39 were in Ontario, 3 in Quebec, and 2 in Alberta. No sites in WBEA network recorded a fourth highest daily maximum 8-h ozone concentration >65 ppb.

4.7   SUMMARY

The Wood Buffalo Environmental Organization (WBEA) is a consensus-based, independent, multistakeholder organization with responsibility to monitor ambient air quality in the RMWB. The 68,000-km^2 municipality includes the Athabasca Oil Sands Region (AOSR). WBEA has used continuous and time-integrated methods to measure air quality at 15 monitoring stations located along a longitudinal gradient of some 250 km. Air quality was measured for compliance, attribution to source, community health, and background data purposes. WBEA achieves this with a highly trained staff of four senior air quality specialists assisted by four air quality technicians. Network performance by the last quarter of 2011 was 98–99%.

In 2011, there were no exceedances recorded of the AAAQOs for the criteria air contaminants (CACs) SO_2, NO_2, and CO. Ammonia (NH_3) concentrations were below the detection limit of the analyzers, and often difficult to measure with a trace-level analyzer. Maximum 1-h H2S/TRS concentrations ranged from 3 to 7 ppb at community stations and 6 to 98 ppb at compliance stations. In 2011, there were 23 exceedances of the 1-h AAAQO based on an odor threshold of 10 ppb. This was many fewer than in 2010 and 2009. There were no exceedances at community stations. However, odor episodes were recorded in some communities, and odors remain a concern.

Twenty-three PAH species were measured at four community stations on a time-integrated basis. Mean 2011 concentrations averaged across stations ranged from 0.01 ng m^{-3} for dibenzo(*a,i*)pyrene to 4.35 ng m^{-3} for phenanthrene. Most frequently measured VOCs in 24-h canister sampling were the BTEX compounds, acetone, isopentane, and isobutane. Carbonyl sulfide (COS), CS_2, and H_2S were also frequently measured.

Air quality in 2011 was dominated by the massive, 700,000-km^2 forest fire complex that burned north of Fort McKay and Fort McMurray throughout May and June. During fire smoke episodes, visibility was severely reduced, and health advisories were issued by regional authorities using WBEA air quality data. Frequently, maximum hourly $PM_{2.5}$ concentrations exceeded 400 µg m^{-3} and analyzers became inoperable pending filter cleanout and recalibration on site. At the time, WBEA was testing new technology which remained operating and recorded short-term $PM_{2.5}$ concentrations up to 900 µg m^{-3}. Due to the fire, the $PM_{2.5}$ and O_3 data were skewed toward higher than historical (1998–2010) values. There were 15 exceedances of the 82 ppb 1-h AAAQO for O_3, all during the fire event. There were 97 exceedances of the 24-h AAAQO for PM2.5, 97% of these occurring during the smoke episodes.

For the first time in the AOSR, ambient mercury measurements were made by Environment Canada-Meteorological Service of Canada. During the period between October 21, 2010 and May 31, 2011, average ambient TGM concentration at Patricia McInnes (AMS 6) in Fort McMurray was 1.40±0.15 ng m^{-3}. This value was similar to TGM concentrations measured elsewhere in Canada.

The year 2011 was the first full year in which WBEA publically reported hourly AQHI values, calculated by AESRD using WBEA continuous data from four community stations. Despite the massive fire event, air quality was in the low risk to health category between 96% and 99.3% (depending on station) of available hours. Most hours in the moderate risk (0.18–2.1%) and all hours in the high or very high risk categories occurred during the most intense smoke episodes.

The WBEA air quality monitoring network has been responsive and has adapted to fit stakeholder needs since 1997. It is expected that the network

will continue to expand to meet compliance and regional monitoring requirements. New technologies will continue to be evaluated for use in the region,
so network performance is maintained at the 99% + level. WBEA data processing and management systems and software are evolving through increased
investment to provide secure, transparent, and high-quality data to stakeholders and the public. WBEA is actively engaged in air monitoring with both
levels of government under the Joint Canada/Alberta Implementation Plan for
Oil Sands Monitoring.

ACKNOWLEDGMENTS

The WBEA air monitoring network infrastructure and its operation are fully supported and
funded by its multistakeholder membership. The authors acknowledge the technical
expertise and dedication shown by WBEA contractors and staff in their work. We
acknowledge the ongoing and important role played by the WBEA Ambient Air Technical
Committee (AATC) in network planning and oversight. The AATC has been ably supported
by the WBEA program manager and science advisors.

The massive forest fire event during May to June 2011 presented huge, ongoing challenges
for analyzer maintenance under extreme fine particulate loadings. During the 6-week long
event, severely reduced visibility due to fire smoke resulted in some 35,000 hits to the
WBEA Web site in 1 week alone. Despite daily shifts in assigning priorities to stations and
analyzers, WBEA was able to successfully maintain a near uninterrupted flow of air quality
data to the public, our members, and regional health authorities.

The Tekran instrument for measurement of ambient TGM at AMS 6 was provided by
Environment Canada, Meteorological Service of Canada, Prairie, and Northern Region,
Edmonton, Alberta. We are most grateful to Rachel Mintz and Monique Lapalme for providing
the TGM data and results contained in Section 4.4.4, along with Table 4.3 and Figures 4.20 and
4.21. We also thank WBEA air technical staff for their assistance in its operation.

REFERENCES

National Atmospheric and Oceanic Administration (NOAA), 2008. Carbon dioxide, methane rise
    sharply in 2007. http://www.noaanews.noaa.gov/stories2008/20080423_methane.html.
Wood Buffalo Environmental Association, 2009. Annual Report 2008. http://www.wbea.org.
    accessed October 4, 2012.
Wood Buffalo Environmental Association, 2010. Annual Report 2009. http://www.wbea.org.
    accessed October 4, 2012.
Wood Buffalo Environmental Association, 2011. Annual Report 2010. htttp://www.wbea.org.
    accessed October 4, 2012.
Environment Canada, 2012. Canadian Smog Science Assessment, http://www.ec.gc.ca/air.
    accessed October 4, 2012.
Dibb, J.E., Talbot, R.W., Scheuer, E., Seid, G., DeBell, L., Lefer, B., Ridley, B., 2003. Strato-
    spheric influence on the northern North American free troposphere during TOPSE: Be-7 as
    a stratospheric tracer. J. Geophys. Res. Atmos. 108 (D4), 11.1–11.8. http://dx.doi.org/
    10.1029/2001JD001347.
Dionex, 2011. Extraction of PAHs from environmental samples by accelerated solvent extraction
    (ASE). Dionex Application Note 313. Sunnyvale, CA.

Alberta Environment and Sustainable Resource Development (AESRD), 2012. http://environment.alberta.ca/03603.html.

U.S. EPA, 1999a. Compendium of methods for the determination of toxic organic compounds in ambient air, second ed. (EPA/625/R-96/010b, January 1999).

U.S. EPA, 1999b. Compendium U.S. EPA Method TO-13A: determination of polycyclic aromatic hydrocarbons (PAHs) in ambient air using gas chromatography/mass spectrometry (GC/MS). U.S. Environmental Protection Agency, January, 1999.

U.S. EPA, 2000. Test method for evaluating solid wastes physical/chemical methods, U.S. EPA SW-846 Method 3545A: pressurized fluid extraction (PFE). U.S. Environmental Protection Agency, Revision 1, November, 2000.

Friedli, H.R., Radke, L.F., Lu, J.Y., Banic, C.M., Leaitch, W.R., MacPherson, J.I., 2003. Mercury emissions from burning of biomass from temperate North American forests: laboratory and airborne measurements. Atmos. Environ. 37, 253–267.

Tekran Instruments Corporation, 2007. Model 2537B ambient mercury vapor analyzer: user manual. Toronto, Tekran.

Jaffe, D., Hafner, W., Chand, D., Westerling, A., Spracklen, D., 2008. Interannual variations in $PM_{2.5}$ due to wildfires in the western United States. Environ. Sci. Technol. 42 (8), 2812–2818.

Jenkin, M.E., Clemitshaw, K.C., 2000. Ozone and other secondary photochemical pollutants: chemical processes governing their formation in the planetary boundary layer. Atmos. Environ. 34 (16), 2499–2527.

Justice, C.O., Giglio, L., Korontzi, S., Owens, J., Morisette, J.T., Roy, D., Descloitres, J., Alleaume, S., Petitcolin, F., Kaufman, Y., 2002. The MODIS fire products. Remote Sens. Environ. 83, 244–262.

Kellerhals, M., Beauchamp, S., Belzer, W., Blanchard, P., Froude, F., Harvey, B., McDonald, K., Pilote, M., Poissant, L., Puckett, K., Schroeder, B., Steffen, A., Tordon, R., 2003. Temporal and spatial variability of total gaseous mercury in Canada: results from the Canadian Atmospheric Mercury Measurement Network (CAMNet). Atmos. Environ. 37 (7), 1003–1011.

Kindzierski, W.B., 2010. Ten year trends in regional air quality for criteria pollutants in the Athabasca Oil Sands Region. Paper 2010-A-1079-AWMA, Proceedings of the Air and Waste Management Association Annual Conference, Calgary, Alberta, June 23–26.

Kindzierski, W.B., Chelme-Ayala, P., Gamal El-Din, G., 2009. Wood Buffalo Environmental Association Ambient Air Quality Data Summary and Trend Analysis. http://www.publichealth.ualberta.ca/research/~/media/publichealth/Research/Research%20Reports/WBEA_summary.ashx.

Lefohn, A.S., Oltmans, S.J., Dann, T., Singh, H.B., 2001. Present day variability of background ozone in the lower troposphere. J. Geophys. Res. 9, 9945–9958.

Lin, C.-J., Pehkonen, S.O., 1999. The chemistry of atmospheric mercury: a review. Atmos. Environ. 33 (13), 2067–2079.

Mazur, M., Mintz, R., Lapalme, M., Wiens, B., 2009. Ambient air total gaseous mercury concentrations in the vicinity of coal-fired power plants in Alberta, Canada. Sci. Total Environ. 408 (2), 373–381.

Monks, P.S., 2000. A review of the observations and origins of the spring ozone maximum. Atmos. Environ. 34, 3545–3561.

National Pollutant Release Inventory (NPRI), 2010, 2012. Environment Canada. http://www.ec.gc.ca/inrp-npri.

Phillips, D., 2010. The WBEA air quality monitoring network: history of operation and current status. Extended Abstract 2010-A-914-AWMA, Proceedings of the Air and Waste Management Association Annual Conference, Calgary, Alberta, June 23–26.

Richardson, G.M., Mitchell, I.A., Mah-Paulson, M., Hackbarth, T., Garrett, R.G., 2003. Natural emissions of mercury to the atmosphere in Canada. Environ. Rev. 11 (1), 17–36.

Schroeder, W.H., Munthe, J., 1998. Atmospheric mercury—an overview. Atmos. Environ. 32 (5), 809–822.

Simpson, I.J., Blake, N.J., Barletta, B., Diskin, G.S., Fuelberg, H.E., Gorham, K., Huey, L.G., Meinardi, S., Rowland, F.S., Vay, S.A., Weinheimer, A.J., Yang, M., Blake, D.R., 2010. Characterization of trace gases measured over Alberta oil sands mining operations: 76 speciated C_2-C_{10} volatile organic compounds (VOCs), CO_2, CH_4, CO, NO, NO_2, NO_y, O_3 and SO_2. Atmos. Chem.Phys. Discuss. 10, 18507–18560.

Spink, D., Dennis, J., 2010. Odor Event Air Quality Monitoring in the Community of Fort McKay. Report to the Fort McKay Industry Relations Corporation, Fort McKay, Alberta.

Steffen, A., Scherz, T., Olson, M., Gay, D., Blanchard, P., 2012. A comparison of data quality control protocols for atmospheric mercury speciation measurements. J. Environ. Monit. 14, 752–765.

Tarasick, D.W., Fioletov, V.E., Wardle, D.I., Kerr, J.B., Davies, J., 2005. Changes in the vertical distribution of ozone over Canada from ozonesondes: 1980–2001. J. Geophys. Res. 11, D02304, 19. http://dx.doi.org/10.1029/2004JD004643.

Temme, C., Blanchard, P., Steffen, A., Banic, C., Beauchamp, S., Poissant, L., Tordon, R., Wiens, B., 2007. Trend, seasonal and multivariate analysis of total gaseous mercury data from the Canadian Atmospheric Mercury Measurement Network (CAMNet). Atmos. Environ. 41 (26), 5423–5441.

Turetsky, M.R., Harden, J.W., Friedli, H.R., Flannigan, M.D., Payne, N., Crock, J., Radke, L.F., 2006. Wildfires threaten mercury stocks in northern soils. Geophys. Res. Lett. 33, L16403. http://dx.doi.org/10.1029/2005GL025595.

Chapter 5

Development and Application of Statistical Approaches for Reducing Uncertainty in Ambient Air Quality Data

M. Nosal[*,1], A.H. Legge[†], E.M. Nosal[‡] and M.C. Hansen[‡]

[*]*Department of Mathematics and Statistics, University of Calgary, Calgary, Alberta, Canada*
[†]*Biosphere Solutions, Calgary, Alberta, Canada*
[‡]*Wood Buffalo Environmental Association, Fort McMurray, Alberta, Canada*
[1]*Corresponding author: e-mail: nosal@ucalgary.ca*

ABSTRACT

Uncertainty estimation in continuous ambient air quality monitoring is one of the most important concepts related to data quality assurance. Even though its significance has been continuously emphasized in the scientific literature, as well as in governmental directives and regulations, its scientific methodology has been only partly developed and published. Uncertainty estimation is in principle a statistical concept based on probability distributions of ambient air pollutant concentrations. The problem stems from very complex distributions due to highly inaccurate estimates and large temporal and spatial variability of ambient air quality. Distributions are always highly skewed, often polymodal and thus noncompliant with the assumptions of standard statistical methodology. This chapter offers an innovative approach based on approximation of distributions of ambient air pollutant concentrations by the truncated Weibull family of probability distributions or their mixtures in cases of polymodal distributions. Such distributions are then used in Monte Carlo simulations to estimate the required uncertainty, which is related to the ambient air quality parameters. These methods have been developed and applied using the data collected by the continuous air quality monitoring network of the Wood Buffalo Environmental Association (www.wbea.org/) in the Athabasca Oil Sands Region of north-eastern Alberta, Canada.

Disclaimer: The content and opinions expressed by the authors in this chapter do not necessarily reflect the views of the WBEA or of the WBEA membership.

Developments in Environmental Science, Vol. 11. http://dx.doi.org/10.1016/B978-0-08-099443-7.00005-3

5.1 INTRODUCTION

Continuous monitoring of ambient air quality results in a large scale database of measurements of fundamental air quality parameters, such as concentrations of various gaseous air pollutants, meteorological parameters, etc. Like all measurements, air quality monitoring measurements are subject to various types of errors. In order to estimate the true but unknown numerical values of the measured parameters, one must evaluate and assess the measurement errors and their impact on the validity, precision, and accuracy of the estimation process. This constitutes a necessary and essential component of quality assurance and quality control (QA/QC) of the resulting data. While we cannot be absolutely sure about the true but unknown value of an air quality parameter, we can use the results of measurements to evaluate a likely range of possible values which the parameter can attain. Such a range then describes the uncertainty of the monitoring measurements. If we can construct such an uncertainty range which contains the true value of the parameter with a very high probability, then we have acquired very valuable information about the actual true value of the parameter.

The ideas discussed in the above paragraph apply virtually to all measurements of physical and/or chemical parameters. Air quality measurements, however, have their own particular challenges that play an important role in the derivation of suitable methods for uncertainty assessment. The first challenge is the fact that it is well known that ambient gaseous air quality concentrations can fluctuate widely near sources of emissions (Finlayson-Pitts and Pitts, 2000; Legge et al., 1990b). They may often occur below the field effective lower detection limit (LDL) of the air quality monitor, which results in erroneous values, even negative values and must be considered as unacceptable. This happens often in remote background areas distant from emissions sources with very low air pollutant concentrations (Finlayson-Pitts and Pitts, 2000; Legge et al., 1990a). Similarly, this may happen at air monitoring locations with topography shielding the receptors from emission sources due to meteorological factors such as prevailing wind direction and wind speed. These facts will be demonstrated later on in this chapter using selected examples of actual WBEA ambient air quality monitoring data collected at two stations in the northern Athabasca Oil Sands Region (AOSR; AMS 1 and AMS 10).

The second challenge of air quality monitoring is the fact that the monitoring equipment is located in the field, often operating under extremely adverse operational conditions. In such circumstances the manufacturer's specification for operating parameters such as temperature range, voltage, etc. cannot be easily maintained. This is particularly true for monitoring in remote areas which lack a stable source of electrical power or in northern latitudes with extremely cold temperatures as will be demonstrated later in this chapter. As a result, the manufacturer's LDL evaluated under ideal laboratory operating conditions may be inapplicable. Therefore, the monitoring process has to consider an actual field effective LDL of the monitor, as WBEA routinely does for its actual field operating conditions. Historically, values below LDL were often entirely removed or replaced by a half LDL value. All such methods can result in incorrect conclusions.

The third challenge of air quality monitoring is the fact that the frequency distributions of the air quality parameters monitored do not follow standard probability distributions such as Gaussian (normal) and related patterns (Student T-distribution, χ^2 distribution, Fisher distribution, etc.). Not only do the air quality parameters (gas concentrations) not follow the simple distributions, even the means of large samples of measurements do not obey the assumptions. This is a consequence of the continuous air quality measurements not representing independent random samples or independent repetitions as is required in standard statistics. Continuous ambient air quality measurements represent a highly autocorrelated time series. The actual frequency distributions of air quality parameters are highly skewed to the right. Fortunately they can be well approximated by the Weibull family of probability distributions. While this observation is quite helpful, finding uncertainty ranges for Weibull distributions is much more complex than in the case of standard statistical distributions.

As a consequence of these challenges, existing published scientific methodology for estimation and assessment of uncertainties related to ambient air quality measurements is rather limited. A search for air quality uncertainty in the ScienceDirect SciVerse (http://www.info.sciverse.com) yielded over 13,000 hits, but virtually all of them discussed only uncertainty related to modeling air quality or other issues, not air quality measurement uncertainty, which is the topic of this chapter. Thus the main objective and goal of this chapter is to develop and describe in full detail robust statistical methods for the definition and the numerical algorithms required for the computations of uncertainties related to air quality measurements. We will use methods based on Weibull probability distributions and Monte Carlo Methods (MCMs). Since these topics are not generally well known by air quality monitoring specialists, they will be described in the following sections.

5.2   RECENT ATTEMPTS RELATED TO UNCERTAINTY

The concept of quantifying uncertainty in air quality data is becoming increasingly important in all jurisdictions responsible for ambient air quality monitoring. The Alberta Environment (AE) *Air Monitoring Directive (AMD)* (Alberta Environment, 1989 and 2006), for example, requires that uncertainty estimation be considered during the air monitoring process. It specifies that when estimating the uncertainty of measurements, all associated components that are of importance must be taken into account. The recent Environment Canada document *Integrated Monitoring Plan for the Oil Sands: Air Quality*

Component (Environment Canada, 2011) also emphasizes in many sections the importance of uncertainty assessment of air quality estimates but is exceedingly short on technical details. Unfortunately, both aforementioned documents are very general and conceptual in nature without providing adequate technical details and references for implementing the required uncertainty assessment. On March 26, 2012, the National Sciences and Engineering Research Council of Canada announced the Climate Change and Atmospheric Research program that is expected to address national priorities of Canada. Among top priorities of the program is the investigation of ambient air quality and air pollution from enhanced industrial activities in the Arctic and adjacent cold regions. A strong emphasis is placed on techniques to reduce model bias and on methods to quantify predictive uncertainty.

The Wood Buffalo Environmental Association (WBEA) has assumed responsibility for the development and implementation of corresponding statistical methodology and procedures in the AOSR of north-eastern Alberta. AE further stipulated that all procedures developed for this purpose should be in line with the recent International Organization for Standardization (ISO) Guideline 20988: *Air Quality—Guidelines for estimating measurement uncertainty* (ISO, 2007). This document is based on the general concept of uncertainty estimation as described in the ISO Guide to the Expression of Uncertainty in Measurement. It should be noted that both ISO documents are based on modern statistical methodology. Many of the formulations in these documents are conceptually confusing. Only a limited amount of work has been done regarding the development of the air quality measurement uncertainty assessment methodology in Canada. The EC document entitled *National Air Pollution Surveillance Network Quality Assurance and Quality Control Guidelines* (Environment Canada, 2004) describes management policies and monitoring protocols which specify how to produce quality-assured data of known and acceptable precision, accuracy, representativeness, and completeness. This document, however, does not deal with uncertainty of ambient air quality data.

The U.S. Environmental Protection Agency (U.S. EPA) has devoted considerable attention to the management of uncertainty of environmental monitoring data and has published the guidance information in three volumes. The U.S EPA volume *Guidance for Data Quality Assessment, Practical Methods for Data Analysis* (U.S. EPA, 2000), is almost entirely devoted to statistical methodology for data quality assessment (DQA) which is the scientific and statistical evaluation of data to determine its uncertainty and to decide if data obtained from environmental data acquisition complies with the data quality objectives (DQO). It is an excellent treatment of fundamental statistical methods and includes descriptive statistics, graphical methods, and confidence intervals (CIs) as measures of uncertainty, an extensive selection of various statistical tests, statistical power functions, etc. The U.S. EPA volume *Guidance on Systematic Planning Using the Data Quality Objectives Process*

(U.S. EPA, 2006) addresses uncertainty in great detail with actual applications of statistical methods for uncertainty assessment described in the previous volume QA/G-9 to various types of environmental data. The U.S. EPA volume *Quality Assurance Handbook for Air Pollution Measurement Systems, Volume II, Ambient Air Quality, Monitoring Program* (U.S. EPA, 2008), Chapter 18 describes in detail the process of evaluating data against the DQOs, a process which has been termed *DQA*. This process is entirely based on complex statistical methodology for quantifying the uncertainty in the data. It should be emphasized that the U.S. EPA materials are very clear and detailed and fully utilize up to date state of the art modern statistical methodology.

One of the first theoretically based attempts regarding air quality uncertainty deals with the concept of the effects of measurement uncertainty on various air quality summary statistics used for compliance with legally based air quality standards (Curran and Suggs, 1986). The authors investigated uncertainty due to bias and imprecision. However, the underlying model is a simple and multiplicative and may be rather limiting in general applications. Since the authors assume a lognormal distribution of the measurements, its expedience is perhaps based more on the mathematical convenience than on practical reality.

5.3   ISO MEASUREMENT UNCERTAINTY ESTIMATION METHODOLOGY

ISO *Guideline 20988: Air Quality—Guidelines for Estimating Measurement Uncertainty* is technically one of the most detailed documents. In this section, we will follow ISO terminology and methodology and comment on its validity and suitability. Here, a measurand is a particular parameter subject to measurement. Measurement uncertainty is a concept associated with the result of a measurement that characterizes the measurand (ISO, 2007: 3.1 and 5.2).

Standard uncertainty of the measurand is the uncertainty $u(y)$ of the result of measurement y expressed as a standard deviation of the measured values (ISO, 2007: 9.2 and eq. 15):

$$u(y) = \sqrt{\mathrm{VAR}(y)}.$$

While the concept of the variance, VAR, treated in the ISO document is much more general than the concept used in standard descriptive statistics, under certain conditions (ISO case A1, simple random sampling) it is reduced to the standard statistical concept of variance. In real world air quality monitoring situations, to get an unbiased estimate $u(y)$ of the standard deviation, we have to use a sample of n independent repetitions y_1, y_2, ..., y_n of the measurement Y. In this case, we can actually use standard statistical procedure (in agreement with descriptive statistical methods and the EPA documents) and estimate the true unknown mean value μ_Y of the measurand Y by calculating the sample mean of the repeated measurements as follows:

$$y = \bar{y} = \frac{1}{n}\sum_{i=1}^{n} y_i.$$

The unbiased estimate of the standard uncertainty is then given as:

$$u(y) = \sqrt{\frac{\sum_{i=1}^{n}(y_i - \bar{y})^2}{n(n-1)}}.$$

Confidence limit $L_\gamma(u(y))$ for the estimated standard uncertainty (deviation) $u(y)$ are obtained in the ISO document by using critical points of the χ^2 distribution (ISO, 2007: eq. 17) as follows:

$$L_\gamma(u(y)) = \sqrt{\frac{(n-1)}{\chi^2_{(n-1,\gamma)}}}u(y).$$

However, use of the χ^2 distribution of the monitored measurements assumes them to be Gaussian. This assumption is most often unrealistic of continuous air quality data. Individual air quality measurements do not exhibit a Gaussian distribution, but rather they follow highly skewed distributions to the right such as the Weibull distribution family. Even the distribution of the means of a large set of samples (e.g., means of 5-min measurements with 1 s frequency yielding a sample size; e.g., $n = 300$) are greatly skewed to the right and hence non-Gaussian. It is essential to note that these facts are in a direct conflict with the basic assumptions in applied statistical methodology. Examples of such distributions of monitoring means obtained by the WBEA are given later in this chapter. It follows from statistical theory that this ISO approach based on the assumption of Gaussian distribution of the measurements is quite questionable.

Expanded uncertainty is a quantity defined as an interval about the result of a measurement y that may be expected to encompass a large proportion p of the distribution of values that could reasonably be attributed to the measurand (ISO, 2007: 9.3.1).

$$[y - U_p(y), y + U_p(y)].$$

The corresponding interval $[y - U_p(y), y + U_p(y)]$ characterizes the range of values within which the true value of the measurand is confidently expected to lie and in statistical terminology it corresponds to the CI for the true value of the measurand. The coverage probability is the proportion of results of measurement expected to be encompassed by the specified interval and, in general, it is equal to $(1-p)$, thus corresponding to the statistical concept of Confidence. The calculation of the limit $U_p(y)$ is covered in detail for the case of Gaussian distribution of the estimated mean y. In this case (ISO, 2007: eqs. 18 and 20),

$$U_p(y) = t(p,v)u(y),$$

where $t(p,v)$ is a critical point of the corresponding Student T-distribution. In this case, the expanded uncertainty of the measurement y is given as follows (ISO 9.3):

$$[y - t(p,v)u(y), \; y - t(p,v)u(y)],$$

where $(1-p)$ is the required coverage probability and $t(p,v)$ is the $p/2$ critical point of the Student T-distribution with v degrees of freedom. The degrees of freedom is $v = n - 1$ and the expanded uncertainty interval can be interpreted as an interval encompassing the unknown true mean μ_Y of the measurand Y with confidence level of $(1-p)100\%$. A very similar approach can be taken when evaluating uncertainty of a calibration process of a particular air quality monitoring apparatus. The problem with this approach, however, is exactly the same as in the case of standard uncertainty: the distribution of the mean value y of the estimated measurand must be approximately Gaussian but this is almost always false for ambient air quality continuous data.

5.4   ALTERNATIVE APPROACH TO UNCERTAINTY USING THE WEIBULL DISTRIBUTION

As stated in the Section 5.1, individual air quality measurements do not follow a Gaussian distribution but follow a distribution skewed to the right such as the Weibull distribution family. Even the distributions of the means of large samples (e.g., 5-min means of measurements with 1-s frequency yielding a large sample size; e.g., $n = 300$) are greatly skewed to the right and hence non-Gaussian (Figures 5.1 and 5.2). Thus, ISO uncertainty methods mentioned above are invalid most of the time.

In the great majority of cases of actual ambient air quality monitoring, repeated measurements of pollutant concentrations constitute a highly autocorrelated time series. For this reason, the central limit theorem is inapplicable (as in traditional statistical theory) and hence large sample means (e.g., $n > 300$) of repeated measurements do not necessarily follow a Gaussian distribution. For this reason, the above-mentioned ISO methods are inadequate. Cases dealing with non-Gaussian distribution are relegated in the ISO document to a rather confusing section ISO A2 on robust estimation of coverage probability (ISO, 2007).

It is well known, however, that a majority of air quality measurements under the conditions of well-mixed air and stable meteorological conditions can be well approximated by probability distributions skewed to the right such

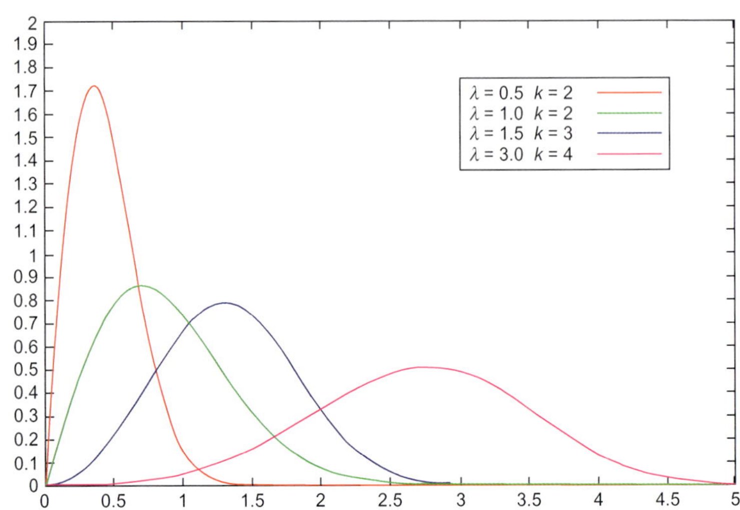

FIGURE 5.1 Examples of Weibull distribution densities.

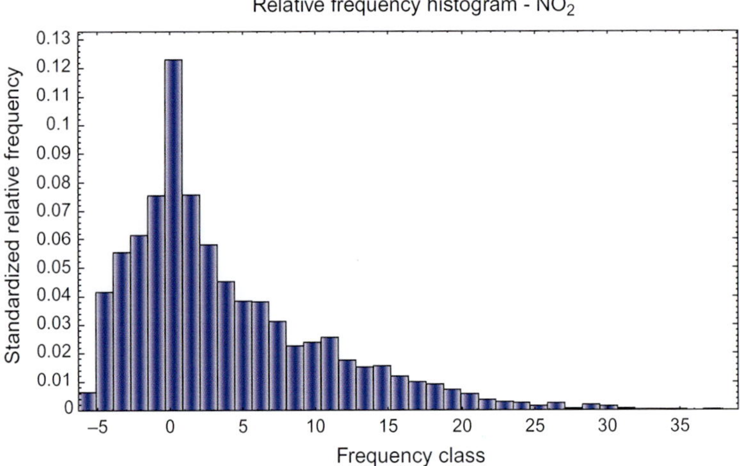

FIGURE 5.2 NO_2 distribution, WBEA AMS 10 (Albian Mine), July 2007, validated and baseline corrected data.

as the Weibull distribution (Berger et al., 1982; Buttazoni et al., 1986; IPCC, 2000; Legge et al., 1990a; Nosal and Nosal, 2008; Nosal et al., 2000, 2008, 2010). Early data analyses applying skewed distributions to air quality data used the lognormal distribution (Larsen, 1969, 1971, 1973, 1974). Further extensive research regarding applications of various skewed distributions to air quality (including the Weibull distribution family) demonstrated why these distributions are so fundamental in this respect (Georgopoulos and Seinfeld, 1982; Jakeman et al., 1986; Taylor et al., 1986). All these authors analyzed the goodness-of-fit tests of particular air quality data to various skewed distributions (lognormal, gamma, and Weibull distributions) and provided recommendations in specific cases.

When using the Weibull family of distributions, the corresponding distribution parameters have to be estimated from the data. In such a case, one can use methods of robust statistical estimation based on MCMs (Bergin and Milford, 1999; Fishman, 1995; Hammersley and Handscomb, 1967; Kuik et al., 1993; Nosal and Nosal, 2008; Nosal et al., 2008; Rubinstein and Kroese, 2008; Shrieder, 1966) to estimate the corresponding critical points required to obtain the expanded uncertainty assessment. The Weibull distribution density has the following form:

$$f(x;\lambda,k,\xi_0) = \frac{k}{\lambda}(x-\xi_0)/\lambda k^{k-1} \exp\left\{[(x-\xi_0)/\lambda]^k\right\}, (\xi_0 < x),$$

where ξ_0 is the location, λ is the scale, and k is the shape parameter. In air quality investigations one usually sets $\xi_0 = 0$, since concentrations of air pollutants are always nonnegative (unless there are negative readings due to the instrument artifacts). The corresponding density is then

$$f(x;\lambda,k) = \frac{k}{\lambda}[x/\lambda]^{k-1} \exp\left\{[x/\lambda]^k\right\}.$$

For any data set consisting of actual continuous air quality pollutant monitoring, a best fitting Weibull distribution can be obtained by estimating the required parameters k and λ. Detailed theoretical background and corresponding numerical algorithms are described in detail by Nosal and Nosal (2008).

The actual frequency distributions of various air pollutants measured by the WBEA continuous air quality monitoring network contain values close to and below the field effective LDL. Many measurements in fact are recorded as negative values. Sometime, the percentage of negative values can be very high, ranging from 50% to over 90%. While this may seem odd, it is a logical consequence of monitoring background air quality in remote areas which are very distant from any emission sources as is frequent in regions located at latitudes of north-eastern Alberta (Fort Chipewyan, WBEA AMS 8).

The following is an example of validated data (after zero correction) frequency distribution of 5-min means of $n = 300$ measurements (once per second) of NO_2 concentrations at one air monitoring station (AMS 10) in July 2007. Figure 5.2 is typical of many of the air quality frequency distributions collected by the WBEA in the northern AOSR.

One can clearly see that this typical pollutant distribution is nonsymmetrical and highly skewed to the right. It is very different from the Gaussian distribution assumed by the ISO methodology in case A1. Distributions as in

Figure 5.2 are common for many air quality species monitored by the WBEA (e.g., NO, NO_x, SO_2, etc.). This demonstrates that the ISO methodology of case A1 for uncertainty estimation described above in Section 5.2 is entirely inappropriate. These skewed distributions can be very well approximated by a Weibull distribution family.

A more convenient and faster method for fitting an empirical air quality frequency distribution by a Weibull distribution is based on minimization of the Kolmogorov–Smirnov distribution distance, KSD (Nosal and Nosal, 2008). Let us denote the observed (empirical) distribution function of the monitored air pollutant concentrations as $E(x)$, and let us denote the Weibull distribution function with parameters λ, k as $F(x; \lambda, k)$.

$$F(x;\lambda,k) = 1 - \exp\left\{[x/\lambda]^{k-1}\right\}.$$

Then the corresponding Kolmogorov–Smirnov distribution distance is defined as

$$\mathrm{KSD}(\lambda,k) = \sup\{\|E(x) - F(x;\lambda,k) : x > 0\}.$$

This KS distance depends on the parameters of scale λ, and shape k. The best fitting Weibull distribution parameters λ_0 and k_0 are obtained by minimization of the Kolmogorov–Smirnov distance:

$$\mathrm{KSD}(\lambda_0,k_0) = \inf \mathrm{KSD}(\lambda,k) \text{ with respect to } \lambda,k.$$

The fact that empirical air quality distributions can be very well approximated by a Weibull distribution family can be used for construction of measurement uncertainty estimates using MCMs (IPCC, 2000; National Research Council, 1994).

5.5   MCMS FOR UNCERTAINTY ESTIMATION

The MCM is based on designing a series of statistical trials which simulate probabilistic properties of a phenomenon which is being studied. By actual realization of these trials, the properties of the investigated phenomenon can be assessed using statistical methodology. This definition may sound rather unclear, general, and diffuse. However, many proponents believe that its generality allows applying MCMs to solve virtually any problem that can be formulated in quantitative terms. It has been successfully applied to perform numerical integration, solve systems of differential equations, optimize management problems, schedule traffic flows, etc. The MCM is not directly tied to computers, but the simulation process of the studied phenomenon is best realized using fast computing technology.

The main advantages of using MCMs for estimation of uncertainty related to ambient air quality monitoring was first mentioned by the Committee on Risk Assessment of Hazardous Air Pollutants (National Research Council,

1994). In the chapter on uncertainty guidelines, the authors noted limitations of the U.S. EPA requirements for complete knowledge of the distributions of all parameters for uncertainty assessment (the same as Alberta Environment, 2006 and ISO, 2007). However, they state that, "complete knowledge is not necessary for Monte Carlo or similar approach to uncertainty analysis." The authors further state that the U.S. EPA (and consequently ISO and AE) requirement of complete knowledge of parameter distributions "is tautology: it is the uncertainty analysis that tells the scientists how their 'lack of complete knowledge' affects the confidence they can have in their estimate." Even though the authors discuss in principle many aspects of the Monte Carlo approach, arguments are presented only on a heuristic, philosophical level, and there are no specific statistical procedures and analytical algorithms provided. Apparently, the authors intended to, but did not include an explanation of algorithms for the methods (National Research Council, 1994).

MCMs have also been used by the Intergovernmental Panel on Climate Change (IPCC, 2000). It is generally accepted that an estimation of the environmental impact of climate change is associated with extremely high levels of uncertainty. It is for this reason that IPCC devoted a great deal of attention to the problem of uncertainty (www.ipcc-nggip.iges.or.jp/public/gp/english/6_uncertainty.pdf) using MCMs (McCann & Associates and Nosal, 1994, Nosal, 1995). An application of the MCM to quantification of uncertainty in air quality modeling and a development of CIs for model forecasts have been discussed by Cheng and Sandu (2009). This method is in principle similar to the ideas presented in this chapter but has been limited to uncertainty related only to air quality model forecasting.

The main purpose of the monitoring process is to estimate some air quality parameter θ which could be, for example, an air pollutant hourly mean, hourly standard deviation, monthly mean, monthly standard deviation, or any other air quality parameter—descriptor. The following derivation of the Monte Carlo uncertainty estimation interval is based on basic statistical concepts. A monitoring process results in a sequence of measurements which is recorded as a sequence of numerical values $x_1, x_2, x_3, \ldots$ representing pollutant concentrations at times $t_1, t_2, t_3, \ldots$ (usually seconds, minutes, 5-min intervals, etc.). Based on these observations, a statistical estimator

$$\hat{\theta} = \hat{\theta}(x_1, x_2, x_3, \ldots) = f(x_1, x_2, x_3, \ldots)$$

of the unknown parameter θ is computed as a suitable function $f(x_1, x_2, x_3, \ldots)$ of the original measurements $x_1, x_2, x_3, \ldots$ This function f could represent an hourly mean, hourly standard deviation, monthly mean, monthly standard deviation, or any other air quality parameter specified. Since measurements $x_1, x_2, x_3, \ldots$ are taken at random (due to the concentrations fluctuating in time and space) and are influenced by random errors (due to the measurement errors), they must be viewed as random variables and consequently the estimator $\hat{\theta}$ is a random variable. An estimation of the uncertainty interval of the parameter θ, estimated by the estimator $\hat{\theta}$ computed on the basis of air quality measurements, requires to find a CI $(\hat{\theta}_{\text{Low}}, \hat{\theta}_{\text{High}})$ such that for a small number α the following condition is satisfied:

$$P(\hat{\theta}_{\text{Low}} < \theta < \hat{\theta}_{\text{High}}) = 1 - \alpha.$$

The number α is usually chosen to be small, such as $0.1, 0.05$, or 0.01. If the probability distribution of the measurements is simple, for example, Gaussian, the problem leads to a simple CI and can be solved using ISO methodology described in Section 5.3 above. However, the main problem is that the probability distributions obtained by continuous air quality monitoring are always very complex, and hence the distribution of $\hat{\theta} = \hat{\theta}(x_1, x_2, x_3, \ldots)$ is always unknown.

As discussed in Section 5.4, the fact that frequency distribution of air pollutant concentrations can be often well approximated by a Weibull distribution and we demonstrate how to find the parameters λ_0 and k_0 of the best fitting distribution. After obtaining these parameters, we can use the resulting Weibull cumulative distribution function $F(x; \lambda_0, k_0)$ to generate pseudorandom observations $\xi_1, \xi_2, \xi_3, \ldots$ which will have exactly the same probability distribution $F(x; \lambda_0, k_0)$ and hence the same statistical properties as the original monitoring measurements $x_1, x_2, x_3, \ldots$ These pseudorandom numbers are obtained by generating uniform pseudorandom variables $u_1, u_2, u_3, \ldots$ on the unit interval $(0, 1)$ and then transforming these using the inverse of the Weibull distribution function as follows:

$$\xi_i = F^{-1}(u_i; \lambda_0, k_0).$$

The generation is a relatively simple matter using a computer. Since the generated pseudorandom observations $\xi_1, \xi_2, \xi_3, \ldots$ have exactly the same statistical properties as the original observations $x_1, x_2, x_3, \ldots$ we can conclude that their functions have also the same statistical properties. Therefore given the required function f (e.g., hourly mean, hourly standard deviation, monthly mean, monthly standard deviation, etc.), the estimators

$$\hat{\theta} = \hat{\theta}(x_1, x_2, x_3, \ldots) = f(x_1, x_2, x_3, \ldots) \text{ and } \hat{\theta}_{\text{MC}} = \hat{\theta}(\xi_1, \xi_2, \xi_3, \ldots)$$
$$= f(\xi_1, \xi_2, \xi_3, \ldots)$$

must have the same statistical properties. However, since the second estimator has been computed, we can repeat such calculations numerous times and then compute the frequency distribution of the estimators $\hat{\theta}_{\text{MC}} = \hat{\theta}(\xi_1, \xi_2, \xi_3, \ldots)$ with arbitrarily high precision. Once such frequency distribution is obtained, one can easily find a suitable CI and hence uncertainty range for the Monte Carlo estimator $\hat{\theta}_{\text{MC}} = \hat{\theta}(\xi_1, \xi_2, \xi_3, \ldots)$ as percentiles of this distribution. Let us say that for some very large n, we have computed n numerical values $\hat{\theta}_{\text{MC}}^1, \hat{\theta}_{\text{MC}}^2, \hat{\theta}_{\text{MC}}^3, \ldots, \hat{\theta}_{\text{MC}}^n$ using n repeated Monte Carlo simulations of the estimator $\hat{\theta}_{MC} = \hat{\theta}(\xi_1, \xi_2, \xi_3, \ldots)$. We order these n values according to their magnitude from the smallest to the largest and denote the ordered values as follows:

$$\hat{\theta}_{MC}^{(1)} \le \hat{\theta}_{MC}^{(2)} \le \hat{\theta}_{MC}^{(3)} \le \cdots \le \hat{\theta}_{MC}^{(n)}.$$

For a given α ($=0.1$, 0.05, 0.01, etc.), in order to get a $(1-\alpha)100\%$ uncertainty (confidence) interval, we find the following two numbers

$$n_{Low} = \left[\frac{\alpha n}{2}\right] \quad \text{and} \quad n_{High} = \left[\frac{(1-\alpha)n}{2}\right] + 1,$$

where $[x]$ denotes the integer (whole) part of the number x. The corresponding lower and upper limit of the required $(1-\alpha)100\%$ uncertainty (confidence) interval are then computed as follows:

$$\hat{\theta}_{Low} = \hat{\theta}_{MC}^{(n_{Low})} \quad \text{and} \quad \hat{\theta}_{High} = \hat{\theta}_{MC}^{(n_{High})}.$$

This is the required uncertainty interval based on $\hat{\theta}_{MC}$. Since both estimators $\hat{\theta}$ and $\hat{\theta}_{MC}$ have exactly same statistical properties, the above interval must be equal also to the required uncertainty CI for the original parameter θ based on the original estimator $\hat{\theta} = \hat{\theta}(x_1, x_2, x_3, \ldots)$.

5.6   ESTIMATION OF UNCERTAINTY IN WBEA MEASUREMENTS

The Monte Carlo uncertainty interval estimation of ambient air quality data is often influenced by measurements. These could be values of concentrations below the field effective LDL or measurements above the upper range limit of the monitor. Values below the LDL are often recorded as negative values or as half the detection limit. Due to prevailing air quality in much of the AOSR, it is common that the percentage of negative values (below effective LDL of the monitor) in the WBEA air quality monitoring data files can be very high (as shown in Figure 5.2), and often exceeds 50%. Sometimes, the percentage of values below LDL may range upward to 90% due to the low concentrations of air pollutants in the ambient air at background remote sites distant from any emission sources. These values introduce a large degree of uncertainty into the monitoring data and greatly complicate QA/QC of such data.

To illustrate this point, the WBEA 5 min continuous air quality data for SO_2 from monitoring site AMS1 (Bertha Ganter–Fort McKay) in January 2010 is considered and shown in Figure 5.3. Of particular interest were the SO_2 concentrations that were rather low at that time. The data consisted of $n=8515$ of which 1300 (15%) values were negative. Furthermore, the field effective LDL of the monitor was found to be 1 ppb and hence $n=4354$

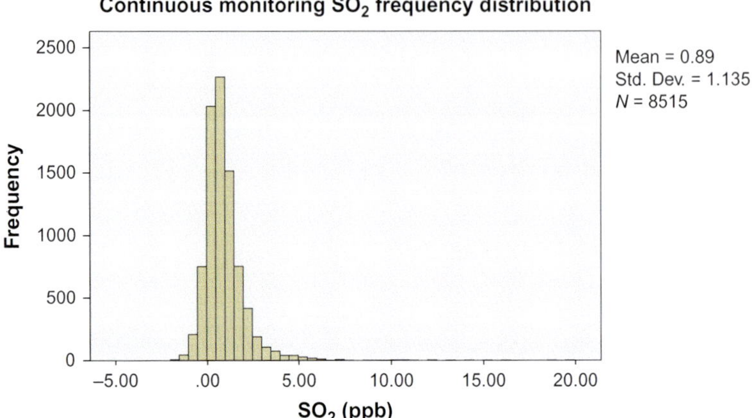

FIGURE 5.3 Frequency histogram of the uncorrected continuous SO_2 data from the WBEA monitoring site AMS 1 (Bertha Ganter–Fort McKay) for January 2010.

between 0 and 1 ppb (51%) must be considered to be also unreliable. Thus, there are altogether $n = 5654$ (66%) unreliable and hence measurements which cannot be used for the data analysis. Figure 5.3 illustrates the frequency histogram for the original uncorrected SO_2 data.

The mean of the original data is 0.89 ppb and the median is 0.70 ppb. The histogram contains a very high frequency of measurements below 1 ppb, including negative values. These measurements have a very significant impact on the data mean and the median, rendering them highly biased. These measurements have to be replaced by synthetically generated new values using a Weibull pseudorandom data generator, hereafter referred to as the Weibull generator. These new values are fully distribution compatible with the original error free measurements above 1 ppb. The frequency histogram of the Weibull corrected SO_2 data is shown in Figure 5.4.

While the Weibull corrected histogram has maintained the overall shape and scale of the original frequency histogram of the uncorrected measurements in Figure 5.3, all values below 1 ppb have been replaced by Weibull generated values compatible with the shape and structure of the original distribution histogram. Importantly, all these corrected measurements are nonnegative.

Thus, the Weibull algorithm described in Section 5.5 allows one to replace these values by corrected values that are homogeneous in the distribution with the shape, scale, and structure of the entire distribution of the original valid (above 1 ppb) measurements. More importantly, however, it also allows one to compute the uncertainty of the estimated air quality parameters such as the mean, the median, the variance, and others. Table 5.1 gives the 90% CIs for these parameters, and provides a clear measure of the uncertainty of each

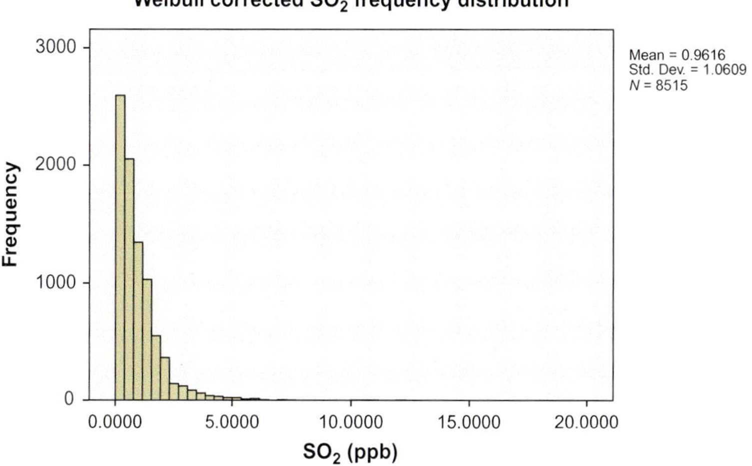

FIGURE 5.4   Frequency histogram of the best Weibull corrected, continuous SO$_2$ data from WBEA monitoring site AMS 1 (Bertha Ganter–Fort McKay) for January 2010.

TABLE 5.1 Uncertainty at 90% Confidence Intervals

Statistic	CI lower limit	CI upper limit
Mean	0.8788 ppb	0.9707 ppb
Median	0.5572 ppb	0.6969 ppb
Variance	1.1156 ppb^2	1.1645 ppb^2

of the estimated parameters. The actual numerical computations were performed using the Weibull algorithm.

Similarly one can obtain 95%, 99%, etc. uncertainty CIs. The first conclusion to be drawn from the Table 5.1 is that all of the 90% CI expressing the uncertainty of the mean, median, and the variance estimation are very narrow and therefore the uncertainty is relatively very small. This is a consequence of the fact that the Weibull corrected data is very homogeneous and does not contain measurements that affect the original monitoring data and thus inflate the uncertainty. Furthermore, the CI for the median (0.5572–0.6969) is located at lower ppb values than the CI for the mean (0.8788–0.9707). This is a necessary consequence of the fact that the SO$_2$ distribution is highly skewed to the right. Furthermore, the mean of the original uncorrected data is 0.8926 ppb and it is therefore inside of the corresponding mean uncertainty CI (0.8788–0.9707). The original uncorrected data median is 0.7 ppb and therefore it just sits on the upper limit of the corresponding uncertainty CI.

Thus, all properties of the uncertainty CI derived using the MCM and the Weibull distributions are in full agreement with the statistical theory.

5.7 CONCLUSIONS

The MCM for numerical evaluation of uncertainty of ambient air quality continuous monitoring described in the Section 5.6 has a number of major advantages compared to the previous historical attempts described in Section 5.2 and are summarized below.

1. The most important advantage of MCM is that it does not depend on unrealistic assumptions about the probability distributions of the concentrations of the monitored air quality. These assumptions are false most of the time and therefore conclusions based on the historical methods discussed in Section 5.2 are inappropriate. MCM also does not permit violations of the basic statistical assumptions as the ISO methodology does. The new MCM introduced in this chapter is also statistically quite simple and straightforward.
2. The advantage of MCM is that the uncertainty estimation and assessment is not dependent on negative and below LDL measurements and their treatment in the course of data QA/QC. The resulting MCM uncertainty assessment is based entirely on correct true measurements above LDL and on the verified and best fitting Weibull distributions.
3. The advantage of the MCM is that the resulting estimation of monitoring uncertainty is obtained in the form of straightforward uncertainty (confidence) intervals for any required parameters of air quality monitoring such as the means, medians, quartiles, percentiles, standard deviations, variances, or any other distribution parameters.
4. The advantage of MCM is that the computations can be very easily arranged to obtain uncertainty intervals with any degree of confidence, be it 90%, 95%, 99%, 99.5%, or any other required value.
5. The advantage of MCM is that the uncertainty (confidence) intervals are very narrow and specific considering very high confidence level which can be achieved. These intervals are not some conservative and inflated approximations of uncertainty based on unrealistic assumptions. These uncertainty (confidence) intervals provide more exact and very accurate estimates.

The main challenge with the MCM is that it requires quite sophisticated computer software for finding the best fitting Weibull distribution for continuous air quality monitoring data and then generating repeated pseudorandom observations from this distribution. Although such software is currently not commercially available in its full functionality, it has been funded, developed, and heavily tested by the authors using WBEA continuous air quality monitoring data. Furthermore, computation of highly accurate intervals with a very high confidence level (95%, 99%, 99.5%) requires very high number of MCM repetitions. However, considering the current advanced state and availability of the computing technology and fast decreasing prices of computer hardware, application of the above described Monte Carlo methodology is very practical, feasible, and relatively inexpensive (compared to the cost of monitoring) and therefore merits very serious consideration by organizations and regulatory bodies involved in ambient air quality monitoring.

Obtaining more accurate and precise information concerning the most likely ranges of ambient air quality and pollutant concentrations resulting from anthropogenic emissions can provide important advantages in environmental impact assessment. When one uses the low and high uncertainty limits of pollutant concentrations obtained from the MCM estimation, this allows for more accurate assessment of the potential effects of air pollutants on the biosphere and their associated risks.

ACKNOWLEDGMENTS

The authors gratefully acknowledge input from the WBEA, Fort McMurray, Alberta, Canada. We particularly appreciate helpful comments from Kevin Percy and Sanjay Prasad.

REFERENCES

Alberta Environment, 1989. Air Monitoring Directive: Monitoring and Reporting Procedures for Industry. AE Environmental Protection Services, Edmonton, AB, Canada.

Alberta Environment, 2006. Amendments to the Air Monitoring Directive, 1989 (AMD2006). AE Environmental Protection Services, Edmonton, AB, Canada Monitoring and Reporting Directive Series, Environmental Monitoring and Evaluation.

Berger, A., Melice, J.L., Demuth, C., 1982. Statistical distributions of daily and high atmosphere SO_2 concentrations. Atmos. Environ. 16, 2863–2872.

Bergin, S., Milford, J.B., 1999. Application of Bayesian Monte Carlo analysis to Lagrangian photochemical air quality model. Atmos. Environ. 34, 781–792.

Buttazoni, C., Lavagnini, I., Marani, A., Grandi, F.Z., Del Rurco, A., 1986. Probability model for atmospheric sulphur dioxide concentrations in the area of Venice. J. Air Pollut. Control Assoc. 36, 1028–1030.

Cheng, H., Sandu, A., 2009. Uncertainty quantification and apportionment in air quality models using the polynomial chaos method. Environ. Model. Softw. 24 (8), 917–925.

Curran, T.C., Suggs, J.C., 1986. Effects of air quality uncertainty on air quality summary statistics. Atmos. Environ. 20 (3), 571–576.

Environment Canada, 2004. National air pollution surveillance network: quality assurance and quality control guidelines. Report No. AAQD 2004-1. Environmental Technology Centre, Analysis and Air Quality Division, Environmental Protection Service, Environmental Technology Advancement Directorate, Analysis and Air Quality Division, Ottawa, ON, Canada.

Environment Canada, 2011. Integrated Monitoring Plan for the Oil Sands: Air Quality Component. Environment Canada, Ottawa, ON ISBN 978-1-100-18813-3; Cat. No.: En14-45/2011E-PDF.

Finlayson-Pitts, B.J., Pitts Jr., J.N., 2000. Chemistry of the Upper and Lower Atmosphere—Theory, Experiments and Applications. Academic Press, San Diego, CA 969 pp.

Fishman, G.S., 1995. Monte Carlo, Concepts, Algorithms, and Applications. Springer, New York, NY 698 pp.

Georgopoulos, P.G., Seinfeld, J.H., 1982. Statistical distributions of air pollution concentrations. Environ. Sci. Technol. 16, 401–416.

Hammersley, J.M., Handscomb, D.C., 1967. Monte Carlo Methods. Methune & Co. Ltd., London 178 pp.

IPCC, 2000. Good practice guidance and uncertainty management in national greenhouse gas inventories, Chapter 6, quantifying uncertainties in practice. Intergovernmental Panel on Climate Change, Montreal. www.ipcc-nggip.iges.or.jp/public/gp/index.html. Last accessed 25 August 2012.

ISO, 2007. Air quality—guidelines for estimating measurement uncertainty. ISO 20988, Geneva, Switzerland. www.iso.org/iso/iso_catalogue/catalogue_tc/catalogue_detail.html. Last accessed 25 August 2012.

Jakeman, A.J., Taylor, R.W., Simpson, R.W., 1986. Modeling distributions of air pollutant concentrations—II. Estimation of one and two parameter statistical distributions. Atmos. Environ. 20, 2435–2447.

Kuik, P., Sloof, J.E., Wolterbeek, H.Th., 1993. Application of Monte-Carlo assisted analysis to large sets of environmental data. Atmos. Environ. 27A, 1975–1993.

Larsen, R.I., 1969. A new mathematical model of pollutant concentration averaging time and frequency. J. Air Pollut. Control Assoc. 19, 24–30.

Larsen, R.I., 1971. A mathematical model for relating air quality measurements to air quality standards. U.S. EPA, Research Triangle Park, NC AP-89.

Larsen, R.I., 1973. An air quality data analysis system for interrelating effects, standards and needed source reductions—II. J. Air Pollut. Control Assoc. 24, 551–558.

Larsen, R.I., 1974. An air quality data analysis system for interrelating effects, standards and needed source reductions. J. Air Pollut. Control Assoc. 23, 933–940.

Legge, A.H., Peake, E., Strosher, M., Nosal, M., McVehil, G.E., Hansen, M., 1990a. Characteristics of the background air quality. In: Legge, A.H., Krupa, S.V. (Eds.), Acidic Deposition, Sulphur and Nitrogen Oxides. Lewis Publishers, Chelsea, MI, pp. 129–248.

Legge, A.H., Peake, E., Strosher, M., Nosal, M., McVehil, G.E., Hansen, M., 1990b. Air quality of an area proximal to anthropogenic emissions. In: Legge, A.H., Krupa, S.V. (Eds.), Acidic Deposition, Sulphur and Nitrogen Oxides. Lewis Publishers, Chelsea, MI, pp. 249–345.

McCann, T.J., & Associates, Nosal, M., 1994. Uncertainties in Greenhouse Gas Emission Estimates. Environment Canada, Ottawa, ON, Canada. K-2019-3-7081, 91 pp.

National Research Council, 1994. Science and Judgment in Risk Assessment, Committee on Risk Assessment of Hazardous Air Pollutants. Board on Environmental Studies and Toxicology, Commission on Life Sciences, National Academy Press, Washington, DC.

Nosal M. (1995), Uncertainty Estimating When Emission is a Difference of Two Sources. Appendix A in Uncertainties in 1990 NO_x SO_2 and CO Emission Estimates by T.T. McCann and Associates, Environment Canada, Ottawa, ON, Canada, 72 pp.

Nosal, M., Nosal, E.M., 2008. Estimating detection limit measurement uncertainty using Weibull distribution. In: Proceedings of the Hawaii International Conference on Statistics, Mathematics, and Related Fields. Honolulu, Hawaii, pp. 353–360.

Nosal, M., Legge, A.H., Krupa, S.V., 2000. Application of a stochastic, Weibull probability generator for replacing missing data on ambient concentrations of gaseous pollutants. Environ. Pollut. 108, 439–446 Elsevier Science.

Nosal, M., Legge, A.H., Nosal, E.M., 2008. Statistical methods for the determination of the effective lower detection limit in measurements from continuous ambient air quality monitors in the remote Athabasca Oil Sands Region of Alberta, Canada. In: Proceedings of the AWMA Symposium on Air Quality Measurement Methods and Technology, Chapel Hill, NC.

Nosal, M., Legge, A.H., Nosal, E.M., Hansen, C.M., 2010. Development and application of statistical approaches for air quality data management. In: AWMA's 103rd Annual Conference, Air & Waste Management Association, Calgary, AB, Canada Paper 2010-A-1068.

Rubinstein, R.Y., Kroese, D.P., 2008. Simulation and the Monte Carlo Method, second ed. Wiley – Interscience, John Wiley & Sons Inc., Hoboken, NJ, p. 372.

Shrieder, Y.A. (Ed.), 1966. The Monte Carlo Method, the Method of Statistical Trials. Pergamon Press, Oxford, p. 381.

Taylor, J.A., Jakeman, A.J., Simpson, R.W., 1986. Modeling distributions of air pollutant concentrations—I. Identification of statistical models. Atmosph. Environ. 20, 1781–1789.

U.S. EPA, 2000. Guidance for Data Quality Assessment, Practical Methods for Data Analysis, EPA QA/G-9, QA00 Update. United States Environmental Protection Agency, Office of Environmental Information, Washington, DC.

U.S. EPA, 2006. Guidance on Systematic Planning Using the Data Quality Objectives Process. EPA QA/G-4Office of Environmental Information, Washington, DC EPA/240/B-06/001.

U.S. EPA, 2008. Quality Assurance Handbook for Air Pollution Measurement Systems. Volume II. Ambient Air Quality Monitoring Program. Office of Air Quality Planning and Standards, Air Quality Assessment Division, Research Triangle Park, NC EPA-454/B-08-003.

Chapter 6

Co-measurement of Volatile Organic and Sulfur Compounds in the Athabasca Oil Sands Region by Dual Detector Pneumatic Focusing Gas Chromatography

R.J. O'Brien[*,1], K.E. Percy[†] and A.H. Legge[‡]

[*]*VOC Technologies & Portland State University, Damascus/Portland, Oregon, USA*
[†]*Wood Buffalo Environmental Association, Fort McMurray, Alberta, Canada*
[‡]*Biosphere Solutions, Calgary, Alberta, Canada*
[1]*Corresponding author: e-mail: obrienr@pdx.edu*

ABSTRACT

Odors are a continuing source of concern to some residents in the Regional Municipality of Wood Buffalo that includes the Athabasca Oil Sands Region (AOSR). Sulfur dioxide (SO_2), fugitive volatile organic compounds (VOCs), and a variety of sulfur-inorganic and -organic compounds, which in total are called total reduced sulfur (TRS), can be a source of this odor. The organic fraction of TRS is, as a general class, the most odiferous. To help understand this issue, the Wood Buffalo Environmental Association (WBEA) of Alberta instituted a program for the dual measurement of these compounds by pneumatic focusing gas chromatography (PFGC). The PFGC, normally equipped with a flame ionization detector (FID) for measurement of VOCs, was fitted with a parallel pulsed flame photometric detector (PFPD), and was deployed in 2009 at a WBEA ambient air monitoring station (AMS-2) near emission sources in the area. The instrument successfully measured a variety of hydrocarbon and sulfur compounds at the ppb level. After 2009, Oil Sands processing procedures were apparently modified

---

Disclaimer: The content and opinions expressed by the authors of this chapter do not necessarily reflect the views of the Wood Buffalo Environmental Association (WBEA) or of the WBEA membership.

Developments in Environmental Science, Vol. 11. http://dx.doi.org/10.1016/B978-0-08-099443-7.00006-5

113

in the AOSR, with a resultant 10- to 100-fold drop in gaseous sulfur compound levels, and a drop in public odor complaints. At that time, the PFGC was moved to the WBEA Bertha Ganter–Fort McKay air monitoring station (AMS-1) in the First Nation community of Ft. McKay. Here, in spite of greatly reduced sulfur compound levels, odor complaints were still received. The concentrations of sulfur compounds at this new location, however, were below the detection limit of the PFPD. To address this challenge, the PFPD was replaced with the more sensitive sulfur chemiluminescence detector (SCD). As documented in this chapter, the newly designed system is now routinely identifying and quantifying individual sulfur compound concentrations well below 100 parts per trillion (ppt). Such greatly enhanced sensitivity is necessary to address odors that still persist in the AOSR, so that odor types can be identified and communicated to stakeholders.

6.1 INTRODUCTION

The Wood Buffalo Environmental Association (WBEA) monitors ambient air quality in the Regional Municipality of Wood Buffalo (RMWB) including the Athabasca Oil Sands Region (AOSR) in northeastern Alberta, Canada. The airshed that WBEA is responsible for monitoring in includes the communities of Fort McMurray, Fort Chipewyan, Fort McKay, Anzac, Conklin, Janvier, and the oil sands operations. WBEA's monitoring network comprises 15 continuous stations measuring sulfur dioxide, total reduced sulfur (TRS), volatile organic compounds (VOCs), particulate matter, hydrocarbons (total, methane, and nonmethane), ozone, oxides of nitrogen, and carbon monoxide (Chapters 4 and 9).

The Alberta Oil Sands, of which the AOSR (Figure 6.1) is the largest deposit, contains oil reserves in the form of bitumen (very heavy crude oil), third only in size to Saudi Arabia and Venezuela. In 2010, these reserves were 169 billion barrels or about 12% of global reserves. Some 80% is deep and recoverable only by drilling (*in situ*) and 20% is recoverable by surface mining. Fort McMurray is more or less centered in the AOSR and the First Nation native community of Ft. McKay lies about 60 km to the north. Both are along the Athabasca River Valley.

One of the environmental concerns in the region is air quality, and an important ongoing subset of these concerns is odor. Odors in ambient air often originate from VOCs, which can comprise hydrocarbons, or hydrocarbons containing heteroatoms such as oxygen, sulfur, and/or nitrogen and perhaps a few other elements. The predominant form of gaseous sulfur in the region is sulfur dioxide (SO_2), which, although odiferous at high-enough concentration, has a much higher odor threshold than the so-called reduced sulfur compounds (RSCs). These include hydrogen sulfide (H_2S), carbon disulfide (CS_2), and a wide range of organic sulfur compounds, such as mercaptans (thiols), organic sulfides and disulfides, thiophenes, and others. The organic sulfur compounds, in particular, are noted for their low olfactory thresholds and adverse human perception.

FIGURE 6.1   The Athabasca Oil Sands Region (AOSR) of NE Alberta and two adjacent bitumen-rich regions of the province (Wikipedia; http://en.wikipedia.org/wiki/Oil_sands).

Based upon the ongoing odor-related concerns of some residents of the AOSR, scientific efforts were initiated by WBEA to characterize and quantify to the extent possible the chemical nature of odors at selected locations in the AOSR. To understand the nature of these odors, a dual-detector pneumatic focusing gas chromatograph (PFGC) (O'Brien, 2005, 2007, 2009) has been operated at two WBEA ambient air monitoring stations for the past 3 years. This single column GC with the two different detectors has the capability to measure both VOCs and inorganic/organic sulfur compounds at the same time. The operation and response of the flame ionization detector (FID) and two sulfur detectors, OI Analytical pulsed flame photometric detector (PFPD) and agilent sulfur chemiluminescence detector (SCD) will be discussed. Either sulfur detector can sample any fraction of the column effluent from the PFGC as split with the FID. The measurements to date encompass both elevated and very low TRS levels. We summarize here the potential VOC or RSC compounds responsible for odor complaints and those we have identified and quantified to date. The likelihood of fully attributing the range of odor issues through ongoing measurement and method development is discussed.

6.2 BACKGROUND REVIEW

6.2.1 VOC Measurements

Atmospheric VOCs are associated with human activities such as transportation, solvent, and other industrial emissions, as well as natural biogenic emissions. VOCs have been the subject of analysis and study for many years (Hester and Harrison, 1995). Traditional methods of ambient VOC measurements can involve canister sampling, followed by transport to a laboratory and analysis by gas chromatography/mass spectrometry (GC/MS) (Sweet and Vermette, 1992). Short-term storage of canisters before chemical analysis is generally considered not to be a problem, but the situation is not entirely clear. This method is well documented and considered accurate but is complex (even though highly automated) and expensive. The expense often limits the number of samples taken and analyzed. Often 24-h integrated samples are employed to minimize expense. These samples give no information about diurnal variation of individual VOC concentrations and allow limited inference about individual sources and their variation.

A more recent method is the collection of VOCs on less cumbersome and less expensive sorbent tubes (Gallego et al., 1999, 2011; Kim and Kim, 2012; Woolfenden, 2010).These are analyzed by GC/MS as with the canister samples. Sampling with these tubes for quantitative analysis requires consideration of "breakthrough" of more volatile components, which would prevent quantitative recovery. Analysis of the tubes requires rapid-volatilization of the adsorbed VOCs into the GC carrier flow, which can be followed by cryogenic freeze out or other methods. The fundamental problem is that a very adsorptive material may prevent breakthrough of the lighter VOCs but then the heaviest VOCs will be very difficult or impossible to desorb. A variety of published methods address the issue of cartridge sample quantification, but these do not concern us here, as our procedures (PFGC) use adsorbent tubes only for compound identification.

6.2.2   Sulfur Measurements

In contrast to a widespread literature of VOC measurements, field measurements of sulfur compounds have been few and have usually been carried out at relatively high concentrations near sources as reviewed by Pandey and Kim (2009).

6.3   EXPERIMENTAL METHODS—PNEUMATIC FOCUSING GAS CHROMATOGRAPHY

6.3.1   VOC Measurements

Outside of measurements at the source, levels of ambient VOCs are usually too low for measurement without some type of concentration step in the analytical process. To date, field operating GCs usually capture analytes onto an absorbent medium, which is then rapidly desorbed (heated) onto the column, very much as with cartridges shipped to a laboratory for analysis. In contrast to traditional methods, we have developed PFGC (O'Brien, 2005) for *in situ* automated measurements. Our detector of choice for these measurements is the FID. Pneumatic focusing uses pressurization as the concentration step, and therefore does not involve any removal of the target VOCs from the gas phase. An air sample of 10–300 cc or more is compressed to 40 bar and injected in entirety onto and through a chromatographic column. Since this instrument runs continuously, typically on a one sample per hour basis, it provides a very complete picture of speciated (peak-forming), and nonspeciated (baseline) VOCs. Depending upon the column chosen, quantified VOCs can range from methane (C_1) out to methyl-isobutylketone, naphthalene, and tetramethyl benzenes (C_{10}).

6.3.2   Gas Chromatography/Mass Spectrometry

In most GC/MS analyses, a sample is submitted to a laboratory where the GC/MS has been programmed to look for a specific set of target compounds with prior knowledge of their retention time on the column of choice, their molecular fragmentation pattern, and their calibration coefficients in terms of a particular fragment ion. This procedure allows coeluting compounds to be separated as long as they have sufficiently distinct fragmentation patterns. This is often not the case for closely related isomers, however. The procedure may ignore (in many cases) the presence of any additional compounds, which have not been included in the target list.

In our application, we trigger an air sample for laboratory analysis only when the field chromatogram indicates an episode "of interest." In this sample, we have no set, *a priori* or "target" list of compounds. Rather we identify all the peaks in the field/laboratory chromatogram. Such a situation can occur if a different peak pattern than the normal one is occurring, or if there is a particular reason to analyze the sample by some legally recognized procedure such as the several US EPA-certified laboratory methods. That is, there is no reason to take samples either at random or at some pre-programmed interval once such ambient VOC profiles are qualitatively understood in terms of the pattern of compound elution orders for a particular site. Thus, once identification of the components of this pattern has taken place, the need is only for quantification, which is accomplished by the field GC through calibrations in the field. This approach can produce a great deal more information than laboratory analysis at a significantly lower cost. Even when a canister sample is taken "on demand" there is usually no need for quantitative GC/MS analysis, but only to determine or confirm compound/peak identities, unless so desired either for legal reasons or to compare with the accuracy of the field GC.

6.3.3 "Baseline" VOCs

In standard GC/MS analysis for a list of target compounds, it may be that some peaks in the chromatogram are not quantified because they are not on the target list. Further, even if all peaks are identified and quantified, it may be that some or many additional compounds remain nonquantified because individually they are not present in sufficient concentration to form a peak and their peak intensity in analysis is mixed with some or many others.

Lewis et al. (2000) published a study of ambient VOCs using two-dimensional GC in which they identified about 500 individual VOCs and suggested that traditional methods of targeted analysis significantly underestimated both the identities and the quantity of total atmospheric VOCs. This observation was important both because some potentially significant (but low concentration) atmospheric VOCs were not being determined and because standard methods of photochemical ozone control often relied upon targeted VOCs *only* to compute photochemical reactivity.

It is unfortunate that the work of Lewis et al. (2000) seems to have been largely ignored by the air pollution community. The exception is the more recent and more expansive work of Myeong et al. (2003) who expended efforts to quantify (but not identify) the contribution of nonresolved VOCs to the total atmospheric VOC content of numerous areas in the Los Angeles (LA) Basin. These researchers found that typical ratios of total/resolved VOCs for the LA Basin were in the range of 1.1–1.5. That is, they found up to 33% of the nonmethane hydrocarbon (NMHC) was not being accounted for by peak-forming VOCs. They associated high ratios for the presence of VOC oxidation products in terms of air mass ageing. An abundance of unresolved VOCs can have other causes as well, as we have found in the AOSR.

6.3.4 Sulfur Gas Measurements

Ambient air, inorganic sulfur gases are rather few in number, including carbonyl sulfide (COS), H_2S, CS_2, and SO_2. The FID does not respond to any of these inorganics but the PFPD does (Tzanani and Amirav, 1995). The

SCD detects all but SO_2 (Shearer, 2002). Organic sulfur compounds could potentially include a very large array of possible species. To the extent that these compounds also qualify as volatile organic compounds (or S-VOCs), the FID does respond to these organic sulfur compounds. Such compounds are also referred to as RSCs and are a subset of TRS essentially all gaseous sulfur-containing compounds except sulfur dioxide SO_2. With parallel FID/ PFPD or SCD detectors, the dual-detector PFGC produces one chromatogram for the sulfur-containing compounds and another for VOCs. The organic RSCs will give peaks at the same retention time in both chromatograms and thus can be unambiguously identified as RSCs and quantified on either or both detectors. Since the RSCs are few in number relative to the VOCs, a dedicated sulfur detector does not have the "baseline" problem that the FID has for the more abundant VOCs. The relative sensitivity of each detector is an inherent property in relation to the noise of each, but can also be manipulated by the fractional diversion of GC column flow to each detector.

In our measurements in the AOSR, since the chief emphasis at the current time is the sulfur compounds, the greater proportion of the flow is diverted to the sulfur detector. This split of column effluent can detect peaks with the sulfur detector that are below the detection limit of the FID detector. A major advantage of either sulfur detector is that by freeing the S atom(s) from a molecule, the detector is specific to the S atom and need not necessarily be calibrated individually for every sulfur-containing compound. However, in calibration, we employ concentrations of the organic sulfur compounds, so they appear on both detectors, a valuable cross-calibration.

Stevens et al. (1971) pioneered the chromatographic measurements of speciated atmospheric sulfur compounds using a Teflon column, chosen due to the reactivity of the RSCs. Nevertheless, such measurements have been infrequent in nonindustrial, ambient air where concentrations are in the low- or sub-ppb levels. Such field and laboratory chromatographic sulfur measurements have been extensively performed and recently reviewed by Pandey and Kim (2009).

6.3.5 Dual-Detector Pneumatic Focusing GC—Principles and Operation

The PFGC carries out *in situ*, near real-time analysis of ambient air samples. The operation of the PFGC can be monitored and controlled over the Internet. In the standard configuration, the gas chromatograph is operated with a helium carrier gas at 600 psi (about 40 bar) and an FID. At this pressure, an air sample is compressed 40-fold, thereby concentrating the analytes by an equivalent amount. Water is condensed in this procedure, most of which remains within the sample loop, not entering the column. The large amount of water present in ambient samples, relative to the VOCs, can be a problem in ambient air chromatography. In addition, upon injection, analytes are further concentrated at the beginning of the column in such a fashion that it is

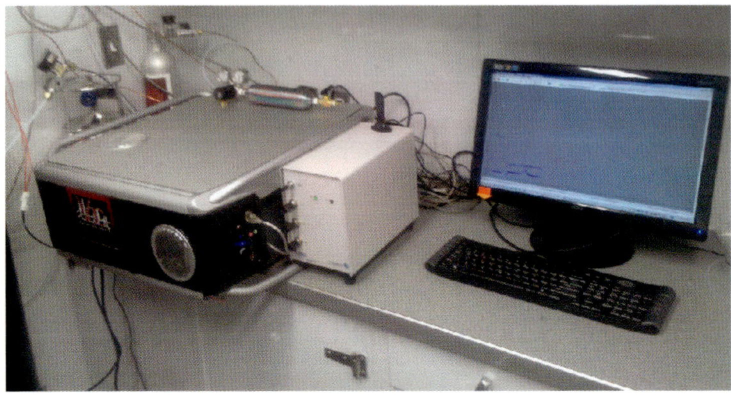

FIGURE 6.2   Photograph showing the PFPD-equipped PFGC. The PFPD control box is bolted to the GC and the burner/photomultiplier assembly is inside, forming a single unit. The monitor and keyboard can be placed on top of the GC for a more compact footprint or eliminated altogether. The gas cylinders are calibration standards and the canister is for GC/MS analysis. With the SCD, the burner is inside the GC, but there are two boxes not bolted to the GC along with a vacuum pump housed in the nearby pump cabinet, forming a considerably larger footprint.

possible to obtain quite respectable resolution in the mid-portion of the chromatogram (Appendix).

The PFGC equipped with the PFPD is shown in Figure 6.2. The SCD has a somewhat larger footprint and a vacuum pump mounted in a nearby cabinet with pumps for other station instruments. The monitor and keyboard are not required in remote operation over the Internet and could be eliminated or can be mounted on top of the GC to reduce the footprint if desired. Calibration gases are included in canisters attached to the GC. Calibrations, blanks, cartridge, or canister samples can be triggered over the Internet "on demand," can be scheduled daily in the case of calibrations, or can be triggered in response to elevated readings on other station instrumentation, such as the TRS reading which the PFGC now acquires continuously.

Both sulfur detectors use a photomultiplier tube to observe emissions from either excited S_2^* (PFPD) or sulfur dioxide (SO_2^*) (SCD). In the PFPD, S_2^* is generated by pulsed combustion of the column effluent in a hydrogen/air flame. S_2^* is formed by recombination of sulfur atoms produced in combustion and the signal is proportional to the square of the analyte concentration. Thus, the signal drops off rapidly at lower sulfur concentration. In the SCD, combustion produces sulfur monoxide (SO), which is then reacted with ozone to form SO_2^*. This signal has the advantage of being linear with the sulfur concentration.

The PFPD is approximately one-half the cost of the SCD, is more compact, and incorporates more easily into a field GC. This design, however, is about a decade old and has technical challenges in a field situation. Flows, for example, are controlled simply by dials on the front of the controller, of which there are four, with no integral flow measurement or control. The large number of adjustable flows makes the instrument difficult to optimize for maximum sensitivity. Tuning consists of adjusting dials controlling the four gas flows into the reaction/emission cell. In these efforts it was observed that the PFPD, although experiencing few problems as a laboratory instrument used at relatively high concentrations, appeared to experience metastable performance (multistationarity) when carefully tuned for maximum sensitivity. In one experiment, turning a single flow knob successively 1/8 of a turn resulted in a gradual 10-fold increase in sensitivity to a calibration H_2S concentration. But the last 1/8 turn resulted in a 10-fold decrease in signal, and the original high sensitivity could not be recovered by simply turning the knob back. This behavior is consistent with chemical hysteresis. Given the flow complexity of this new instrument, this is not unexpected. As a remote field instrument, for which it was never designed, it would benefit from electronic flow measurement and/or control.

While the SCD was also probably not designed as a field instrument, it has the advantage of integral flow measurement. Flows are adjusted with dials as in the PFPD but flow rates can be read out on a digital front panel. To assist in remote operation, the PFGC continuously records these flows so that their stability can be assessed. Further, as there are fewer flows than in the PFPD, detector response does not appear prone to multistationarity in flow rates and the flows appear more stable. Nevertheless, we do observe as yet unexplained variation in SCD response even when flows are constant, as described later. This variation is much less than in the PFPD. The SCD requires a vacuum pump, has a larger footprint than the PFPD and has an additional disadvantage of not measuring SO_2, which does not combust to the reactive intermediate sulfur monoxide, SO.

6.3.6   Chromatographic Separation

Experience over the past 10 years has shown that the Agilent GasPro column performs well over the long term for the separation of VOCs in Pneumatic Focusing. It performs equally well for sulfur-containing compounds as illustrated in Figure 6.3.

The dual-detector PFGC instrument shows a number of unique features:

1. First, methane is quantitatively measured, which is not possible with other concentration methods in an ambient air measurement, using either cryogenic or adsorbent focusing, for which methane is too volatile for quantitative recovery.
2. The baseline in the FID chromatogram contains valuable information about the range of nonresolved total nonmethane hydrocarbons (TNMHC), which are not sufficiently high in concentration to form individual peaks in the chromatogram.

Sulfur gas analysis in light hydrocarbon streams

Column: GS-GasPro
          30 m x 0.32 mm I.D.
P/N: 113-4332
Injector: 200 °C, 1:20 split
Carrier: Helium, 10 psig, 2.0 mL / min @ 60 °C
Oven: 60 °C for 2 min, 20 °C /min to 260 °C and hold

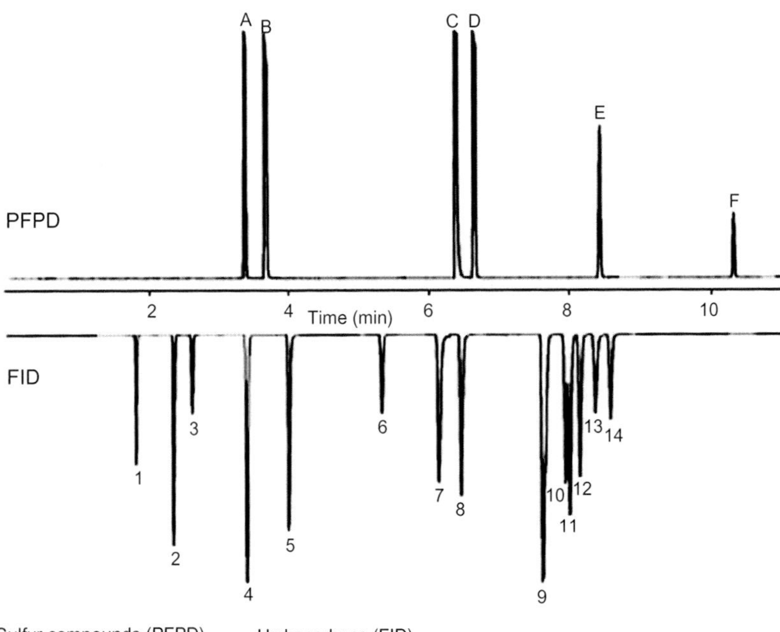

Sulfur compounds (PFPD)	Hydrocarbons (FID)	
A. Carbonyl sulfide	1. Methane	8. *n*-Butane
B. Hydrogen sulfide	2. Ethane	9. 1-Butene/methyl acetylene
C. Sulfur dioxide	3. Ethylene	10. *trans*-2-Butene
D. Carbon disulfide	4. Acetylene	11. 1,3-Butadlene
E. Methyl mercaptan	5. Propane	12. cis-2-Butene
F. Ethyl mercaptan	6. Propylene	13. iso-Pentane
	7. iso-Butane	14. *n*-Pentane

FIGURE 6.3   Elution pattern of VOCs and sulfur-containing compounds (Agilent GasPro column). The horizontal axis is chromatographic elution time (min), and the vertical axis (reversed for the FID) is the signal from each compound. This graphics format is standard for chromatograms. From: Agilent literature.

3. Heavy VOCs are more completely transferred to the column because they never leave the gas phase as they do in adsorption/desorption or cryogenic focusing.

4. Although the FID easily measures so-called S-VOCs since they give a "VOC" signal, it can be difficult to "pick them out" from the more numerous VOCs especially if they are low in concentration. For their potential role in odor

determination, RSCs must be resolved with high sensitivity. The PFGC/SCD has this sensitivity.

5. The calibration RSCs can be quantified on both detectors, providing considerable confidence in accurate measurement.

6. Sulfur compounds of unknown but potentially significant lability may be difficult to recover quantitatively from either a canister (Traube et al., 2008) or an adsorbent. Nevertheless, the lability of some RSCs, especially H_2S, presents significant measurement challenges for the PFGC as well.

7. The PFGC (the GC-in-a-PC^{TM}) is completely compatible with operation over the Internet so that performance may be monitored continuously and potential problems diagnosed based upon extensive past experience. PFGC parameters may also be changed remotely to meet changing situations at the site.

8. By connecting the GC computer to a variety of peripheral instruments and devices, a wide range of flexibilities become easily implemented, in real time, in the field.

9. Since all software is written in-house, new features can be continually added or improved over the Internet or with the help of only moderately trained technicians.

Many of these features have been added over the course of this study in spite of the 1200-mile distance between our laboratory in Oregon and the field site in northern Alberta.

6.4   CURRENT LOCATIONS FOR PFGC MONITORING IN THE AOSR

The AOSR comprises a large area in NE Alberta (Figure 6.1) where oil sand bitumen is known to be present. Figure 6.4A and B shows two locations, labeled A and B where the PFGC has operated.

6.5   RESULTS AND DISCUSSION

This work has required continuing improvement in measurement capability for sulfur and organic compounds, the identification of these compounds, and improvements in technology. We report here the sulfur compounds of greatest interest at present with respect to odors, document the need for improved sensitivity, correlate the GC measurements with other atmospheric parameters acquired continuously by station instrumentation, examine the PFPD's sensitivity as correlated with SO_2 levels, and describe the improvements achieved with the SCD. Then we conclude with a prognosis for routine measurement of those trace atmospheric constituents, which even at very low concentrations are a source of odors.

N
Google earth
Eye alt 56.18 mi
B
+A
Fort McMurray
© 2012 Google
© 2012 Cnes/Spot Image
57°15'49.69" N 112°00'20.51" W elev 1275 ft

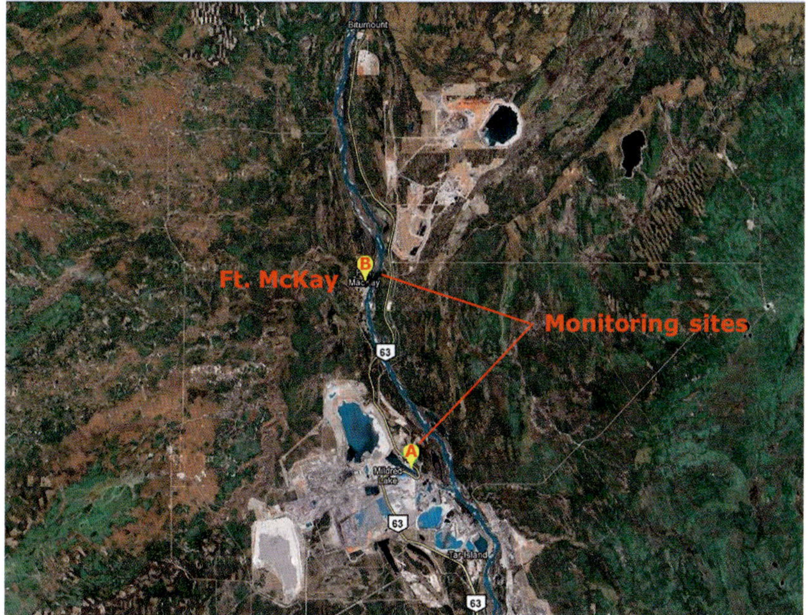

FIGURE 6.4   (A) A satellite image showing the Alberta Athabasca Oil Sands Region north of the city of Ft. McMurray which is indicated at the bottom center of the image. A number of Oil Sands developments are seen. VOC and sulfur compound measurements have been carried out at locations A (2009) and B (2010–2012). Highway 63 approximately parallels the Athabasca River on its west side. Location A represents one "source" region for emissions, while location B represents a downwind "receptor" area, the First Nation community of Ft. McKay. Site A is 43 km north of Ft. McMurray and site B is an additional 16 km north. (Copyright Google Earth, used with permission). (B) An enlargement of the monitoring areas shown in (A), illustrating the extent of development and the community of Ft. McKay (Copyright Google Earth, used with permission).

6.5.1   Sulfur Compounds

6.5.1.1   Sulfur Chromatograms

When the PFPD was first deployed in 2009 (site A in Figure 6.4), RSC concentrations were elevated in some areas of the AOSR, sometimes exceeding the 10-ppb Alberta Air Quality Objective (http://environment.gov.ab.ca/info/library/5726.pdf) and occasionally exceeding 100 ppb for very short periods. When this occurred, there were frequent odor complaints by residents of the area. During this period, the PFPD instrument measured several RSC compounds up to the C_{5-8} region. An "episode" set of three consecutive, hourly chromatograms is shown in Figure 6.5. The upper set are VOCs recorded by the FID, and the lower set contains RSCs recorded by the PFPD. Note that the two RSC peaks appear in both FID and PFPD chromatograms. While the

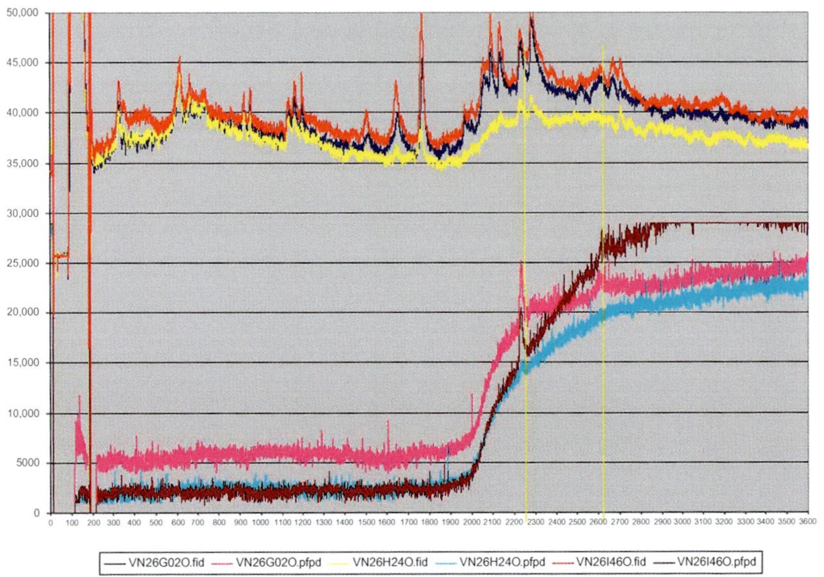

FIGURE 6.5 Three, successive, characteristic chromatogram sets recorded by the FID (upper) and the PFPD (lower) during a high-concentration episode in fall, 2009 at Site A in Figure 6.4. Vertical lines connecting the FID and PFPD chromatograms indicate two organic sulfur compounds. Axes are in standard chromatographic terminology described in Figure 6.3.

FID has linear response between signal and concentration, the PFPD has a quadratic response. Thus, with the Gas Pro GC column (Figure 6.3), if the concentration represented in a FID peak were reduced 10-fold, the peak would decrease 10-fold but still be detectable. However, the PFPD peak would decrease 100-fold in intensity and would in this case be below the detection limit. The PFPD is not capable of extreme sensitivity because of this quadratic dependence.

6.5.1.2 Distribution of Sulfur Compounds in 2009

Figure 6.6 gives the measured distribution of sulfur compounds by the PFPD during the 2009 period. At that time, the organic RSC compounds had not been characterized, however, identified now by their retention times in the chromatogram, and quantified by their sulfur response.

6.5.1.3 Correlation of TRS with Wind Direction and with Total Hydrocarbon

In spite of the large drop in ambient sulfur levels from 2009 to 2010, the authors ((while at the WBEA Bertha Ganter–Ft. McKay air monitoring station (AMS-1)) observed a correlation between the TRS readings and perceived

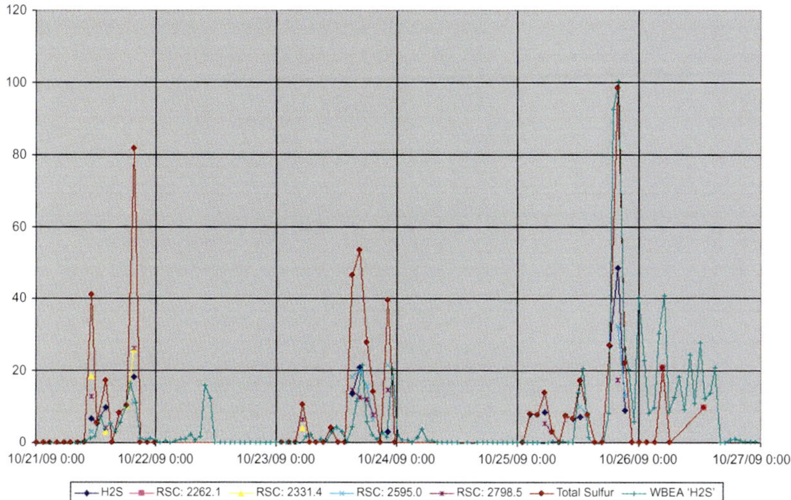

FIGURE 6.6   An example of the distribution of sulfur compounds during a 2009 episode as measured by the PFPD–PFGC at Site A. The concentrations shown are in parts per billion by volume (ppbv). The unidentified RSC compounds are labeled by their chromatographic retention times (seconds). The "H_2S" shown signifies the signal from the continuous ambient H_2S monitor at AMS-2 also located at Site A. It is important to note that this instrument also partially responds to a variety of additional RSC compounds.

odor. Analysis of the AMS-1 continuous data for TRS and THC versus wind direction is shown in Figure 6.7. The THC data in Figure 6.7 has the global background methane concentration of 1.8 ppm subtracted and the maximum of about 1.5 ppm above background THC and TRS is seen to occur at this location when the wind is from the south, although northerly winds contribute a broad, secondary maximum. Refer to Figure 6.4A and B to better understand these wind directions relative to site geography. Note that TRS rarely exceeded 4 ppb over this 6-month period (Chapter 4).

6.5.1.4   Maximum Achieved PFPD Sensitivity

In light of the great reduction in levels of TRS, considerable effort was expended in the attempt to "tune" (by flow adjustment) the PFPD to high-enough sensitivity to detect the individual sulfur compounds, as described in the experimental section. In summary, the PFPD appears to have a "plateau" of moderate sensitivity in which it is not overly sensitive to flows and the potential for much higher, but unstable sensitivity. Careful tuning produced a relatively stable detection limit for SO_2 of 2–3 ppb as shown by comparison of the PFPD–PFGC with the AMS-1 pulsed-fluorescence SO_2 analyzer data in Figure 6.8. With this detection limit, it is not surprising that the PFPD failed to detect any RSC during the period of 2010 onward.

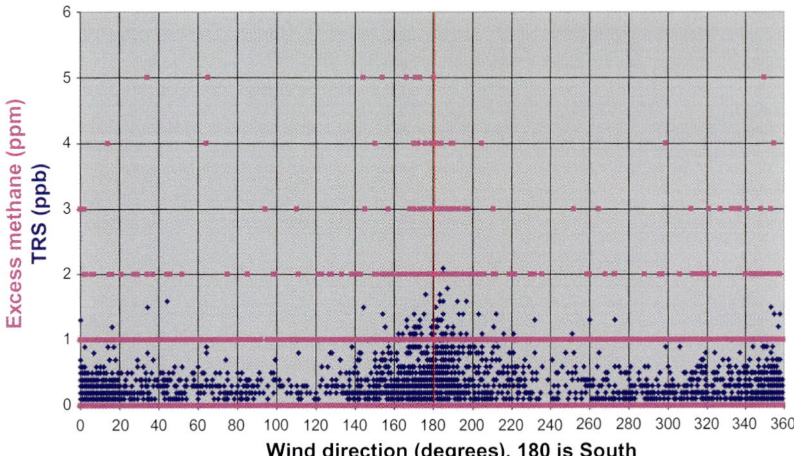

FIGURE 6.7   An example of excess total hydrocarbon (ppm) and total reduced sulfur (ppb) versus wind direction at AMS-1 (hourly averages). Excess THC refers to levels above the 1.8 ppb global methane background, which has been subtracted. Wind direction is shown in degrees, with south as 180. Data for AMS-1 (Ft. McKay), January through July 2011. Refer to Figure 6.4A or B for directions.

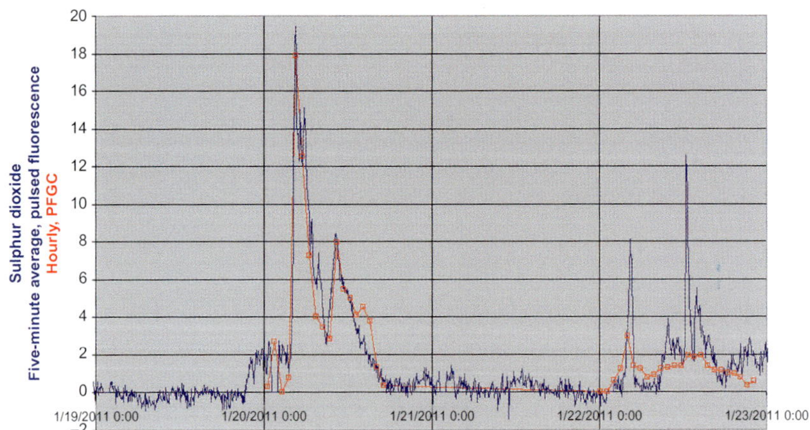

FIGURE 6.8   A comparison of the hourly SO_2 concentrations measured by the PFPD–PFGC (an integration of SO_2 chromatographic peak) and the SO_2 fluorescence-analyzer (5-min average). January 2011 at AMS-1 (site B). These data indicate that the PFPD–PFGC is capable of minimum detection limits (MDLs) in the region of 2–3 ppb for sulfur compounds with optimal flow tuning. The vertical axis is in ppb.

The goal of the PFGC measurements was not to measure SO_2. The SO_2 sensitivity achieved, coupled with the failure to detect RSC chromatogram peaks at any time at the downwind Site B (AMS-1) in Figure 6.4A and B, indicated that the putative sulfur compounds contributing to odor at that site were below 1–2 ppb. This was consistent with the TRS readings that rarely exceeded 3–4 ppb. This turned out to be consistent with later SCD measurements, which were successful in quantifying individual RSC compounds at concentrations well below 1 ppb. It should be remembered that the PFGC measures individual sulfur compounds, so that even when TRS is 2–4 ppb, individual sulfur compounds may be much less than these levels.

6.5.1.5   Correlation Between TRS and Odor Reports

Although it is not certain that sulfur compounds are solely responsible for odor events, there is a coincidence of complaints with elevated TRS readings. These odor episodes are reported by local residents (data supplied by Alberta Environment and Sustainable Resource Development) and then correlated with TRS readings from the nearby WBEA monitoring station. A 5-month period of such events is illustrated in Figure 6.9. The PFGC–PFPD detected no RSCs during this period, clearly because of insufficient sensitivity of the detector.

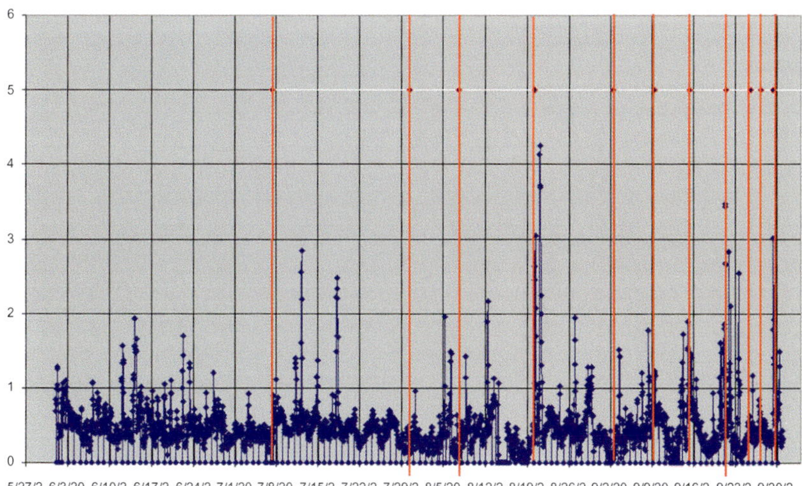

FIGURE 6.9   Hourly average TRS levels (ppb, blue) and reported "odor events" (red and also indicated by the dots/points at 5 ppb) for a 5-month period at AMS-1 (Ft. McKay). TRS is measured continuously and reported as a 5-min average. Local residents report odor events to either Alberta Environment and Sustainable Resource Development or WBEA (odor hotline).

6.5.2 Compound Identification, Calibration, and Quantification

6.5.2.1 VOC Calibration

Quantitative calibration of the PFGC for VOCs is relatively straightforward and is carried out by periodic injection of a multicomponent VOC calibration gas mixture of 7 ppb per compound. Calibration procedures for sulfur compounds are more problematic because of their potential instability in gas cylinder storage.

An example of characterizing VOCs in the AOSR is illustrated in Figure 6.10, which shows a one-day accumulation of approximately hourly chromatograms. Note the buildup of a VOC episode starting from the bottom chromatogram at midnight (red). The 10th chromatogram from the bottom is an instrument blank (green), performed by the GC operator over the Internet. This blank (no methane peak) quantifies contamination in the carrier gas line

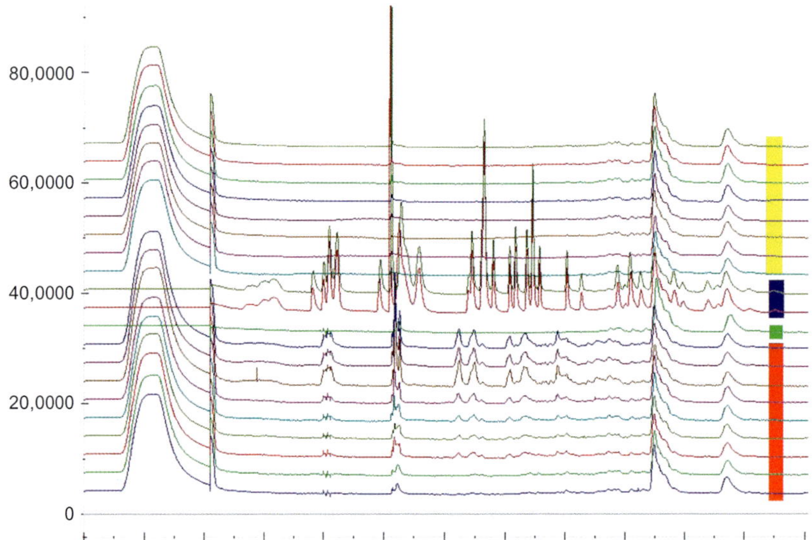

FIGURE 6.10   Set of daily chromatograms from August 31, 2010, which indicate the course of an episode building and then ending toward midday. The figure axes use standard chromatographic notation, see Figure 6.3. The bottom chromatogram is taken just after midnight and the top chromatogram just before the subsequent midnight. The large, wide peak at start is Methane. After the Methane peak, the signal is amplified 10-fold. The color bars at the right indicate: pollution episode building (red), blank run showing (known) contamination due to inappropriate helium carrier gas line (soon replaced) (green), two calibration runs (blue), episode has ended (yellow). (In monochrome, these coloured indicators are consecutive from bottom to top). The PFGC was manipulated over the Internet to produce this experimental sequence, which occurred soon after the PFGC was set up at Site B (AMS-1).

that shows two wide peaks near the end of the chromatogram. These contaminant VOCs did not interfere with the measured VOCs because they eluted after them. They were later eliminated. The next two chromatograms (blue) are calibrations with known VOCs at 7 ppb each. After the calibration, the chromatograms (yellow) indicate that the episode had passed. These episodic VOCs were identified as naphtha hydrocarbons (alkanes) detected by GC/MS analysis of a canister sample also triggered over the Internet.

6.5.2.2 Resolved (Identified) VOCs Versus Unresolved (Baseline) VOCs

Figures 6.11 and 6.12 illustrate another episode as a series of 17 "offset-stacked" chromatograms for November 30, 2009 at Site A. Note the baseline difference is zero up to 1900 s, indicating negligible unresolved VOCs at low-molecular weight. Unresolved peaks run to $\sim C_{12}$, at which point the baseline returns to "background" levels. This indicates that the TNMHCs do not include significant components "heavier" than about 12 carbons. The total integrated areas of the resolved and unresolved VOCs for the entire chromatogram are roughly equal at 190 ppb C. But, for compounds eluting after about C_{7-8}, the "unresolved" baseline hump is greater in area that the resolved VOCs which ride upon it. These peaks are quite significant in total, although not individually, and are usually ignored in standard methods of VOC analysis.

Further, on that day (Nov 30, 2009, Figs. 6.11 and 6.12) the unresolved hydrocarbons were not simply proportional to the resolved VOCs. The baseline hump at the bottom of Figure 6.13 represents the change in the

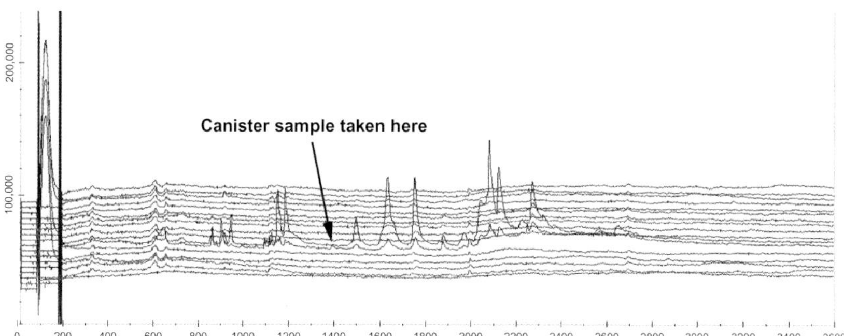

FIGURE 6.11   VOC-FID chromatograms for November 30, 2009. First (bottom) chromatogram is a blank, which is automatically run after midnight, wherein no sample is injected and the baseline was determined solely by column "bleed," which is a cumulative result of previously injected samples. Each succeeding chromatogram is progressively ~ 1.25 h later. Note the two chromatograms with elevated VOC concentrations and a large increase in the baseline due to a large number of unresolved (baseline) VOCs in midmorning.

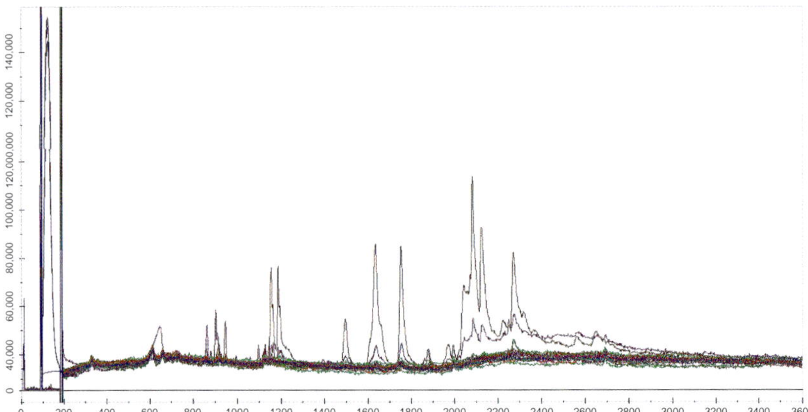

FIGURE 6.12   This figure is the same as Figure 6.11, except the artificial spacing between chromatograms has been removed to show the relative constancy of the "baseline," except for the two elevated VOC episodes. This baseline contains varying amounts of nonpeak forming VOCs and the change from one run to the next allows for the change in these unresolved VOC changes to be quantified. This behavior is further elucidated in Figure 6.13, which illustrates the subtraction of successive chromatograms in the Nov 30 episode.

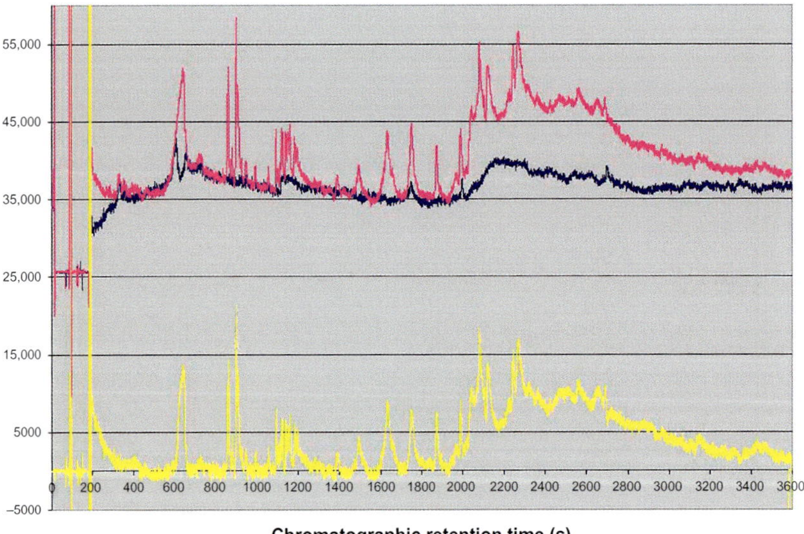

FIGURE 6.13   The first of the two elevated VOC chromatograms and previous chromatogram along with a difference chromatogram indicating very many unresolved VOCs are present in the air sample relative to the previous one at the onset of the episode.

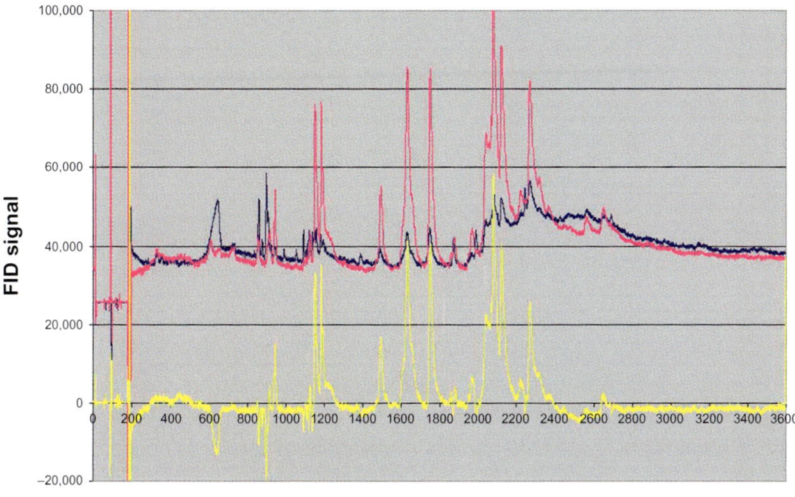

FIGURE 6.14   The second and first of the two elevated VOC chromatograms and their difference is shown. Note that there is no change in the unresolved VOC concentration in spite of a 10-fold increase in most resolved VOCs. The negative peak at 600 s retention time is methyl chloride, which is higher in the 2nd chromatogram.

unresolved VOCs as the episode starts, with a large increase over the previous chromatogram. But in the subtraction of Figure 6.14, there was no baseline hump at all. In the second chromatogram of the episode, the resolved VOCs increased by 10-fold, but there was no significant increase in the (previous) unresolved (baseline) VOCs over the previous sample. In fact, the baseline difference is slightly negative.

The large "negative" peak at 650 s retention time is methyl chloride that decreased between the GC samples giving a negative difference. A number of the C_4 compounds at 900 s retention time also decreased. The unresolved peaks did not increase with the resolved, and therefore could not be closely coupled in terms of their source(s), nor could some sort of proportionality between resolved and unresolved VOCs be even remotely correct.

This baseline behavior is easily interpreted. At early elution times, low carbon numbers allow only a limited number of isomers (C2 has 3: ethane, ethene, ethyne; C3 has 5: propane, propene, propyne, propadiene, cyclo-propane; etc.). The column resolves them all, so there is no raise in baseline due to unresolved VOCs. At successively larger carbon numbers, as more and more isomers are possible, it is not possible to resolve them all, only the predominant (in concentration) compounds are "speciated." Finally, the baseline discrepancy disappears at the end of the run, indicating virtually all the heavier VOCs have eluted in each chromatogram, whether speciated or not. Thus, we know the carbon number range of the "unresolved"

VOCs. Had the baseline not returned to the previous value it would be clear that "heavier" VOCs had not fully eluted and would appear on the next chromatogram. In this situation, if so desired, the chromatograms could be run long enough to attain the true baseline for the current air sample, and this lengthening of the run time could be implemented over the Internet.

6.5.2.3 Compound Identification

The separated compounds can be quantified in the field chromatograms by means of a calibration standard as illustrated in Figure 6.10. This standard is typically injected on a daily basis automatically but can be injected "on demand" as shown in Figure 6.10. Although this commercially available standard contains about 40 VOCs, it can never contain all atmospheric VOCs and many of those it contains may not be present in a field chromatogram. Further, even among these 40, several peak groups coelute to form a single, larger, resolved (multicomponent) peak. Therefore, compound identification is carried out by laboratory GC/MS analysis of an air sample, usually triggered over the Internet at an appropriate time. This sample is then analyzed on the same column as the field chromatograph, by GC/MS. Rather than subjecting the sample to a programmed search for target compounds, we attempt to identify all peaks by their characteristic fragmentation pattern using the chromatographic libraries provided with the GC/MS. It is important to note that even with known coelutions, this procedure cannot be exact.

An example of the identification of field-measured VOCs by remote triggering of a Silcosteel^{TM} coated canister during an episode, and then GC/MS analysis in the laboratory is shown in Figure 6.15. The All peaks were identified by their characteristic fragmentation pattern and knowledge of the elution order of a VOC standard on the GasPro column as determined by analysis of a standard VOC mixture shown as the calibration in Figure 6.9, and by reference to Agilent literature for the GasPro. Time of canister sample is indicated, but this PFGC sample was injected into PFGC about 30 min earlier. During a prolonged episode, the canister sample could be taken over the Internet at the time of injection for exact correspondence.

6.5.2.4 Sulfur Calibration

Although VOCs calibration is relatively straightforward as VOCs are generally quite stable in Silcosteel^{TM} Calibration Cylinders, sulfur-containing standards are more problematic. After considerable effort, a SCD calibration standard with four compounds so far shows good stability at 1–10 ppb per compound (Figure 6.16). A second standard exhibits good stability for two compounds at about 50 ppt concentration. By including several S-VOCs, which elute throughout the chromatogram, we can obtain peaks on both the FID and the SCD for the same compound. This cross-calibration removes many potential ambiguities in instrument response versus calibration standard

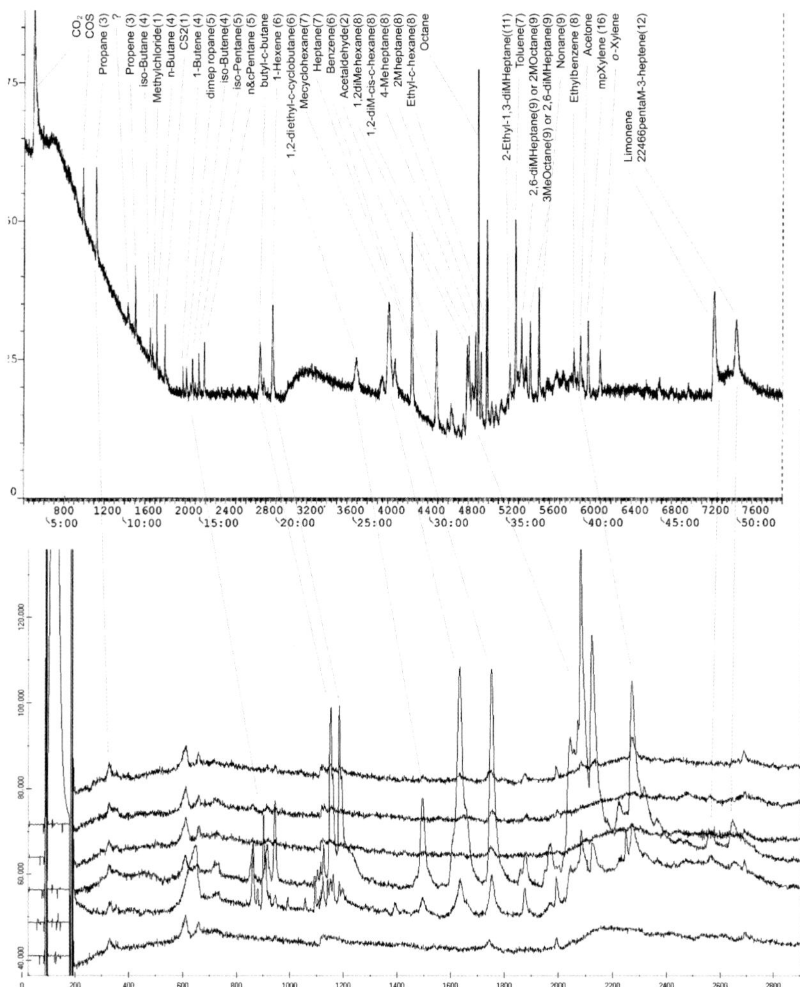

FIGURE 6.15   Comparison of field PFGC chromatogram (lower) and GC/MS analysis (upper) of a canister sample taken at the same time, both on the Agilent GasPro column. Horizontal axis is chromatographic retention time (seconds).

stability. This is illustrated in Figure 6.16 for a 3–4 week period. Thiophene is present at 12 ppb and its FID peak area along with that of benzene is relatively constant over this period. This indicates stability of both compounds in the calibration cylinder as well as stability of the FID response. The drop in both peak areas near the midpoint of the graph is due to the hydrogen fuel gas delivery pressure for the FID and SCD dropping as the cylinder runs empty. Response returns when this cylinder is replaced. In contrast, the SCD thiophene peak area drops considerably to a minimum and then rises

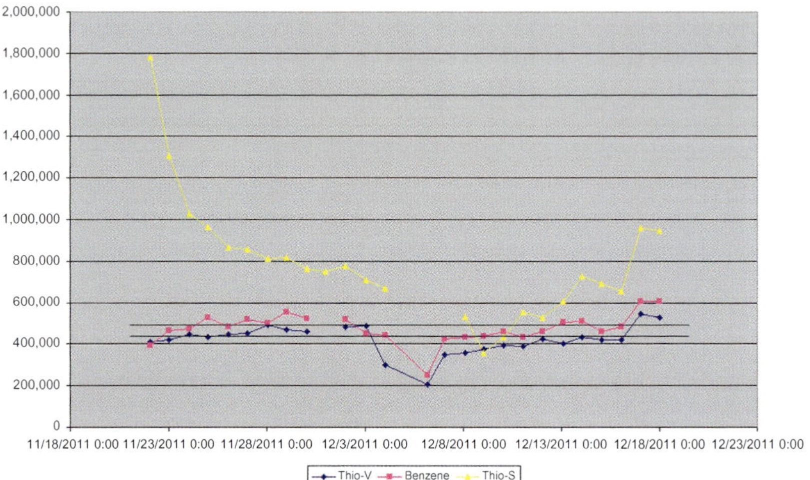

FIGURE 6.16   Variation of the thiophene and benzene daily calibration peak areas over a 3–4 week period. Thiophene response initially drops on the SCD while both thiophene and benzene response are constant within error limits on the FID. This indicates that thiophene is stable in the calibration cylinder, but the SCD response is varying (discussed in text). The apparent calibration drop in the middle of the graph was due to the hydrogen cylinder running empty. The calibration returns when the cylinder was replaced.

again. Continuous monitoring of all SCD gas flows indicates that they were stable over this period, so the cause of this drop in sensitivity is unknown.

Nevertheless accurate calibration is maintained because this drop in sensitivity can be corrected in data analysis. So with a systematic time-varying SCD response (calibration factor), a curve of response versus time is fit to the SCD data of Figure 6.16 and then the appropriate calibration factor is used for each sample. However, decreasing response is still a problem, in that it worsens the detection limit. Work is ongoing to identify the source of this variation.

6.6   RECENT SULFUR MEASUREMENTS

6.6.1   RSC Chromatograms

The increased sensitivity of the SCD has allowed a variety of compounds not previously detected at Station AMS-1 to be quantified in the field PFGC. Several chromatograms for a typical odor episode are shown in Figure 6.17. This figure was prepared before peak identification, which is discussed next.

6.6.2   Cartridge Versus Canister Samples for GC/MS Analysis

Identification of the peak was carried out by GC/MS by shipping a canister taken at an appropriate time (Figure 6.15). However, with increased SCD sensitivity, sulfur compound identification became problematic because RSC peaks did not appear in canister samples taken for GC/MS analysis, probably

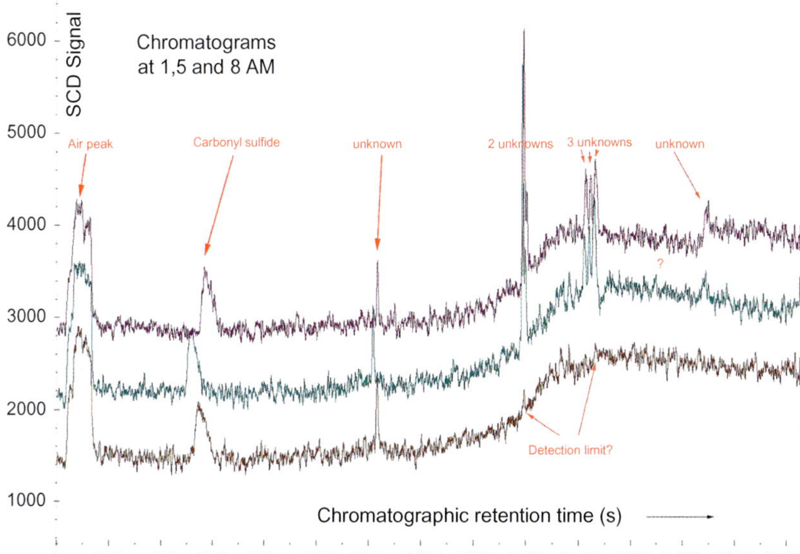

FIGURE 6.17   Three recent, selected SCD field chromatograms during a typical TRS episode at Station AMS-1 (Site B in Figure 6.4A and B). The RSC compounds of unknown identity at the time were subsequently identified by GC/MS (see Figure 6.19). Only a few chromatograms for the extended episode are shown.

because their concentration was too low. So, recently the canister samples have been augmented with cartridge samples taken concurrently.

As discussed previously, one problem with cartridge samples is breakthrough of more volatile components leading to nonquantitative samples. In our case, we do not use the cartridges for quantification, so breakthrough is not an issue. To automate taking of cartridge samples for GC/MS analysis, the PFGC continuously monitors the signal from the TRS machine and is programmed to activate a cartridge sampler during periods when TRS > 2 ppb. Figure 6.18 illustrates the capture of a TRS episode by the automated cartridge sampler over a period of several days.

The total ion chromatogram (TIC) taken from a cartridge sample by the GC/MS is shown in Figure 6.19A and B for both the identified RSCs and selected non-sulfur-containing VOCs. The rough, relative preponderance of the RSCs versus "VOCs" is apparent from the relative peak sizes in the TIC, although TIC signal does vary significantly among various compounds.

6.7   SUMMARY AND CONCLUSIONS

6.7.1   Current Measurement Status

The dual-detector PFGC has been demonstrated to measure a variety of RSCs at concentrations below 100 ppt during odor episodes at remote locations in the AOSR. RSCs identified in chromatograms to date include COS,

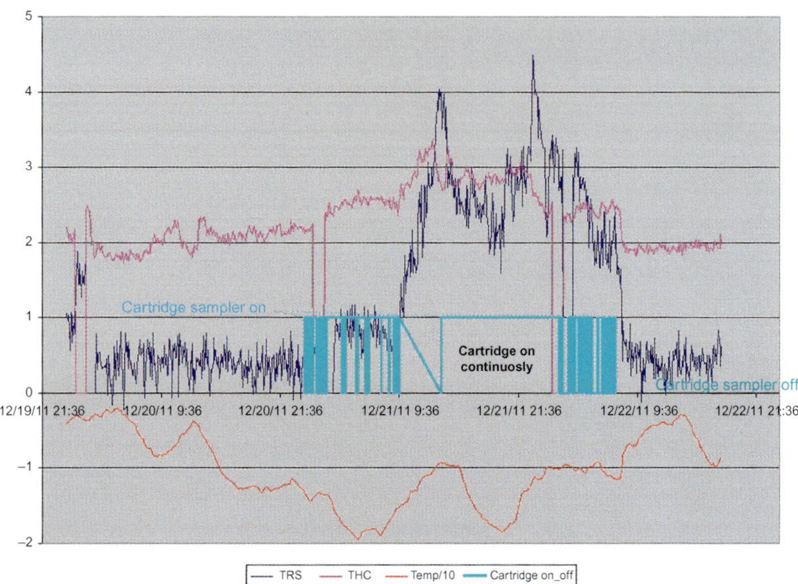

FIGURE 6.18 An example of the automated cartridge sampler being triggered by signal from the AMS-1 continuous TRS air monitor whenever TRS > 2 ppb. The cartridge was sampled continuously during the period indicated, and turned on and off during the oscillatory regions as TRS rose above the preset trigger level of 2 ppb before and after the continuous sampling. Station temperature (Centigrade/10) and THC (ppm) are shown as well. THC shows correlation with TRS, and the low temperature during the episode (−2 to −20 °C) is significant.

carbon disulfide, 2-methylthiophene, 3-methylthiophene, 2-ethylyhiophene, 2,5-dimethylthiophene, 2,4-dimethylthiophene. A few additional RSCs may have been present in the laboratory mass spectrum of the cartridge sample at intensities too low for identification. One of these was identified by retention time (only) as methyl–ethyl disulfide. These were below the current detection limit in the field PFGC, but effort is ongoing to continually improve sensitivity.

Automated cartridge sampling allows identification of the compounds, which can appear as peaks on both the FID and SCD traces, depending upon the split of the column effluent between the detectors. Currently, due to the importance of odors, most of the flow is directed to the SCD and small peaks on the SCD do not appear on the FID. Daily cross-calibration (FID/SCD) with organic RSCs provides for accurate atmospheric measurements and stable calibration gas concentrations at ∼ 10 ppb. The SCD, although recorded to have stable flows, has been found to vary significantly in sensitivity (calibration). Measured daily calibration allows for this variation to be applied to the calibration factor so that accurate concentrations are obtained. However, further work is required to identify the source of this variation so that maximum sensitivity can be maintained. Work is ongoing to extend the overall PFGC–SCD sensitivity a little further to capture compounds apparently just below the current detection limit.

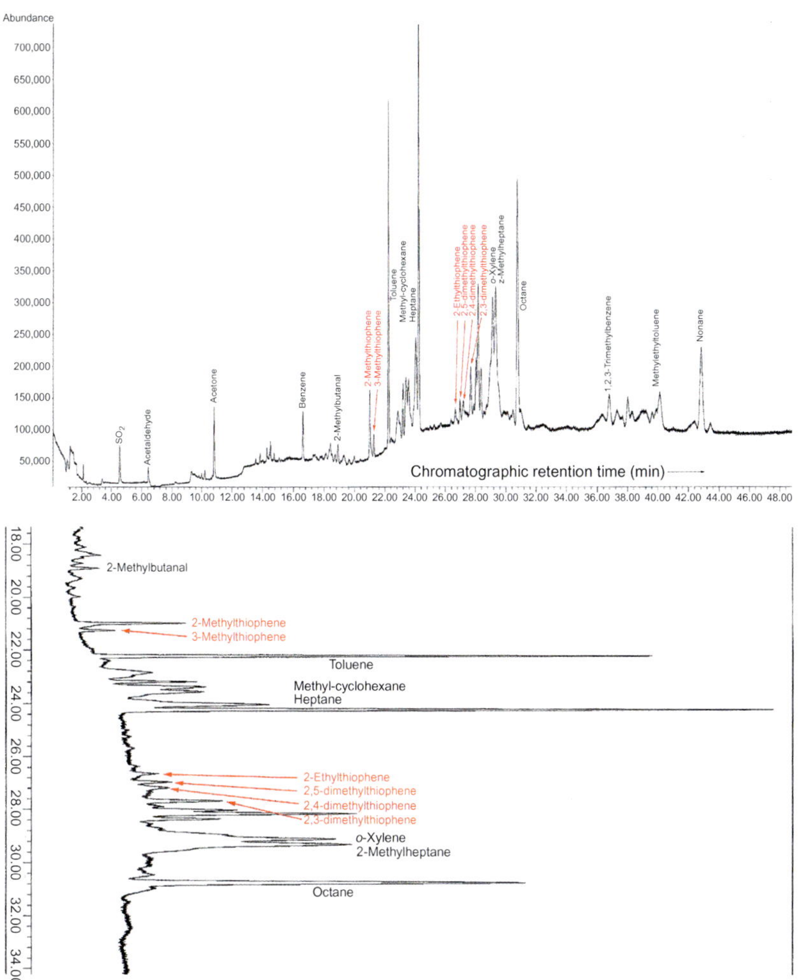

FIGURE 6.19   (A) Total (GC/MS) ion chromatogram taken from cartridge sample during an odor episode. Identified RSCs are indicated along with a few selected VOCs. This laboratory chromatogram (GC/MS–TIC) corresponds to the field SCD chromatogram in Figure 6.17 using the automated sampler of Figure 6.18. (B) RSC region of the GC/MS chromatogram taken from Figure 6.19A, which has been rotated for legibility. Several additional RSC compounds were suggested in this region, too weak for compound identification or for detection in the field chromatogram.

6.7.2   Potential Odor Compounds

Odors are likely, *but not certainly*, at this point to be caused largely by sulfur compounds. With this in mind, we have compiled a list of available candidate VOC and sulfur compounds from a critically reviewed assessment of odor thresholds by Devos et al. (1990). These are sorted by odor threshold in

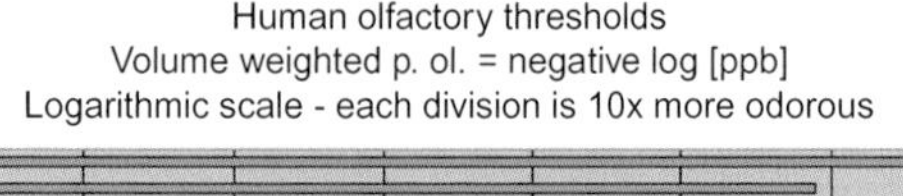

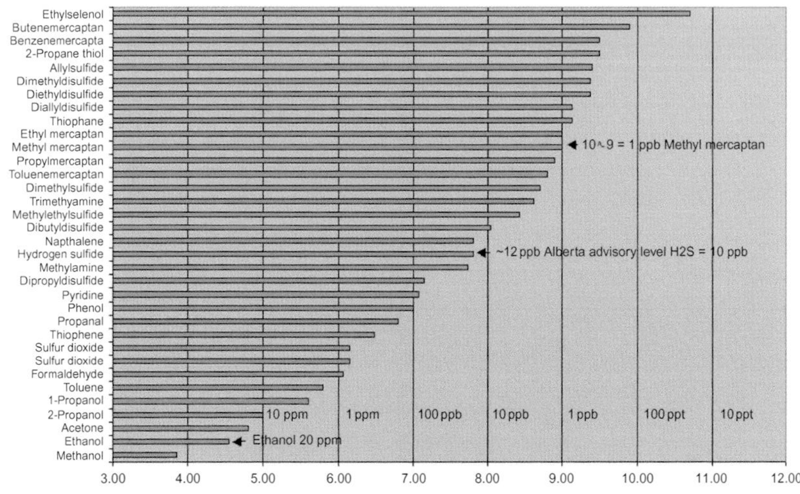

FIGURE 6.20　A diagram listing the standardized human olfactory thresholds of selected compounds, which were summarized by Devos et al. (1970). The negative log of the concentration at which the odor is just detected by an odor panel was determined by a weighted averaging of previous studies dating back over 100 years. Here, odor detection limits have been transferred to ppb.

Figure 6.20. Note that methyl mercaptan (methane thiol) has a threshold of 1 ppb; hydrogen sulfide has a threshold of about 12 ppb (vs. the Alberta Ambient Air Quality Objective of 10 ppb), and, for reference, ethanol, at the very high level of 20 ppm.

Since January 1, 2012, a wide range of substituted thiophenes present during TRS episodes of 2 ppb have been measured and identified at the Bertha Ganter–Ft. McKay air monitoring station (Figure 6.20). Individually, each compound is on the order of 50 ppt but in total they have ranged up to about 1 ppb at their highest. Interestingly, thiophene itself is never present, and we have not found odor thresholds for substituted thiophenes. If collectively they have a similar threshold to thiophene itself, we can correlate with the recommended threshold of Figure 6.20, which is about 200–300 ppb. It seems unlikely that these thiophenes are responsible for odor complaints. It is possible, however, from the GC/MS cartridge analysis that other RSCs are present at concentrations somewhat lower than the thiophenes and are therefore not yet detected by the field PFGC. One such suspected compound is methylethyl-disulfide (tentative identification based upon retention time in GC/MS but not confirmed by spectral analysis). This compound's odor threshold is well below 1 ppb. Further modest improvement in SCD sensitivity is recommended.

6.7.3   On-Going Efforts and Goals

The on-going goal of these measurements in the AOSR is to accurately measure the concentrations of VOCs, and especially sulfur-containing compounds, whose presence in the air leads to odor complaints. With analytical sensitivity now below 100 ppt, and based upon mass spectral analysis of cartridge samples, a further (modest) reduction in field GC detection limits may quantify an additional suite of compounds, now below the detection limit, which likely contribute to odor. We have several for this improvement in sensitivity.

The PFGC–SCD uses a variety of compressed gases whose cost and replacement at remote sites would benefit from reduction. Second generation technology is being developed which will significantly reduce this gas consumption, making the PFGC even more practical as a field instrument. The SCD has now run continuously at the field site for more than 6 months with little or no attention and in past studies we have found the FID and column to require little or no maintenance for periods of a year or more so the prognosis appears good for a viable, cost-effective remote field instrument requiring little maintenance or upkeep, monitored and manipulated if necessary over the Internet.

This work continues so that measurements of such compounds can become a routine part of the extensive WBEA monitoring network, where appropriate.

ACKNOWLEDGMENTS

We are grateful to the Wood Buffalo Environmental Association, Fort McMurray, Alberta, for funding this work.

APPENDIX

TABLE A1 CONDITIONS FOR THE CHROMATOGRAM COMPARISONS

Parameter	Agilent laboratory chromatogram	PFGC field chromatogram	Units
Benzene concentration	100	0.007	ppm
Chromatogram starting temperature	80	21	Centigrade
Benzene retention time	11	16	Minutes
Sample volume	0.5	200	Standard cc (NTP)
Sample pressure upon injection	1	40	Atmospheres

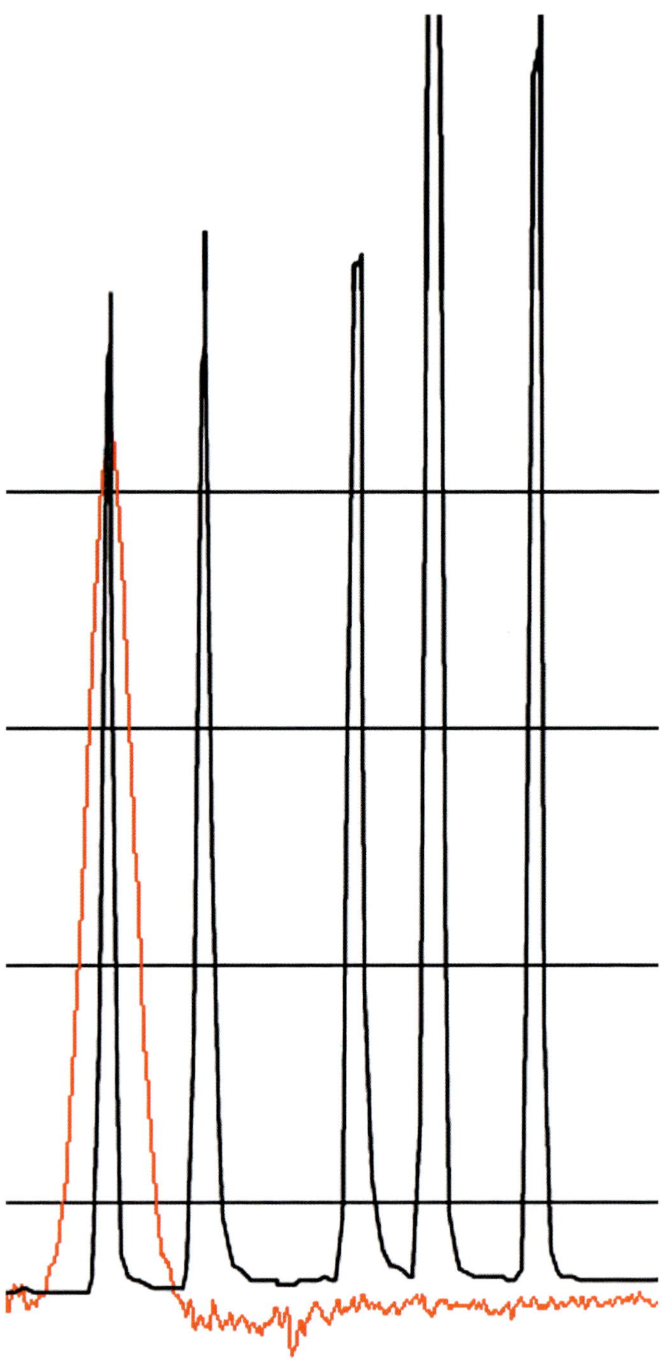

FIGURE A1   Comparison of "standard" laboratory and "field" pneumatic focusing chromatograms as described in Table A1. The Agilent chromatogram (blue) was of a mixture of ozone precursors including benzene, the PFGC was of an aromatic hydrocarbon mixture (BTEX) including benzene (red). The portion shown for each chromatogram is 2 min wide. It is seen that pneumatic focusing of a $400\times$ larger sample volume increases the peak width of benzene by only about a factor of 2–3. The "eyeball" PFGC signal-to-noise ratio appears to be about 1000. Further reduction in the pneumatically focused benzene peak width could be accomplished by lowering the injection temperature but this would require additional complexity in the field PFGC instrument, which can operate at ambient temperatures from 0 to 40 C without surrounding temperature control.

REFERENCES

Devos, M., Patte, F., Rouault, J., Laffort, P., Van Gemert, L.J., 1990. Standardized Human Olfactory Thresholds. IRL Press at Oxford University Press, Oxford, UK.

Gallego, E., Roca, F. J., Perales, J. (1999), US EPA Determination of volatile organic compounds in ambient air using active sampling onto sorbent tubes; Method TO-17.

Gallego, E., Roca, F.J., Perales, J.F., Guardino, X., 2011. Assessment of chemical hazards in sick building syndrome situations: Determination of concentrations and origin of VOCs in indoor air environments by dynamic sampling and TD-GC/MS analysis. In: Abdul-Wahab, S.A. (Ed.), Sick Building Syndrome: Public Buildings and Workplaces. Springer-Verlag, Berlin Heidelberg, Chapter 16, 591 pp.

Hester, R.E., Harrison, R.M. (Eds.), 1995. Volatile Organic Compounds in the Atmosphere. In: Issues in Environmental Science and Technology, vol. 4. Royal Society of Chemistry, Cambridge, UK.

Kim, Y.-H., Kim, K.-H., 2012. Novel approach to test the relative recovery of liquid-phase standard in sorbent-tube analysis of gaseous volatile organic compounds. Anal. Chem. 84, 4126–4139.

Lewis, A., Carslaw, N., Marriott, P., Kinghorn, R., Morrison, P., Lee, A., Bartle, K., Pilling, M., 2000. A larger pool of ozone-forming carbon compounds in urban atmospheres. Nature 405, 778–781.

Myeong, Y.C., Maris, C., Krischke, U., Meller, R., Paulson, S.E., 2003. An investigation of the relationship between total non-methane organic carbon and the sum of speciated hydrocarbons and carbonyls measured by standard GC/FID: Measurements in the Los Angeles air basin. Atmos. Environ. 37 (Suppl. 2), S159–S170.

O'Brien, R. J. (2005). Method and apparatus for concentrating samples for analysis. US Patent Number: 6952945.

O'Brien, R. J. (2007). Method and apparatus for concentrating samples for analysis. US Patent Number: 7257987.

O'Brien, R. J. (2009). Method and apparatus for concentrating samples for analysis. US Patent Number: 6952945.

Pandey, K.P., Kim, K.-H., 2009. A review of methods for the determination of reduced sulfur compounds (RSCs). Environ. Sci. Technol. 43, 3020–3029.

Shearer, R.L., 2002. Development of flameless sulfur chemiluminescence detection: Application to gas chromatography. Anal. Chem. 64 (18), 2192–2196.

Stevens, R.K., Mulik, J.D., O'Keeffe, A.E., Krost, K.J., 1971. Gas chromatography of reactive sulfur gases in air at the parts-per-billion level. Anal. Chem. 43, 827–831.

Sweet, C.W., Vermette, S.J., 1992. Toxic volatile organic compounds in urban air in Illinois. Environ. Sci. Technol. 26, 165–173.

Traube, S., Scoggin, K., Mitloehner, F., Hong, L., Burns, R., Xin, H., 2008. Consideration on the broad quantification range of gaseous reduced sulfur compounds with the combined application of gas chromatography and thermal desorber. Atmos. Environ. 42, 3332–3341.

Tzanani, N., Amirav, A., 1995. Combined pulsed flame photometric ionization detector. Anal. Chem. 67 (1), 167–173.

Wikipedia. http://en.wikipedia.org/wiki/Athabasca_oil_sands (assessed 03-Mar-2012).

Woolfenden, E., 2010. Sorbent-based sampling methods for volatile and semi-volatile organic compounds in air: Part 1: Sorbent-based air monitoring options. J. Chromatogr. A 1217 (16), 2674–2684.

Overview of Real-World Emission Characterization Methods

J.G. Watson[1], J.C. Chow, X.L. Wang, S.D. Kohl, L.-W.A. Chen and V. Etyemezian

Division of Atmospheric Sciences, Desert Research Institute, Reno, Nevada, USA
[1]Corresponding author: e-mail: john.watson@dri.edu

ABSTRACT

Real-world emissions do not necessarily correspond with those derived from certification tests, owing to changes in equipment, fuels, and operating cycles. Therefore, emission rates from sources that affect ambient air quality are needed to drive air quality models and to provide accountability for air quality management strategies. Since the early 1960s, source characterization methods have been established that quantify emission rates to certify sources and determine their compliance over time. However, these certification and compliance methods have not been adapted to changes in emission processes and controls nor have they incorporated advances in measurement technology. This results in source tests that are incompatible with each other and with ambient measurement methods. Source characterization methods need to be improved to better represent real-world hardware, operating conditions, and feedstocks and to obtain more information at lower costs. Hot-ducted exhaust can be cooled to ambient temperatures prior to measurement to better approximate emissions as they appear in the atmosphere. Engine exhaust can be characterized *in situ* with portable monitoring systems. A portable wind tunnel can characterize fugitive dust threshold suspension velocities, reservoir sizes, particle size distributions, and chemical profiles.

7.1 INTRODUCTION

Air pollutant emission estimates are a fundamental component of effective air quality management (AQM) (Bachmann, 2007; Chow et al., 2007). They are

Developments in Environmental Science, Vol. 11. http://dx.doi.org/10.1016/B978-0-08-099443-7.00007-7

used to regulate polluters, establish accountability through long-term trends, and evaluate the effectiveness of different emission reduction strategies. Air pollutant emitters are divided into large stationary, mobile, and area-source categories, with different testing methods applicable to each one. Large stationary sources are those with tall stacks, usually associated with industrial facilities such as power stations, refineries, steel mills, and cement plants. Mobile sources include those using spark ignition (gasoline), compression ignition (diesel), and turbine (jet) engines in both on-road and non-road applications, including cars, trucks, buses, haulers, portable generators, locomotives, ships, and aircraft. Area sources include everything else, such as fugitive dust, fires, evaporating fuels and solvents, and aggregates of smaller ducted emissions such as wood stoves, fireplaces, and small industries. The most common species of interest correspond to the criteria contaminants of sulfur dioxide (SO_2), oxides of nitrogen (NO_x), carbon monoxide (CO), and particulate matter (PM). PM is divided into $PM_{2.5}$ and PM_{10} size fractions (mass of particles with aerodynamic diameters <2.5 and 10 μm, respectively). Also of interest are emissions of volatile organic compounds (VOCs) owing to their participation in secondary ozone (O_3) and PM formation and ammonia (NH_3) for its neutralizing effects on sulfuric and nitric acids (H_2SO_4 and HNO_3). Emissions of additional components become important when considering multipollutant/multieffect AQM (Chow and Watson, 2011). Although adverse health effects are the most important motivators for reducing emissions, ecosystem and crop damage, odors, visibility degradation, cultural heritage deterioration, and climate change are of concern in many communities.

Because it is not possible to directly measure emissions for all pollutants on all emitters all the time, emission models are used to create a weighted sum of emission measurements from similar source types:

$$ER_{ijk} = F_{ij}EF_{jk}A_{jk}\left(1 - P_{jk}\right) \tag{7.1}$$

ER_{ijk}, mass emission rate (μg/s, tons/yr, or other mass per unit time) of pollutant i from source type j corresponding to time period k. F_{ij}, fractional quantity of pollutant i in source type j. For PM measurements, these source profile abundances are normalized to mass emissions from a source in the desired size range and averaged over several tests on similar sources (Watson et al., 2008). For VOCs, these are normalized to the sum of the species measured or to the sum of the 55 Photochemical Assessment Monitoring Station (PAMS) species (Watson et al., 2001). $F_{ij}=1$ for single constituents such as SO_2, NO_x, etc. EF_{jk}, mass rate of emissions (emission factor; g/unit of activity) for source type j corresponding to time period k. A_{jk}, activity related to emissions (unit of activity/s) for source type j over corresponding time period k. P_{jk}, fractional reduction (unitless) due to emission controls applied to source type j over time period k.

Each of the components of ER_{ijk} is empirically derived from a limited number of tests. ER_{ijk} are summed over j for period k to obtain total emissions for species i. Some emission factors and activities (e.g. cold engine starts, roadway deicing, evaporating fuels, and solvents) may change with season or even time of day, and time and temperature dependence is an important consideration.

Air pollution emission characterization, often referred to as "source testing," is performed periodically to determine EF_{jk} and F_{ij}. A_{jk} is usually derived from databases on quantities of fuel used, vehicle kilometers traveled, amounts of product produced, and population densities. The derived emissions are used for the following purposes:

- *Certification* intends to verify that a process or hardware design has emissions within the legal limit set by an emission standard. These tests are performed by procedures specified in the regulation on a small number of representative units to demonstrate attainment. The U.S. and Canadian Tiers and European EURO engine emission limits exemplify the certification process (U.S. EPA, 2004).
- *Compliance* intends to determine that emissions from an individual unit are within acceptable ranges, as specified in an operating permit. These include periodic tests of stacks at stationary sources (Cooper and Rossano, 1974) and inspection and maintenance tests (e.g., smog checks) for on-road vehicles (Lloyd and Cackette, 2001; St. Denis and Lindner, 2005; Van Houtte and Niemeier, 2008).
- *Emissions trading* compares annual emissions that are measured continuously (Jahnke, 2000) with the allowances owned by a given source. These allowances can be bought and sold at market prices. The most successful example of emissions trading is the SO_2 allowances originally established in the U.S. 1990 Clean Air Act Amendments (U.S. Congress, 1990).
- *National emission inventories* (Environment Canada, 2012; U.S. EPA, 2012a) evaluate annual rates to compare emissions among different regions and long time periods. These trends represent first step in assessing AQM accountability (HEI, 2003).
- *Dispersion modeling inventories* are inputs for urban- to regional-scale chemical transport models (Zannetti, 2003, 2005) that simulate the effects of actual emissions and emission changes on ambient concentrations.
- *Source apportionment receptor models* (Watson et al., 2001, 2008; Watson and Chow, 2012) use the multipollutant pattern of the source profiles (F_{ij}) to deconvolute mixtures of pollutants measured at a receptor to derive source contribution estimates (SCEs). The SCEs provide an independent check on emission inventories and dispersion modeling results.

As thousands of emission tests are conducted each year for certification and compliance purposes, these emission factors (EF_{jk}) are often used for the other purposes. However, these test methods do not adequately represent the real-world emissions that derive from hardware, fuels, and operating conditions differing from those specified by the test methods. These differences can result in large discrepancies, by factors of two or more, between real-world and certification/compliance emissions. This chapter intends to: (1) describe certification and compliance source testing methods in current use, (2) identify alternatives and improvements that better approximate real-world emissions, and (3) provide examples relevant to emitters from oil sands operations.

7.2 STATIONARY SOURCE EMISSIONS

Large industrial stacks are noticeable emitters, and many have installed PM, SO_2, and NO_x control devices. Since the early 1960s, stack effluents have been isokinetically removed through a buttonhook nozzle, drawn through a filter, and bubbled through gas-absorbing solutions contained in cooled glass impingers (Figure 7.1). For PM, the sample is drawn through a filter maintained at or near stack temperature, then through distilled water to collect condensable vapors. When in-stack PM levels are high, the impinger catch is a small fraction of that on the filter at the in-stack temperature. After PM controls, however, the impinger catch often exceeds the mass on the front filter. Most of the impinger fraction derived from dissolved SO_2 and VOCs rather than condensable particles (England et al., 2000).

Figure 7.2 illustrates how the stationary source test method affects the tracking of emissions trends. In the early 1970s, electric utilities contributed 15% of total U.S. nonmiscellaneous PM_{10} emissions, declining each year (along with other categories) to 9% of the 1998 total. Starting in 1999, however, the electric utilities' share increased to 23% of the total and to 24% in 2011. This increase was due to the addition of the impinger catch in 1999 and does not represent the real primary emissions, which are somewhere between the hot filter and hot filter/impinger emissions.

Separate compliance sampling procedures are used for PM (U.S. EPA, 1996, 1997, 2000a), SO_2 (U.S. EPA, 2000d), NO_x (U.S. EPA, 2000e), and other pollutants requiring multiple trips up and down the stack with ice buckets,

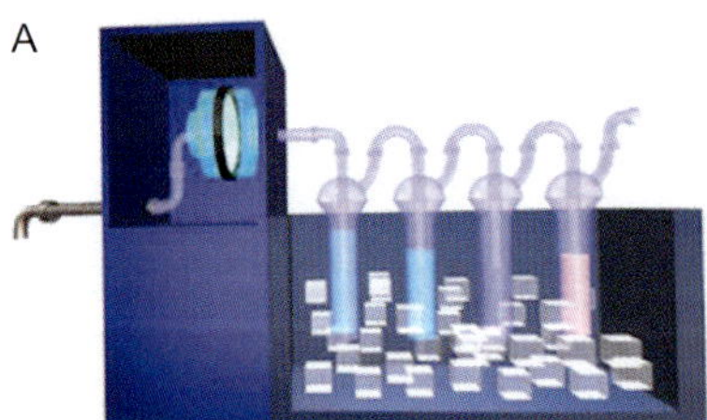
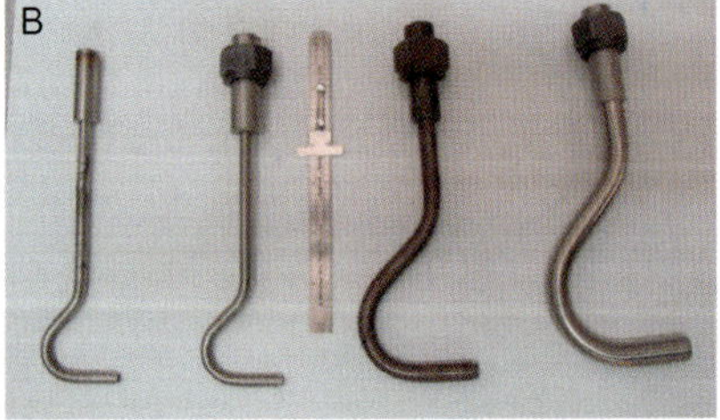

FIGURE 7.1   (A) U.S. EPA Method 5 particulate matter (PM) stack sampler and (B) buttonhook nozzles. In this method, the filter is contained in a heated compartment ($<150\,^\circ$C). Method 201 for PM_{10} places the filter inside the stack following a size-selective inlet.

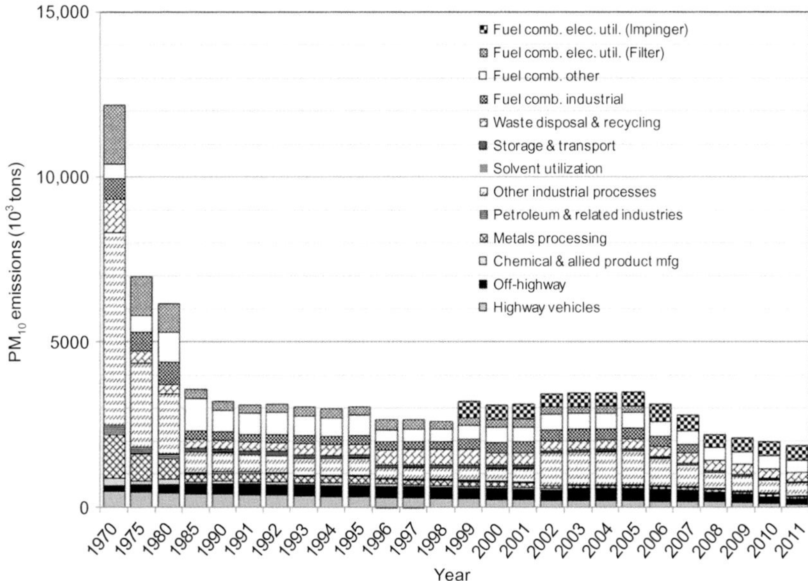

FIGURE 7.2   U.S. PM_{10} emission trends from major source categories (U.S. EPA, 2012a,b). The miscellaneous category (wildfires, windblown dust, etc.) was excluded. The hot filter (filterable) and impinger (condensable) fractions are reported separately for electric utility emissions from 1999 to 2005. From 2006 onward, the sum of the two is reported as the impinger catch. Only the hot filter results are reported prior to 1999. Environment Canada's (2012) National Pollutant Release Inventory (NPRI) uses the hot filter PM_{10} emissions for electric utility emissions.

glassware, and chemicals to obtain multipollutant emissions. The glass-fiber filters are not compatible with most analyses to determine PM source profiles, and there is no provision for stack VOC emission rates or profiles.

A more modern approach, similar to that applied to mobile sources described below, is to extract a portion of the stack gases into a chamber where it can be diluted with filtered air to near ambient temperatures, then draw air from this chamber through substrates and sensors compatible with those used at ambient air monitoring stations (Chang et al., 2004; England et al., 2007a,b; Hildemann et al., 1989; Lipsky and Robinson, 2005), as illustrated in Figure 7.3 and further demonstrated by Wang et al. (2012a,b). When hot exhaust is cooled and aged, some of the vapors nucleate or condense on preexisting particles. Condensational growth of particles in a diluted plume depends on temperature, relative humidity, aging time, mixing rate, and partitioning of species between the gas and particle phases (Chang et al., 2004; Lipsky et al., 2002, 2004; Lipsky and Robinson, 2006); a 2- to 10-s aging time for a well-mixed diluted plume was found to achieve a stable ultrafine size distribution for coal, residual oil, and natural gas combustion effluents (Chang et al., 2004).

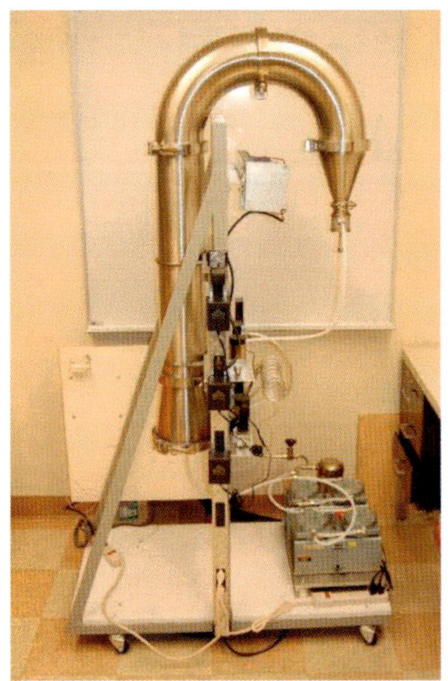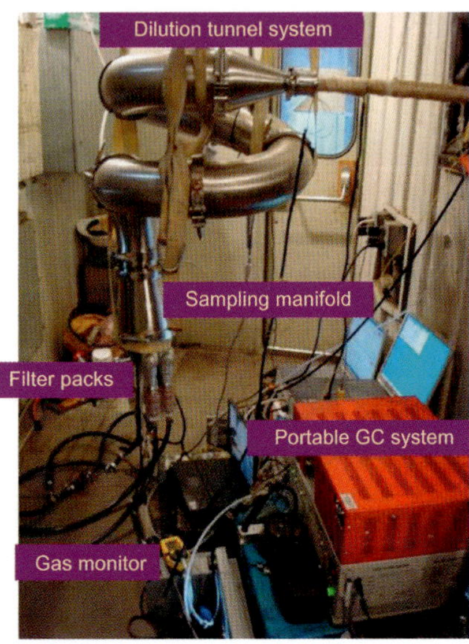

FIGURE 7.3 Examples of PM stack dilution sampling systems. The spiral design on the right allows for greater adjustment of the residence time by adding additional sections to the spiral. Different instruments and filter packs can be used to obtain gas and particle measurements simultaneously that are compatible with those from ambient air quality monitors.

Figure 7.4 shows the differences between tests using the standard compliance methods and a dilution method applied to an oil sands flue gas desulfurization (FGD) unit. Some of the differences are due to changes in feedstocks and operating conditions that occurred between the tests. The SO_2 and NO_x emissions are reasonably comparable, given these changes. The Method 5 procedure (Alberta Environment, 1995) shows that most of the PM is from the impinger fraction. The winter 2010 dilution measurement shows a value somewhere between the hot filter and hot filter/impinger values, while the summer 2008 value is closer to the hot filter value from the compliance tests.

ASTM (2008) provides dilution sampling guidance for stationary source certification that better reconciles the current discrepancy between stationary and mobile source emissions and ambient PM measurements. With an effective baghouse or electrostatic precipitator, most stack PM is in the $PM_{2.5}$ fracton, and isokinetic extraction is less critical. However, with the use of wet SO_2 scrubbers, PM can dissolve in water droplets with up to 20 μm diameters, too large to negotiate the bend in the buttonhook nozzle. A potential solution to this is an in-stack nozzle that slows the droplets as they are rammed through the inlet so they can make the bend into the dilution chamber where the water evaporates, leaving much smaller particles for tranport to the sensing devices (Baldwin et al., 2012). As a dilution chamber cannot completely simulate conditions in the atmosphere, there is disagreement among

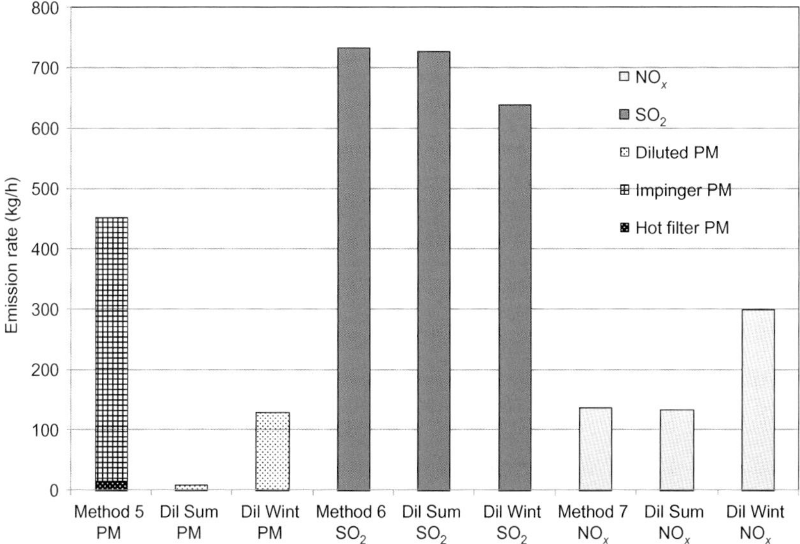

FIGURE 7.4   Comparison of average PM, SO$_2$, and NO$_x$ emission rates from an oil sands flue gas desulfurization stack for winter 2007 compliance sampling (Lehder Environmental Services, 2007), summer 2008 dilution sampling (Watson et al., 2010b), and winter 2010–2011 dilution sampling (Watson et al., 2011b). Some of the variability is due to changes in feedstocks and processes between the source testing periods as discussed by Wang et al. (2012b). NO$_x$ is expressed as equivalent NO$_2$. Procedures for Methods 5, 6, and 7 are specified in U.S. EPA 2000a, d, and e, respectively.

practitioners on the precise conditions that should be simulated. Nevertheless, there is general agreement that some form of dilution and cooling, similar to that used for mobile source engine testing, would provide more realistic emission factors than the hot filter/impinger method. An example is the use of two methods for certifying residential wood stove designs in the U.S. Method 5H (U.S. EPA, 2000c), which uses the approach in Figure 7.1, and Method 5G (U.S. EPA, 2000b) that mimics the dilution approach in Figure 7.3. All of the certification tests done to date have used the 5G approach as being a more realistic representation of the semivolatile biomass burning emissions.

7.3   ENGINE EXHAUST EMISSIONS

Engine certifiation tests operate an engine through different cycles (DieselNet, 2012b) and dilute the exhaust to $\sim 50\,^\circ$C, draw air through NO/NO$_2$, CO, and nonmethane hydrocarbon (NMHC, a subset of VOC) monitors, and sample PM onto filters that are weighed before and after sampling. Most on-road gasoline engines are tested within the vehicle on a chassis dynamometer while larger diesel engines and small non-road gasoline engines are tested outside the vehicle on an engine dynamometer (Figure 7.5). Emphasis in recent years has been on diesel engines, and especially non-road applications, owing to their high PM

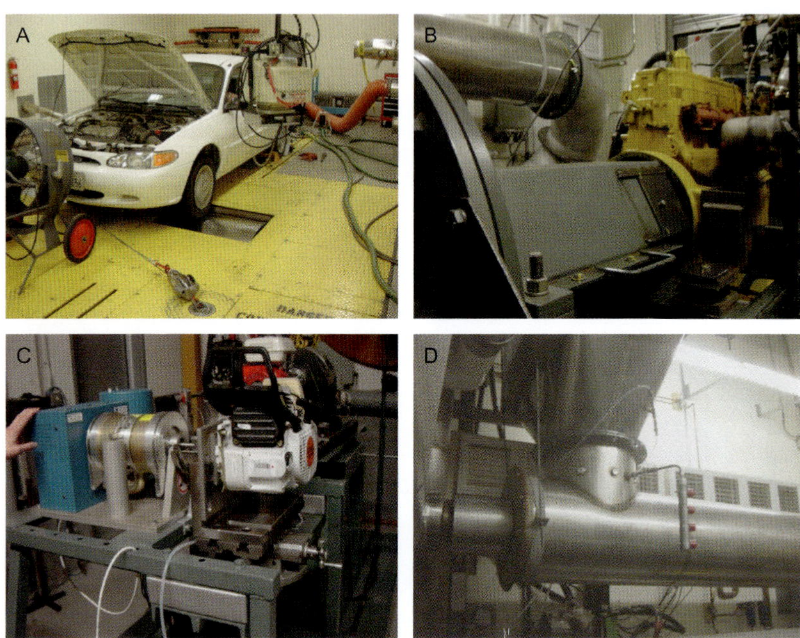

FIGURE 7.5 Examples of engine certification test hardware: (A) the chassis dynamometer intends to simulate on-road driving conditions with the drive wheels on a roller that provides selected resistances; (B) large and (C) small engine dynamometers use a water brake or resistance cell to apply loads; (D) the dilution tunnel has sampling ports which allow extraction of cooled exhaust to continuous monitors, filters, and canisters for multipollutant quantification.

and NO_x emissions (DieselNet, 2012a). Diesel engines are usually tested at various speeds and loads in steady state modes. Cold start and transient emissions are not included. Certification cycles for non-road sources are multimode, steady state, and depend on the application. The only fuel certification is a sulfur (S) content <5000 ppmw for Tiers 1, 2, and 3 engines, and <15 ppmw for Tier 4 and beyond. For non-road engines, such as those in the mine fleet, the ISO (2007) 8178 test cycle is run with C-1 weightings for engine certification against the appropriate tier. Diesel emission factors are reported in grams per brake horsepower-hour (g/bhp-h), which is not a useful unit for AQM. Distance-based EFs are related to g/bhp-h using brake-specific fuel consumption (BSFC in lb fuel/bhp-h), fuel density (ρ in lb/gal), and fuel economy (FE in miles/gal) (Machiele, 1989):

$$EF(g/mile) = EF(g/bhp - h) \times (\rho/[BSFC \times FE]) \tag{7.2}$$

Manufacturers usually select the engine and the testing laboratory when applying for certification of a new engine design. Certification emissions using chassis and engine dynamometers are intended only to determine compliance with emission standards and are not intended to represent real-world emissions encountered in practice (Sawyer et al., 2000). Periodic workshops have been conducted for nearly two decades to determine how real-world emissions differ from the certification values (Cadle et al., 2009).

Different methods are used by Alberta oil sands companies to estimate their diesel exhaust emission rates (ER in g/yr) for compliance submissions. One facility used fuel-based EFs derived by U.S. EPA (1995) multiplied by annual fuel consumption (Method 1). This facility changed to emission estimates in 2004 using the product of emission standards, gross operating hours, rated vehicle power, and a conservative load factor (Method 2). Another facility uses emission standards or a weighted sum of certification emissions, whichever is larger, and adds a 10% safety margin (Method 3). The three calculation methods are summarized below:

- Method 1: ER (g/yr) = EF (g/kg fuel) × annual fuel consumption (kg) with EF selected for pre-1985 engines from U.S. EPA (1985) AP-42.
- Method 2: ER (g/yr) = $EF_{standard}$ (g/kWh) × operating hours (h/yr) × rated vehicle power (kW) × load factor. Load factors are taken conservatively (59% for heavy haulers and 100% for supporting equipment as recommended by U.S. EPA).
- Method 3: ER (g/yr) = max($EF_{standard}$, $EF_{cert\ composite}$) (g/kWh) × 1.1/BSFC (kg fuel/kWh) × fuel consumption (kg fuel/yr).

These are all good engineering approaches that err on the side of overestimating emissions. As illustrated in Figure 7.6, however, these approaches differ among each other and from real-world measurements on representative trucks as described below.

As engine exhaust becomes cleaner, it is more difficult to measure the pollutants for certification, especially PM by the filter-weighing technique. Concerns about ultrafine particle health effects (Brugge et al., 2007; Mauderly and Chow, 2008; McDonald et al., 2007) and black carbon (BC) contributions to global warming (U.S. EPA, 2012c) have stimulated efforts in Europe (Johnson et al., 2009b; Wang et al., 2010) and in California (CARB, 2011) to revise engine certification methods to better reflect real-world conditions.

Cross-plume and in-plume measurement systems have been developed and applied to better quantify real-world engine emissions and the factors that influence them. These methods derive fuel-based EFs (g emitted/kg fuel consumed) based on carbon mass balance (Dreher and Harley, 1998; Fraser et al., 1998; Kean et al., 2000; Moosmüller et al., 2003) as:

$$EF_i = CMF_{fuel}\ \frac{C_i}{C_{CO_2}\left(\frac{M_C}{M_{CO_2}}\right) + C_{CO}\left(\frac{M_C}{M_{CO}}\right)} \tag{7.3}$$

where CMF_{fuel} is the carbon mass fraction of the fuel, which is 86.2% for diesel, assuming it has an average formula of $C_{12}H_{23}$. C_i is the concentration of

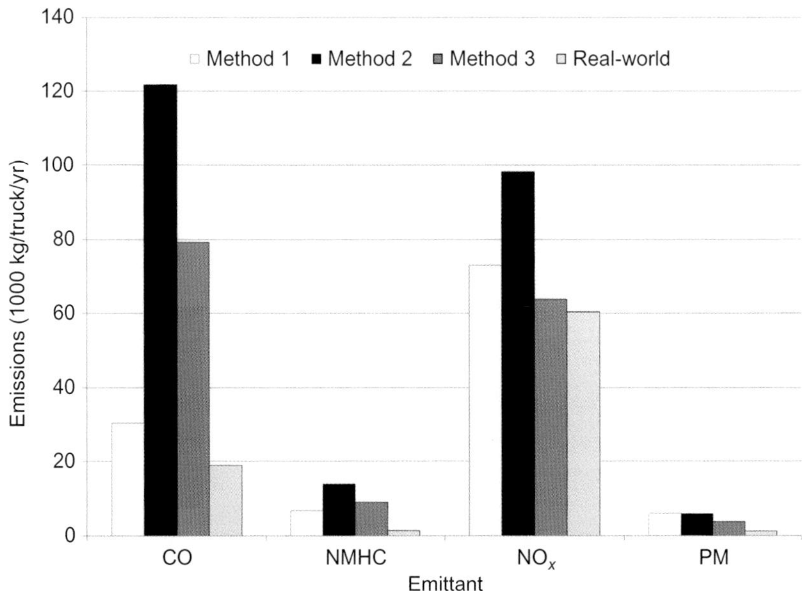

FIGURE 7.6 Comparison of annual emissions calculated by three different methods used by different oil sands operating units for annual emissions from a CAT 797B heavy hauler. Real-world emission rates were derived from the average of on-board measurements of four CAT 797Bs at each of the operating units (Watson et al., 2011c).

pollutant i in g/m^3 or particle number/m^3, and C_{CO_2} and C_{CO} are the concentrations of CO_2 and CO in g/m^3, respectively. M_C, M_{CO_2}, and M_{CO} are the atomic or molecular weights of C, CO_2, and CO in g/mol. Equation (7.3) assumes that the carbon content in CH_4, NMHC, and PM is negligible compared to carbon in CO and CO_2. CO is often <1% of CO_2 and may be dropped from the calculation. Use of Equation (7.3) avoids the need to capture all of the exhaust to determine the dilution factor and opens many possibilities for cost-effective real-world emission characterization.

Figure 7.7 illustrates cross-plume and in-plume configurations for quantifying real-world fuel-based emissions. Cross-plume measurements consist of remote sensing instruments that measure radiation transmission and scattering in the infrared (IR) and ultraviolet (UV) portions of the spectrum across an exhaust plume (Barber et al., 2004; Bishop et al., 1989; Kuhns et al., 2004; Mazzoleni et al., 2004a,b, 2010). Gaseous cross-plume sensors use IR absorption for CO_2, CO, and some VOCs, UV absorption is used for NO, and UV-lidar backscatter is used for PM. Cross-plume sensors can measure gaseous EFs for large numbers of individual vehicles (>1000/h), albeit under a limited variety of operating conditions largely determined by monitoring location. These measurements have a high temporal resolution (~10 ms) resulting in 20–50 measurements before, during, and after vehicle passage through the

FIGURE 7.7   Different configurations for cross-plume and in-plume sampling: (A) a vehicle crosses a transmitter and detector that remotely senses gases and PM; (B) exhaust is sampled at some distance from the outlet to allow for cooling; (C) a probe is placed across the roadway that encounters the down-mixed plume from a passing vehicle; and (D) exhaust is extracted from a CAT 797B heavy hauler exhaust pipe and directed to a small dilution chamber for cooling.

measurement path. Remote sensing studies have shown that comparatively few vehicles cause a majority of the emissions, that is, EFs do not follow a symmetric frequency distribution (Zhang et al., 1994), as is assumed by the averaging technique used to estimate most EFs. A better approach might be to average within percentiles and report separate emissions for low, median, and high emitters.

In-plume measurements extract a portion of exhaust and draw it directly into the measurement devices when the plume has been diluted in ambient air, or into a small dilution chamber if the plume is hot, as illustrated in Figure 7.7. Several mobile emission laboratories (MELs) (Beckerman et al., 2008; Bukowiechi et al., 2002; Canagaratna et al., 2004; Cocker et al., 2004; Kittelson et al., 2004; Morawska et al., 2007; Nussbaum et al., 2009; Pirjola et al., 2004) have been constructed that contain continuous monitors and substrate samplers. MELs can be stationed downwind of roadways to capture the plumes of passing vehicles (as identified by a CO_2 spike) or to follow vehicles as they transition through idling, accelerating, slowing, and cruising. MELs require an on-board power source, usually a gasoline or diesel generator, and their large size and limited maneuverability limit their ability to sample in many locations, especially for characterizing non-road emissions. Their biggest advantage is that they can accommodate some of the most advanced air quality measurement technologies.

The ongoing development of small, battery-powered microsensors creates new possibilities for portable emission monitoring systems (PEMS). PEMS are under consideration for engine certification tests as their data would better simulate real-world conditions. Most commercial PEMS (Abolhasani et al., 2008; Bishop et al., 2009; Cooper and Ekstrom, 2005; Johnson et al., 2009a, 2011; Zhang and Frey, 2008) address only criteria contaminants, which limits their utility for a broader multipollutant analysis for multiple effects. As a result, several research systems (Barrios et al., 2011; Bishop et al., 2006; Boughedaoui et al., 2008; Brunet et al., 2008; Collins et al., 2007; Wang et al., 2012a) have been assembled for on-board measurements of the criteria contaminants plus black and brown carbon, PM size distribution, ultrafine particle number concentration, non-criteria gases (e.g. H_2S), as well as PM and VOC source profiles. The on-board system can be customized for a specific application, placed in the trunk or back seat of a vehicle and powered by the vehicle or a deep cycle 12 V battery, and draw the sample from a probe located $\sim$30 cm behind the exhaust pipe. Accompanied by a Global Position System tracker and an On-Board Diagnostics monitor (Supnithadnaporn et al., 2011), emissions can be related to engine operating conditions.

Figure 7.8 illustrates the continuous output from an on-board system applied to a heavy hauler for a typical cycle in an Alberta oil sands mine. These continuous measurements can be used to divide emissions among different processes. During idling, most pollutant concentrations were stable with lower emission concentrations than when the truck was moving. However, NO_2 concentrations increased at the beginning of the test cycle when the truck idled for more than 30 min. The background subtracted CO/CO_2 ratio was low, indicating near-complete fuel combustion. When the truck started moving, all pollutant concentrations increased and reached local maxima. As the truck accelerated toward the loading pit, the pollutant concentration gradually decreased. While the truck waited for a load (idling), $PM_{2.5}$, BC, and CO concentrations, as well as the CO/CO_2 ratio decreased while the NO and NO_2 concentrations increased. There was not much change to the particle numbers. When the truck moved forward in the waiting line, the $PM_{2.5}$, BC, and CO concentrations increased, but the NO, NO_2 levels, and particle numbers decreased. When the truck backed to the shovel, the CO/CO_2 ratio reached its maximum. Other pollutants did not follow this pattern. When the loaded truck left the pit, engine speed and load were close to their maximum values, NO, NO_2 concentrations, and particle numbers decreased while $PM_{2.5}$, BC, and CO levels increased. When the truck accelerated uphill out of the pit, all pollutant concentrations reached local maxima. Cruising on a level road at $\sim$32 km/h produced slightly elevated $PM_{2.5}$, particle numbers, BC, and CO levels, but NO, NO_2 stayed at lower concentrations. $PM_{2.5}$ and BC approximately followed the trend of CO/CO_2 ratio. When the truck dumped its cargo, the engine load was $\sim$85% and the engine speed

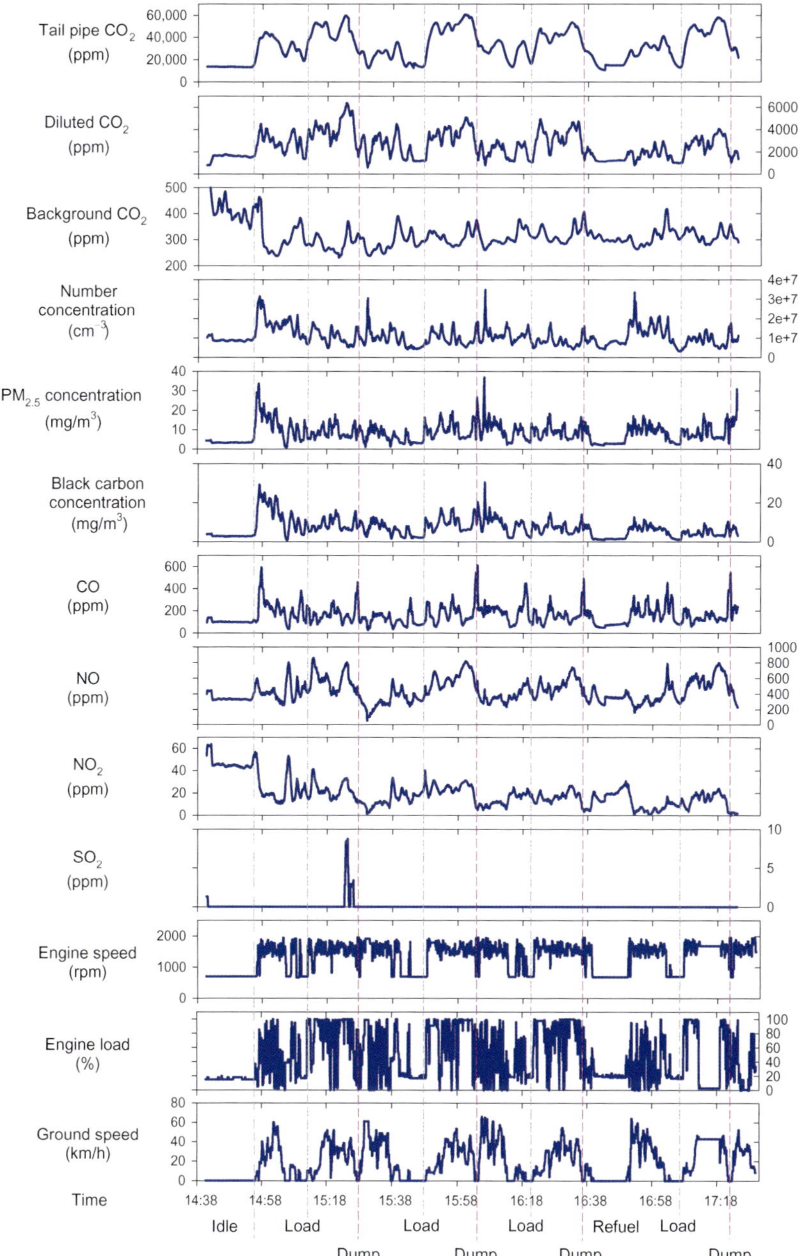

FIGURE 7.8   Example of continuous emission concentrations from on-board measurements on a CAT 797B heavy hauler (Watson et al., 2010a).

almost reached its maximum ($\sim$2000 rpm). The CO/CO_2 ratio had a sharp peak, which produced small peaks of $PM_{2.5}$ and BC, and CO concentrations. After dumping, the truck accelerated downhill, and all pollutant concentrations dropped. However, the particle numbers reached a maximum when the truck was cruising at 61 km/h. The CO/CO_2 ratio was higher when the empty truck was driving downhill to the loading pit than when it was moving uphill to the dumping area with its cargo.

Overall, PEMS and on-board monitors are adaptable to other source characterization applications, especially for area sources. Wildfire and wood stove plumes can be sampled using a probe similar to that shown in Figure 7.7B, while passing vehicles can use an arrangement such as that in Figure 7.7C.

7.4   FUGITIVE DUST EMISSIONS

Fugitive dust (Cowherd, 2001; Merritt et al., 2003; Watson et al., 2000) emissions constitute large fractions of most $PM_{2.5}$ and PM_{10} emission inventories, yet they are the most poorly characterized. Fugitive dust is derived from windblown and mechanical suspension from traffic on paved and unpaved roads and shoulders, industrial materials handling, construction and demolition, mining, tailings ponds, and high winds passing over these and other disturbed soil surfaces. These emissions depend on the amount of dust in the suspendable reservoir, the $\sim$0.3 to $\sim$30 µm size distribution, and the suspension force related to wind-shear or mechanical force (e.g. tires on the road). Current dust emissions are based on the silt content of suspendable soil. Surface dust is swept or troweled from several locations, dried, weighed, and sieved to a geometric diameter of 75 µm (equivalent to $\sim$120 µm aerodynamic diameter for $\rho = 2.5$ g/cm^3), which is divided by the total mass. Silt has little relevance to $PM_{2.5}$ and PM_{10} mass emissions.

Source profiles (F_{ij}) are measured more simply than ER (Chow et al., 1994), and these are often used in receptor models to estimate fugitive dust SCEs at receptors (Watson et al., 2012) as demonstrated in Figure 7.9. Geological materials are trowelled, swept, or vacuumed from locations that are believed to represent the surface-suspendable material. Ten or more samples are taken over the surface and mixed to represent the average composition. This is often done three or more times to obtain an average and standard deviation for the source profile that incorporates the within-source variability (Chow et al., 2003). Approximately 0.5 kg is needed because the amount of suspendable $PM_{2.5}$ and PM_{10} is relatively small compared to the mass of the bulk soil. The samples are dried, sieved, puffed into a chamber, and sampled through $PM_{2.5}$ and PM_{10} inlets onto filters for chemical characterization (Chow and Watson, 2012).

Upwind/downwind measurements (Chakradhar, 2004; Gillies et al., 1999; Watson et al., 2011a) locate an array of samplers downwind of a fugitive dust source and subtract the upwind concentrations to estimate the horizontal flux.

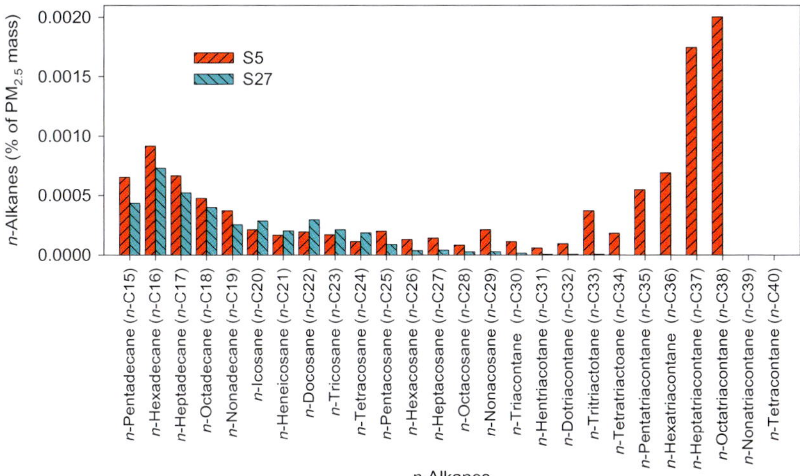

FIGURE 7.9   Comparison of $PM_{2.5}$ source profiles for resuspended $PM_{2.5}$ fugitive dust samples from a mine haul road (S5) and bare forest (S27) soil for *n*-alkanes. The haul road dust has a bimodal or trimodal distribution, with main peaks at C15–C17 and C35–C37, and a smaller peak at C22–C28. For the forest soil, *n*-alkanes are concentrated at C15–C26, with high-molecular-weight *n*-alkanes near or lower than the detection limits. Contributions from these sources would be easily distinguished from each other at receptors.

These fluxes are related to variables such as wind speed, traffic speed, silt loading, vehicle weight, soil moisture content, and others to develop EFs for emission models. Owing to the large costs and logistical requirements, these experiments are limited in scope, but most of the U.S. EPA AP-42 fugitive dust emission factors are derived by this method. These factors do not adequately represent transportable dust, the fraction of suspendable dust that is likely to travel more than a few hundred meters from the emitter. Vertical flux estimates might be more appropriate for determining fugitive dust impacts over urban and larger spatial scales. Vertical flux is an alternative to horizontal flux as a method to estimate fugitive dust emissions (Gillette, 1977). Vertical flux is proportional to particle density, surface friction velocity, and the difference between particle concentrations at different elevations above ground level. Only a fraction of the horizontal flux moves upward, depending on meteorological, surface, and dust composition variables.

Portable wind tunnels (Gillette, 1978a,b; Gillette et al., 1990; McKenna-Neuman and Nickling, 1989) have been used to measure threshold suspension velocities, reservoir sizes, and size distributions. Figure 7.10A illustrates how these measurements have been made in the past, with large heavy systems requiring generators and transport trailers. The Portable *In Situ* Wind Erosion Laboratory (PI-SWERL) (Etyemezian et al., 2007; Goossens and Buck, 2009; Kavouras et al., 2009; King et al., 2011; Kuhns et al., 2010; Sweeney et al.,

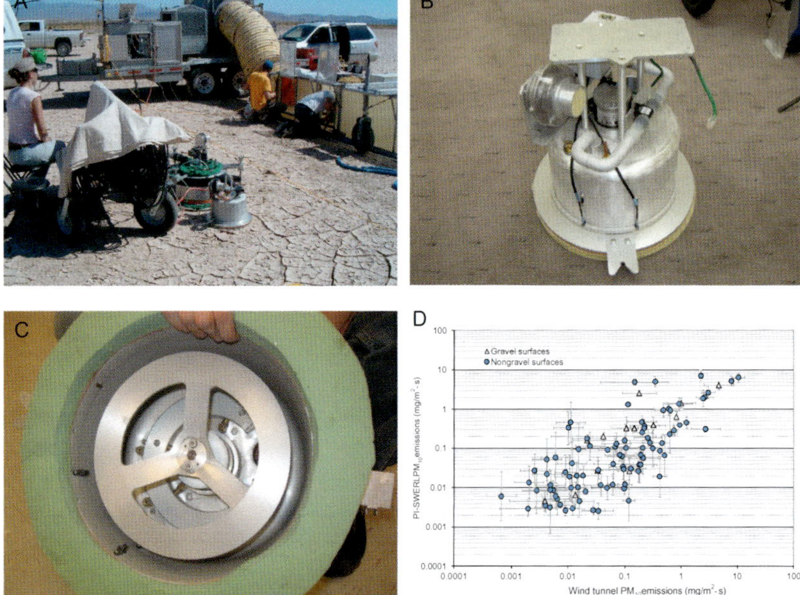

FIGURE 7.10 Wind tunnels to determine suspension properties and emission factors for different soil surfaces: (A) University of Guelph wind tunnel collocated with PI-SWERL; (B) the PI-SWERL portable wind tunnel; (C) the PI-SWERL rotating blade that creates a variable shear wind force over the soil surface; and (D) a comparison between PM_{10} emission factors from the large wind tunnel and the PI-SWERL.

2008, 2011) operates on battery power and can be rapidly moved to various locations on an exposed dust-disturbed surface. When equipped with particle sizing monitors similar to those used on the on-board system, it can estimate the dust reservoir associated with different wind shear velocities. The filter module of the on-board system can be attached to the chamber to obtain a source profile. Figure 7.10D shows that comparable EFs can be obtained by the PI-SWERL and the larger wind tunnel. PI-SWERL results can be incorporated into a vertical flux emission model and can also be used to rank different surfaces for their potential to emit under different meteorological conditions. It has also been used to evaluate the short- and long-term effectiveness of different dust suppression measures.

Vehicular movement creates a reservoir of particles as well as providing the energy to inject them into the atmosphere. Vehicle shape, speed, weight, number of wheels as well as previous history (e.g. dust acquisition for track-out and carry-out) interact with different road surfaces to change the particle size, surface loading, and wind effects. Vehicular traffic adds to particle suspension because tire contact creates a shearing force with the road that lifts particles into the air (Nicholson et al., 1989; Nicholson and Branson, 1990). Moving vehicles also create turbulent wakes that act much like natural winds to raise particles (Moosmüller et al., 1998). Natural crusts are often disturbed by vehicular movement, increasing the reservoir available for wind erosion. Paved roads are limited reservoirs, as the dust deposit must be continually replenished. Dust loadings on a paved road surface build up by being tracked out from unpaved areas, spilled from haul trucks, deposited from vehicle undercarriages, wear of tires, brakes, clutches, and exhaust systems, wear of the pavement surface, and windblown dust from adjacent areas. Unpaved roads and other unpaved areas with vehicular activity are unlimited reservoirs when vehicles are moving, as they are always being disturbed, and wind erosion seldom has an opportunity to decrease surface loadings or increase the surface roughness sufficiently to attenuate particle suspension.

The Testing Reentrained Aerosol Kinetic Emissions from Roads (TRAKER; Etyemezian et al., 2003a,b, 2006; Kuhns et al., 2001, 2005; Zhu et al., 2009) locates a probe behind one of a test-vehicle's tires and samples the suspended dust through the particle sizing instrumentation available from the on-board monitoring system described above. Surface loadings and size distributions are then determined as a function of location and vehicle velocity. $PM_{2.5}$ and PM_{10} source profiles can also be obtained by including the filter system from the on-board monitor.

7.5   EMERGING TECHNOLOGIES FOR SOURCE CHARACTERIZATION

Given the large number and types of sources that need to be characterized, that many of the most important emitters are non-road or fugitive sources, and that the cost of past characterization are too high, new technologies need to be developed.

The adaptation of smart phone technology shows promise. Ramanathan et al. (2011) adapted a smart phone camera to measure the darkness of filter paper acquired from solid-fuel stoves in India to obtain a large data base of burning conditions to reduce human exposure during cooking and heating. In addition to optically analyzing the sample, the smart phone immediately transmits data to a central database. Smart phones are also being adapted to recognize microscopic particles (Beiser, 2012) that might serve as part of a source profile to better determine source contributions.

Unmanned aerial vehicles (UAVs) are being used to map surfaces that can be related to area source activity levels (Bryson et al., 2010; Dunford et al., 2009; Rango et al., 2009). Several UAVs have been loaded with micro-sensor payloads to characterize elevated plumes and vertical distributions (Pollanen et al., 2009; Spiess et al., 2007; Techy et al., 2010).

Microsensor development is proceeding rapidly and is a prerequisite for using UAVs and on-board monitoring systems. BC (Hansen and Mocnik, 2010), NO_x (Prabhakar et al., 2010; Sitnikov et al., 2005; Sluis et al., 2010), and O_3 (Vallejos et al., 2007) can be detected, along with several other gases (Do and Chen, 2007). Small mass spectrometers and gas chromatographs (Contreras et al., 2008; Gao et al., 2008; Ouyang et al., 2009) open many possibilities for multipollutant characterization as their costs and detection limits are improved.

ACKNOWLEDGMENTS

This work was sponsored by the Wood Buffalo Environmental Association (WBEA), Alberta, Canada (www.wbea.org). The authors thank Drs. Allan Legge and Kevin Percy and the environmental officers at each facility for their efforts to coordinate the field-testing for the examples illustrated here. Ms. Jo Gerrard of DRI assisted with editing the chapter.

REFERENCES

Abolhasani, S., Frey, H.C., Kim, K., Rasdorf, W., Lewis, P., Pang, S.H., 2008. Real-world in-use activity, fuel use, and emissions for nonroad construction vehicles: a case study for excavators. J. Air Waste Manage. Assoc. 58 (8), 1033–1046.

Alberta Environment, 1995. Alberta stack sampling code 1995. Report Number ISBN# 0773214062. Prepared by Alberta Environment, Edmonton, AB, Canada. http://www.qp.alberta.ca/570.cfm. Accessed 21 August 12.

ASTM, 2008. WK752 Test method for determination of $PM_{2.5}$ mass and species emissions from stationary combustion sources by dilution sampling. Prepared by American Society for Testing Materials International, Conshohocken, PA.

Bachmann, J.D., 2007. Will the circle be unbroken: a history of the US national ambient air quality standards—2007 critical review. J. Air Waste Manage. Assoc. 57 (6), 652–697. http://pubs.awma.org/gsearch/journal/2007/6/10.3155-1047-3289.57.6.652.pdf. Accessed 21 August 12.

Baldwin, T.A., Watson, J.G., Neulicht, R., 2012. New test method for measuring fine particulate matter ($PM_{2.5}$) in wet stacks. In Proceedings, Air Quality Measurement Methods and Technology, Air & Waste Management Association: Piitsburgh, PA, 88-1-88-6.

Barber, P.W., Moosmüller, H., Keislar, R.E., Kuhns, H.D., Mazzoleni, C., Watson, J.G., 2004. On-road measurement of automotive particle emissions by ultraviolet lidar and transmissometer: theory. Meas. Sci. Technol. 15, 2295–2302.

Barrios, C.C., Dominguez-Saez, A., Rubio, J.R., Pujadas, M., 2011. Development and evaluation of on-board measurement system for nanoparticle emissions from diesel engine. Aerosol Sci. Technol. 45 (5), 570–580.

Beckerman, B., Jerrett, M., Brook, J.R., Verma, D.K., Arain, M.A., Finkelstein, M.M., 2008. Correlation of nitrogen dioxide with other traffic pollutants near a major expressway. Atmos. Environ. 42 (2), 275–290.

Beiser, V., 2012. Turning cellphones into mobile microscopes. Pacific Standard. http://www.psmag.com/health/turning-cellphones-into-mobile-microscopes-36611/. Accessed 21 August 12.

Bishop, G.A., Starkey, J.R., Ihlenfeldt, A., Williams, W.J., Stedman, D.H., 1989. IR long-path photometry: a remote sensing tool for automobile emissions. Anal. Chem. 61 (10), 671A–677A.

Bishop, G.A., Burgard, D.A., Dalton, T.R., Stedman, D.H., Ray, J.D., 2006. Winter motor-vehicle emissions in Yellowstone National Park. Environ. Sci. Technol. 40 (8), 2505–2510.

Bishop, G.A., Stadtmuller, R., Stedman, D.H., Ray, J.D., 2009. Portable emission measurements of Yellowstone Park snowcoaches and snowmobiles. J. Air Waste Manage. Assoc. 59 (8), 936–942.

Boughedaoui, M., Kerbachi, R., Joumard, R., 2008. On-board emission measurement of high-loaded light-duty vehicles in Algeria. J. Air Waste Manage. Assoc. 58, 45–54.

Brugge, D., Durant, J.L., Rioux, C., 2007. Near-highway pollutants in motor vehicle exhaust: a review of epidemiologic evidence of cardiac and pulmonary health risks. Environ. Health 6, 23. http://dx.doi.org/10.1186/1476-069X-6-23.   http://www.ehjournal.net/content/6/1/23. Accessed 21 August 12.

Brunet, J., Pauly, A., Varenne, C., Lauron, B., 2008. On-board phthalocyanine gas sensor microsystem dedicated to the monitoring of oxidizing gases level in passenger compartments. Sens. Actuators B Chem. 130 (2), 908–916.

Bryson, M., Reid, A., Ramos, F., Sukkarieh, S., 2010. Airborne vision-based mapping and classification of large farmland environments. J. Field Robot. 27 (5), 632–655.

Bukowiechi, N., Dommen, J., Prevot, A.S.H., Richter, R., Weingartner, E., Baltensperger, U., 2002. A mobile pollutant measurement laboratory-measuring gas phase and aerosol ambient concentrations with high spatial and temporal resolution. Atmos. Environ. 36 (36–37), 5569–5579.

Cadle, S.H., Ayala, A., Black, K.N., Graze, R.R., Koupal, J., Minassian, F., Murray, H.B., Natarajan, M., Tennant, C.J., Lawson, D.R., 2009. Real-world vehicle emissions: a summary of the 18th Coordinating Research Council on-road vehicle emissions workshop. J. Air Waste Manage. Assoc. 59 (2), 130–138.

Canagaratna, M.R., Jayne, J.T., Ghertner, D.A., Herndon, S., Shi, Q., Jimenez, J.L., Silva, P.J., Williams, P., Lanni, T., Drewnick, F., Demerjian, K.L., Kolb, C.E., Worsnop, D.R., 2004. Chase studies of particulate emissions from in-use New York City vehicles. Aerosol Sci. Technol. 38 (6), 555–573.

CARB, 2011. Appendix P: LEV III PM technical support document—development of particulate matter mass standards for future light duty vehicles. Prepared by California Air Resources Board, Sacramento, CA. http://www.arb.ca.gov/regact/2012/leviiighg2012/levappp.pdf. Accessed 21 August 12.

Chakradhar, B., 2004. Fugitive dust emissions from mining areas. J. Environ. Syst. 31 (3), 279–288.

Chang, M.-C.O., Chow, J.C., Watson, J.G., Hopke, P.K., Yi, S.M., England, G.C., 2004. Measurement of ultrafine particle size distributions from coal-, oil-, and gas-fired stationary combustion sources. J. Air Waste Manage. Assoc. 54 (12), 1494–1505. http://pubs.awma.org/gsearch/journal/2004/12/chang.PDF. Accessed 21 August 12.

Chow, J.C., Watson, J.G., 2011. Air quality management of multiple pollutants and multiple effects. Air Quality Climate Change 45 (3), 26–32.

Chow, J.C., Watson, J.G., 2012. Chemical analyses of particle filter deposits. In: Ruzer, L., Harley, N.H. (Eds.), Aerosols Handbook: Measurement, Dosimetry, and Health Effects 2, CRC Press/Taylor & Francis, NewYork, NY, pp. 177–202. http://www.taylorandfrancis.com/books/details/9781439855102/. Accessed 21 August 12.

Chow, J.C., Watson, J.G., Houck, J.E., Pritchett, L.C., Rogers, C.F., Frazier, C.A., Egami, R.T., Ball, B.M., 1994. A laboratory resuspension chamber to measure fugitive dust size distributions and chemical compositions. Atmos. Environ. 28 (21), 3463–3481.

Chow, J.C., Watson, J.G., Ashbaugh, L.L., Magliano, K.L., 2003. Similarities and differences in PM_{10} chemical source profiles for geological dust from the San Joaquin Valley, California. Atmos. Environ. 37 (9–10), 1317–1340. http://dx.doi.org/10.1016/S1352-2310(02) 01021-X. Accessed 21 August 12.

Chow, J.C., Watson, J.G., Feldman, H.J., Nolan, J., Wallerstein, B.R., Hidy, G.M., Lioy, P.J., McKee, H.C., Mobley, J.D., Bauges, K., Bachmann, J.D., 2007. 2007 Critical review discussion—will the circle be unbroken: a history of the U.S. National Ambient Air Quality Standards. J. Air Waste Manage. Assoc. 57 (10), 1151–1163. http://pubs.awma.org/gsearch/ journal/2007/10/10.3155-1047-3289.57.10.1151.pdf. Accessed 21 August 12.

Cocker, D.R., Shah, S.D., Johnson, K., Miller, J.W., Norbeck, J.M., 2004. Development and application of a mobile laboratory for measuring emissions from diesel engines. 1. Regulated gaseous emissions. Environ. Sci. Technol. 38 (7), 2182–2189.

Collins, J.F., Shepherd, P., Durbin, T.D., Lents, J., Norbeck, J., Barth, M., 2007. Measurements of in-use emissions from modern vehicles using an on-board measurement system. Environ. Sci. Technol. 41 (18), 6554–6561.

Contreras, J.A., Murray, J.A., Tolley, S.E., Oliphant, J.L., Tolley, H.D., Lammert, S.A., Lee, E.D., Later, D.W., Lee, M.L., 2008. Hand-portable gas chromatograph-toroidal ion trap mass spectrometer (GC-TMS) for detection of hazardous compounds. J. Am. Soc. Mass Spectrom. 19 (10), 1425–1434.

Cooper, D.A., Ekstrom, M., 2005. Applicability of the PEMS technique for simplified NOX monitoring on board ships. Atmos. Environ. 39, 127–137.

Cooper, H.B.H., Rossano, A.T., 1974. Source Testing for Air Pollution Control. McGraw Hill, Columbus, OH.

Cowherd, C., 2001. Fugitive dust emissions. In: Baron, P., Willeke, K. (Eds.), Aerosol Measurement: Principles, Techniques, and Applications. second ed. John Wiley & Sons, New York, NY, pp. 845–857.

DieselNet, 2012a. Emission standards: summary of worldwide diesel emission standards. Prepared by DieselNet. http://www.dieselnet.com/standards/#asia. Accessed 21 August 12.

DieselNet, 2012b. Emission test cycles: Summary of worldwide engine and vehicle test cycles. Prepared by Ecopoint Inc., Mississauga, ON, Canada. http://www.dieselnet.com/standards/ cycles/. Accessed 21 August 12.

Do, J.S., Chen, P., 2007. Amperometric sensor array for NOx, CO, O_2 and SO_2 detection. Sens. Actuators B Chem. 122 (1), 165–173.

Dreher, D.B., Harley, R.A., 1998. A fuel-based inventory for heavy-duty diesel truck emissions. J. Air Waste Manage. Assoc. 48 (4), 352–358.

Dunford, R., Michel, K., Gagnage, M., Piegay, H., Tremelo, M.L., 2009. Potential and constraints of Unmanned Aerial Vehicle technology for the characterization of Mediterranean riparian forest. Int. J. Remote Sens. 30 (19), 4915–4935.

England, G.C., Zielinska, B., Loos, K., Crane, I., Ritter, K., 2000. Characterizing $PM_{2.5}$ emission profiles for stationary sources: comparison of traditional and dilution sampling techniques. Fuel Process. Technol. 65, 177–188.

England, G.C., Watson, J.G., Chow, J.C., Zielinska, B., Chang, M.-C.O., Loos, K.R., Hidy, G.M., 2007a. Dilution-based emissions sampling from stationary sources: Part 1. Compact sampler, methodology and performance. J. Air Waste Manage. Assoc. 57 (1), 65–78. http://pubs. awma.org/gsearch/journal/2007/1/england.pdf. Accessed 21 August 12.

England, G.C., Watson, J.G., Chow, J.C., Zielinska, B., Chang, M.-C.O., Loos, K.R., Hidy, G.M., 2007b. Dilution-based emissions sampling from stationary sources: Part 2. Gas-fired combustors compared with other fuel-fired systems. J. Air Waste Manage. Assoc. 57 (1), 79–93. http://pubs.awma.org/gsearch/journal/2007/1/england2.pdf. Accessed 21 August 12.

Environment Canada, 2012. National Pollutant Release Inventory (NPRI) online data search. Prepared by Environment Canada, Ottawa, Ontario, Canada. http://ec.gc.ca/pdb/websol/emissions/ap/ap_query_e.cfm. Accessed 21 August 12.

Lehder Environmental Services, 2007. Compliance survey 26-1 FGD Stack. Report Number 071260.9. Prepared by Lehder Environmental Services, Inc., Edmonton, AB, Canada.

Etyemezian, V., Kuhns, H.D., Gillies, J.A., Chow, J.C., Hendrickson, K., McGown, M., Pitchford, M.L., 2003a. Vehicle-based road dust emissions measurement (III): effect of speed, traffic volume, location, and season on PM_{10} road dust emissions in the Treasure Valley, ID. Atmos. Environ. 37 (32), 4583–4593.

Etyemezian, V., Kuhns, H.D., Gillies, J.A., Green, M.C., Pitchford, M.L., Watson, J.G., 2003b. Vehicle-based road dust emission measurement. I. Methods and calibration. Atmos. Environ. 37 (32), 4559–4571.

Etyemezian, V., Kuhns, H.D., Nikolich, G., 2006. Precision and repeatability of the TRAKER vehicle-based paved road dust emission measurement. Atmos. Environ. 40 (16), 2953–2958.

Etyemezian, V., Nikolich, G., Ahonen, S., Pitchford, M.L., Sweeney, M., Purcell, R., Gillies, J.A., Kuhns, H.D., 2007. The Portable In Situ Wind Erosion Laboratory (PI-SWERL): a new method to measure PM_{10} potential for windblown dust properties and emissions. Atmos. Environ. 41 (18), 3789–3796.

Fraser, M.P., Cass, G.R., Simoneit, B.R.T., 1998. Gas-phase and particle-phase organic compounds emitted from motor vehicle traffic in a Los Angeles roadway tunnel. Environ. Sci. Technol. 32 (14), 2051–2060.

Gao, L., Sugiarto, A., Harper, J.D., Cooks, R.G., Zheng, O.Y., 2008. Design and characterization of a multisource hand-held tandem mass spectrometer. Anal. Chem. 80 (19), 7198–7205.

Gillette, D.A., 1977. Fine particulate emissions due to wind erosion. Trans. ASAE (St. Joseph, MI) 20, 890–897.

Gillette, D.A., 1978a. A wind tunnel simulation of the erosion of soil: effect of soil texture, sandblasting, wind speed, and soil consolidation on dust production. Atmos. Environ. 12, 1735–1743.

Gillette, D.A., 1978b. Tests with a portable wind tunnel for determining wind erosion threshold velocities. Atmos. Environ. 12, 2309–2313.

Gillette, D.A., Adams, J.B., Endo, A., Smith, D., Kihl, R., 1990. Threshold velocities for input of soil particles into the air by desert soils. J. Geophys. Res. 85, 5621–5630.

Gillies, J.A., Watson, J.G., Rogers, C.F., DuBois, D.W., Chow, J.C., Langston, R., Sweet, J., 1999. Long term efficiencies of dust suppressants to reduce PM_{10} emissions from unpaved roads. J. Air Waste Manage. Assoc. 49 (1), 3–16. http://pubs.awma.org/gsearch/journal/1999/1/gillies.pdf. Accessed 21 August 12.

Goossens, D., Buck, B., 2009. Dust dynamics in off-road vehicle trails: measurements on 16 arid soil types, Nevada, USA. J. Environ. Manage. 90 (11), 3458–3469.

Hansen, A.D.A., Mocnik, G., 2010. The "Micro" Aethalometer(R)—an enabling technology for new applications in the measurement of aerosol black carbon. In: Chow, J.C., Watson, J.G., Cao, J.J. (Eds.), Proceedings, Leapfrogging Opportunities for Air Quality Improvement. Air & Waste Management Association, Pittsburgh, PA, pp. 984–989. http://www.dri.edu/conference-proceedings. Accessed 21 August 12.

HEI, 2003. Assessing health impact of air quality regulations: concepts and methods for accountability research. Report Number HEI Communication 11. Prepared by Health Effects Institute, Boston, MA. http://pubs.healtheffects.org/getfile.php?u=262. Accessed 21 August 12.

Hildemann, L.M., Cass, G.R., Markowski, G.R., 1989. A dilution stack sampler for collection of organic aerosol emissions: design, characterization and field tests. Aerosol Sci. Technol. 10 (10–11), 193–204.

ISO, 2007. ISO 8178: Reciprocating internal combustion engines—exhaust emission measurement—Part 4: steady-state test cycles for different engine applications. Prepared by International Organization for Standardization, Geneva, Switzerland. http://www.iso.org/iso/iso_catalogue/catalogue_tc/catalogue_detail.htm?csnumber=42275. Accessed 21 August 12.

Jahnke, J.A., 2000. Continuous Emission Monitoring, second ed. John Wiley & Sons, Inc., New York, NY.

Johnson, K.C., Durbin, T.D., Cocker, D.R., Miller, W.J., Bishnu, D.K., Maldonado, H., Moynahan, N., Ensfield, C., Laroo, C.A., 2009a. On-road comparison of a portable emission measurement system with a mobile reference laboratory for a heavy-duty diesel vehicle. Atmos. Environ. 43 (18), 2877–2883.

Johnson, K.C., Durbin, T.D., Jung, H., Chaudhary, A., Cocker, D.R., Herner, J.D., Robertson, W.H., Huai, T., Ayala, A., Kittelson, D., 2009b. Evaluation of the European PMP methodologies during on-road and chassis dynamometer testing for DPF equipped heavy-duty diesel vehicles. Aerosol Sci. Technol. 43 (10), 962–969.

Johnson, K.C., Durbin, T.D., Jung, H.J., Cocker, D.R., Bishnu, D., Giannelli, R., 2011. Quantifying in-use PM measurements for heavy duty diesel vehicles. Environ. Sci. Technol. 45 (14), 6073–6079.

Kavouras, I.G., Etyemezian, V., Nikolich, G., Gillies, J., Sweeney, M., Young, M., Shafer, D., 2009. A new technique for characterizing the efficacy of fugitive dust suppressants. J. Air Waste Manage. Assoc. 59 (5), 603–612.

Kean, A.J., Sawyer, R.F., Harley, R.A., 2000. A fuel-based assessment of off-road diesel engine emissions. J. Air Waste Manage. Assoc. 50 (11), 1929–1939.

King, J., Etyemezian, V., Sweeney, M., Buck, B.J., Nikolich, G., 2011. Dust emission variability at the Salton Sea, California, USA. Aeolian Res. 3 (1), 67–79.

Kittelson, D.B., Watts, W.F., Johnson, J.P., Remerowki, M.L., Ische, E.E., Oberdörster, G., Gelein, R.A., Elder, A., Hopke, P.K., Kim, E., Zhao, W., Zhou, L., Jeong, C.H., 2004. On-road exposure to highway aerosols. 1. Aerosol and gas measurements. Inhal. Toxicol. 16 (Suppl. 1), 31–39.

Kuhns, H.D., Etyemezian, V., Landwehr, D., Macdougall, C.S., Pitchford, M.L., Green, M.C., 2001. Testing Re-entrained Aerosol Kinetic Emissions from Roads (TRAKER): a new approach to infer silt loading on roadways. Atmos. Environ. 35 (16), 2815–2825.

Kuhns, H.D., Mazzoleni, C., Moosmüller, H., Nikolic, D., Keislar, R.E., Barber, P.W., Li, Z., Etyemezian, V., Watson, J.G., 2004. Remote sensing of PM, NO, CO, and HC emission factors for on-road gasoline and diesel engine vehicles in Las Vegas, NV. Sci. Total Environ. 322, 123–137.

Kuhns, H.D., Gillies, J.A., Etyemezian, V., DuBois, D.W., Ahonen, S., Nikolic, D., Durham, C., 2005. Spatial variability of unpaved road dust PM_{10} emission factors near El Paso, Texas. J. Air Waste Manage. Assoc. 55 (1), 3–12.

Kuhns, H.D., Gillies, J.A., Etyemezian, V., Nikolich, G., King, J., Zhu, D.Z., Uppapalli, S., Engelbrecht, J.P., Kohl, S.D., 2010. Effect of soil type and momentum on unpaved road particulate matter emissions from wheeled and tracked vehicles. Aerosol Sci. Technol. 44 (3), 187–196.

Lipsky, E.M., Robinson, A.L., 2005. Design and evaluation of a portable dilution sampling system for measuring fine particle emissions from combustion systems. Aerosol Sci. Technol. 39 (6), 542–553.

Lipsky, E.M., Robinson, A.L., 2006. Effects of dilution on fine particle mass and partitioning of semivolatile organics in diesel exhaust and wood smoke. Environ. Sci. Technol. 40 (1), 155–162.

Lipsky, E.M., Stanier, C.O., Pandis, S.N., Robinson, A.L., 2002. Effects of sampling conditions on the size distribution of fine particulate matter emitted from a pilot-scale pulverized-coal combustor. Energy Fuel 16 (2), 302–310.

Lipsky, E.M., Pekney, N.J., Walbert, G.F., O'Dowd, W.J., Freeman, M.C., Robinson, A., 2004. Effects of dilution sampling on fine particle emissions from pulverized coal combustion. Aerosol Sci. Technol. 38 (6), 574–587.

Lloyd, A.C., Cackette, T.A., 2001. Critical review—diesel engines: environmental impact and control. J. Air Waste Manage. Assoc. 51 (6), 809–847.

Machiele, P.A., 1989. Heavy duty vehicle emission conversion factors II (1962–2000). Report Number EPA-AA-SDSB-89-01. Prepared by U.S. Environmental Protection Agency, Research Triangle Park, NC.

Mauderly, J.L., Chow, J.C., 2008. Health effects of organic aerosols. Inhal. Toxicol. 20 (3), 257–288. http://dx.doi.org/10.1080/08958370701866008. Accessed 21 August 12.

Mazzoleni, C., Kuhns, H.D., Moosmüller, H., Keislar, R.E., Barber, P.W., Robinson, N.F., Watson, J.G., 2004a. On-road vehicle particulate matter and gaseous emission distributions in Las Vegas, Nevada, compared with other areas. J. Air Waste Manage. Assoc. 54 (6), 711–726. http://pubs.awma.org/gsearch/journal/2004/6/mazzoleni.PDF. Accessed 21 August 12.

Mazzoleni, C., Moosmüller, H., Kuhns, H.D., Keislar, R.E., Barber, P.W., Nikolic, D., Nussbaum, N.J., Watson, J.G., 2004b. Correlation between automotive CO, HC, NO, and PM emission factors from on-road remote sensing: implications for inspection and maintenance programs. Transport. Res. D9, 477–496.

Mazzoleni, C., Kuhns, H.D., Moosmüller, H., 2010. Monitoring automotive particulate matter emissions with LiDAR: a review. Remote Sens. 2 (4), 1077–1119. http://www.mdpi.com/2072-4292/2/4/1077.

McDonald, J.D., Reed, M.D., Campen, M.J., Barrett, E.G., Seagrave, J.C., Mauderly, J.L., 2007. Health effects of inhaled gasoline engine emissions. Inhal. Toxicol. 19 (Suppl. 1), 107–116.

McKenna-Neuman, C., Nickling, W.G., 1989. A theoretical and wind tunnel investigation of the effects of capillary water on the entrainment of sediment by wind. Can. J. Soil Sci. 69 (1), 79.

Merritt, W.S., Letcher, R.A., Jakeman, A.J., 2003. A review of erosion and sediment transport models. Environ. Model. Software 18 (8–9), 761–799.

Moosmüller, H., Gillies, J.A., Rogers, C.F., DuBois, D.W., Chow, J.C., Watson, J.G., Langston, R., 1998. Particulate emission rates for unpaved shoulders along a paved road. J. Air Waste Manage. Assoc. 48 (5), 398–407. http://pubs.awma.org/gsearch/journal/1998/5/moosmull.pdf. Accessed 21 August 12.

Moosmüller, H., Mazzoleni, C., Barber, P.W., Kuhns, H.D., Keislar, R.E., Watson, J.G., 2003. On-road measurement of automotive particle emissions by ultraviolet lidar and transmissometer: instrument. Environ. Sci. Technol. 37 (21), 4971–4978.

Morawska, L., Ristovski, Z.D., Johnson, G.R., Jayaratne, E.R., Mengersen, K., 2007. Novel method for on-road emission factor measurements using a plume capture trailer. Environ. Sci. Technol. 41 (2), 574–579.

Nicholson, K.W., Branson, J.R., 1990. Factors affecting resuspension by road traffic. Sci. Total Environ. 93, 349–358.

Nicholson, K.W., Branson, J.R., Giess, P., Cannell, R.J., 1989. The effects of vehicle activity on particle resuspension. J. Aerosol Sci. 20 (8), 1425–1428.

Nussbaum, N.J., Zhu, D., Kuhns, H.D., Mazzoleni, C., Chang, M.-C.O., Moosmüller, H., Watson, J.G., 2009. The in-plume emissions test-stand: a novel instrument platform for the real-time characterization of combustion emissions. J. Air Waste Manage. Assoc. 59 (12), 1437–1445. http://pubs. awma.org/gsearch/journal/2009/12/10.3155-1047-3289.59.12.1437.pdf. Accessed 21 August 12.

Ouyang, Z., Noll, R.J., Cooks, R.G., 2009. Handheld miniature ion trap mass spectrometers. Anal. Chem. 81 (7), 2421–2425.

Pirjola, L., Parviainen, H., Hussein, T., Valli, A., Hameri, K., Aalto, P., Virtanen, A., Keskinen, J., Pakkanen, T.A., Makela, T., Hillamo, R.E., 2004. "Sniffer"—a novel tool for chasing vehicles and measuring traffic pollutants. Atmos. Environ. 38 (22), 3625–3635.

Pollanen, R., Toivonen, H., Perajarvi, K., Karhunen, T., Smolander, P., Ilander, T., Rintala, K., Katajainen, T., Niemela, J., Juusela, M., Palos, T., 2009. Performance of an air sampler and a gamma-ray detector in a small unmanned aerial vehicle. J. Radioanal. Nucl. Chem. 282 (2), 433–437.

Prabhakar, A., Iglesias, R.A., Wang, R., Tsow, F., Forzani, E.S., Tao, N.J., 2010. Ultrasensitive detection of nitrogen oxides over a nanoporous membrane. Anal. Chem. 82 (23), 9938–9940.

Ramanathan, N., Lukac, M., Ahmed, T., Kar, A., Praveen, P.S., Honles, T., Leong, I., Rehman, I.H., Schauer, J.J., Ramanathan, V., 2011. A cellphone based system for large-scale monitoring of black carbon. Atmos. Environ. 45 (26), 4481–4487.

Rango, A., Laliberte, A., Herrick, J.E., Winters, C., Havstad, K., Steele, C., Browning, D., 2009. Unmanned aerial vehicle-based remote sensing for rangeland assessment, monitoring, and management. J. Appl. Remote Sens. 3 (1), 033542.

Sawyer, R.F., Harley, R.A., Cadle, S.H., Norbeck, J.M., Slott, R.S., Bravo, H.A., 2000. Mobile sources critical review: 1998 NARSTO assessment. Atmos. Environ. 34 (12–14), 2161–2181.

Sitnikov, N.M., Sokolov, A.O., Ravengnani, F., Yushkov, V.A., Ulanovsky, A.E., 2005. A chemiluminescent balloon-type nitrogen dioxide meter for tropospheric and stratospheric investigations (NaDA). Instrum. Exp. Tech. 48 (3), 400–405.

Sluis, W.W., Allaart, M.A.F., Piters, A.J.M., Gast, L.F.L., 2010. The development of a nitrogen dioxide sonde. Atmos. Meas. Tech. 3 (6), 1753–1762.

Spiess, T., Bange, J., Buschmann, M., Vorsmann, P., 2007. First application of the meteorological Mini-UAV 'M(2)AV'. Meteorol. Z. 16 (2), 159–169.

St. Denis, M., Lindner, J., 2005. Review of light-duty diesel and heavy-duty diesel gasoline inspection programs. J. Air Waste Manage. Assoc. 55 (12), 1876–1884.

Supnithadnaporn, A., Noonan, D.S., Samoylov, A., Rodgers, M.O., 2011. Estimated validity and reliability of on-board diagnostics for older vehicles: comparison with remote sensing observations. J. Air Waste Manage. Assoc. 61 (10), 996–1004.

Sweeney, M., Etyemezian, V., Macpherson, T., Nickling, W.G., Gillies, J.A., Nikolich, G., McDonald, E., 2008. Comparison of PI-SWERL with dust emission measurements from a straight-line field wind tunnel. J. Geophys. Res. Atmos. 113 (F1).

Sweeney, M.R., McDonald, E.V., Etyemezian, V., 2011. Quantifying dust emissions from desert landforms, eastern Mojave Desert, USA. Geomorphology 135 (1–2), 21–34.

Techy, L., Schmale, D.G., Woolsey, C.A., 2010. Coordinated aerobiological sampling of a plant pathogen in the lower atmosphere using two autonomous unmanned aerial vehicles. J. Field Robot. 27 (3), 335–343.

U.S. Congress, 1990. Clean Air Act. Prepared by U.S. Congress, Washington, DC. http://www. epa.gov/air/caa/. Accessed 21 August 12.

U.S. EPA, 1995. Compilation of air pollutant emission factors (fifth ed.). Report Number AP-42. Prepared by U.S. Environmental Protection Agency.

U.S. EPA, 1996. Method 202. Condensable particulate matter—determination of condensable particulate emissions from stationary sources. Prepared by U.S. Environmental Protection Agency,

Office of Air Quality Planning and Standards, Technical Support Division, Research Triangle Park, NC. http://www.epa.gov/ttn/emc/promgate/m-202.pdf. Accessed 21 August 12.

U.S. EPA, 1997. Method 201A—Determination of PM_{10} emissions (constant sampling rate procedure). Prepared by U.S. Environmental Protection Agency, Office of Air Quality Planning and Standards, Technical Support Division, Research Triangle Park, NC. http://www.epa.gov/ttn/emc/promgate/m-201a.pdf. Accessed 21 August 12.

U.S. EPA, 2000a. Method 5. Particulate matter (PM). Determination of particulate matter emissions from stationary sources. Prepared by U.S. EPA, Research Triangle Park, NC. http://www.epa.gov/ttn/emc/promgate.html. Accessed 21 August 12.

U.S. EPA, 2000b. Method 5G—PM wood heaters from a dilution tunnel. Prepared by U.S. EPA, Research Triangle Park, NC. http://www.epa.gov/ttn/emc/promgate/m-05g.pdf. Accessed 21 August 12.

U.S. EPA, 2000c. Method 5H—PM wood heaters from a stack. Prepared by U.S. EPA, Research Triangle Park, NC. http://www.epa.gov/ttn/emc/promgate/m-05h.pdf. Accessed 21 August 12.

U.S. EPA, 2000d. Method 6. Determination of sulfur dioxide emissions from stationary sources. prepared by U.S. EPA, Research Triangle Park, NC. http://www.epa.gov/ttn/emc/promgate/m-06.pdf. Accessed 21 August 12.

U.S. EPA, 2000e. Method 7. Determination of nitrogen oxide emissions from stationary sources. Prepared by U.S. EPA, Research Triangle Park, NC. http://www.epa.gov/ttn/emc/promgate/m-06.pdf. Accessed 21 August 12.

U.S. EPA, 2004. Control of emissions of air pollution from nonroad diesel engines and fuel: final rule. Federal Register 69 (124), 38958–39273. http://epa.gov/nonroad-diesel/2004fr.htm. Accessed 21 August 12.

U.S. EPA, 2012a. 2008 National Emissions Inventory Data. Prepared by U.S. Environmental Protection Agency. Research Triangle Park, NC. http://www.epa.gov/ttn/chief/net/2008inventory.html. Accessed 21 August 12.

U.S. EPA, 2012b. National Emissions Inventory (NEI) air pollutant emissions trends data. Prepared by U.S. Environmental Protection Agency, Research Triangle Park, NC. http://www.epa.gov/ttn/chief/trends/index.html. Accessed 21 August 12.

U.S. EPA, 2012c. Report to congress on black carbon. Report Number EPA-450/D-12-001. Prepared by U.S. Environmental Protection Agency, Washington DC. http://www.epa.gov/blackcarbon/2012report/fullreport.pdf. Accessed 21 August 12.

Vallejos, S., Khatko, V., Aguir, K., Ngo, K.A., Calderer, J., Gracia, I., Cane, C., Llobet, E., Correig, X., 2007. Ozone monitoring by micro-machined sensors with WO3 sensing films. Sens. Actuators B Chem. 126 (2), 573–578.

Van Houtte, J., Niemeier, D., 2008. A critical review of the effectiveness of I/M programs for monitoring PM emissions from heavy duty vehicles. Environ. Sci. Technol. 42 (21), 7856–7865.

Wang, X.L., Caldow, R., Sem, G.J., Hama, N., Sakurai, H., 2010. Evaluation of a condensation particle counter for vehicle emission measurement: experimental procedure and effects of calibration aerosol material. J. Aerosol Sci. 41 (3), 306–318.

Wang, X.L., Watson, J.G., Chow, J.C., Gronstal, S., Kohl, S.D., 2012a. An efficient multipollutant system for measuring real-world emissions from stationary and mobile sources. Aerosol Air Quality Res. 12 (1), 145–160. http://aaqr.org/VOL12_No2_April2012/1_AAQR-11-11-OA-0187_145-160.pdf. Accessed 21 August 12.

Wang, X.L., Watson, J.G., Chow, J.C., Kohl, S.D., Chen, L.-W.A., Sodeman, D.A., Legge, A.H., Percy, K.E., 2012b. Measurement of real-world stack emissions with a dilution sampling system. In: Percy, K.E. (Ed.), Alberta Oil Sands: Energy, Industry, and the Environment. Elsevier Press, Amsterdam, The Netherlands, pp. 167–188.

Watson, J.G., Chow, J.C., 2012. Source apportionment. In: El-Shaarwi, A.H., Piegorsch, W.W. (Eds.), The Encyclopedia of Environmetrics. John Wiley & Sons, Hoboken, NJ.

Watson, J.G., Chow, J.C., Pace, T.G., 2000. Fugitive dust emissions. In: Davis, W.T. (Ed.), Air Pollution Engineering Manual. second ed. John Wiley & Sons, Inc., New York, pp. 117–135.

Watson, J.G., Chow, J.C., Fujita, E.M., 2001. Review of volatile organic compound source apportionment by Chemical mass balance. Atmos. Environ. 35 (9), 1567–1584. ftp://ftp.cgenv.com/pub/downloads/Watson.pdf. Accessed 21 August 12.

Watson, J.G., Chen, L.-W.A., Chow, J.C., Lowenthal, D.H., Doraiswamy, P., 2008. Source apportionment: findings from the US Supersite Program. J. Air Waste Manage. Assoc. 58 (2), 265–288. http://pubs.awma.org/gsearch/journal/2008/2/10.3155-1047-3289.58.2.265.pdf. Accessed 21 August 12.

Watson, J.G., Chow, J.C., Wang, X., Kohl, S.D., Gronstahl, S., 2010a. Measurement of in-use emissions from non-road diesel trucks. Report Number 010109-123109. Prepared by Desert Research Institute, Reno, NV, for Wood Buffalo Environmental Association, Ft. McMurray, AB, Canada.

Watson, J.G., Chow, J.C., Wang, X., Kohl, S.D., Sodeman, D.A., 2010b. Measurement of real-world stack emissions with a dilution sampling system. Report Number 010109-123109. Prepared by Desert Research Institute, Reno, NV, for Wood Buffalo Environmental Association, Ft. McMurray, AB, Canada.

Watson, J.G., Chow, J.C., Chen, L., Wang, X.L., Merrifield, T.M., Fine, P.M., Barker, K., 2011a. Measurement system evaluation for upwind/downwind sampling of fugitive dust emissions. AAQR (Aerosol and Air Quality Research) 11 (4), 331–350. http://dx.doi.org/10.4209/aaqr.2011.03.0028. http://aaqr.org/VOL11_No4_August2011/1_AAQR-11-03-OA-0028_331-350.pdf. Accessed 21 August 12.

Watson, J.G., Chow, J.C., Wang, X.L., Kohl, S.D., Gronstal, S., 2011b. Winter stack emissions measured with a dilution sampling system. Prepared by Desert Research Institute, Reno, NV, for Wood Buffalo Environmental Association, Ft. McMurray, AB, Canada.

Watson, J.G., Chow, J.C., Wang, X.L., Lowenthal, D.H., Kohl, S.D., Gronstal, S., 2011c. Real-world emissions from non-road mining trucks. Report Number 010109-123109. Prepared by Desert Research Institute, Reno, NV, for Ft. McMurray, AB, Canada, Wood Buffalo Environmental Association.

Watson, J.G., Chow, J.C., Chen, L.-W.A., Kohl, S.D., Casuccio, G., 2012. Elemental analysis of filter tape deposits from beta attenuation monitors. Atmos. Res. 106, 181–189.

Zannetti, P., 2003. Air Quality Modeling: Theories, Methodologies, Computational Techniques, and Available Databases and Software, vol. 1, Fundamentals. Air & Waste Management Association, Pittsburgh, PA.

Zannetti, P., 2005. Air Quality Modeling: Theories, Methodologies, Computational Techniques, and Available Databases and Software, vol. 2, Advanced Topics. Air & Waste Management Association, Pittsburgh, PA.

Zhang, K., Frey, C., 2008. Evaluation of response time of a portable system for in-use vehicle tailpipe emissions measurement. Environ. Sci. Technol. 42 (1), 221–227.

Zhang, Y., Bishop, G.A., Stedman, D.H., 1994. Automobile emissions are statistically g-distributed. Environ. Sci. Technol. 28 (7), 1370–1374.

Zhu, D.Z., Kuhns, H.D., Brown, S., Gillies, J.A., Etyemezian, V., Gertler, A.W., 2009. Fugitive dust emissions from paved road travel in the Lake Tahoe Basin. J. Air Waste Manage. Assoc. 59 (10), 1219–1229.

Chapter 8

Measurement of Real-World Stack Emissions with a Dilution Sampling System

X.L. Wang[*][1], J.G. Watson[*], J.C. Chow[*], S.D. Kohl[*], L.-W.A. Chen[*], D.A. Sodeman[†], A.H. Legge[‡] and K.E. Percy[§]

[*]*Division of Atmospheric Sciences, Desert Research Institute, Reno, Nevada, USA*

[†]*County of San Diego Air Pollution Control District, San Diego, California, USA*

[‡]*Biosphere Solutions, Calgary, Alberta, Canada*

[§]*Wood Buffalo Environmental Association, Fort McMurray, Alberta, Canada*

[1]*Corresponding author: e-mail: xwang@dri.edu*

ABSTRACT

Stationary sources with emissions ducted through tall stacks are among the largest point sources for greenhouse gases and air pollutants in the Athabasca Oil Sands Region (AOSR) of northern Alberta, Canada. A dilution sampling system was used to quantify multipollutant gaseous and particulate emissions from three stacks (A, B, and C) in the AOSR under real-world operations. The major particle component was found to be ammonium sulfate for Stacks A and B, and sulfuric acid for Stack C. Mass distributions of particles from Stacks A and B were bimodal, with $\sim$50% mass in PM_1, $\sim$30% mass in $PM_{1-2.5}$, and $\sim$20% mass in $PM_{2.5-10}$, while particles from Stack C were in a single submicron mode, with 96% mass in PM_1 and $\sim$99% mass in $PM_{2.5}$. Emission rates of major air pollutants were lower than the emission guidelines for each stack during the test period.

8.1 INTRODUCTION

The Athabasca Oil Sands Region (AOSR) in northern Alberta, Canada contains about 170 billion barrels of recoverable bitumen. Crude oil derived from this bitumen is projected to increase from 1.5 to $\sim$3.5 million barrels/day from 2010 to 2025 (OAP, 2010). This large-scale oil production and

---

Disclaimer: The content and opinions expressed by the authors in this chapter do not necessarily reflect the views of WBEA or of the WBEA membership.

Developments in Environmental Science, Vol. 11. http://dx.doi.org/10.1016/B978-0-08-099443-7.00008-9

processing has raised environmental concerns, including potential adverse effects in air and water quality, ecosystems, human health, and climate change (Charpentier et al., 2009; Dowdeswell et al., 2010; Schindler, 2010; Tenenbaum, 2009). Tall stacks are used in upgrading facilities to vent process and combustion gases, most of which have passed through pollution control devices. The tall stacks allow primary emittant concentrations to be diluted when they reach nearby ground locations, but they also facilitate transport of pollutants over long distances. An emission inventory (TMAC WG, 2003) estimated large stationary source emissions contributing 97.6% of sulfur dioxide (SO_2), 30.2% of oxides of nitrogen (NO_x), 43.9% of carbon monoxide (CO), nearly 100% of ammonia (NH_3), nearly 100% of sulfuric acid (H_2SO_4), and 80% of $PM_{2.5}$ (mass of particles with aerodynamic diameters $<2.5\ \mu m$) from anthropogenic sources in the northern AOSR (Fort Chipewyan to Fort McMurray). Described here is the application of a dilution sampling system to measure real-world multipollutant emission rates (ERs) and source profiles from three stacks in two AOSR facilities (hereafter referred to as Facilities A and B) during the summer of 2008 (Watson et al., 2010b).

Major stationary sources in Alberta are tested for total suspended particulate (TSP) matter ER compliance using hot filter-impinger methods following Method 5 of the Alberta Stack Sampling Code (Alberta Environment, 1995), which was adapted from U.S. EPA Method 5 (U.S. EPA, 2000a). This method employs a heated ($120\pm14\ ^\circ C$) glass-fiber filter external to the stack to collect particles with the intent to minimize interferences from condensation of water vapor. The air passing through the filter is then routed through a set of solutions contained in glass impingers immersed in an ice bath to capture condensed particulate matter (PM) that penetrates the hot filter. Some applications of U.S. EPA Method 5 allow heating the filter up to 146–$174\ ^\circ C$ to minimize the collection of H_2SO_4 (Myers and Logan, 2002). The PM_{10} (mass of particles with aerodynamic diameters $<10\ \mu m$) compliance method (U.S. EPA, 2010) includes an in-stack size-selective inlet and by filter at the stack temperature, and by impingers. U.S. EPA (2000b) Method 17 also uses an in-stack filter to capture solid and liquid particles present at the stack temperature.

Hot filter-impinger configurations (Cooper and Rossano, 1974) were adequate in the early 1960s when industrial effluents were uncontrolled and contained large quantities primary particle emissions. The impingers were intended to determine condensable PM, which was expected to be small compared to the large particle masses captured on the front filter. Most gases were measured using wet chemistry (i.e. pulling the gas through absorbing solutions with subsequent laboratory analyses). These five-decade-old test methods are not sufficient to test emissions from today's industrial processes that are equipped with efficient control devices (Watson et al., 2012). High temperatures of the filter preclude collection of condensable organic and inorganic compounds in hot-stack emissions, resulting in a PM underestimation. On the other hand, the downstream impinger catch overestimates actual PM emissions because soluble gases (e.g. SO_2, NO_x, volatile organic compounds [VOCs]) as well as condensable particles are captured in the solutions. Gas phase reactions among NH_3, SO_2, and/or hydrogen chloride (HCl) in the impingers generate particulate ammonium sulfate [$(NH_4)_2SO_4$] and/or ammonium chloride (NH_4Cl) which should not be included as part of the primary $PM_{2.5}$ emissions (Richards et al., 2005).

Many Canadian provinces (including Alberta) and U.S. states do not report the impinger "condensables" (Corio and Sherwell, 2000). Purging of the absorbed SO_2 from the impinger solution for 1 h after sampling may reduce positive bias, but the effectiveness of purging is pH-dependent and varies among stacks (Corio and Sherwell, 2000). These compliance tests are also labor intensive, since separate sampling trains are needed for SO_2, NO_x, and PM, and the sampling and analytical methods are not consistent with more modern ambient monitoring approaches. Glass impingers, solutions that can be easily contaminated, and ice buckets would not be tolerated in a modern ambient monitoring station, yet these practices still persist for stationary source testing. Most important for determining multipollutant effects (Chow and Watson, 2011), these compliance testing do not represent real-world emission factors and chemical compositions needed for speciated emission inventories used in source models and receptor-oriented source apportionment.

Industrial stationary source PM emissions result from incomplete combustion and contaminants in the feedstocks (Lighty et al., 2000). As emissions exit the stack, the hot exhaust rapidly mixes with ambient air and cools, resulting in aggregations of gas molecules (nucleation) and condensation onto these nuclei or preexisting particles. Particle nucleation and growth in a diluted plume depends on temperature, relative humidity (RH), mixing and cooling rate, aging time, and partitioning between gas and particle phases (Chang et al., 2004). Dilution sampling methods are used for compliance testing of engine exhaust (Code of Federal Regulations, 2001a,b, 2002) and have been applied to stationary sources developed for real-world source characterization (ASTM, 2010; Chang et al., 2005; England et al., 1998, 2000, 2007; Heinsohn et al., 1980; Hildemann et al., 1989; Lee et al., 2004; Li et al., 2011; Lipsky and Robinson, 2005; McDonald et al., 2000; Myers and Logan, 2002; U.S.EPA, 2004; Wang et al., 2011, 2012; Watson and Chow, 2001; Watson et al., 2002). Hot stack emissions are diluted with clean air in a mixing and residence chamber that simulates the dilution, cooling, and aging when the flue gas exits the stack. Sufficient residence time is provided for gas-particle phase equilibrium while minimizing sampling losses and contamination (Hildemann et al., 1989).

TABLE 8.1 Physical and Operating Parameters of the Three Stacks

	Facility A		Facility B
Stack physical characteristics	Stack A	Stack B	Stack C
Stack inside diameter (m)	7.94	6.10	7.01
Stack height (m)	183	94	91
Elevation of sampling platform (m)	91	70	61
Stack temperature at the sampling port (°C)	258.3±0.6	79.6±0.2	54.5±0.1
Flue gas velocity (m/s)	24.7±0.3	12.7±0.1	9.6±0.2
Stack standard flow rate (m³/s)	634.0±6.8	290.7±2.7	311.0±5.6

8.2 MATERIAL AND METHODS

8.2.1 Source Description and Sampling Conditions

Stack A at Facility A exhausts flue gas from two CO boilers that burn overhead gas from two fluid cokers, sour water treating units, and sometimes the sulfur recovery units; effluents pass through two electrostatic precipitators (ESPs) to remove particles. Stack B at Facility A exhausts flue gas from a CO boiler that burns the overhead gas from a third fluid coker and sulfur recovery units; it is equipped with an ESP and an SO_2 scrubber (i.e. a diluted slurry of $(NH_4)_2SO_4$ containing an excess of NH_3 to produce $(NH_4)_2SO_4$ fertilizer) as part of a flue gas desulfurization (FGD) system. Flue gas flow rate and SO_2 concentration are monitored by Continuous Emissions Monitoring Systems (CEMS) in Stacks A and B by Facility A and the hourly ER is reported to Alberta Environment. Stack C at Facility B vents emissions from a coke-burning powerhouse and is equipped with a FGD system. CEMS data are not available for comparison. Key parameters for the three stacks are listed in Table 8.1.

8.2.2 Dilution Sampling System

The dilution sampling system is illustrated in Figure 8.1. A sample of approximately five standard liter per minute (L/min) flue gas was drawn isokinetically from the stack with a buttonhook nozzle (inserted 1.5 m inside of the stack), transferred through a 2.4 m-long heated probe (i.e. 5 °C above stack temperature), and mixed with clean dilution air generated by pumping ambient air (125 L/min) through an activated charcoal capsule filter followed by a high efficiency particulate air filter to remove volatile gas species and particles, respectively. To enhance turbulent mixing, the flue gas was pulled through a tube surrounded by a multihole diffuser plate supplying the clean

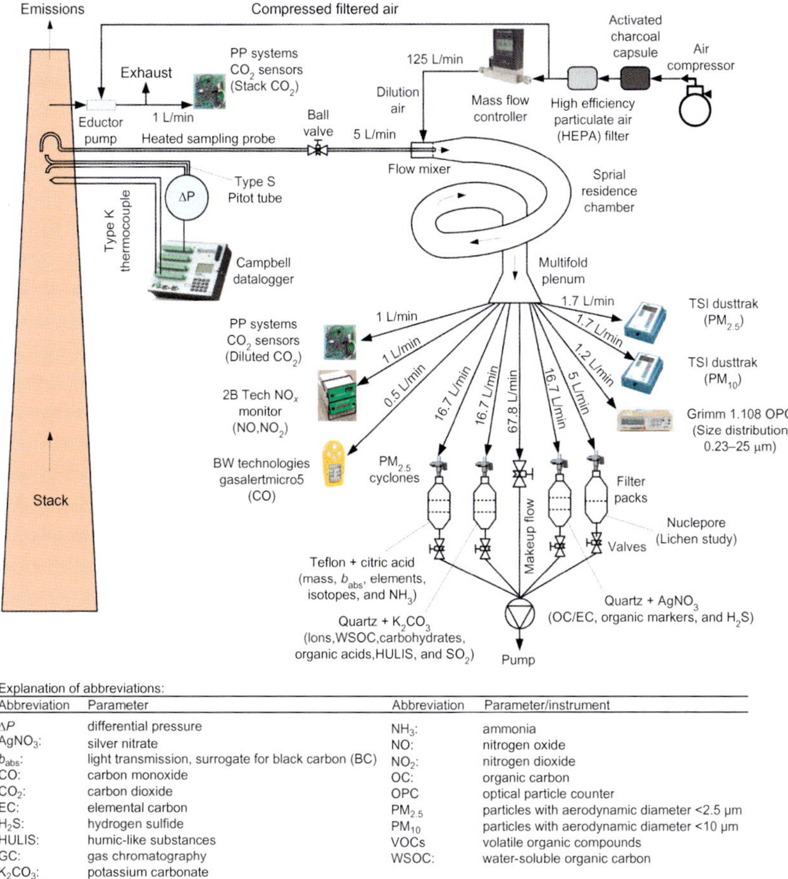

Explanation of abbreviations:

Abbreviation	Parameter	Abbreviation	Parameter/instrument
ΔP	differential pressure	NH_3:	ammonia
$AgNO_3$:	silver nitrate	NO:	nitrogen oxide
b_{abs}:	light transmission, surrogate for black carbon (BC)	NO_2:	nitrogen dioxide
CO:	carbon monoxide	OC:	organic carbon
CO_2:	carbon dioxide	OPC	optical particle counter
EC:	elemental carbon	$PM_{2.5}$	particles with aerodynamic diameter <2.5 µm
H_2S:	hydrogen sulfide	PM_{10}	particles with aerodynamic diameter <10 µm
HULIS:	humic-like substances	VOCs	volatile organic compounds
GC:	gas chromatography	WSOC:	water-soluble organic carbon
K_2CO_3:	potassium carbonate		

FIGURE 8.1   Schematic diagram of the dilution sampling system.

dilution air (England et al., 2007; Wang et al., 2012). This diluted effluent entered a spiral residence chamber for ~ 28 s aging to attain equilibrium. This was nearly three times the ~ 10 s deemed sufficient by Chang et al. (2004) in similar tests. The sectional stainless steel spiral-shaped residence chamber (made of six quick-clamp $90°$ elbow tube fittings with an inner diameter of 14.7 cm and arranged into a spiral) offered flexibility in chamber size and portability. The aged effluent entered a manifold, which was connected to real-time instruments (see Table 8.2) and filter packs (Figure 8.2).

Particle losses through the dilution system excluding the 2.4 m-long sampling probe were evaluated prior to the field measurement using monodisperse polystyrene latex spheres (PSL) of five sizes (0.5, 1.0, 2.5, 5.0, and 10.0 µm) entrained in sample flow heated to $70\ °C$ and diluted by a factor of ~ 20. The measured transmission efficiencies were $\sim 100\%$ for 0.5–5 µm PSL and

TABLE 8.2 Real-Time Instruments Applied to the Stack Emission Testing in the AOSR

Instrument	Parameter of interest and measurement principles	Measurement range	Time resolution	Nominal precision/ accuracy
PP System CO_2 analyzers Model SBA-4	CO_2 by nondispersive infrared (NDIR)		$\sim$1.5 s	<1% of span concentration
	• Stack gas	0–20,000 ppmv		
	• Diluted sample	0–5000 ppmv		
2B Tech NO Monitor and NO_2 Converter Model 400 and 401	NO, NO_2, and NO_x by reaction with O_3	2–2000 ppbv	10 s	Higher of 3 ppbv or 3% of reading
BW Technologies Multi-Gas Detector Model GasAlertMicro 5	CO by electrochemical sensor	0–500 ppmv	30 s	1 ppmv
TSI DustTrak Model 8520	PM mass concentration by light scattering ($PM_{2.5}$ or PM_{10})	Size: $\sim$0.1–2.5 μm or 0.1–10 μm Mass: 0.001–100 mg/m^3	1 min	±20% (for calibration aerosol)
Grimm Optical Particle Counter (OPC) Model 1.108	Particle size distribution by light scattering	Size: 0.23–25 μm Number: 0.001–2000 particle/cm^3 Mass: 0.0001–100 mg/m^3	1 min	±2.5%
Pitot tube	Stack velocity	1–100 m/s	1 min	±2.5%
Thermocouple	Stack temperature	−200 to 1250 °C	1 min	Higher of 2.2 °C or 0.75%

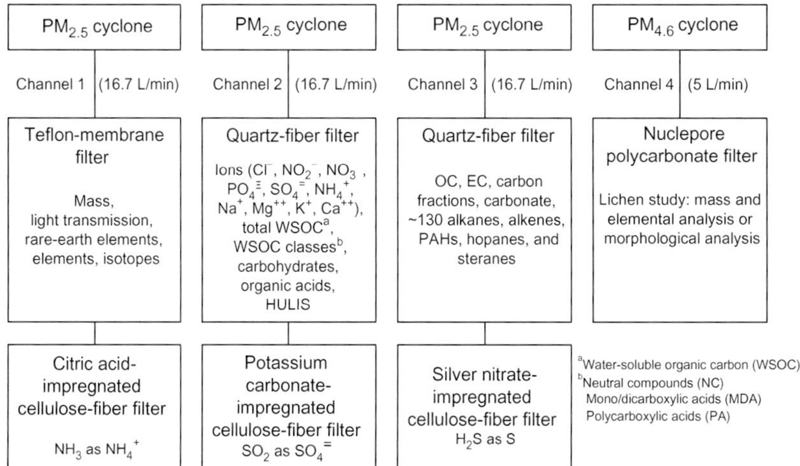

FIGURE 8.2 Four-channel filter pack configuration used for stack samples. Chemical analyses are documented by Chow and Watson (2012).

$86.2 \pm 18.6\%$ for 10 μm PSL. Particle losses through the 2.4 m sampling probe were estimated for Brownian diffusion and gravimetric setting under stack sampling conditions (Kulkarni et al., 2011). The overall transmission efficiencies are estimated to be $\sim 100\%$ for 0.1–1 μm, 96–98% for 2.5 μm, 46–56% for 10 μm, and 0% for ≥ 18 μm.

Parallel to the sampling probe, a type S pitot tube measured the pressure difference (ΔP) between the two openings, from which the stack flow velocity and flow rate were calculated (U.S. EPA, 1991). Flow velocities measured by this method across the stack diameter demonstrated $<3\%$ variation from the average velocity at the sampling point, so it is assumed that pollutant concentrations at the sampling location represent those across the sampling cross section. The stack temperature (T) was measured by a type K thermocouple. The ΔP and T were recorded by a data logger (Model 21X, Campbell Scientific, Logan, Utah). To obtain in-stack CO_2 concentrations without water vapor interference, a separate sample was extracted and diluted with an eductor pump by a factor of 21.4 with filtered air.

Nineteen sets of samples were collected over 10 days with 6–7 sample runs per stack. Sample durations ranged from 120 to 180 min per run and dilution ratios (DR) varied from 22:1 to 50:1 with ~ 25:1 dilutions for most runs. One 120-min system blank run was conducted for each stack with the valve on the sampling probe closed and all instruments sampling only the dilution air. At the beginning of each test day, all real-time instruments sampled diluted air for 10 min to check their readings for filtered dilution air; background CO_2 concentrations (380–420 ppm) obtained during this period were used for DR calculation.

8.2.3 Data Reduction

ERs for each species i (ER_i in µg/h) are expressed in mass emitted per unit time:

$$ER_i = 3600 \times C_i \times V_{Stk} \times A \times \frac{T_{std}}{T_{Stk}} \times \frac{P_{Stk}}{P_{std}} \tag{8.1}$$

where C_i is the stack concentration of the pollutant i in µg/m^3 under standard temperature ($T_{std} = 293$ K) and pressure ($P_{std} = 101{,}325$ Pa) conditions, V_{Stk} is the average stack velocity in m/s, A is the stack cross-section area in m^2, and T_{Stk} and P_{Stk} are stack temperature and pressure in K and Pa, respectively.

The diluted concentration ($C_{i,Dil}$ in µg/m^3) is used to calculate the stack concentration (C_i):

$$C_i = C_{i,Dil} \times DR \tag{8.2}$$

The DR is calculated using measured stack, diluted, and background CO_2 concentrations (i.e. $CO_{2,Stk}$, $CO_{2,Dil}$, and $CO_{2,Bkg}$, respectively) by

$$DR = \frac{CO_{2,Stk} - CO_{2,Bkg}}{CO_{2,Dil} - CO_{2,Bkg}} \tag{8.3}$$

8.3 RESULTS AND DISCUSSION

8.3.1 PM$_{2.5}$ Source Profiles

Each species abundance, including major gases, is normalized to the gravimetric PM$_{2.5}$ mass from the Teflon-membrane filter. Organic constituents are also normalized to PM$_{2.5}$ organic carbon (OC). Table 8.3 shows that Stack B had high abundances of NH_3 and SO_2 ($1025 \pm 637\%$ and $9205 \pm 4662\%$ of PM$_{2.5}$, respectively). NH_3 and SO_2 abundances differed by two to four orders of magnitude among the three stacks. H_2S had a low abundance in all three stacks

TABLE 8.3 Abundances (Mass Percentage of PM$_{2.5}$) for NH_3, SO_2, and H_2S Measured from Backup Filters

Chemical species	Stack A	Stack B	Stack C
NH_3	34.2 ± 10.5	1025 ± 637	0.43 ± 0.27
SO_2	$>2226^a$	9205 ± 4662	472 ± 152
H_2S	0.038 ± 0.041	0.026 ± 0.042	0.008 ± 0.006

[a] *The K_2CO_3-impregnated cellulose-fiber filters were designed for ambient measurements, with a maximum SO_2 holding capacity of 730 µmol/filter. Ion analysis of the filter extract shows that the K_2CO_3 was totally consumed for Stack A samples and therefore, the actual SO_2 concentrations was higher than those measured from filters.*

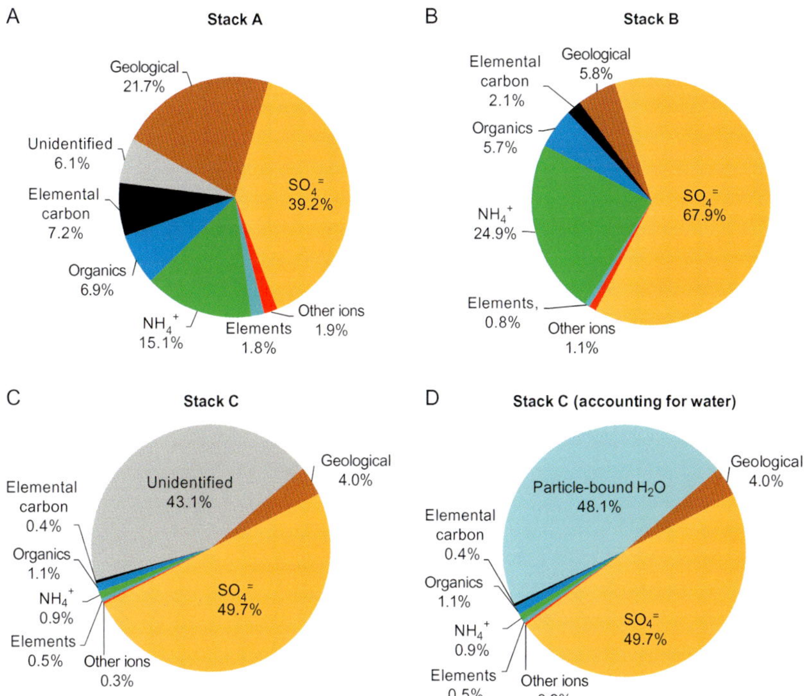

FIGURE 8.3   PM$_{2.5}$ material balance for: (A) Stack A, (B) Stack B, (C) Stack C without accounting for particle-bound water, and (D) Stack C after accounting for water associated with un-neutralized H$_2$SO$_4$. This accounting assumed thermodynamic equilibrium under the laboratory filter-conditioning environment of $21.5 \pm 1.5\ °C$ temperature and $35 \pm 5\%$ RH. Geological material is assumed to be composed of Al$_2$O$_3$, SiO$_2$, CaO, TiO$_2$, FeO, and Fe$_2$O$_3$ (geological material $= 2.2 \times$ [Al] $+ 2.49 \times$ [Si] $+ 1.63 \times$ [Ca] $+ 1.94 \times$ [Ti] $+ 2.42 \times$ [Fe]). Other soluble ions include Cl$^-$, NO$_2^-$, NO$_3^-$, PO$_4^=$, Na$^+$, Mg^{2+}, K$^+$, and Ca^{2+}. Other elements include those measured by XRF from Na to U, excluding Na, Mg, Al, Si, Cl, S, Ca, Ti, and Fe. Organics are assumed to be organic carbon [OC] $\times 1.2$. The sum of all species was 108% and 105% for (B) and (D), respectively. The excess mass is within the precision of the measurements and assumptions for the material balance.

($<0.038\%$ of PM$_{2.5}$). The PM$_{2.5}$ material balance in Figure 8.3A–C shows that SO$_4^=$ was the most abundant species, accounting for 39–68% of PM$_{2.5}$ mass. A high abundance of NH$_4^+$ (15–25% of PM$_{2.5}$ mass) was also found in Stacks A and B, with molar ratios of NH$_4^+$ versus SO$_4^=$ of ~ 2 (Figure 8.4A). The balanced anions and cations (regression slope ≈ 1 in Figure 8.4B) are consistent with fully neutralized (NH$_4$)$_2$SO$_4$ for Stacks A and B, accounting for 53.9% and 91.3% of PM$_{2.5}$, respectively. The low NH$_4^+$ content ($0.9 \pm 1.0\%$; Figures 8.3C and 8.4A) and cation deficiency (Figure 8.4B) in Stack C indicates the presence of unmeasured hydrogen ions (H$^+$), most likely in the form of H$_2$SO$_4$, accounting for 47.8% of PM$_{2.5}$. Under laboratory filter equilibrium conditions

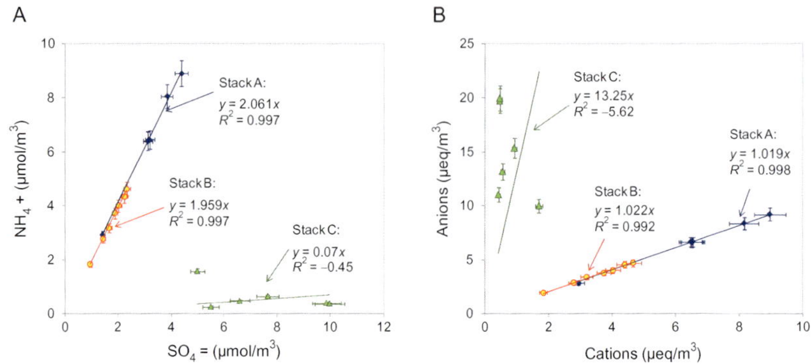

FIGURE 8.4 Comparison of molar concentrations for: (A) ammonium (NH_4^+) versus sulfate ($SO_4^=$) and (B) anions versus cations in $PM_{2.5}$ from Stacks A, B, and C. The slopes of $\sim$2 for Stacks A and B are consistent with sulfate as $(NH_4)_2SO_4$. This is not the case for Stack C.

($21.5 \pm 1.5\ ^\circ$C and $35 \pm 5\%$ RH) for gravimetric analysis, H_2SO_4 would be in the thermodynamic equilibrium form of $H_2SO_4 \cdot 5.48H_2O$ (Clegg et al., 1998). The particle-bound liquid water mass would be 48.1% of $PM_{2.5}$, which accounts for the unidentified mass in Figure 8.3C and completes the $PM_{2.5}$ mass balance for Stack C (Figure 8.3D).

$PM_{2.5}$ chemical abundances are summarized in Table 8.4. Total carbon (TC; sum of organic and elemental carbon, OC and EC) was most abundant in Stack A (12.9% of $PM_{2.5}$), and accounted for 6.9% and 1.3% of $PM_{2.5}$ in Stacks B and C, respectively. OC and EC abundances in Stack A were 5.7% and 7.2%, respectively. The Fe abundance in Stack A ($2.9 \pm 0.4\%$) was higher than that in Stacks B ($1.6 \pm 0.5\%$) and C ($0.3 \pm 0.2\%$). Abundances for other elements and ions (except for NH_4^+ and $SO_4^=$) were low and variable. Abundances of rare earth elements were <0.01% of $PM_{2.5}$ with three to eight times higher levels found in Stack A. $PM_{2.5}$ in Stack C had higher stable lead isotope ratios (19.91, 15.99, and 40.91 for ^{206}Pb/^{204}Pb, ^{207}Pb/^{204}Pb, and ^{208}Pb/^{204}Pb, respectively) than Stacks A (18.92, 15.33, and 38.55, respectively) and B (18.85, 15.57, and 38.85, respectively), providing a potential composition marker to differentiate $PM_{2.5}$ in Stack C from those in Stacks A and B. For $\sim$100 nonpolar organic compounds, n-alkanes (i.e., n-tricosane (n-C23), n-tetracosane (n-C24), n-pentacosane (n-C25), n-hexacosane (n-C26), and n-heptacosane (n-C27)) were the most abundant (>0.01% of OC). Polycyclic aromatic hydrocarbons (PAHs), hopanes, and steranes were low but detectable in Stacks A and B; they were below detection limits in Stack C. The abundance patterns were similar (peaking at n-C24) between Stacks A and B with lighter n-alkanes (peaking at n-C22) in Stack C. Most carbohydrates, organic acids, and water-soluble organic carbon (WSOC) were below detection limits.

TABLE 8.4 Abundances of $PM_{2.5}$ Constituents in-Stack Emissions

Stack	Distribution of chemical abundances greater than their uncertainties in % of $PM_{2.5}$			
	<0.1%	0.1–1%	1–10%	>10%
Stack A	Cr, Y, Pb, Br, Mo, Sr, Zr, Mg^{++}, Zn, K^+, Na^+, NO_3^-, Ni, Mn	Ca^{++}, Mg, V, K, Ca, Ti, OC4[a], EC2[a], Na	Al, Cl, Cl^-, OC2[a], Fe, Si, OC, EC1[a], EC[a]	TC, S, NH_4^+, $SO_4^=$
Stack B	Br, Sr, Pb, K^+, Mn, Na^+, Mg^{++}, V, K	Ti, Ca^{++}, Ca, Mg, NO_3^-, Cl, Ni, Al, EC2[a], Si, OC4[a], OC3[a]	Na, Fe, EC1, EC, OC2[a], OC[a], TC	S, NH_4^+, $SO_4^=$
Stack C	Rb, Y, Sr, Mn, Zr, Se, Zn, EC3[a], Mg^{++}, Na^+, Ni, Mo, Ti, $PO_4^\equiv$, K^+, Ca^{++}	Mg, K, Ca, EC2[a], Fe, Al, V, EC[a], Na	Si, S, TC	$SO_4^=$

[a]$OC1$–$OC4$ are organic carbon evolved at 140, 280, 480, and 580 °C, respectively, in a 100% helium (He) atmosphere. $EC1$–$EC3$ are elemental carbon evolved at 580, 740, and 840 °C in a 98% He/2% oxygen (O_2) atmosphere, following the IMPROVE_A protocol (Chow et al., 2007, 2011).

8.3.2   PM Size Distributions and Optical Properties

$PM_{2.5}$ mass concentrations were acquired in three ways: gravimetry by Teflon[®]-membrane filters, light scattering by DustTrak, and single particle optical sizing by an optical particle counter (OPC). Figure 8.5 shows that these three measurements were correlated for Stacks A ($R^2 = 0.81$–0.92) and B ($R^2 = 0.95$–0.99), with poor correlations for Stack C ($R^2 = 0$–0.6). Figure 8.5C shows that both optical instruments (i.e. DustTrak and OPC) had high correlation ($R^2 = 0.92$ and 0.95 for Stacks A and B, respectively), with twice the readings (2.1–2.2) from DustTrak compared with OPC. This is due to different calibration factors that convert light scattering signals to particle mass and illustrates why a collocated filter measurement is needed. Regression slopes in Figure 8.5C are consistent with the similar particle compositions [$(NH_4)_2SO_4$] and size distributions (Wang et al., 2009) for both Stacks A and B, in contrast with a much higher regression slope (4.22) and lower correlation ($R^2 = 0.6$) found for Stack C. Particles in Stack C (mainly $H_2SO_4 \cdot xH_2O$) contain different amounts of water at different RHs (Seinfeld and Pandis, 2006), resulting in changes of mass concentration and particle size distribution, thereby contributing to the poor correlation.

The mass calibration factor for the DustTrak was obtained by taking the ratio of the $PM_{2.5}$ mass concentration by gravimetry to that by the DustTrak for each run. The same approach was taken for the OPC for particles smaller

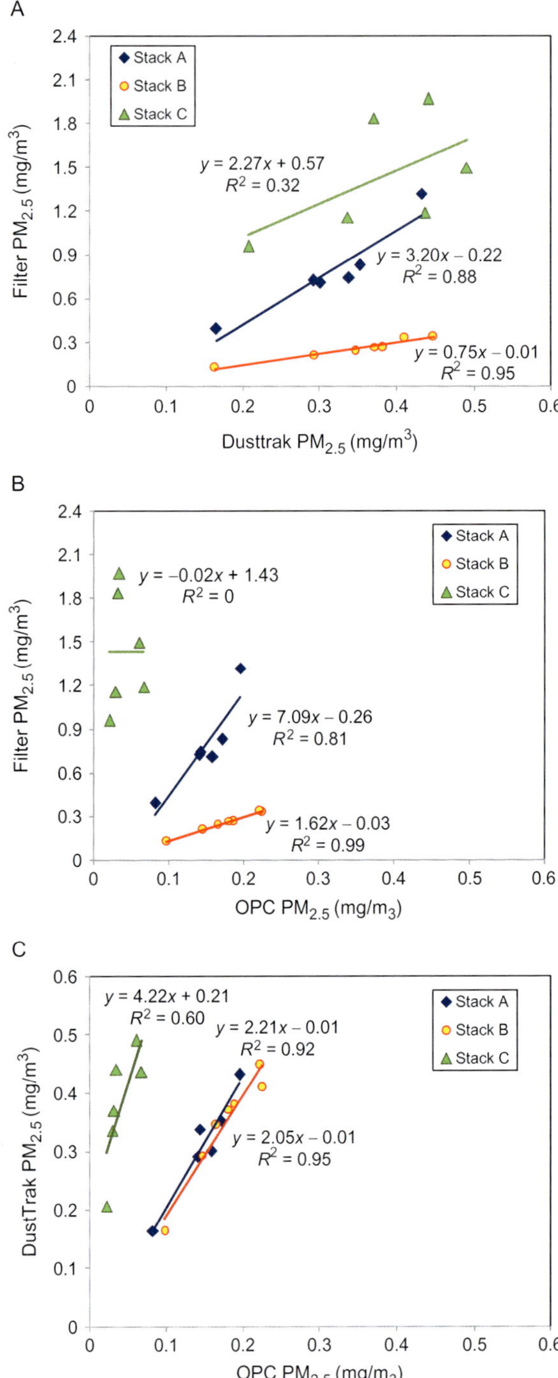

FIGURE 8.5 Correlations between diluted PM$_{2.5}$ concentrations from: (A) gravimetric filter versus DustTrak; (B) gravimetric filter versus OPC; and (C) DustTrak versus OPC. The reported concentrations by the DustTrak and OPC use default internal calibration factors (i.e. without normalization to the filter masses).

than 2.5 μm. The $PM_{2.5}$ calibration factors were also applied to remaining OPC size fractions, recognizing the potential for larger uncertainties in these fractions. The OPC sizes particles according to light scattering, and hence, the reported geometric particle diameters differ from aerodynamic particle diameters. Optical sizes were assumed to be close to aerodynamic sizes, and the calibration factor-corrected particle size distributions measured by the OPC were grouped into PM_1, $PM_{2.5}$, PM_{10}, and TSP (PM_{25}) and used for reporting ERs for size segregated PM. Particle losses in the dilution sampling system were corrected for the size range of 1–15 μm. Losses for >15 μm particles were not corrected because most measured concentrations were zero and the estimated transmission efficiencies were nearly zero. Therefore, the reported TSP by OPC would underestimate true values.

Figure 8.6 shows the mass median diameter (MMD) were 1.15, 0.72, and 0.34 μm for Stacks A, B, and C, respectively. Although the PM mass

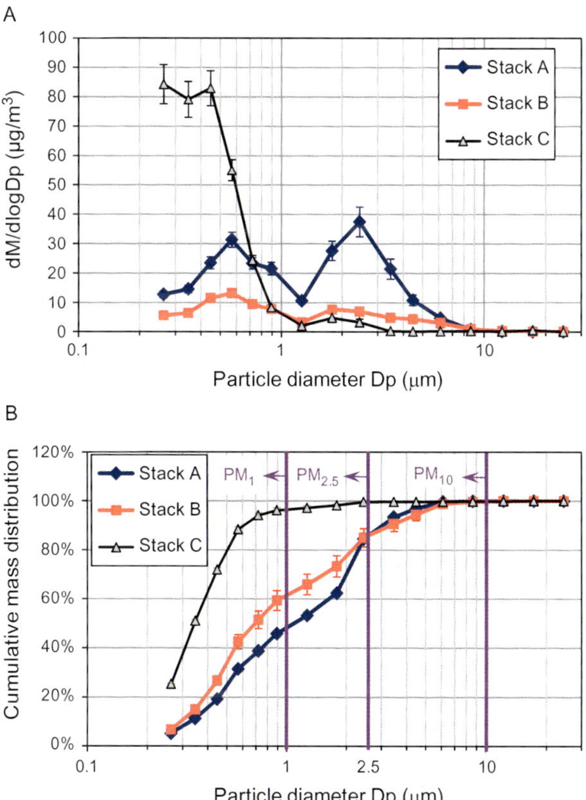

FIGURE 8.6   Particle mass distribution measured by the optical particle counter (OPC) for: (A) differential and (B) cumulative distribution.

concentration in Stack A was higher, particles from Stacks A and B had similar bimodal mass distributions, peaking at 0.5–0.6 μm and 1.5–2.5 μm. This size distribution similarity is consistent with the comparable regression slopes found in Figure 8.5C. The cumulative size distributions in Figure 8.6B show $\sim 50\%$ mass in PM_1, $\sim 30\%$ in $PM_{1-2.5}$, and $\sim 20\%$ in $PM_{2.5-10}$ for Stacks A and B. Particles from Stack C were much smaller, with a single mode peaking at 0.45 μm, 96% of mass in PM_1, and $\sim 99\%$ of mass in $PM_{2.5}$. The two FGD stacks (B and C) show distinct differences in particle size distribution despite comparable stack physical and operational parameters such as velocity, flow rate, and temperature (Table 8.1). Differences in feedstocks, processes, and pollution controls all affect the particle size distributions.

8.3.3 Gas and PM Concentrations and Emission Rates

The continuous measurement devices detected short-duration stack emission changes indicative of process and control device variations, as illustrated in Figure 8.7. Stack parameters and gas concentrations were relatively stable over the test periods, in contrast to real-time concentrations from heavy haulers in mining operations (Watson et al., 2010a). The PM spikes in Stack C corresponded to NO spikes, probably due to short-duration changes. Such correspondence was not found in Stack A or B samples during this study. Detection of short-term variations might be useful for explaining test-to-test differences in ERs and compositions.

Table 8.5 shows that CO_2 had the highest ERs among all pollutants, ranging from 1.8×10^5 to 2.7×10^5 kg/h. While H_2S ERs (0.003–0.038 kg/h) were low, large variations were found in NH_3 ERs. Stack C emitted 1–2 orders of magnitude lower NH_3, only 1.1% and 0.2% of levels in Stacks A and B, respectively, which explains the higher acidity of $PM_{2.5}$. ERs ranged from 500–1599 kg/h for CO, 201–>1050 kg/h for SO_2, 132–295 kg/h for NO_x, and 8–49 and 11–68 kg/h for $PM_{2.5}$ and PM_{10}, respectively.

Table 8.6 shows that for Stack A, the SO_2 ER from dilution tests was one-eighth that from CEMS during the same sampling period because the potassium carbonate (K_2CO_3) on the backup filter was completely consumed by SO_2. SO_2 from CEMS and a 2007 compliance test differed <2%. NO_x and TSP ERs from dilution tests were 52% and 17%, respectively, of ERs from 2007 compliance tests. ERs from dilution sampling (except the unquantified SO_2), CEMS, and compliance were well within Alberta's emission guidelines for these species. For Stack B, SO_2 from dilution tests was $\sim 25\%$ lower than the CEMS but was similar to the 2007 compliance tests. NO_x by dilution sampling was 45% higher than that from the 2007 compliance test. The TSP by dilution sampling was 21% of the hot filter catch and 3% of the total TSP from the compliance tests. The Stack B total TSP would exceed the emission guideline value by 51%, if the impinger catch was included. For Stack C, TSP by dilution sampling was 16% lower than the 2010 compliance test result, and

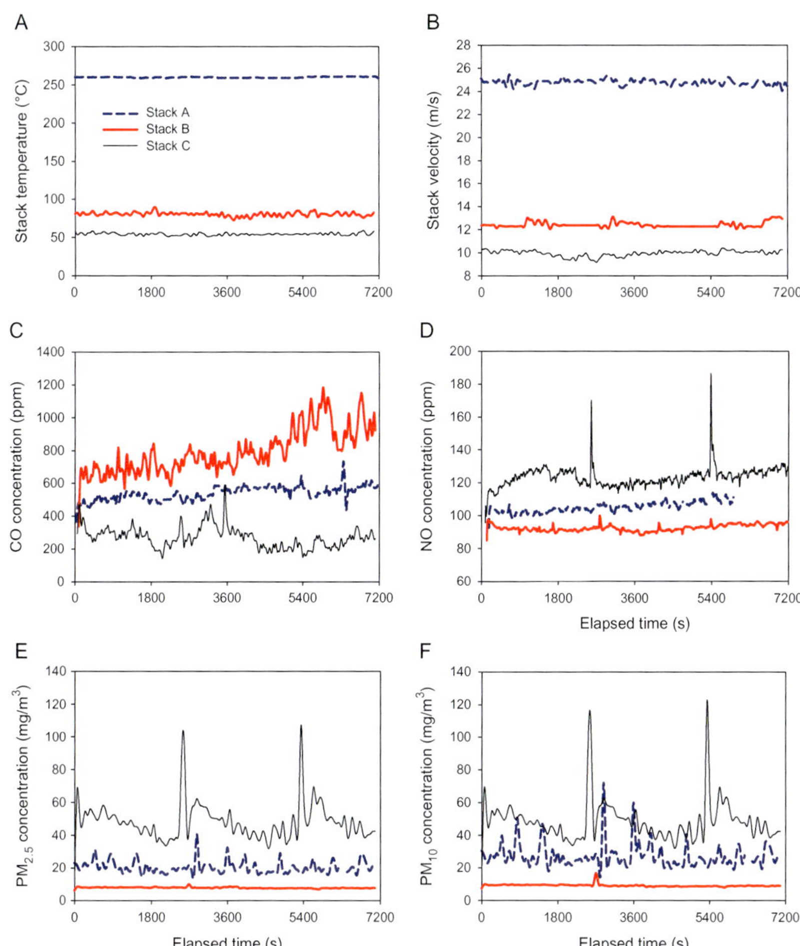

FIGURE 8.7   Examples of real-time data for: (A) stack temperature, (B) velocity, (C) CO, (D) NO, (E) $PM_{2.5}$, and (F) PM_{10} concentrations from one run at each stack.

both NO_x and TSP were <16% of emission guidelines. The discrepancy in TSP between dilution sampling and compliance tests might be partially attributed to losses of >15 μm particles in dilution sampling, positive artifacts from the impinger catch, and emission variations between the 2007 and 2008 testing periods.

Table 8.7 compares particle size distributions from this study with those measured from a 2002 in-stack survey. The methods are different, and the processes are probably better controlled for the 2008 tests, so a precise correspondence is not expected. The $PM_{2.5}$ and PM_{10} ERs from the in-stack survey were, respectively, 63% and 15% lower than those by the dilution method,

TABLE 8.5 Average Gas and PM Concentrations (Under Standard Conditions) and Emission Rates for the Three Stacks

Measured species		Concentration (mg/m^3)			Emission rates (kg/h)		
		Stack A	Stack B	Stack C	Stack A	Stack B	Stack C
Gases	CO	698 ± 27	929 ± 43	451 ± 74	1599 ± 54	980 ± 44	500 ± 73
	CO_2	$(1.14 \pm 0.02) \times 10^5$	$(1.68 \pm 0.03) \times 10^5$	$(2.40 \pm 0.03) \times 10^5$	$(2.61 \pm 0.05) \times 10^5$	$(1.77 \pm 0.02) \times 10^5$	$(2.70 \pm 0.05) \times 10^5$
	NO	128 ± 4	126 ± 2	167 ± 11	295 ± 11	132 ± 2	187 ± 12
	NO_2	0.0 ± 1.2	0.0 ± 0.7	0.0 ± 1.3	0.0 ± 2.7	0.0 ± 0.7	0.0 ± 1.4
	NO_x	128 ± 3	126 ± 2	166 ± 10	295 ± 11	132 ± 2	186 ± 11
	NH_3	7.3 ± 1.1	82.0 ± 21.3	0.16 ± 0.01	16.6 ± 2.4	86.4 ± 22.9	0.18 ± 0.01
	SO_2	>461	689 ± 122	177 ± 22	>1050	727 ± 132	201 ± 28
	H_2S	0.017 ± 0.008	0.005 ± 0.001	0.003 ± 0.001	0.038 ± 0.017	0.005 ± 0.002	0.003 ± 0.001
PM	PM_1	13.3 ± 1.1	5.6 ± 0.1	36.7 ± 2.7	30.6 ± 2.7	5.9 ± 0.1	41.3 ± 3.4
	$PM_{2.5}$	21.5 ± 2.1	7.6 ± 0.3	38.0 ± 3.0	49.1 ± 5.0	8.0 ± 0.3	42.7 ± 3.8
	PM_{10}	29.6 ± 3.5	10.1 ± 1.2	38.2 ± 3.2	68.0 ± 8.4	10.7 ± 1.4	43.0 ± 4.0
	PM_{25}	29.6 ± 3.5	10.3 ± 1.4	38.2 ± 3.2	68.1 ± 8.4	10.9 ± 1.6	43.1 ± 4.0

TABLE 8.6 Comparison of Emission Rates (kg/h) from Dilution Sampling, Continuous Emissions Monitoring Systems (CEMS) During the Dilution Sampling Period, Compliance Tests (Conducted in 2007 for Stacks A and B, and 2010 for Stack C), and Alberta Environment's Emission Guidelines for Each Stack

Stacks/species		NO_x[a]	SO_2	TSP-front	TSP-total[b]
Stack A	Dilution	452±17	>1050	NA[c]	68±8
	CEMS	NA	8699±157	NA	NA
	Compliance	870±69	8843±431	NA	392±48
	Guideline	1500	16,400	NA	600
Stack B	Dilution	203±4	727±132	NA	10.9±1.6
	CEMS	NA	951±288	NA	NA
	Compliance	140±19	741±239	52±10	378±50
	Guideline	NA	NA	NA	250
Stack C	Dilution	284±17	201±28	NA	43.1±4.0
	CEMS	NA	NA	NA	NA
	Compliance	NA	NA	NA	51.09
	Guideline	1800	NA	NA	340

[a]Compliance NO_x was measured by Alberta Method 7A, where NO_x in the flue gas sample was oxidized to nitrate, measured by ion chromatography (IC), and reported in term of NO_2. To be comparable to the compliance tests, NO_x from dilution sampling is also reported as NO_2 by: ER $(NO_2) = ER\ (NO) \times 46/30$.
[b]The compliance tests measured TSP emissions based on Alberta Method 5: TSP-front: hot filter catch; TSP-total: hot filter and impinger catches. The dilution sampling TSP uses PM_{25} by the OPC.
[c]Data not available.

TABLE 8.7 Comparison of Particle Size Distributions Measured from Stack A by an In-Stack Survey Test and the Dilution Sampling in This Study

Test name[a]	Test date	$PM_{2.5}$ ER (kg/h)	PM_{10} ER (kg/h)	TSP ER (kg/h)
In-stack survey	5/1/2002–5/2/2002	18±2	58±10	103±27
This study	8/9/2008–8/11/2008	49±5	68±8	68±8

[a]The in-stack sampling followed a setup similar to the U.S. EPA Method 201A by installing both a PM_{10} and $PM_{2.5}$ in-stack cyclones (U.S. EPA, 2010). The condensable fraction captured in the impingers was not accounted for in the in-stack test data.

while TSP was 51% higher. The in-stack hot filter did not collect particles that would nucleate and grow upon cooling and therefore underestimated fine particle concentrations. This is probably why the hot filter $PM_{2.5}$ was so much lower than the dilution sample. The in-stack TSP ($>10\ \mu m$) fraction is uncertain and subject to contamination, as it was recovered by washing the sampling probe and cyclone with acetone (U.S. EPA, 2010). However, losses for particles $>15\ \mu m$ in the dilution sampling method were not accounted for, and the OPC only measures particles $\lesssim 25\ \mu m$.

8.4 SUMMARY

A dilution sampling system was used for real-world emission characterization from three stacks in two AOSR facilities. Real-time gases (i.e. CO, CO_2, NO, and NO_x), particles (i.e. $PM_{2.5}$ and PM_{10}), and particle size distributions were measured. Precursor gases (i.e. SO_2, H_2S, and NH_3) as well as $PM_{2.5}$ mass and chemical composition were collected with stacked filter packs. Continuous measurements generated data to calculate average pollutant concentrations and ERs, and also detected short-duration emission variability, while integrated filter samples allowed for comprehensive laboratory analyses. Chemical source profiles and ERs were derived from these measurements.

H_2S abundances were low ($<0.04\%$ of $PM_{2.5}$) for the three stacks. NH_3 and SO_2 abundances ($1025 \pm 637\%$ and $9205 \pm 4662\%$ of $PM_{2.5}$, respectively) were high for Stack B. The major $PM_{2.5}$ constituent was $(NH_4)_2SO_4$ for Stacks A and B, accounting for 53.9% and 91.3% of $PM_{2.5}$, respectively; while 95.9% of $PM_{2.5}$ in Stack C was hydrated sulfuric acid, with H_2SO_4 and H_2O accounting for 47.8% and 48.1% of $PM_{2.5}$, respectively. $PM_{2.5}$ in Stack A contained 21.7% geological material. Carbon accounted for a minor fraction of $PM_{2.5}$, with the TC being $12.9 \pm 2.3\%$, $6.9 \pm 0.6\%$, and $<1.4\%$ of $PM_{2.5}$ for Stacks A, B, and C, respectively. Low abundances of nonpolar organic compounds were found for Stacks A and B, peaking at n-C24, while Stack C peaked at n-C22. Abundances of rare earth elements and other species were low ($<0.01\%$ of $PM_{2.5}$).

Bimodal mass distributions were found for Stacks A and B, with $\sim 50\%$ mass in PM_1, $\sim 30\%$ mass in $PM_{1-2.5}$, and $\sim 20\%$ mass in $PM_{2.5-10}$. Particles from Stack C were unimodal and much smaller, with 96% mass in PM_1 and $\sim 99\%$ mass in $PM_{2.5}$. Good correlations were found between $PM_{2.5}$ by gravimetry and DustTrak or OPC for Stacks A and B, while the correlation was poor for Stack C due to the dependence of H_2SO_4 droplet size on the temperature and RH of the measurement environment.

CO_2 had the highest ERs among all pollutants, ranging 1.8–2.7×10^5 kg/h, with 1–2 orders of magnitude variations found for criteria pollutants (i.e. CO, SO_2, NO_x, and PM). ERs from both dilution sampling and compliance tests were found to be lower than the emission guidelines for each stack, except that the TSP including impinger catches from 2007 compliance tests would exceed the emission guideline by 51% for Stack B.

Several cautions need to be noted when interpreting or using the dilution sampling results and their comparisons with compliance tests:

- These data represented stack conditions during the test period. Dilution sampling was carried out over a period of 2–3 days, with each test lasting 120–180 min. Variations in upstream processes and stack operation conditions need to be taken into account when extrapolating the current data to annual averages.
- Compliance test data reported here were not taken at the same time. Temporal variations of emissions may contribute to the observed differences.
- Larger particles ($>$15 µm) were not recovered by this system and were also poorly recovered by TSP compliance measurements.
- Wet FGD stacks present other difficulties in extracting large droplets through the buttonhook nozzle and transferring the water-saturated droplets to the dilution chamber. An in-stack dilution system is being tested by U.S. EPA to more effectively capture and transport these droplets to the dilution chamber (Baldwin et al., 2012).

ACKNOWLEDGMENTS

This work was sponsored by the Wood Buffalo Environmental Association (WBEA, www.wbea.org). The authors thank Dr. Yu-Mei Hsu, Ms. Carna MacEachern, Ms. Simone Balaski, and the environmental officers at each facility for their efforts to coordinate the field testing and supply necessary stack information. Mr. Tom Baldwin, Don House, and Allan Budd assisted with the particle loss tests. Ms. Jo Gerrard of DRI assisted with editing the chapter.

REFERENCES

Alberta Environment, 1995. Alberta stack sampling code. Publication Number: REF.89; ISBN: 0-7732-1406-2. Prepared by Alberta Environment, Edmonton, AB Canada.

ASTM International, 2010. Standard Guideline for Determining $PM_{2.5}$ and PM_{10} Mass in Stack Gases: D2203-WK8124_Guideline.

Baldwin, T.A., Watson, J.G., Neulicht, R., 2012. New test method for measuring fine particulate matter ($PM_{2.5}$) in wet stacks. In: Proceedings, Air Quality Measurement Methods and Technology. Air & Waste Management Association, Pittsburgh, PA, pp. 88-1–88-6.

Chang, M.-C.O., Yi, S.M., Hopke, P.K., England, G.C., Chow, J.C., Watson, J.G., 2004. Measurement of ultrafine particle size distributions from coal-, oil-, and gas-fired stationary combustion sources. J. Air Waste Manage. Assoc. 54 (12), 1494–1505.

Chang, M.-C.O., Chow, J.C., Watson, J.G., Glowacki, C., Sheya, S.A., Prabhu, A., 2005. Characterization of fine particulate emissions from casting processes. Aerosol Sci. Technol. 39 (10), 947–959. http://www.tandfonline.com/doi/pdf/10.1080/02786820600623711. Accessed 22 August 12.

Charpentier, A.D., Bergerson, J.A., MacLean, H.L., 2009. Understanding the Canadian oil sands industry's greenhouse gas emissions. Environ. Res. Lett. 4 (1), 1–11. http://stacks.iop.org/1748-9326/4/i=1/a=014005. Accessed 22 August 12.

Chow, J.C., Watson, J.G., 2011. Air quality management of multiple pollutants and multiple effects. Air Qual. Clim. Change J. 45 (3), 26–32.

Chow, J.C., Watson, J.G., 2012. Chemical analyses of particle filter deposits. In: Ruzer, L., Harley, N.H. (Eds.), Aerosols Handbook: Measurement, Dosimetry, and Health Effects, vol. 2. CRC Press/Taylor & Francis, New York, NY, pp. 177–202.

Chow, J.C., Watson, J.G., Chen, L.-W.A., Chang, M.C.O., Robinson, N.F., Trimble, D.L., Kohl, S.D., 2007. The IMPROVE_A temperature protocol for thermal/optical carbon analysis: maintaining consistency with a long-term database. J. Air Waste Manage. Assoc. 57 (9), 1014–1023. http://pubs.awma.org/gsearch/journal/2007/9/10.3155-1047-3289.57.9.1014.pdf. Accessed 22 August 12.

Chow, J.C., Watson, J.G., Robles, J., Wang, X.L., Chen, L.-W.A., Trimble, D.L., Kohl, S.D., Tropp, R.J., Fung, K.K., 2011. Quality assurance and quality control for thermal/optical analysis of aerosol samples for organic and elemental carbon. Anal. Bioanal. Chem. 401 (10), 3141–3152. http://dx.doi.org/10.1007/s00216-011-5103-3. Accessed 22 August 12.

Clegg, S.L., Brimblecombe, P., Wexler, A.S., 1998. Thermodynamic model of the system H^+-NH_4^+-Na^+-SO_4^{2-} NO_3^- Cl H_2O at 298.15 K. J. Phys. Chem. 102 (12), 2155–2171.

Code of Federal Regulations USA, 2001a. 40 CFR Part 86, Control of emissions from new and in-use highway vehicles and engines.

Code of Federal Regulations USA, 2001b. 40 CFR Part 89, Control of emissions from new and in-use nonroad compression-ignition engines.

Code of Federal Regulations USA, 2002. 40 CFR Part 1065, Engine Testing Procedures.

Cooper, H.B.H., Rossano, A.T., 1974. Source Testing for Air Pollution Control. McGraw-Hill, Columbus, OH, USA.

Corio, L.A., Sherwell, J., 2000. In-stack condensible particulate matter measurements and issues. J. Air Waste Manage. Assoc. 50 (2), 207–218.

Dowdeswell, L., Dillon, P., Ghoshal, S., Miall, A., Rasmussen, J. and Smol, J.P., 2010. A foundation for the future: building an environmental monitoring system for the oil sands. Prepared by Canadian Minstry of the Environment, Ottawa, ON, Canada.

England, G.C., Toby, B. and Zielinska, B., 1998. Critical review of source sampling and analysis methodologies for characterizing organic aerosol and fine particulate source emission profiles. Report number 344; prepared by American Petroleum Institute, Washington, DC.

England, G.C., Zielinska, B., Loos, K., Crane, I., Ritter, K., 2000. Characterizing $PM_{2.5}$ emission profiles for stationary sources: comparison of traditional and dilution sampling techniques. Fuel Proc. Technol. 65–66, 177–188. http://www.sciencedirect.com/science/article/B6TG3-40FGBMD-H/2/012c269ab472a9e4716aedc4e0c76ba5. Accessed 22 August 12.

England, G.C., Watson, J.G., Chow, J.C., Zielinska, B., Chang, M.-C.O., Loos, K.R., Hidy, G.M., 2007. Dilution-based emissions sampling from stationary sources: part 1. Compact sampler, methodology and performance. J. Air Waste Manage. Assoc. 57 (1), 65–78. http://pubs.awma.org/gsearch/journal/2007/1/england.pdf. Accessed 22 August 12.

Heinsohn, R.J., Davis, J.W., Knapp, K.T., 1980. Dilution source sampling systems. Environ. Sci. Technol. 14 (10), 1205–1209.

Hildemann, L.M., Cass, G.R., Markowski, G.R., 1989. A dilution stack sampler for collection of organic aerosol emissions: design, characterization and field tests. Aerosol Sci. Technol. 10 (10–11), 193–204.

Kulkarni, P., Baron, P.A., Willeke, K., 2011. Aerosol Measurement Principles, Techniques and Applications, third ed. John Wiley & Sons, Inc., Hoboken, NJ, USA.

Lee, S.W., He, I., Young, B., 2004. Important aspects in source $PM_{2.5}$ emissions measurement and characterization from stationary combustion systems. Fuel Proc. Technol. 85, 687–699.

Li, X., Wang, S., Duan, L., Hao, J., Long, Z., 2011. Design of a compact dilution sampler for stationary combustion sources. J. Air Waste Manage. Assoc. 61 (11), 1124–1130.

Lighty, J.S., Veranth, J.M., Sarofim, A.F., 2000. Critical review: combustion aerosols: factors governing their size and composition and implications to human health. J. Air Waste Manage. Assoc. 50 (9), 1565–1618.

Lipsky, E.M., Robinson, A.L., 2005. Design and evaluation of a portable dilution sampling system for measuring fine particle emissions from combustion systems. Aerosol Sci. Technol. 39 (6), 542–553.

McDonald, J.D., Zielinska, B., Fujita, E.M., Sagebiel, J.C., Chow, J.C., Watson, J.G., 2000. Fine particle and gaseous emission rates from residential wood combustion. Environ. Sci. Technol. 34 (11), 2080–2091.

Myers, R.E., Logan, T., 2002. Progress on developing a federal reference PM fine source test method. In: 11th International Emission Inventory Conference—"Emission Inventories—Partnering for the Future", 15 April 2002; U.S. EPA.

Oil Sands Advisory Panel (OAP), 2010. A foundation for the future: building an environmental monitoring system for the oil sands. Prepared by Canadian Minstry of the Environment, Ottawa, ON, Canada.

Richards, J., Holder, T., Goshaw, D., 2005. Optimized method 202 sampling train to minimize the biases associated with method 202 measurement of condensable particulate matter emissions. In: Proceedings, Specialty Conference on Hazardous Waste Combustion. Air & Waste Management Association, Pittsburgh, PA, USA, pp. 1–9.

Schindler, D., 2010. Tar sands need solid science. Nature 468 (7323), 499–501. http://dx.doi.org/10.1038/468499a. Accessed 22 August 12.

Seinfeld, J.H., Pandis, S.N., 2006. Atmospheric Chemistry and Physics: From Air Pollution to Climate Change. John Wiley & Sons, New York, NY.

Tenenbaum, D.J., 2009. Oil sands development: a health risk worth taking? Environ. Health Perspect. 117 (4), A150–A156. http://dx.doi.org/10.1289%2Fehp.117-a150. Accessed 22 August 12.

TMAC, W.G., 2003. A priority ranking of air emissions in the oil sands region. Prepared by Clearstone Engineering Ltd. and Golder Associates Ltd. for Trace Metals and Air Contaminants Working Group, Cumulative Environmental Management Association, Fort McMurray, AB, Canada.

U.S. EPA, 1991. Method 2—Determination of Stack Gas Velocity and Volumetric Flow Rate (Type S Pitot Tube)(40 CFR 60. Appendix A to Part 60).

U.S. EPA, 2000a. Method 5. Particulate matter (PM). Determination of particulate matter emissions from stationary sources. Prepared by U.S. EPA, Research Triangle Park, NC. http://www.epa.gov/ttn/emc/promgate/m-5.pdf. Accessed 22 August 12.

U.S. EPA, 2000b. Method 17. In-Stack Particulate-Determination of particulate matter emissions from stationary sources. Prepared by U.S. EPA, Research Triangle Park, NC. http://www.epa.gov/ttn/emc/promgate/m-17.pdf. Accessed 22 August 12.

U.S. EPA, 2004. Conditional test method (CTM) 039: Measurement of $PM_{2.5}$ and PM_{10} emissions by dilution sampling (constant sampling rate procedures). Prepared by U.S. Environmental Protection Agency, Research Triangle Park, NC. http://www.epa.gov/ttn/emc/ctm/ctm-039.pdf.

U.S. EPA, 2010. Methods for measurement of filterable PM_{10} and $PM_{2.5}$ and measurement of condensable PM emissions from stationary sources, Final Rule (40 CFR Part 51). Federal Register. 75(244), 80118–80172.

Wang, X.L., Chancellor, G., Evenstad, J., Farnsworth, J.E., Hase, A., Olson, G.M., Sreenath, A., Agarwal, J.K., 2009. A novel optical instrument for estimating size segregated aerosol mass concentration in real time. Aerosol Sci. Technol. 43, 939–950.

Wang, X.L., Robbins, C., Hoekman, S.K., Chow, J.C., Watson, J.G., Schuetzle, D., 2011. Dilution sampling and analysis of particulate matter in biomass-derived syngas. Front. Environ. Sci. Eng. China 5 (3), 320–330.

Wang, X.L., Watson, J.G., Chow, J.C., Gronstal, S., Kohl, S.D., 2012. An efficient multipollutant system for measuring real-world emissions from stationary and mobile sources. AAQR—Aerosol Air Qual. Res. 12 (1), 145–160.http://aaqr.org/VOL12_No2_April2012/1_AAQR-11-11-OA-0187_145-160.pdf.

Watson, J.G., Chow, J.C., 2001. Source characterization of major emission sources in the Imperial and Mexicali valleys along the U.S./Mexico border. Sci. Total Environ. 276 (1–3), 33–47.

Watson, J.G., Chow, J.C., Lowenthal, D.H., Robinson, N.F., Cahill, C.F., Blumenthal, D.L., 2002. Simulating changes in source profiles from coal-fired power stations: use in chemical mass balance of $PM_{2.5}$ in the Mt. Zirkel Wilderness. Energy Fuel 16 (2), 311–324.

Watson, J.G., Chow, J.C., Wang, X., Kohl, S.D., and Gronstahl, S. 2010a. Measurement of in-use emissions from non-road diesel trucks. Report number 010109-123109; prepared by Desert Research Institute, Reno, NV, for Wood Buffalo Environmental Association, Ft. McMurray, AB, Canada.

Watson, J.G., Chow, J.C., Wang, X., Kohl, S.D., and Sodeman, D.A. 2010b. Measurement of real-world stack emissions with a dilution sampling system. Report number 010109-123109; prepared by Desert Research Institute, Reno, NV, for Wood Buffalo Environmental Association, Ft. McMurray, AB, Canada.

Watson, J.G., Chow, J.C., Wang, X.L., Kohl, S.D., Chen, L.-W.A., Etyemezian, V., 2012. Overview of real-world emission characterization methods. In: Percy, K.E. (Ed.), Alberta Oil Sands: Energy, Industry, and the Environment. Elsevier Press, Amsterdam, The Netherlands, pp. 141–166.

Chapter 9

Applying the Forest Health Approach to Monitoring Boreal Ecosystems in the Athabasca Oil Sands Region

K.E. Percy*,[1], D.G. Maynard[†] and A.H. Legge[‡]

*Wood Buffalo Environmental Association, Fort McMurray, Alberta, Canada
[†]Canadian Forest Service, Pacific Forestry Centre, West Victoria, British Columbia, Canada
[‡]Biosphere Solutions, Calgary, Alberta, Canada
[1]Corresponding author: e-mail: kpercy@wbea.org

ABSTRACT

The increased development in the Athabasca Oil Sands Region (AOSR) has raised concerns about elevated emissions of air pollutants and the potential for negative effects on terrestrial ecosystems. A forest health monitoring program was established in 1998 by the Wood Buffalo Environmental Association (WBEA). Field sites were sampled in 1998, 2004, and 2011. To date, while there is evidence of increased elemental concentrations in plant foliage with increasing predicted deposition levels, there was no evidence of a negative effect on nutrient cycling processes or forest productivity. However, differences in site factors confounded interpretations of the potential effects of air emissions and bioassays suggested that modeled PAI (potential acid input) values used in data analysis were insufficient to link cause and effect. As a result, the network of monitoring sites was expanded and enhanced in 2011 with updated science-based monitoring concepts to better support decision making and regulatory processes. A forest health approach to terrestrial monitoring was adopted and built upon the existing terrestrial monitoring network in order to determine cause/effect relationships between air pollution and forest ecosystem health in the AOSR. This enhanced design will also serve Alberta government regulatory expectations under cumulative effects management, and regional land-use planning.

Developments in Environmental Science, Vol. 11. http://dx.doi.org/10.1016/B978-0-08-099443-7.00009-0

9.1   INTRODUCTION

There has been a general pattern across North America of decreasing levels of sulfur, oxidized nitrogen species (NO_x), mercury deposition, and ozone (O_3) since the 1980s, although important exceptions exist. There is evidence for increasing NO_x deposition across western North America, and stable or increasing ammonium (NH_4), while intercontinental transport of mercury from Asia to North America is rising (Environment Canada and Health Canada 2010; Hidy et al., 2011). In north-eastern Alberta, there was no statistically significant trend in ambient sulfur dioxide (SO_2) concentrations between 1998 and 2007 (Kindzierski, 2010). However, there is evidence that ambient concentrations of NO_2 (nitrogen dioxide) have increased during 2005–2010 (McLinden et al., 2011).

The North American Research Strategy for Tropospheric Ozone (NARSTO) published an assessment of scientific resources available to address options for air quality improvement. NARSTO included in this assessment a state-of-science evaluation consequent improvement in public health and welfare, and ecosystem health (Hidy et al., 2011). Clair et al. (2011) provided a review of the major air pollutants which have widespread and documented effects on ecosystems in North America. The major pollutant stressors affecting ecosystems have been identified as acidic deposition, ecosystem acidification, nitrogen deposition, and O_3, with mercury and persistent organic pollutants considered important stressors of aquatic systems. There is now an increasing policy-level focus in North America on the need to address multipollutant air quality management, despite the considerable technical challenges of so doing.

In North America, natural areas are important interfaces between air quality, the public, science, and regulation (Percy and Karnosky, 2007). All forests are shaped by disturbance regimes driven by climate variability in temperature, wind, and moisture, which in turn affect fire, herbivory, and other ecosystem processes. Forest structures, landscapes, and functions at any time are in dynamic disequilibria between maturation processes (e.g., tree growth) and disturbances at various spatial and temporal scales. Forests are strongly influenced by tree growth rates (via slow processes) and disturbance regimes (via rapid processes). Slow processes and rapid processes can be influenced simultaneously by a complex array of factors that includes several dimensions of climate (drought, temperature, wind, etc.) (Lucier et al., 2009). Nonclimatic factors (air pollution) can influence forest disturbance regimes via interactions with climatic effects. In the case of air pollution, such interaction can take the form of feedback systems that can tend either to stabilize or destabilize forest ecosystems by acting on sensitive genotypes (shorter-term) or species (longer-term) (Karnosky et al., 2003a; Percy, 2002), most often by predisposing trees to drought, and then inciting insect attack (Miller and McBride, 1999).

North-eastern Alberta now represents such an interface between air quality, the public, science, and regulation. The Athabasca Oil Sands Region

(AOSR) has a growing number of fixed, fugitive, and mobile sources emitting a wide range of primary pollutants in a changing industrial landscape. Regulatory decisions on air quality must, therefore, be made with clear understanding of source emissions, deposition modes/patterns, and cause–effect relationships need to be scientifically demonstrated. The Wood Buffalo Environmental Association (WBEA) has recently enhanced its level and scale of monitoring within the AOSR in order to better support decision making and regulatory policies (Percy, Introduction to the volume).

9.2   TERRESTRIAL ENVIRONMENTAL MONITORING IN THE ATHABASCA OIL SANDS PRIOR TO 2008

Canada's oil sands are found in three deposits—the Athabasca, Peace River, and Cold Lake deposits in Alberta and Saskatchewan (see Figure 1, in Preface). The oil sands are at the surface near Fort McMurray, but deeper underground in the other areas (CAPP, 2011). The Regional Municipality of Wood Buffalo (RMWB) is located in the northeast portion of the province of Alberta, Canada, extending northwards from latitude 55° to 59°. The RMWB is 68,454 km^2 in area, and includes the AOSR. The WBEA is a community-based, multistakeholder, not-for-profit association based in Fort McMurray. WBEA (www.wbea.org) "…monitors air quality, and air quality related environmental impacts to generate accurate and transparent information which enables stakeholders to make informed decisions." It does this through an extensive network of continuous, time-integrated, and passive air quality monitoring techniques (Chapter 4), as well as through its Terrestrial Environmental Effects Monitoring (TEEM) program.

The AOSR bitumen deposits lie beneath the Boreal Plains Ecozone consisting of upland jack pine, aspen, mixed forest, and wetlands (Figure 9.1). Monitoring of air pollutant effects on forests in the AOSR has been underway to varying degrees since the first oil sands operation began in the late 1960s (Addison, 1980; Addison and Puckett, 1980; Addison et al., 1986). The WBEA TEEM program was established in the mid 1990s. The objective of TEEM was to determine if anthropogenic emissions of acidifying compounds such as SO_2 and NO_x gases from oil sands operations are having a long-term adverse effect on the regional terrestrial environment and if so, to determine the magnitude of this effect. In 1998, TEEM initiated measurement and sampling at a network of 10 jack pine (*Pinus banksiana* Lamb.) dominated interior forest stand plots (AMEC, 2000). Five additional plots were added between 1999 and 2003, although two of these were lost to development. Another cycle of measurement and sampling of soils and vegetation occurred at the remaining 13 plots in 2004 (eight were previously sampled in 1998 and one in 2001).

The 2004 resampling resulted in a report that summarized some aspects of forest condition at that time and compared vegetation and soil status to those reported from the 1998 sampling (Jones and Associates, 2007). Monitoring of

A

B

FIGURE 9.1   (A) Location of the Boreal Plains Ecozone (NRCan, www.atlas.nrcan.gc.ca). (B) Aerial view of the ecozone northwest of Fort McMurray.

foliar vigor and stand condition in 2004 revealed no emissions-related effects on either needle retention or condition, and no anomalous damage/health issues in any of the 13 study sites (Jones and Associates, 2007). Analysis of foliar chemistry data showed that local industrial emissions were evident in increased concentrations of total sulfur, inorganic sulfur, iron, and nickel, all of which are known to be components of oil sands emissions. However, despite evidence of increasing elemental concentrations in foliage with increasing predicted deposition levels, there was no demonstrated evidence of a negative effect on forest productivity.

A parallel study was carried out in 2004 at the 13 jack pine monitoring sites to assess the soil biological and chemical status of the surface organic horizon and 0–5 cm mineral soil (Visser, 2006). Several soil chemical response variables were related to the modeled potential acid input (PAI) at these sites (Table 9.1). The variables significantly correlated with PAI are shown in Table 9.1. No biological indicators were correlated to the PAI. Increases in total S, available S, and ammonium (NH_4^+) in the surface organic horizon suggested that measurable deposition of S and possibly N had occurred, but there were no indications that deposition had impaired the microbial biomass and nutrient cycling processes in either the surface organic horizon or the 0–5 cm mineral soil (Visser, 2006). Other site-specific studies (Cheng et al., 2011; Laxton et al., 2012) assessing N and S deposition in the AOSR found that nitrate leaching was negligible with no indication of N saturation, providing further evidence that the nutrient cycling processes have not been affected; however, the potential risk dictates that monitoring continues (Addison et al., 1986).

Potential soil acidification effects were also assessed in a long-term soil acidification monitoring program initiated in 1981 by Alberta Environment (Abboud et al., 2012). There are nine locations in Alberta including one site near oil sands mining and extraction facilities (adjacent to jack monitoring site 104 of the TEEM monitoring program). Two subsites per location were established, and they were to be sampled every 4 years; however, logistical issues limited the sampling after1993. Soils were sampled by depth (0–2 cm, 2–5 cm, 5–10 cm, 10–15 cm, 15–30 cm, 30–45 cm, and 45–60 cm). Since 2004, only the upper three depths and one of the two subsites at the Fort McMurray site were sampled (Abboud et al., 2012). Results of three potential indicators of acidification trends in soils; pH, exchangeable base saturation percentage, and the soil solution base cation:aluminum (Al) ratio at the subsite sampled up to 2008 are presented in Table 9.2 (adapted from Abboud et al., 2012).

Changes in the pH, base saturation percentage, and base cation:Al ratio were statistically significant among sampling times; however, there was no consistent downward or upward trend. For example, the pH of the surface three soil horizons was similar in 2008 to the initial pH measurements in 1981 (Table 9.2). Base saturation percentage and base cation:Al ratio were

TABLE 9.1 Soil Chemical and Biological Response Variables Significantly Correlated with Modeled Potential Acid Input (PAI)

Response variable	Forest floor	0–5 cm Mineral soil
Organic matter mass	Increase with PAI	Not applicable
Total S	Increased with PAI	
NH_4—N and SO_4—S	Increased significantly with PAI	
pH, electrical conductivity (EC)	pH increases significantly with PAI	EC increased significantly with PAI
Exchangeable Ca and effective cation exchange capacity (ECEC)	Positive significant relationship between PAI and exch. Ca and EC	
Soil Solution Fe and S	Fe and S increase significantly with PAI	Positive significant correlation between PAI and S
"Total" acid extractable Al, B, Ca, Cu, Fe, K, Mg, Mn, S, Zn	All positively correlated with PAI	
Microbial biomass/ microbially mediated processes	No correlation with PAI	No correlation with PAI
Jack pine seedling growth potential	No correlation with PAI	No correlation with PAI
Soil fauna (mites/springtails)	No correlation with PAI	No correlation with PAI
Jack pine ectomycorrhizae	No correlation with PAI	No correlation with PAI

Adapted from Visser (2006).

lower in 2008 compared to 1981; however, there was considerable variability within a sampling year (e.g., coefficient of variation 30–50% for base cation: Al ratio) and in 2004 and 2008 both base saturation percentage and base cation:Al ratios increased from the previous sampling year (1989 or 1993) (Abboud et al., 2012).

Analyses conducted on 2004 data in all applicable program parameters indicated that effects were either not present (given nonsignificant deposition terms), or were present but not cumulative over the sampling interval. Importantly, Jones and Associates (2007) revealed that the 13 sites in 2004 were imperfect ecological analogues. In other words, differences in site factors confounded interpretations of the potential effects of local industrial emissions.

TABLE 9.2 Soil pH, Base Saturation Percentage and Base Cation: Aluminum Ratio at the Fort McMurray Long-Term Soil Acidification Monitoring Site from 1981 to 2008

Year	pH (CaCl$_2$)	% Base saturation	Base cation: Al ratio
0–2 cm ($n=12$)			
1981	4.1±0.2 (5.4)	60±13 (22)	3.3±1.2 (37)
1985	4.2±0.2 (4.5)	54±5 (10)	3.5±0.8 (23)
1989	4.3±0.3 (6.1)	46±11 (24)	2.6±0.6 (22)
1993	4.3±0.4 (9.2)	46±14 (31)	Not determined
2004	3.9±0.2 (6.1)	51±9 (18)	3.2±0.6 (19)
2008	4.2±0.1 (3.3)	54±5 (9)	3.2±1.1 (36)
2–5 cm ($n=12$)			
1981	4.5±0.3 (5.8)	57±21 (38)	4.4±2.5 (56)
1985	4.4±0.2 (4.1)	52±8 (15)	3.9±0.9 (24)
1989	4.5±0.3 (6.7)	46±14 (31)	3.3±0.8 (24)
1993	4.4±0.3 (7.0)	44±17 (38)	Not determined
2004	4.4±0.2 (5.4)	49±12 (24)	3.9±1.4 (35)
2008	4.4±0.2 (3.6)	51±8 (16)	3.4±1.0 (30)
5–10 cm ($n=12$)			
1981	4.4±0.2 (3.9)	52±14 (27)	5.6±2.6 (47)
1985	4.4±0.1 (2.5)	51±6 (12)	4.0±0.5 (13)
1989	4.5±0.1 (2.9)	44±11 (24)	3.5±0.6 (18)
1993	4.5±0.2 (5.6)	43±14 (33)	Not determined
2004	4.6±0.3 (5.9)	53±14 (27)	6.9±6.3 (90)
2008	4.5±0.2 (4.2)	56±12 (21)	4.4±1.4 (31)

Values are means±standard deviation (coefficient of variation).
Adapted from Abboud et al. (2012).

Importantly, examination of lichen tissue elemental concentrations as a bioassay of air quality monitoring data suggested that modeled PAI values used in data analysis were insufficient to link cause and effect.

For credible attribution of change (i.e., increased foliar concentrations), comeasurement of inputs and receptor response in time and space is required

(Karnosky et al., 2003b). Future sampling cycles using the 2004 protocols were concluded as having little opportunity of relating air emissions to forest ecosystem change. It was, therefore, apparent that the existing approach to assessing these parameters, while founded on accepted designs and concepts available at the time in 1997 (D'Eon et al., 1994), required review.

9.3 DEFINING FOREST HEALTH

The Oxford Dictionary defines health as "the state of being free from illness or injury." In his comprehensive treatise of the subject, Innes (1993) indicated that there are many dimensions to forest health. O'Laughlin et al. (1994) have defined forest health as "...a condition of forest ecosystems that sustains their complexity while providing for human needs." This definition stresses the utilitarian endpoint, which is defined in anthropocentric terms. Kolb et al. (1994) differentiated between the utilitarian view of forest health, which stresses timber production as the primary endpoint and the ecosystem perspective, which views forest health in terms of measures of longer-term forest function. Function includes properties such as resilience, diversity, and the flow of carbon, water, and nutrient resources required for tree resistance to natural stresses and maintenance of biogeochemical cycles, as has been shown to hold true in the case of O_3 (Karnosky et al., 2003a).

Building upon Kolb et al. (1994), McLaughlin and Percy (1999) proposed a less anthropocentric, and more function based definition: "A capacity to supply and allocate water, nutrients, and energy in ways that increase or maintain productivity while maintaining resistance to biotic and abiotic stresses." Subsequently, that definition was endorsed by the United Nations Forum on Forests (2003). Under this definition, less healthy forests are ones in which trees lose productive capacity and/or become more sensitive to environmental stresses. Of course, we recognize that unhealthy trees and stands occur naturally as a part of successional processes by which a balance between forest production, site resources, and climate are attained.

Utilitarian approaches to assess forest condition are usually inadequate to detect change, and certainly for attribution of natural (stand dynamics, climate, pests) and anthropogenic (air pollutants) influences. Where air quality is a serious concern, retrospective analysis (Percy, 2002) of monitoring programs has demonstrated air pollution to be an important factor in forest ecosystems only:

- When scales of stressors have been considered;
- When monitoring has been succeeded by process-oriented research;
- When appropriate indicators and endpoints were measured;
- When investigations on physical/chemical cycles were coupled with biological cycles; and
- When there was continuity in investigation.

9.4   TEEM FOREST HEALTH NETWORK DESIGN

Following scientific review of results and recommendation in Jones and Associates (2007), a new way forward was conceived and proposed (Percy et al., 2007). The existing Acid Deposition Monitoring Program (AMP) objective centered on indirect effects over the long term. It was replaced with a new design to "…implement an approach for establishing/determining cause-effect relationships between air pollutants and forest ecosystem health in the Oil Sands Region." Six key elements of the new design were:

1. Adopt the forest health approach;
2. Change the conceptual design;
3. Relax the stand area restriction;
4. Maintain adaptive capacity under rapid development;
5. Incorporate ecologically analogous sites; and
6. Comeasure inputs (predictors) and responses (indicators) in space and time.

9.4.1   Conceptual Design, Area Restriction, and Adaptive Capacity

A number of conceptual designs have been used in forest monitoring (Percy and Ferretti, 2004). In Europe, forest monitoring plots are arrayed on a uniform pan-European grid. In Canada, plots in the Acid Rain National Early Warning System were allocated by ecoregion. Classical monitoring around strong point sources (i.e., Sudbury Case Study) has typically used a gradient approach to distribute monitoring plots along plume dispersion paths, with sites located both upwind (control) and downwind at varying distances (e.g., Hogan and Wotton, 1984; Maynard et al., 1994).

In the boreal forest of west-central Alberta, the West Whitecourt Case Study was carried out over a 25-year period in the vicinity of a point source of sulfur gas emissions from 1976 to 2001 and used a modified gradient approach (Legge et al., 1988). The forest sampling plots were positioned at locations that were progressively downwind from the point source of sulfur emissions with each additional sampling plot located further from the plume dispersion center line defined by the prevailing direction of the wind. The forest sampling plots, however, were selected to be ecologically analogous (Chapter 10). This combined modified gradient and ecological approach enhanced the opportunity to relate change in terrestrial receptors (vegetation and soils) to sour gas plant air emissions. The classic southern California San Bernardino Case Study is another variant on this approach, where pine tree plots were arrayed by combining a range of oxidant exposure levels with consideration of stand ecology (Miller and McBride, 1999).

Both the West Whitecourt and San Bernardino Case Studies were designed in the late 1960s and 1970s respectively to conform to scientific principles known at the time, and enunciated more explicitly much later following retrospective analysis of the North American experience. Monitoring and understanding the relative roles of natural and anthropogenic stress in influencing forest health will require programs that are structured to evaluate responses at appropriate frequencies across gradients of forest resources that sustain them. Such programs must be accompanied by supplemental process-oriented investigations that more thoroughly test cause and effect relationships among stresses and responses of both forests and the biogeochemical processes that sustain them (McLaughlin and Percy, 1999).

Where gradients are not easily identifiable, and multiple, dispersed sources are present, a categorical approach has been used. This is often employed when no ambient air measurement data exist in a natural area to stratify sites. In the case of the 1998–2004 AMP, there was only very limited passively measured air data in the boreal forest during this period, so modeled PAI (Alberta Environment, 1999; Cheng et al., 1994) was used to assign plots to high and low deposition areas as it was the only option available at the time. Subsequently, TEEM established a more extensive passive sampler monitoring network in the boreal forest so that improved regional deposition data were available in 2007 to better inform the forest health network design.

A forest health approach to terrestrial monitoring was adopted and it was decided to build upon the existing AMP network of 11 sites (two lost post-2004 due to development). This was done in order to retain data, particularly soil and foliar chemical concentrations, from the 1998 and 2004 measurements for assessing the potential change with time. The enhancement of the network since 2008 was centered on locating, validating, and bringing into the network a suite of new ecologically analogous sites. Previous AMP design restrictions for the minimum jack pine stand area (2 ha) in which a monitoring plot could be placed was relaxed. While there are many definitions of a forest, the Canadian Forest Service (www.carbon.cfs.nrcan.gc.ca) defines a forest as "a 1 ha minimum area, 25% canopy cover of trees that have the potential to reach 5 m height at maturity." Upland jack pine grows on sandy soils that are thought to be highly sensitive to acid deposition (Holowaychuk and Fessenden, 1987); however, this is not the dominant ecotype in the AOSR and large stands are difficult to find. Analogous plots selected into the enhanced network are interior-to-stand edges.

There has been, and will continue to be increasing pressures on the land resource within the AOSR. In consideration of this fact, sampling with partial replacement was selected as the sampling method (Scott, 1998) most adaptable for estimating change in the forest resource. This method features a combination of permanent and temporary plots. Most of the effort is directed towards repeated measurement of permanent plots, although some of the temporary plots are routinely measured to some level. Temporary plots can then be brought into the network on relatively short notice, and with less expense, when permanent plots are lost to development. Together, permanent and temporary plots provide continuing efficient estimates of current status and future change.

9.4.2   Ecologically Analogous Plots, Indicators, and Endpoints

A complete description of the theory and application of analogous plots within the AOSR can be found in Chapter 10. With analogous plots (Figure 9.2) classified into the Pj/bearberry/lichen plant community of the boreal mixed-wood lichen–jack pine ecosite phase (Beckingham and Archibald, 1996) making up a majority of the new network, fewer plots are required for comparison across the region. Secondly, Krupa and Legge (1998) have convincingly demonstrated for boreal pine systems, that ecological analogues minimize within site variability, thus, maximizing the potential to detect change over time. The approach is also more cost-effective.

Elzinga et al. (2001) identified three stages in the monitoring of forest ecosystem response to air pollutants. The first stage is *detection*; is there any measurable effect of air pollution on forest productivity or health? The second stage is *quantification*; if there is, how serious is the effect and how will it progress? The third stage is to *understand the processes and resilience*. If air pollutant exposure leading to the effect is diminished, will forest health improve and in the future? Terrestrial ecosystems response to environmental

FIGURE 9.2   An ecologically analogous jack pine plot. Note the dominant *Cladina* sp. ground cover.

stresses in a complex, hierarchical process occurring over time scales ranging from, seconds (cellular), to minutes (physiological), to years (individuals, genotype), to multiple years (stands), to decades (ecosystems) (Hinckley et al., 1992). Therefore, the choice of indicators and endpoints to be measured is critical to success.

Within the forest health context, there are physical, biochemical, chemical, or biological measures of change. Change in an indicator level is simply that; a percentage increase or decrease since the last measurement. Change can be either positive or negative. In fact, the theory of *hormesis* underlying toxicological response stipulates that for the response to be real, a stimulatory response can be measured until such time as the dose is sufficient to inhibit a process (Jäger and Krupa, 2009). Change does not necessarily lead to an effect on forest function or process, nor does an effect on an ecosystem component necessarily lead to an impact, or change in an endpoint.

An effective indicator requires attention to several key constraints (Hunsaker, 1993). Clear objectives and endpoints must be elaborated. Secondly, a conceptual model identifying the linkage between the issue examined and responses expected is needed. In this respect, defoliation could in fact be considered as an indicator of overall tree condition. Although defoliation is an indicator common to some programs, it should not be considered a specific indicator of air pollution effects on trees nor used as a diagnostic tool to detect these effects (Ferretti, 1997; Percy and Ferretti, 2004). To provide an improved interface between stakeholders, science, and regulation, receptor response indicators must be: (1) specific to pollutants of concern; (2) supported by dose–response science; (3) responsive; (4) representative across the region; and (5) measurable within defined detection limits.

Indicators in use or being validated through field trials are shown in Table 9.3. Several indicators are being retained from the 1998 and 2004 measurements and others introduced into the network have been or are being validated in practice, or by external scientific peer review. All forest ecosystems are the product of geology, climate, disturbance, and genetics. Physiochemical and ecological characteristics interact to confer a degree of sensitivity and susceptibility to perturbation by air pollutants. System stressors such as wet and dry deposition tend to a have primary role when critical levels/loads are exceeded affecting change in soil or foliar indicators. Factors such as climate (e.g., temperature, precipitation, radiation, wind speed, and direction) and plant genotype can modify the degree and direction of change. Sufficient change derived from sufficient accumulation (S) or exposure (O_3) can singly or together with modifying factors, lead to an effect on forest function, such as reducing base saturation or predisposing trees to insect attack. Cumulative effects may then cascade to a change in endpoint status.

An endpoint is defined as "the point marking the completion of a process or stage of a process: *the final point*" (Merriam-Webster, 1993). The endpoint selected for use in the jack pine-based forest health network is productivity.

TABLE 9.3 Indicators of Air Pollutants Being Used or are Being Validated in Practice, or by External Scientific Peer Review in the AOSR

Vegetation	Soil	Lichens
Plant community assessment	pH	Epiphytic lichen community composition
Needle retention	Exchangeable cations and cation exchange capacity	Total N and S
Foliar analysis • Total C, N, and S • Total Ca, Mg, K, and Na • Inorganic SO_4S • Inorganic SO_4S/ organic S ratio • Total Fe, Al, and Mn • Micronutrients Cu, Zn, B, and Mo • Potential metal contaminants Ni, V	Total C, N, and S	Potential metal contaminants
Bark deposition	Available SO_4S, NH_4N, NO_3N, and PO_4P	
Cuticular wax structure and chemistry	Calculated C:N and base cation: Al ratios, % base saturation	
	Litter decomposition	
	Ecotmycorrhizal associations	
	Phospholipid fatty acids	

It addresses the six questions outlined by Percy and Karnosky (2007) for endpoint selection:

1. Is it the final point in a key ecosystem process?
2. Can it be measured accurately and precisely with time?
3. Is it supported by published exposure–response science?
4. Does it have social, economic, and ecological relevance?
5. Is it in a form that can be understood and utilized by air quality regulators? and
6. Can it be used in the longer-term to provide scientific input to a criterion setting processes, such as air quality management (protection)?

9.4.3   Deployment and Comeasurement

9.4.3.1   Long-Term Monitoring

In 2011, the enhanced WBEA forest health network structured to evaluate responses at appropriate frequencies across gradients of forest resources that sustain them (McLaughlin and Percy, 1999) were in place. The plots are dispersed across zones of sulfur and nitrogen deposition (Chapter 12). The network consists of 23 forested, interior-to-stand plots in Boreal Plains and Boreal Shield ecozones (Figure 9.3). The plots extend out 150 km to the north, west, and east from Fort McMurray, with two in Saskatchewan. Nine of the plots are equipped with towers holding passive monitors above the tree canopy and measuring monthly average concentrations of SO_2, NO_2, O_3, HNO_3, and NH_3. Eleven other nonforested sites contribute monthly passive data for the region.

Fulfilling another requirement in monitoring forest health, a system for comeasurement of predictors and indicators, and supplemental process-oriented investigations that more thoroughly test cause and effect relationships among stresses and responses is being deployed. As of 2011, four forest health plots (104, 107, 201, and 213) were equipped with 30 m tall towers (Figure 9.4). Four more plots will be added between 2012 and 2014 to complete the grid. With no power available, and a need for continuous meteorological measurements to account for the influence of interannual climate differences, a second tower is fitted at each plot with a solar array and ground-installed batteries to provide year-round power supply. Data from 10 m and 2 m above canopy height, and 2 m below canopy/3 m above ground on wind direction/speed, relative humidity, temperature, radiation (global/PAR) are continuously measured and uploaded through cell modem/satellite connection daily to the WBEA database in Calgary (Figure 9.5). Automated precipitation measurements are collected and uploaded, along with continuous O_3 at several of the sites. Below-ground soil moisture and temperature sensors operate seasonally at two depths and these data are also transmitted to the database. Plant root simulator (PRS^{TM}) probes (ion exchange membranes) are installed to determine the soluble ion concentrations at two soil depths (10 and 50 cm).

The PRS probes were designed to act as an ion sink, constantly accumulating ions as they become available in solution. A preliminary study to determine the suitability of the PRS probes to assess soil solution chemistry was initiated at two of the towered sites. Site 104 is in an area receiving S deposition (PAI from CALPUFF model for the period 1971–2002 was estimated at 0.35 keq H ha^{-1} year^{-1}; Jones and Associates, 2007) and site 201 is considered a background site west of Fort McMurray. At each site, three soil pits were excavated and within each pit four sets of PRS probes were inserted into the soil at two depths; 10 and 50 cm. The four sets within each pit were composited and extracted to give three replicates per site. Probes were buried over winter from October to May in 3 years and during the growing season in 2011

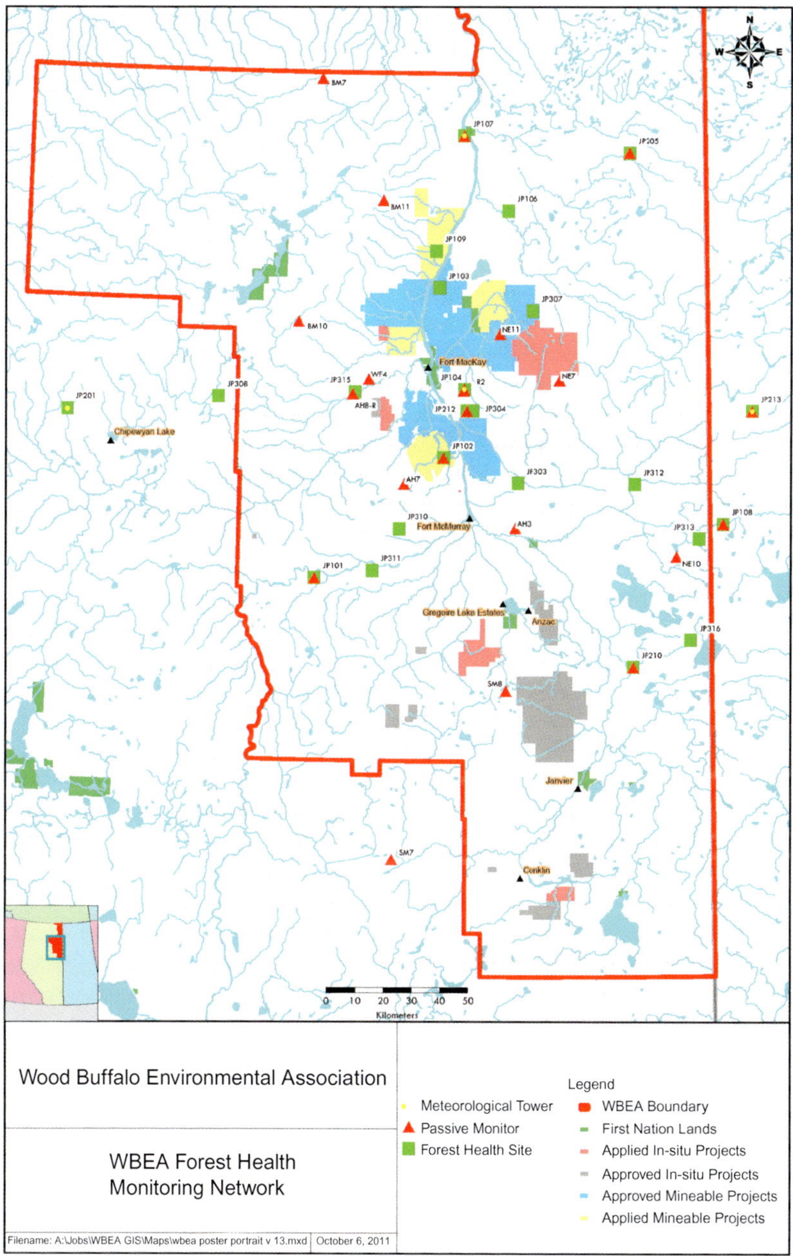

FIGURE 9.3   Map of WBEA forest health network plot locations. Note colocation of passive and solar-powered continuous meteorological-pollutant measurement towers.

FIGURE 9.4 Photograph of 30 m tall measurement tower at plot 104 beside tower holding the solar panels.

(Table 9.4). The results indicate that S in soil solution at both depths is consistently higher at site 104 compared to the background site (Table 9.4). There also appears to be an effect of time. The PRS probe S at site 104 has considerably higher PRS probe S (3–6 times higher) during the over winter and early spring sampling period than site 201. Higher sulfur levels are consistent with the results of the TEEM monitoring program and the study of Visser (2006).

In 2011, the network of 23 plots was intensively sampled over a 6-week period as part of the routine cycle for above- and below-ground measurements begun in 1998, and repeated in 2004. Tree allometry was measured, and tree cores taken on off-plot trees beside the 0.04 ha permanent plots. Foliar, soil, and lichen samples were collected for chemical analysis (see Table 9.3 for list of measurements) from the upper-third crown of the numbered trees. Canopy cover, frequency of occurrence, and composition by canopy cover was evaluated in microplots for the ground cover and for species. Soils were sampled at various depths in four subplots as outlined in Table 9.5.

Tree condition was assessed by Forest Health Specialists with the Alberta Sustainable Resource Department on each plot and insect/diseases presence and severity scored because of the potential for air pollutant–insect interactions (Miller and McBride, 1999; Percy, 2002). This will be the only annual measurement made on the permanent plots.

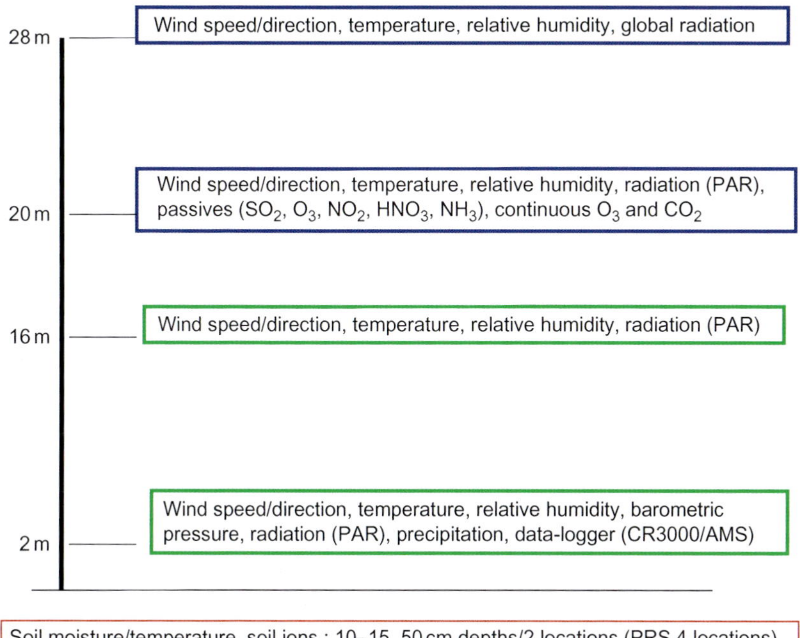

FIGURE 9.5   Schematic of measurements taken 10 m above canopy height, 2 m above canopy, 2 m into the canopy, and 2 m above ground.

9.4.3.2   Early Warning

The longer-term forest health monitoring in stand interiors is being coupled where possible with jack pine edge plots to provide early warning of indicator change. This new addition to the terrestrial monitoring program is driven by experience in case studies such as West Whitecourt. Concentrations of air pollutants are at their maximum when the air impinges on the stand edge, and concentrations decrease as the air moves through the stand. This is best evidenced by the steep drop in foliar inorganic S concentrations from fully exposed trees at the edge to a few *m* into the stand (Chapter 10). Beginning in 2012, up to 25 edge plots, many already located near new ecologically analogous interior-to-stand plots, will be deployed into the TEEM forest health monitoring system. Many of these sites will have mini-towers installed into the fen in front of the edge trees. The mini-tower will be equipped with continuous, solar-powered meteorology, and passive samplers. Indicators will be measured in the early warning edge plots on a three-year cycle. Each second suite of measurements will coincide with the intensive, six-year cycle for interior plots. It is anticipated that the early warning indicators will provide a window of opportunity for detection well in advance of eventual quantification within stands.

TABLE 9.4 Plant Root Simulator (PRS^{TM}) Sulfur Values (μg 10 cm^{-2} time^{-1}) from a Site With Known S Deposition (JP104 Within Site and Edge) and a Background Site (JP201) at Two Soil Depths (10 and 50 cm)

Site	October 2009–June 2010	October 2010–June 2011	June 2011–October 2011	October 2011–June 2012
10 cm depth				
JP104	71.96±25.00	43.53±17.63	74.67±5.24	52.83±6.91
JP104–edge	n.d.	n.d.	73.30±7.66	44.47±9.26
JP201	13.13±1.56	12.00±2.69	61.00±6.37	15.57±1.60
50 cm depth				
JP104	44.33±21.80	21.47±3.65	66.97±11.77	51.37±1.18
JP104–edge	n.d.	n.d.	53.17±5.37	49.60±12.90
JP201	7.13±0.96	6.80±0.69	48.53±6.38	11.77±0.30

Values are means±standard deviation ($n=3$).
n.d., not determined.

TABLE 9.5 Routine Soil Monitoring Program—Samples Per Site and Layer/Depth (Foster et al., 2011)

Depth	Number of plots (a)	Number of subplots (b)	Number of samples by depth (a × b)	Total number of samples
Surface organic horizon (forest floor)	4	4	16	64
0–5 cm	4	4	16	
5–15 cm	4	4	16	
15–30 cm	4	4	16	
15–30 cm[a]	4	4	16	80[a]

[a]*Establishment year only.*

9.5   INVESTIGATIVE STUDIES TO ENHANCE THE TEEM PROGRAM

The review of the jack pine AMP led to a coincidental review by science advisors of the TEEM program. Science advisors recommended a significant science enhancement to program, with a more integrated suite of measurements at key points along the established pollutant pathway. As has been previously stated, terrestrial ecosystems respond to air pollutants in a hierarchical manner, beginning with exposure and potential accumulation of some pollutants in plants and other ecosystem components. Integrated approaches linking the air component with pattern-oriented monitoring along defined pollution gradients are required to define cause–effect linkage (McLaughlin and Percy, 1999), a key objective of the enhanced monitoring program. A general description of the science enhancement within the WBEA TEEM program can be found in Percy et al. (2010).

In 2008, TEEM adopted the source-to-sink approach for terrestrial effects monitoring by measuring at key components along the emissions, chemical transformation, deposition, effects, and value endpoint pathway (Figure 9.6). With industry member cooperation, "real-world" characterization of emissions from several large stacks has been completed (Chapter 8). For the first time, "real-world" emissions from the world's largest 400 ton mine haulers

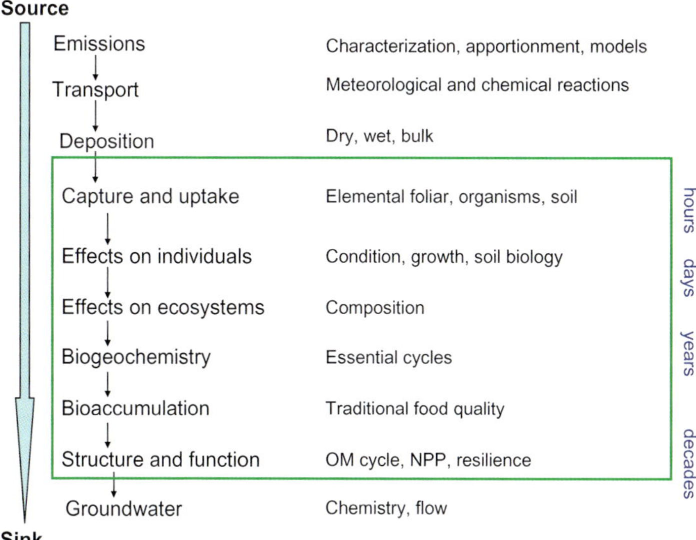

FIGURE 9.6   The pathway for air pollutant interaction with a boreal forest pine stand (modified from Legge et al., 1988).

have been measured and characterized using an innovative, onboard portable emissions monitoring system (Chapter 7). "Real-world" emission factors derived from this work will be compared to certification data with a view to increasing accuracy in emissions databases.

The deployment of an ambient ion monitor provides time-resolved measurements of particulate anions/cations, acidic gases, and aerosols in a mobile, field laboratory setting. Forest health monitoring plots located up to 175 km from sources are equipped with tower-mounted passive samplers measuring time-integrated ambient concentrations of primary (SO_2, NH_3) and secondary (NO_2, HNO_3, and O_3) air pollutants. Four bogs, located in a range of deposition zones are instrumented to measure bulk deposition, meteorology, net primary production, and monitor hydrology and water chemistry. Indicators will be validated for application within the AOSR and monitoring tools developed for future deployment within the region.

Atmospheric deposition to terrestrial receptors is being measured and mapped. The conditional time-averaged gradient (COTAG) measurements above fens (Figure 9.7) provide site-specific deposition velocities for S and N compounds that can be used to validate number inputs in regional dispersion modeling. Ion exchange resins (Fenn et al., 2009) collect bulk deposition in the open and beneath jack pine canopies in order to *estimate* regional SO_4 and NO_3 loadings near source, and at remote sites at forest health plots. A three-year stable isotope tracer study of the transfer (capture, uptake) of S and N to vegetation and soils has been completed (Chapter 11).

Ultimately, the TEEM objective of cause–effect linkage between industrial emissions and ecosystem health requires some confirmation of relative contributions to the terrestrial environment from natural and anthropogenic sources. Lichens are frequently used in monitoring studies to determine spatial and temporal gradients in air pollution. In 2008 during a 6-week period, two epiphytic lichen species were sampled at 359 sites arrayed on a geospatial

FIGURE 9.7 A fen with open fetch in front of jack pine plot 104. Note the conditional time-averaged gradient (COTAG) instrument for measuring flux in the foreground, the black ion exchange resin collectors for measuring bulk deposition, and the tripod for continuous meteorology and passive pollutant measurements in front of the transition to the jack pine stand.

grid. Patterns for S and N in the lichens have been identified (Berryman et al., 2010). Further analysis was then completed using state of the art preparation and analytical techniques for 42 trace elements (Chapter 14) and lead and mercury isotopes (Chapters 15 and 16). Apportionment techniques and receptor modeling were then used to attribute elemental concentrations to source type (Chapters 13 and 18).

In the AOSR, the CALPUFF regional dispersion model is used to estimate deposition from industrial emissions. The geospatial array of 2008 lichen elemental concentrations is being used as an input to CALPUFF to verify accuracy (over-, under-prediction) patterns in modeled deposition with distance from main fixed and mobile emission sources (Chapter 12). The instrumentation of towers outside of the valley will produce real time, regional-scale meteorological data as input to the CALMET module of CALPUFF in order to improve accuracy of dispersion modeling.

9.6   SUMMARY

The Dene definition for monitoring is "Watching, listening, learning" the "things that are changing" (A. Garibaldi, personal communication). Significant investment in funding has been made to enhance the WBEA TEEM program. Key goals were to increase capacity for detection of change, and to elucidate the role of industrial air emissions as a modifier of forest ecosystem health. The TEEM program has been enhanced scientifically since 2008 and will contribute to better decision making through improved and expanded effects monitoring, new measurements of emissions, transfer, and deposition, and receptor modeling that relates emissions to measured pollutant concentrations in terrestrial receptors across the region. A multidisciplinary, integrated program has been deployed to the landscape using a forest health approach. This will provide stakeholders with more accurate, timely, and scientifically credible information upon which to base regional air quality and emissions management decisions.

In the RMWB, monitoring of terrestrial effects is a regulatory requirement that is required for some industry approvals to operate. Effects from atmospheric industrial emissions in the RMWB are managed under cumulative management frameworks and regional plans. The enhanced design that WBEA members supported in 2008 has built upon positive attributes of the existing AMP in order to determine cause/effect relationships between air pollution and forest ecosystem health in the AOSR. The now deployed design will be essential for *tracking and evaluating* the impacts of Government of Alberta policies/actions on airshed management, a key component of any Air Quality Management System (Bachmann, 2007). *Integrated into the full science-enhanced TEEM program, the WBEA forest health network will be essential in meeting the Government of Alberta's regulatory and management requirements under Regional Land Use Planning, and Cumulative Effects Management.*

ACKNOWLEDGMENTS

The authors thank the WBEA membership and funders for their support that has enabled the design and deployment of a more scientific, and holistic approach to ecosystem monitoring in the AOSR. We are very appreciative of the sound and constructive science advice that we have received since 2006 from our colleagues Drs. Sagar Krupa, Dale Johnson, Tom Nash III, Sandy McLaughlin, Mike Miller, Ted Hogg, Ken van Rees, Suzanne Visser, and Neil Cape (see Preface, this volume for affiliations). The advice and input provided by WBEA TEEM, and TEEM-SSC members is gratefully acknowledged. We also thank current and past TEEM program managers for facilitating our work.

REFERENCES

Abboud, S.A., Schwarz, C.J., Dinwoodie, G.D., Byrtus, G.A., Turchenek, L.W., 2012. Trends in Soil Acidification in Alberta Based on Long Term Soil Acidification Monitoring from 1981 to 2002. Alberta Environment and Water, Land Monitoring Team, Edmonton, AB.

Addison, P.A., 1980. Baseline condition of jack pine bio-monitoring plots in the Athabasca Oil Sands area, 1976 and 1977. Alberta Environ. Alberta Oil Sands Environmental Research Program, Edmonton, AB, Canada. AOSERP Rep 98.

Addison, P.A., L'Hirondelle, S.J., Maynard, D.G., Malhotra, S.S., Khan, A.A., 1986. Effects of oil sands processing emissions on the boreal forest. Agriculture Canada, Canadian Forestry Service, Northern Forestry Centre, Edmonton, AB, Canada. Information Report NOR-X-284.

Addison, P.A., Puckett, K.J., 1980. Deposition of atmospheric pollutants as measured by lichen element content in the Athabasca Oil Sands area. Can. J. Bot. 58, 2323–2334.

Alberta Environment, 1999. Application of Critical, Target, and Monitoring Loads for the Evaluation and Management of Acid Deposition. Environmental Service, Environmental Sciences Division, Edmonton, AB Alberta Environment Publication No. T/472, 67 pp.

AMEC (Earth and Environmental Limited), 2000. Monitoring long-term effects of acid emissions in northeast Alberta—1998 annual report. Prepared for the Wood Buffalo Environmental Association, Calgary, AB, Canada.

Bachmann, J.D., 2007. Will the circle be unbroken: a history of the US national ambient air quality standards-2007 critical review. J. Air Waste Manag. Assoc. 57 (6), 652–697.

Beckingham, J.D., Archibald, J.H., 1996. Field guide to ecosites of northern Alberta. Natural Resources Canada, Canadian Forest Service, Northwest region, Northern Forestry Centre, Special Report 5, Edmonton, AB, Canada.

Berryman, S., Straker, J., Krupa, S., Davies, M., Ver Hoef, J., Brenner, G., 2010. Geospatial deposition mapping using lichens in the Athabasca Oil Sands Region. Extended Abstract 2010-A-1002-AWMAIn: Proceedings of the Air and Waste Management Association Annual Conference, June 22–25, 2010, Calgary, AB, Canada.

CAPP, 2011. The Facts on: Oil Sands 2010. Canadian Association of Petroleum Producers, Calgary, AB.

Cheng, L., Angle, R.P., Peake, E. And, Sandhu, H., 1994. Effective acidity modelling to establish deposition objectives and manage emissions. Atmos. Environ. 29, 383–392.

Cheng, Y., Cai, Z., Zhang, J., Chang, S.X., 2011. Gross N transformations were little affected by 4 years of simulated N and S depositions in an aspen-white spruce dominated boreal forest in Alberta, Canada. For. Ecol. Manage. 262, 571–578.

Clair, T.A., Burns, D.A., Rosas, I.R., Blais, J., Percy, K.E., 2011. Ecosystems. In: Hidy, G.M., Brook, J.R., Demerjian, K.L., Molina, L.T., Pennell, W.T., Scheffe, R.D. (Eds.), Technical Challenges of Multipollutant Air Quality Management. Springer Publication, Dordrecht, Germany, pp. 139–230.

D'Eon, S.P., Magasi, L.P., Lachance, D., DesRochers, P., 1994. ARNEWS: Canada's national forest health monitoring plot network: manual on plot establishment and monitoring (revised). Natural Resources Canada, Canadian Forestry Service, Petawawa National Forestry Institute., Chalk River, ON, Canada. Inf. Rep. PI-X-117.

Elzinga, C.L., Salzer, D.W., Willoughby, J.W., Gibbs, J.P., 2001. Monitoring Plant and Animal Population. Blackwell Science, Malden, MA.

Environment Canada and Health Canada, 2010. Risk management strategy for mercury. Ottawa ON, Canada. Available from: http://www.ec.gc.ca/Publications/9B24BD24-7D0B-4A1E-BFE0-53DC4137ED90/Risk_Management_Strategy_for_Mercury_e.COM1241.pdf.

Ferretti, M., 1997. Forest health assessment and monitoring. Issues for consideration. Environ. Monit. Assess. 48, 45–72.

Fenn, M.E., Sickman, J.O., Bytnerowicz, A., Clow, D.W., Molotch, N.P., Pleim, J.E., Tonnesen, G.S., Weathers, K.C., Padgett, P.E., Campbell, D.H., 2009. Methods for measuring atmospheric nitrogen deposition inputs in arid and montane ecosystems of western North America. In: Legge, A.H. (Ed.), Air Quality and Ecological Impacts: Relating Sources to Effects. Elsevier, Amsterdam, The Netherlands, pp. 179–228.

Foster, K. R., Baines D., Percy, K.E., Legge, A.H., Maynard, D.G., Chisholm, V., 2011. Wood Buffalo environmental association, terrestrial environmental effects monitoring. Forest health monitoring program procedures manual. Wood Buffalo Environmental Association Report, Fort McMurray, AB, Canada.

Hidy, G.M., Brook, J.R., Demerjian, K.L., Molina, L.T., Pennell, W.T., Scheffe, R.D. (Eds.), 2011. Technical Challenges of Multipollutant Air Quality Management. Springer Publication, Dordrecht, Germany.

Hinckley, T., Ford, D., Segura, G., Sprugel, D., 1992. Key processes from stand to tree level. In: Wall, G. (Ed.), Implications of Climate Change for Pacific Northwest Forest Management. University of Waterloo, Department of Geogrpahy, Waterloo, ON, pp. 33–43 Paper 15.

Hogan, G.D., Wotton, D.L., 1984. Pollutant distribution and effects in forests adjacent to smelters. J. Environ. Q. 13 (3), 377–3881.

Holowaychuk, N., Fessenden, R.J. 1987. Soil sensitivity to acid deposition and the potential of soils and geology in Alberta to reduce the acidity of acidic inputs. Earth Sciences Rep. 87-1, Terrain Sciences Department, Alberta Research Council, Edmonton, AB, Canada.

Hunsaker, C.T., 1993. New concepts in environmental monitoring: the question of indicators. Sci. Total Environ. Suppl. Part 1, 77–96.

Innes, J.L., 1993. Forest Health: Its Assesment and Status. Commonwealth Agricultural Bulletin, Wallingford, UK.

Jäger, H.-J., Krupa, S.V., 2009. Hormesis—its relevance in phytotoxicology. In: Legge, A.H. (Ed.), Air Quality and Ecological Impacts: Relating Sources to Effects. Elsevier, Amsterdam, The Netherlands, pp. 137–152.

Jones, C.E., Associates., 2007. Terrestrial environmental effects monitoring—acidification monitoring program: 2004 sampling event report for soils, lichen, understory vegetation and forest health and productivity. Prepared for Wood Buffalo Environmental Association, Fort McMurray, AB, Canada.

Karnosky, D.F., Pregitzer, K.S., Hendrey, G.R., Percy, K.E., Zak, D.R., Lindroth, R.L., Mattson, W.J., Kubiske, M., Podila, G.K., Noormets, A., McDonald, E., Kruger, E.L.,

King, J., Mankovska, B., Sober, A., Awmack, C., Callan, B., Hopkin, A., Xiang, B., Hom, J., Sober, J., Host, G., Riemenschneider, D.E., Zasada, J., Dickson, R.E., Isebrands, J.G., 2003a. Impacts of interacting CO_2 and O_3 on trembling Aspen: results from the Aspen FACE experiment. Funct. Ecol. 17, 289–304.

Karnosky, D.F., Percy, K.E., Chappelka, A.H., Percy, K.E., 2003b. Air pollution and global change impacts on forest ecosystems: monitoring and research needs. In: Karnosky, D.F., Percy, K.E., Chappelka, A.H., Simpson, C., Pikkarainen, J. (Eds.), Air Pollution, Global Change and Forests in the New Millenium. Elsevier Ltd., Oxford, UK, pp. 447–459.

Kindzierski, W., 2010. Ten-year trends in regional air quality for criteria pollutants in the Athabasca Oil Sands Region. Extended Abstract 2010-A-1079In: Proceedings of the Air & Waste Management Association Annual Conference, June 22–25, 2010, Calgary, AB, Canada.

Kolb, T.E.M.R., Wagner, M.R., Covington, W.W., 1994. Concepts of forest health: utilitarian and ecosystem perspectives. J. Forestry 92, 10–15.

Krupa, S.V., Legge, A.H., 1998. Sulphur dioxide, particulate sulphur and its impacts on a boreal forest ecosystem. In: Ambasht, R.S. (Ed.), Modern Trends in Ecology and Management. Backhuys Publishers, Leiden, The Netherlands, pp. 285–306.

Laxton, D.L., Watmough, S.A., Aherne, J., 2012. Nitrogen cycling in Pinus banksiana and Populus tremuloides stands in the Athabaska oil sands region, Alberta, Canada. Water Air Soil Pollut. 223 (1), 1–13.

Legge, A.H., Corbin, J., Bogner, J., Strosher, M., Krouse, H.R., Laishley, E.J., Bryant, R.D., Cavey, M.J., Prescott, C.E., Nosal, M., Schellhase, H.U., Weidensaul, T.C., Mayo, J.M., 1988. Acidification in a temperate forest ecosystem: the role of sulphur gas emissions and sulphur dust. A final report submitted to the Whitecourt Environmental Study Group. University Calgary, Kananaskis Centre for Environmental Research, Calgary, AB, Canada.

Lucier, A., Ayres, M., Karnosky, D., Thompson, I., Loehle, C., Percy, K., Sohngen, B., 2009. Forest responses and vulnerabilities to recent climate change In: Seppala, R., Buck, A., and Katila, P. (Eds.). Adaptation of forests and people to climate change—A global assessment report. International Union of Forest research Organizations World Series Volume 22. Helsinki, Finland, pp. 31–52.

Maynard, D.G., Stadt, J.J., Mallett, K.I., Volney, W.J.A., 1994. Sulfur impacts on forest health in west-central Alberta. Natural Resources Canada, Canadian Forestry Service, Northwest Region, Northern Forestry Centre, Edmonton, AB, Canada. Inf. Rep. NOR-X-334.

McLaughlin, S.B., Percy, K.E., 1999. Forest health in North America: some perspectives on actual and potential roles of climate and air pollution. Water Air Soil Pollut. 116, 151–197.

McLinden, C.A., Fioletov, V., Boersma, K.F., Krotkov, N., Sioris, C.E., Veefkind, J.P., Yang, K., 2011. Air quality over the Canadian oil sands: a first assessment using satellite observations. Geophys. Res. Lett. 39, L04804.

Merriam-Webster, 1993. Webster's third new international dictionary. Meriiam-Webster, Inc., Springfield, MA.

Miller, P.R., McBride, J.R. (Eds.), 1999. Oxidant Air Pollution Impacts in the Montane Forests of Southern California. A Case Study of the San Bernardino Mountains. Springer Verlag, New York, NY.

O'Laughlin, J., Livingston, R.L., Thier, R., Thornton, J.P., Toweill, D.E., Morelan, L., 1994. Defining and measuring forest health. J. Sustain. For. 2 (1–2), 65–85.

Percy, K.E., 2002. Is air pollution an important factor in international forest health? In: Szaro, R.C., Bytnerowicz, A., Oszlanyi, J. (Eds.), Effects of Air Pollution on Forest Health and Biodiversity in Forests of the Carpathian Mountains. IOS Press, Amsterdam, The Netherlands, pp. 23–42.

Percy, K.E., Ferretti, M., 2004. Air pollution and forest health: toward new monitoring concepts. Environ. Pollut. 130 (1), 113–126.

Percy, K.E., Karnosky, D.F., 2007. Air quality in natural areas: interface between the public, science and regulation. Environ. Pollut. 149 (3), 256–267.

Percy, K., Legge, A., Maynard, D., Jaques, D., Straker, J., Visser, S., 2007. TEEM new design. Presentation at the WBEA-TEEM/CEAM NSMWG Integration Workshop, October 16–19, Banff, AB, Canada.

Percy, K.E., Maynard, D.G., Legge, A.H., 2010. Going forward: enhancing the WBEA Terrestrial Effects Monitoring Program. Extended Abstract 2010-A-1009-AWMAIn: Proceedings of the Air & Waste Management Association Annual Conference, June 22–25, 2010, Calgary, AB, Canada.

Scott, C.T., 1998. Sampling methods for estimating change in forest resources. Ecol. Appl. 8, 228–233.

United Nations Forum on Forests, 2003. E/CN.18/2003/13 ● E/2003/42.

Visser, S., 2006. Oil sands plant emissions in north-eastern Alberta: terrestrial environmental effects monitoring (TEEM) program. Monitoring soil chemistry, soil biology and ectomycorrhizae in jack pine stands—2004 results. Report prepared for Wood Buffalo Environmental Association, Fort McMurray, AB, Canada.

Chapter 10

Ecological Analogues for Biomonitoring Industrial Sulfur Emissions in the Athabasca Oil Sands Region, Alberta, Canada

D.R. Jaques[*,1] and A.H. Legge[†]

[*]Ecosat Geobotanical Surveys Inc., North Vancouver, British Columbia, Canada
[†]Biosphere Solutions, Calgary, Alberta, Canada
[1]Corresponding author: e-mail: drjaques@shaw.ca

ABSTRACT

An ecological analogue system for biomonitoring the chronic and long-term effects of anthropogenic atmospheric emissions in the Alberta Oil Sands Region (AOSR) is described. This system has shown to be an efficient adjunct to ambient air quality measurements and has been previously applied successfully in western Canada. The essence of an ecological analogue system is the classification and identification of plant associations that are most sensitive to the atmospheric emissions of concern. An ecosystem classification and ordination was applied to sites of the most sensitive plant associations to identify detailed ecological analogue types (EATs). The EATs were then selected for use in locating field sites for the WBEA Forest Health Monitoring Program.

Twenty-one major plant associations were identified within the AOSR with jack pine (*Pinus banksiana*)/bearberry (*Arctostaphylos* sp.)/green reindeer lichen (*Cladina mitis*) communities considered most sensitive. Among those, nine EATs most sensitive to atmospheric emissions were identified by classification and ordination techniques. These EATs possessed 10 specific ecological parameters necessary for field identification and mapping. Field sites were located near major AOSR emission sources, radiating outwards from ~ 18 to 130 km. A significant and high, nonlinear negative correlation ($r = -0.98$) was determined between the foliar inorganic/organic sulfur ratios in first year jack pine needles and the distance from the SO_2 sources. This foliar sulfur ratio metric coupled with other growth parameters provided a robust measure for deploying the ecological analogue system to monitor for the biological effects from the atmospheric chemical species of concern.

Disclaimer: The content and opinions expressed by the authors do not necessarily reflect the views of the WBEA or of the WBEA membership.

Developments in Environmental Science, Vol. 11. http://dx.doi.org/10.1016/B978-0-08-099443-7.00010-7

10.1 INTRODUCTION

While equipment to measure ambient air quality from industrial emissions forms the backbone of any air quality monitoring/regulatory system, those equipment-based systems are expensive to obtain, maintain, and analyze the resulting data. Those factors limit the number of such sites deployed throughout a region of the size (56,400 km^2) of the Athabasca Oil Sands (AOSR), particularly when access by land and availability of electrical power are limiting factors. Biomonitoring methods offer an adjunct, cost-effective, and biologically meaningful support (effects/impact assessment) for such equipment-based ambient air quality monitoring systems. Optimum biomonitoring methods make use of living organisms to monitor the relative effects of atmospheric emissions from industrial activity.

An ecological analogue type (EAT) biomonitoring system produces a cost-effective method to monitor large areas over long periods of time. The EATs have been used very successfully for monitoring the long-term biological effects of sour gas emissions on boreal forest trees for 25 years from 1976 to 2001 as part of the West Whitecourt Case Study in west-central Alberta and other locations in western Canada (Jaques, 2002; Krupa and Legge, 1998; Legge and Bogner, 1983; Legge et al., 1977, 1981, 1988a,b; Prietzel et al., 2004). These systems successfully utilized a specific biogeo-climatic ecosystem type (*sensu* Braun-Blanquet, 1965) that possesses a minimum of variability both within and between tree stands across the landscape. This chapter documents the ecological analogue system which has been developed and applied in the AOSR.

The process of defining and locating a suitable EAT most sensitive to acid-forming emissions and found in close proximity to those emissions is described below for the reader to understand the application on a regional scale. The identification and classification of EATs have been developed from many years of foundational work. Some of the earliest work in establishing the patterns and factors controlling vegetation is by von Humboldt (1805, 1806). He used plant taxonomic principles along with plant geographic features to define ecosystem types. Schouw (1823) in Denmark was the first biologist to clearly define numerous plant communities. In the mid-nineteenth century, Grisebach (1866, 1872) expanded these concepts to include vegetation characteristics and substrate information and termed the new field "geobotany." Subsequent study of geobotanical units was conducted in the late nineteenth and early twentieth centuries; most notably by Kerner (1863), Gray and Hooker (1880), Macoun (1882), Merriam (1890, 1899), Cowles (1899), Schimper (1903), Clements (1905, 1916), Cajander (1909), Warming (1909), Moss (1910), Raunkiaer (1913), Vestal (1913), Visher (1916), Weaver (1917), Nichols (1917), and Livingstone and Shreve (1921).

From the 1920s through to the present day, thousands of synecological and autecological studies have produced findings on major factors controlling vegetation development. Syntheses of these many individual studies led to the development of geobotanical classification systems in Europe and North America. The most significant geobotanical classification systems have been developed by Sukachev (1928, 1932, 1945), Braun-Blanquet (1928, 1932, 1965), and Daubenmire (1952, 1966, 1968). These systems led directly to the "biogeoclimatic classification system" established by Krajina (1933, 1960, 1965, 1969) in Europe and British Columbia, Canada, based largely on the work of Braun-Blanquet but modified and enhanced by the works of Sukachev (1928) and Daubenmire (1968).In the northern oil sands region of Alberta, an ecological classification system based directly on the Braun-Blanquet system was developed by Beckingham and Archibald (1996).

10.2   OVERVIEW OF METHODS

A classification and ordination of ecosystem types was conducted using 176 sites located throughout the AOSR. Quantitative vegetation composition–cover–relative frequency-stand structure (on 15×50 m macroplots), soils conditions, bedrock and surficial parent materials types, mesoclimatic conditions, and inferred microclimatic conditions were analyzed at each site. Vegetation cover was measured at each site using either a line-point sampling method (200 points along each macroplot long axis) or 15×50 cm microplots (30 cm along each long axis). The Sorensen similarity index (Sorenson, 1948) was calculated for each plot; the resulting index metrics were applied into an ordination program (Henderson and Davis, 1977) for vegetation mapping procedures. These analyses were done to identify the driest and most nutrient poor EAT.

The sites were sampled in 2008, 2009, and 2010. The EAT most sensitive to acidifying emissions (i.e., most acidic parent material, highest soils sand content, lowest cation-exchange capacity, pH, macronutrient, and micronutrient contents in recent jack pine foliage) was identified, and a site of this type was selected as close to the source(s) of anthropogenic atmospheric emissions as possible. Then, sites of the same EAT were located in distance gradients away from the sources, both parallel and perpendicular to the major wind rose axes identified using local meteorological data. A search of the AOSR was conducted based upon digital prestratification of the designated study region ranging away from the emissions point as follows: 77 km to the north, 110 km to the west, 100 km to the east, and 118 km to the southeast. The sites were prestratified using digital ArcView Shape files over UTM-corrected LANDSAT TM satellite imagery mosaics of the AOSR. This prestratification used the following delineations:

1. Ecological region and subregion—Boreal Region and Central Mixed-wood Subregion (Downing and Pettapiece, 2006);
2. Ecological districts—the Mackay Plain, Embarras Plain, Steepbank Plain, Lake Athabasca Plain, Wabasca Plain, Garson Lake Plain, and Firebag

Hills ecodistricts of the Central Mixed-wood Subregion of the Boreal Forest Region (Canada Committee on Ecological Classification, 1999);
3. Bedrock geology—noncalcarerous bedrock types (A.E.U.B., 2001; A.G. S./A.E.U.B., 1999);
4. Surficial materials—aeolian or glaciofluvial sand deposits (Andriashek, 2003; Bayrock, 1971; Bayrock and Reimchen, 1973);
5. Soils types—the Marguerite, Harrison, Firebag, Kearl, and Mildred Series (Turchenek and Lindsey, 1982); and
6. Elevation contours—areas above 280 m and below 640 m (Natural Resources Canada, 2009).

Once the prestratification was accomplished, potential EAT sites with the constraints identified above were located by stereo air-photo interpretation of high-resolution (1:15,000 scale) 9×9 in. color infrared aerial photography (Mission 98-052, Alberta Air-photo Archives, Department of Natural Resources, Edmonton, Alberta).

The ecological characteristics of each potential biomonitoring site were analyzed in detail in August of the year of sampling, 2008, 2009, and 2010 to insure optimum plant phenological development.

Based upon the classification and ordination of these 176 macroplots, analyses of a few selected measures of the impacts of acidifying emissions on the dominant plant species (jack pine-*Pinus banksiana*-first year needle tissues, blueberry-*Vaccinium myrtilloides*-leaf tissue, green reindeer lichen-*Cladina mitis*-whole tissue) at each ecological analogue site were conducted as follows: sulfur analyses including inorganic sulfate (Johnson and Nishita, 1952), total sulfur by combustion (Jackson et al., 1985) with LECO SC132 analysis (Pacific Soil Analysis Inc., Richmond, B.C.), and organic sulfur by subtraction of inorganic S from total S and the ratio of inorganic to organic S.

Apical branchlets from the upper crown of five jack pine trees were collected at each site in late September–early October of each year. Each of the frozen branch samples were age-classed into first-, second-, and third-year needle tissues and then refrozen until air-dried at 70 °C for 48 h in a ventilated drying oven The dried samples were finely ground using a stainless steel grinder and submitted for chemical analyses. Blueberry and green reindeer lichen samples were collected from open-edge microsites and composited into one sample per site. Tests for significant differences between measures were conducted using Duncan's new multiple range Test (Duncan, 1955).

10.3   RESULTS

10.3.1   Atmospheric Emissions

A summary of 2009 emissions per AOSR source is listed in Table 10.1. Sources 1, 2, and 3 are located in close proximity to one another, while Source 4 is located 79 km southeast of the other three. The three upgrading

TABLE 10.1 Summary Listing of Air Emissions in 2009 from Major Sources in the AOSR

Air emissions	Plant #1 (tons)	Plant #2 (tons)	Plant #3 (tons)	Plant #4 (tons)	Others (tons)	Totals (tons)
SO_2	81,470.7	19,027.1	10,571.6	7150.6	19,570.5	137,790.5
NO_x	13,931.8	11,692.6	1759.7	1653.2	8612.0	37,649.3
VOCs	8402.0	27,885.7	27,839.8	161.3	34,962.8	99,251.6
PM 2.5	899.5	965.0	142.3	23.0	538.3	2568.1
Totals	104,704.0	59,570.4	40,313.4	8988.1	63,683.6	277,259.5

Source: National Pollution Release Inventory (NPRI), Environment Canada.

sources located in close proximity to one another produced 81% of the total SO_2 emissions, and 74% of all emitted species reported and listed by the AOSR (NPRI, 2009).

10.3.2   Wind Patterns

Wind rose data from the Wood Buffalo Environmental Association (WBEA) meteorological station nearest to three major industrial emission sources were acquired from WBEA. These data for the growing season of 2007 are shown in Figure 10.1. It was assumed that wind rose data for 2007 are generally characteristic of most years on record.

Figure 10.1 shows that during the growing season of 2007 (April 1–September 15), the major wind directions at the 10 m height were predominantly from the north quadrant ($315°–45°$) followed by the south-southeast quadrant ($135°–225°$). These two quadrants account for 66% of the directional wind patterns. Westerly winds account for another 27% of the total. If these patterns apply to the growing seasons of most years, then one would expect major impingement of the industrial emission plumes onto ecosystems located to the north, south, and east of the emissions. As the north and south wind patterns largely follow the major Athabasca River Valley, and this valley airflow appears to extend several thousand feet above the surrounding plains based upon limited data (Martin Hansen, WBEA, personal communication), the effects from the industrial emissions should be most evident at ecological analogue sites along the Athabasca River Valley.

10.3.3   EAT Classification

EATs were defined based upon those found immediately in close proximity to the major sources of atmospheric emissions, and at major cardinal directions

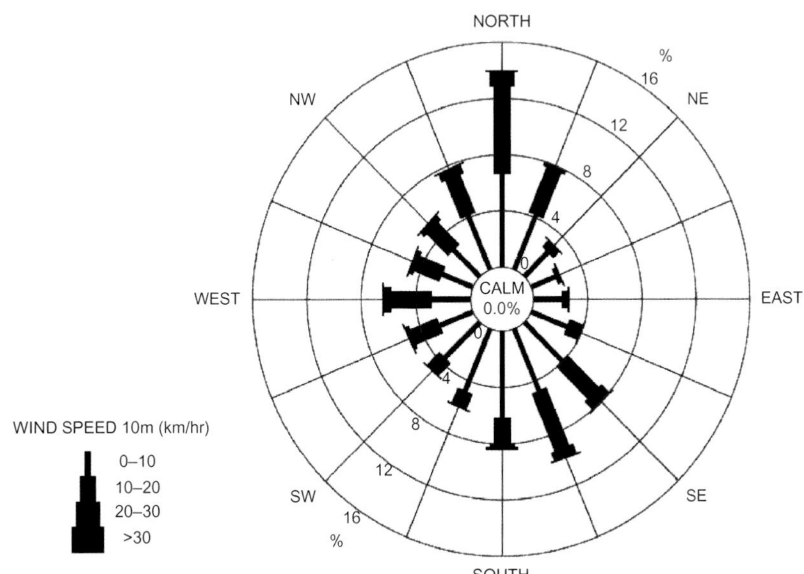

FIGURE 10.1 Wind rose data at the WBEA Mildred Lake air monitoring station (AMS-2) for the 2007 growing season.

extending outward from these sources. The 176 sites surveyed in the AOSR produced the set of Braun-Blanquet plant associations as shown in Table 10.2.

The ordination of these 176 sample sites grouped into their respective biogeoclimatic association types is shown in Figure 10.2. One can see that the biogeoclimatic association types sorted themselves along the nutrient (X-axis) and moisture (Y-axis) regimes as defined (Figure 10.2). While this classification, and the resulting ordination of ecological association types, is not complete for the AOSR, it is a comprehensive summary of the major associations found near the main AOSR emissions sources.

As this ordination shows, Association Type #21 (i.e., the jack pine/bearberry/ lichen type) represents the driest and most nutrient-poor ecosystem type found within close proximity to the three major oil sands processing facilities. This was the ecosystem type chosen for subsequent detailed biomonitoring analysis.

Association Type #21 (jack pine/bearberry/green reindeer lichen), when found in areas where the following prestratification characteristics overlap, identified optimum locations to search for EAT sites of Association Type #21, namely:

1. Areas of the Central Mixed-wood Boreal Ecological Subregion;
2. Elevations between 280 m and 940 m;

TABLE 10.2 Plant Associations of the Boreal Central Mixed-Wood Subregion, NE Alberta

Association Type	1	5	4	2	3	6	7	10	9	8	15	14	11	12	13	18	17	16	20	19	21	21 (3)
# Plots	4	2	8	6	8	1	4	2	3	1	3	2	1	2	8	5	12	12	10	10	3	69
Charcteristic spp.																						
Carex aquatilis	**17.0**	**5.0**	**1.0**	**3.0**	**5.0**																	
Sphagnum spp.			**12.0**	**9.0**	**7.0**	0.3		**7.0**														
Salix planifolia	**32.0**	**9.0**		**6.0**	**4.0**	**35.0**	3.0		3.0		1.0						0.3	0.1				
Salix bebbiana	3.0	2.0	0.1		2.0	**15.0**	0.1			**25.0**	0.1	0.1	**6.0**		1.0	0.1	2.0	1.0	0.1	2.0	2.0	
Larix laricina	0.3		**0.1**	**10.0**	**5.0**		3.0	0.1														
Betula spp.			**3.0**	**23.0**	**10.0**		0.1	3.0	0.1													
Picea mariana	0.3		**20.0**	**5.0**	**21.0**	**8.0**	**38.0**	**25.0**	**21.0**	**3.0**	**22.0**	**7.0**			**4.0**		**2.0**	**1.0**	**20.0**	**6.0**		
Picea glauca	0.3	**9.0**					1.0			**10.0**	**5.0**	**22.0**	**35.0**	**15.0**	**13.0**	**14.0**	**25.0**	**21.0**	**3.0**	**18.0**	**6.0**	
Populus tremuloides	0.1	**22.0**				0.1	1.0			**10.0**	**8.0**	**11.0**	**21.0**	**6.0**	**15.0**	**23.0**	**28.0**	**15.0**	**22.0**	**7.0**	**20.0**	**30.0**
Pinus banksiana		0.1	4.0							0.1	0.1				2.0	2.0	0.1	0.1	**15.0**	3.0	**28.0**	**23.0**
Cladina mitis																		4.0			**2.0**	**48.1**
Vaccinium myrtilloides									1.0								0.1	0.1	1.0		**8.0**	**11.0**
Arctostaphylos uva-ursi																		0.1			**18.0**	**9.9**

Continued

TABLE 10.2 Plant Associations of the Boreal Central Mixed-Wood Subregion, NE Alberta—Cont'd

Association Type	1	5	4	2	3	6	7	10	9	8	15	14	11	12	13	18	17	16	20	19	21	21 (3)
# Plots	4	2	8	6	8	1	4	2	3	1	3	2	1	2	8	5	12	12	10	10	3	69
Differential spp.																						
Ledum groenlandicum			**14.0**	1.0	**10.0**	0.3	**14.0**	**7.0**	**6.0**		**5.0**	0.1	1.0	2.0	0.1	1.0	1.0	0.1	**10.0**	**4.0**		0.1
Carex spp.	**5.0**		**2.0**	1.0	**2.0**		**3.0**	0.1	**3.0**	1.0	**2.0**	0.1					1.0		**3.0**	**2.0**		
Cladonia scabriuscula								**12.0**														
Betula papyrifera	0.3				0.1		0.1			1.0				**3.0**	**0.1**	**1.0**	**2.0**	**2.0**	0.1	0.1	0.1	
Calamagrostis canadensis	**8.0**	**7.0**	2.0	2.0	2.0					0.3		0.1	**8.0**	**8.0**	**2.0**	**4.0**	**3.0**	**3.0**		0.1		
Populus balsamifera	3.0	**8.0**			0.1		**4.0**	2.0		**4.0**		**4.0**	**4.0**	**15.0**	**7.0**	0.1	**7.0**	**2.0**		0.3		
Alnus tenuifolia	1.0		2.0			0.1	3.0				**5.0**		**8.0**	**3.0**	0.1		0.1	0.1		0.3		
Shepherdia canadensis								0.1			**3.0**		**1.0**	**1.0**	**2.0**	**2.0**	0.1	**2.0**	0.1	**2.0**	**2.0**	
Viburnum edule									2.0	1.0				**6.0**	**5.0**	**3.0**	**8.0**	**8.0**	0.1	**5.0**	**3.0**	
Elymus innovatus														0.1	1.0	0.1	0.1	0.1		1.0	0.1	**0.3**
Cornus stolonifera														**2.0**	**3.0**	**0.1**	**1.0**	**0.1**				
Lonicera involucrata					0.3									**1.0**	**0.1**	**1.0**	**0.1**	**1.0**	0.1			
Rosa acicularis									0.1	0.1				**1.0**	**3.0**	**3.0**	**2.0**	**3.0**	**1.0**	**3.0**	**2.0**	0.4
Hylocomium splendens				1.0										**5.0**	**3.0**	**5.0**	**2.0**	**2.0**				

Association Type	1	5	4	2	3	6	7	10	9	8	15	14	11	12	13	18	17	16	20	19	21	21 (3)
# Plots	4	2	8	6	8	1	4	2	3	1	3	2	1	2	8	5	12	12	10	10	3	69
Pleurozium schreberi			**16.0**		**5.0**	0.3	**20.0**		**20.0**		**14.0**	0.1		0.1	**2.0**	**3.0**	0.1	**12.0**	**20.0**	**7.0**	**12.0**	**2.6**
Cornus canadensis							**1.0**		**1.0**		**3.0**		**1.0**	0.1		**2.0**	**2.0**	**2.0**		0.1	**3.0**	0.3
Alnus crispa																**2.0**		1.0	**2.0**		**9.0**	0.2
Solidago simplex																						**0.1**
Carex siccata																						**0.2**
Oryzopsis pungens																						**0.3**
Cladina stellaris																						**0.7**

The 21 association types are defined by their characteristic and differential species: #, Name; 1, Sedge/Willow; 2, Tamarack/Dwarf birch; 3, Black spruce-Tamarack/Dwarf birch; 4, Black spruce/Labrador tea; 5, Poplar-White spruce/Willow/Sedge; 6, Tall willow; 7, Black spruce-Poplar/Labrador tea; 8, White & Black spruce-Poplar/Willow; 9, Black spruce-Aspen/Labrador tea; 10, Black spruce/Labrador tea/Lichen; 11, White spruce-Aspen/Willow/Reedgrass; 12, Aspen-Poplar-White spruce/Reedgrass; 13, Aspen-Poplar-White spruce/Dogwood; 14, Poplar-White spruce/Mountain alder; 15, Black-White spruce-Aspen/Labrador tea; 16, White spruce-Aspen/Highbush cranberry; 17, White spruce-Aspen/Feathermoss; 18, Aspen-White spruce/Reedgrass; 19, White & Black spruce-Aspen/Labrador tea; 20, Black spruce-Aspen-Jack pine/Labrador tea; 21, Jack pine/Bearberry/Green Reindeer lichen. The underlined/bold values indicate dominant cover Characteristic or Differential species and bold values indicate abundant but not dominant Characteristic or Differential species.

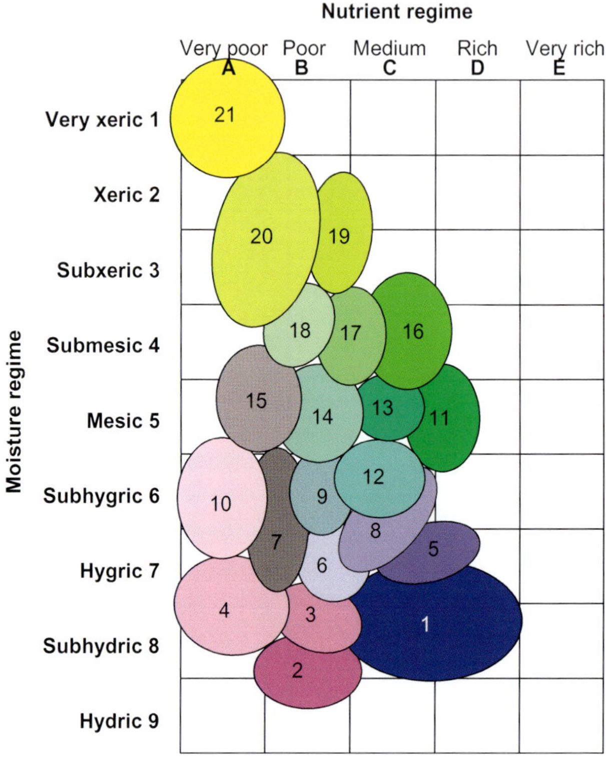

FIGURE 10.2 Major Biogeoclimatic plant association types within the Central Mixed-wood Subregion of the Boreal Forest of Northern Alberta.

3. Bedrock of the calcareous Devonian Waterways Formation is not found in the areas of interest; if it is, then it is covered by at least 10 m of noncalcareous surficial materials;

4. Surficial materials of noncalcareous glaciofluvial or aeolian sands at least 3 m thick; and

5. Soil types are of the Marguerite, Harrison, Firebag, Kearl, and Mildred Series.

10.3.4 Air-Photo Interpretation of EAT #3 Sites

Based upon the above prestratification restrictions, high-resolution color infrared aerial photography was used to locate potential EAT #3 sites having the following characteristics:

1. Jack pine (*Pinus banksiana*) dominant with little, if any, black spruce (*Picea mariana*), aspen (*Populus tremuloides*), white spruce (*Picea glauca*), or birch (*Betula papyrifera*);

2. Jack pine canopy closure of between 24% and 45%;

3. Dominant jack pine tree densities of between 720 and 1360 stems/ha;
4. Stand age classes of between 54 and 96 years; and
5. A rapid transition from the optimum Association Type #21 to the wetter Association Types #20, 19, 15, 10, 9, 7, 4, 3, 2, and 1. This ecotonal transition should be less than 5 m in horizontal distance and should be bounded on the downslope topographic position by Association Types #4, #7, #10, and possibly also Association Type #15.

From the air-photo interpretation, 392 potential sites were identified in the area of study. This area of study covers about 20,500 km^2 (42.4% of the total AOSR area of 48,400 km^2). These potential sites were surveyed by helicopter and on the ground in 2008, 2009, and 2010. The results of these surveys determined that of the 392 potential ecological analogue sites, 157 represent high-quality Association Type #21 sites. Of these 157, 31 sites proved to be optimum for use in this study based upon the following restriction criteria:

1. Association type #21 (jack pine/bearberry/green reindeer lichen);
2. Age class of site is between 54 and 96 years (as of 2012);
3. Maximum dominant jack pine tree height is between 14.5 and 20.3 m;
4. Jack pine stand density and canopy cover range between 720 and 1360 stems/ha and 24–45%, respectively;
5. Dominant jack pine diameter at breast height is between 15.8 and 21.7 cm;
6. No evidence of dwarf mistletoe (*Arceuthobium americanum*) in conifer cover;
7. Ground surface height of the outer edge of the Association Type #21 stand above the adjacent bog or poor fen mean spring water level is >1.8 m and <2.9 m;
8. Distance from the upper edge of the mean spring water level at the bog or poor fen adjacent to Association Type #21 is no more than 5 m in horizontal distance;
9. Ae and B soil horizons contain a minimum of 92% sand;
10. Ae and B soil horizon pH between 3.7 and 4.7;
11. Ae and B soil horizon cation exchange capacity between 1.25 and 4.82 meq/100 g;
12. B soil horizon exchangeable calcium below 450 ppm;
13. Black and white spruce, aspen, and birch overstory trees and seedlings are rare to nonexistent within the jack pine stand;
14. The following plant species exhibit a high relative frequency (>67%) in the sites: jack pine, bearberry (*Arctostaphylos uva-ursi*), blueberry (*V. myrtilloides*), yellow reindeer lichen (*C. mitis*), northern reindeer lichen (*Cladina stellaris*), and short-awned ricegrass (*Oryzopsis pungens*);
15. The following plant species exhibit very low canopy cover (0 to <0.3%) within the sites: Labrador tea (*Ledum groenlandicum*), green alder (*Alnus sinuata*), *Pleurozium schreberi*, ground dogwood (*Cornus canadensis*), hairy wildrye (*Elymus innovatus*), rough-leaved ricegrass (*Oryzopsis asperifolia*), willow (*Salix* spp.), and saskatoon (*Amelanchier alnifolia*);

16. Understory tree regeneration is composed of jack pine, with little or no black spruce, white spruce, aspen, or birch regeneration seedlings or saplings;
17. Sites that experienced a major fire which produced the dominant overstory age-class of jackpine and no more than one (1) understory fire since then; and
18. Association Type #21 site edge with the adjacent bog/poor fen site faces directly toward the major industrial emissions locations.

Characteristics 7 and 8 were important in the location and layout of an ecological analogue site. They related to the site being located where soil moisture was limited to annual precipitation rather than being augmented to any significant degree by groundwater. In addition, these two characteristics were also important in selecting the sampling location as close to the "edge" of the upland jack pine stands as possible. It is at the forest stand edge where exposure to air pollutants is maximized and where the air pollution plume tends to be most concentrated and possesses maximum impingement upon the trees at the edge of the stand.

Floristic data from the 176 sites were used to produce a refined Braun-Blanquet classification that identified EATs classified within Association Types #19, #20, and #21. Table 10.3 shows the plant species composition, cover, and relative frequency data for each EAT. Figure 10.3 shows the ordination of the nine EATs defined by this process. EATs #1, 2, 3, and 5 are included within Association Type #21 (jack pine/bearberry/lichen); EATs #6, 8, 9, and 10 are included within Association Type #20 (Black spruce–Aspen–Jack pine/Labrador tea); and EAT #4 is included within association type #19 (White & Black spruce-Aspen/Labrador tea).

EAT #3 shown in (Figure 10.3) is the driest and most nutrient poor of the nine identified. Any biomonitoring site should fall within this EAT and possess the stand characteristics listed above. Sites made up of EAT #2 could also serve as biomonitoring sites if they are located in important monitoring areas where sites of EAT #3 do not exist.

However, caution must be exercised in the use of EAT #2 for several reasons. EAT #2 sites possess the following:

- A significantly greater soil moisture regime than EAT #3 sites;
- A significant cover of seedlings and pole-sized advanced regeneration of black spruce;
- A higher cover of *Cornus canadensis*, *Pleurozium* sp., *Lycopodium complanatum*, and *Ledum groenlandicum* indicating their higher soil moisture and/or greater nutritional conditions;
- Ae soil horizon silt and clay content of at least 15% in EAT #2 as opposed to 8.0% in EAT #3 which contributes somewhat to the higher soil moisture availability; and
- "Control" ecological analogue sites of EAT #2 exhibit inorganic to organic sulfur ratios significantly greater than those of EAT #3, that is, 0.16 versus 0.09.

TABLE 10.3 Characteristics of Ecological Analogue Types (EATs) for Biomonitoring in the AOSR

Species (% cover/frequency)	Type #1	Type #2	Type #3	Type #4	Type #5	Type #6	Type #8	Type #9	Type #10
Characteristic species									
Pinus banksiana (overstory)	38.0/100	34.0/100	32.2/100	49	38	31.5/100	31.0/100	22.0	35.0
Pinus banksiana (seedlings)	2.1/50	0.1/100	0.1/33	0.5	1.5	0.1/100	0.5/100	0.5	0.1
Arctostaphylos uva-ursi	10.8/100	9.0/100	3.5/100	9.5	**1.7**	2.7/100	3.2/100	**1.0**	14.0
Vaccinium myrtilloides	**6.0/100**	**6.1/100**	3.4/100	4.3	4.3	**11.8/100**	**16.0/100**	**19.0**	**7.0**
Cladina mitis	39.0/100	**14.5/100**	**52.5/100**	24.7	37.4	**28.5/100**	49.0/100	**29.0**	**4.7**
Total characteristic spp.	**95.9**	**63.7**	**91.7**	**88.0**	**82.9**	**74.6**	**99.7**	**71.5**	**60.8**
Differential species									
Alnus crispa		0.1/100	0.1/33	**7.0**		**3.8/100**		3.0	
Amelanchier alnifolia	**0.6/100**				0.1				0.1
Anemone multifida	**0.4/100**								
Betula papyrifera				**0.5**					
Cornus canadensis	0.2/50	**2.7/100**	0.1/40	0.5			0.5/100		**3.2**
Elymus innovatus	0.1/50		0.1/100	0.3	2.4		**0.8/100**	2.0	**1.3**
Ledum groenlandicum		**1.0/100**				3.0/50	**0.1/50**	**0.1**	**0.1**
Linnaea borealis	1.2/100	2.2/100		0.4		0.5/50	0.5/50	**8.0**	2.4
Lycopodium complanatum	0.5/50	1.8/100		**3.5**		**6.0/50**			
Oryzopsis asperifolia				**0.2**	**0.3**				

Continued

TABLE 10.3 Characteristics of Ecological Analogue Types (EATs) for Biomonitoring in the AOSR—Cont'd

Species (% cover/frequency)	Type #1	Type #2	Type #3	Type #4	Type #5	Type #6	Type #8	Type #9	Type #10
Oryzopsis pungens	0.2/50	0.1/100	0.2/67	0.5	0.3	0.1	2.2/100	0.5	0.1
Picea glauca				**0.1**				**0.5**	
Picea mariana		**1.2/100**		**0.1**			**0.1/100**		
Populus tremuloides	**0.1**			**0.3**				**2.0**	
Salix sp.						**0.1/50**			
Shepherdia canadensis		0.1/25	0.3/18						3.0
Vaccinium vitis-idaea	2.7/100	**8.0/100**	2.3/67	2.5	0.1	**7.2/100**	3.0/100	7.0	**5.0**
Pleurozium schreberi	11.6/100	**18.9/100**	9.4/100	3.4	**37.2**		2.8/100	9.0	11.0
Cladonia gracilis				0.2	**1.5**	**13.5/100**	**1.8/100**	**0.5**	
Dicranum polysetum	1.5/100	2.7/100	2.0/100	4.1	**5.3**	0.1/100		0.5	
Total differential species	**19.3**	**38.8**	**14.7**	**23.4**	**47.2**	**34.3**	**12.0**	**33.1**	**26.1**
Common species									
Maianthemum canadense	0.5/100	2.8/100	2.5/100	2.0	4.0/100	0.5/100	3.3/100	0.5	
Rosa acicularis	0.7/50	0.3/100	0.5/33	0.1	0.1/100	0.2/50	0.5/100	0.5	0.1
Polytrichum juniperinum			0.4/67	0.6	0.1/100	0.4/100			
Cladina stellaris	1.3/100	0.1	5.2/100	**7.0**	2.0/100	0.1/50	0.5/100	0.5	0.8
Total common species	**2.5**	**3.2**	**8.6**	**9.7**	**6.2**	**1.2**	**4.3**	**1.5**	**0.9**
Total species cover	**117.7**	**105.7**	**115.0**	**121.3**	**136.3**	**110.1**	**116.0**	**106.2**	**87.9**

The bold values are a tool to assist in visual sorting of the Braun-Blanquet tabular data.

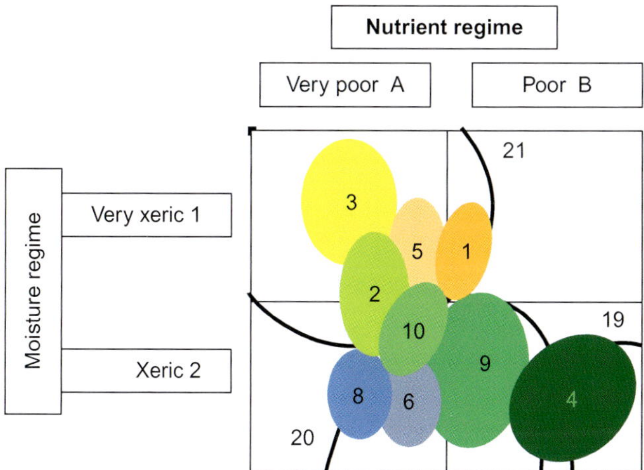

FIGURE 10.3 Ordination of ecological analogue types (EATs) within Plant Association Types #21, #20, and #19.

Although, EAT #2 sites are not directly comparable to EAT #3 sites, they can be used with caution when no other sites exist in a particular region of the AOSR where biomonitoring is required. Similarly, the other EATs with greater moisture and nutrient regimes than EAT #3 should not be used for biomonitoring (i.e., EATs #1, #4, #5, #6, #9, and #10). EAT #8 should not be used in biomonitoring work due to its greater moisture regime conditions than EAT #3.

10.3.5   Jack Pine Forest Stand Edges and Early Warning for Pollutant Effects

Numerous studies have documented and demonstrated that trees located at the edge of a forest stand and adjacent to a low physiognomic stature vegetation type (e.g., bog or fen) receive greater deposition of atmospheric emissions than trees located internal to a forest stand (Beier, 1991; Beier and Gunderson, 1989; Draaijers et al., 1988, 1994; Hager and Kazda, 1985; Hasselrot and Grennfelt, 1987; Krouse et al., 1984; Legge et al., 1977; Lester et al., 1987; Prietzel et al., 2004; Weathers et al., 2001). Foliage from branchlets from the upper crown of five jack pine trees (R2-Trees #1, #2, #3, #4, and #5) were sampled for inorganic and total sulfur at a site located 13.4 km northeast of two upgrading plant stacks (Jaques, 2007). Figure 10.4 shows that a jack pine tree located within the bog/poor fen fronting the upland jack pine stand (i.e., "R2-Tree#1") had a low foliar sulfur ratio of 0.1357; this tree is located 5 m into the fen and was only 5.8 m tall. The next tree (R2-Tree#2) was located at the closest edge of the jack pine stand and was 5 m from R2-Tree#1. This tree at the edge towers 10 m above the fen jack pine tree. Each of the

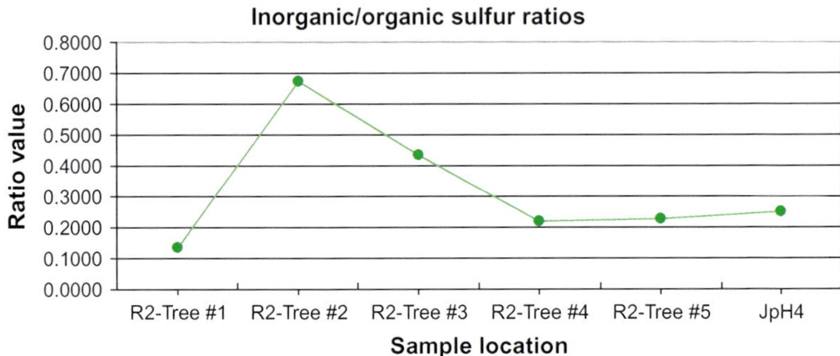

FIGURE 10.4   Inorganic//organic sulfur ratios of 2007 needles along the gradient from the Bog (Tree #1) to the Bog/Jackpine stand edge (Trees #2 and #3) to the Jackpine stand of Site R2 (Trees #4 and #5) and the extensive Jackpine stand at Site JpH4 (JP104 in Chapter 9).

remaining three trees (i.e., "R2-Tree#3," "R2-Tree#4," and "R2-Tree#5") was progressively located 5 m into the jack pine stand from the edge tree.

As Figure 10.4 shows, the foliar sulfur ratio within the current year needles of each pine tree is highest at the edge of the stand (0.67 for "R2-Tree#2") and progressively declines as one moves into the stand to 0.44 in "R2-Tree#3," 0.22 in "R2-Tree#4," and 0.23 in "R2-Tree#5." The foliar sulfur ratio at Site JPH4 (JP104 in Chapter 9) is 0.25, and this site is located 0.51 km north of these five jack pine forest stand edge trees. The sulfur ratio value at JPH4, located 0.51 km from the edge is not significantly different than that at the two trees located 10 and 15 m from the edge of the jack pine stand with the open bog/poor fen edge. This underscores the importance of correct location of trees for maximum early warning of the potential for detrimental foliar sulfur loading at forested sites.

10.4   APPLICATION EXAMPLES

An analysis of two studies of air pollution effects in the AOSR illustrates the importance of utilizing EAT classification to establish sites for sampling and analyses. The first study (Jones and Associates, 2007) did not use the EAT system, while the data collected for this present study did. Sites located close to two major industrial emissions sources in the AOSR (i.e., Jones and Associates. 2007—Site JPH4 and this study—Site 08186) were compared with sites located quite far to the north of the sources (i.e., Jones and Associates, 2007—Site JPL9 (JP109 in Chapter 9) and this study—Site 08395).

Results show that jack pine foliage from the two sites from the 2004 (Jones and Associates, 2007) sampling did not have significantly different ($p=0.10$) foliar inorganic/organic sulfur ratios, while our 2009 study showed the same site nearest to the source of industrial SO_2 emissions did have significantly elevated foliar sulfur ratios. The foliar sulfur ratios from these two separate studies are summarized below:

	Control site	Elevated sulfur exposure site
TEEM 2004 Data (JPL9 vs. JPH4)	0.1308^{ns}	0.2465^{ns}
EAT Ecosat 2009 Data (08395 vs. 08186)	0.0766^*	0.2622^*

ns, values in rows not significantly different $(p=0.10)$; $n=5$.
*values in rows not significantly different $(p=0.10)$; $n=5$.

As a result of the findings from the Jones and Associates (2007) study, no effects could be documented on jack pine forest stands; whereas, the 2009 study has documented that statistically significant sulfur loading has occurred at the Site JPH 4, 13.4 km away from the sources of the sulfur emissions. These data also indicate that this foliar sulfur loading, as of 2009, was at a "low" level (see Table 10.4).

Another pairing of past studies shows a similar problem when ecologically nonanalogous sites are used for analyses. In 1994, a study in the sour gas region of northeastern British Columbia (Case, 1994) measured total sulfur in lodgepole pine stands randomly located throughout the region. In 2002, the same region was studied by the senior author (Jaques, 2002) using ecologically analogous stands. The 1994 study did not find significant differences in total sulfur within lodgepole pine needles in any stands regardless of their distance from the major emissions source in northeastern British Columbia. In fact, this 1994 study found the highest total sulfur at great distances from the major source of industrial SO_2 emissions.

On the other hand, the Jaques (2002) ecological analogue study in this same region of northeastern British Columbia was able to detect a statistically significant $(p=0.10)$ elevated total sulfur concentration in lodgepole pine located 1 km from the emissions source. The Jaques (2002) study also determined that ecologically analogous sites showed a significant negative nonlinear regression relationship $(r=-0.93)$ between total pine needle sulfur levels with distance from the industrial source of SO_2 emissions. This ecological analogue study was also able to determine that the level of foliar sulfur loading was quite low and posed no apparent risk for the lodgepole pine. The nonecological analogue study by Case (1994), however, could not be used to draw any conclusions one way or the other regarding foliar sulfur loading near the sour gas processing plant.

The analyses of ecologically analogous jack pine stands in the AOSR have documented the robust and accurate capabilities of using the EAT method. As Figure 10.5 shows, the foliar sulfur ratios within the Athabasca River valley

TABLE 10.4 Summary Listing of Foliar Inorganic/Organic Sulfur Ratios at Sites North and South of the Major Industrial Atmospheric Emissions Sources for SO_2 in the AOSR

Sites	Distance	S ratio
10014	10.4	0.4471[a]
10011	13.1	0.4090[ab]
10013	21.9	0.3856[ab]
10046	21.8	0.3470[b]
10020	32.3	0.3048[b]
09306	39.7	0.1921[c]
09065	41.2	0.1582[cd]
09221	46.5	0.1254[cd]
09214	49.6	0.1511[cd]
09090	56.1	0.1049[de]
09195	77.0	0.0766[e]

Sites followed by the same letter are not significantly different ($p=0.10$).

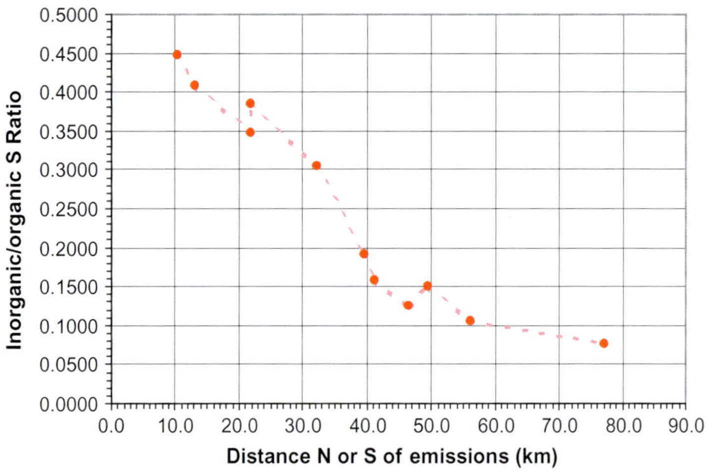

FIGURE 10.5 Correlation ($r=-0.98$) of foliar inorganic/organic sulfur ratios in first year jack pine needle with distance (North, N or South, S) from the major industrial emissions of SO_2 in the Athabasca oil sands region of Alberta.

corridor exhibit a highly significant ($p=0.01$) negative correlation ($r=-0.98$), from sites at 7 km from the two major industrial emissions sources progressively outward to 77 km. The data in Table 10.4 show that significantly elevated foliar inorganic sulfur is found extending 32.3 km away from the emissions sources, while from about 40–50 km the sulfur levels are also elevated but to a much lesser degree. At distances greater than 56 km, foliar inorganic sulfur levels were near or at background concentrations as of 2009.

The data generated from this ecological analogue study allows preliminary sulfur loading levels (%) in jack pine needles to be suggested for the AOSR as follows:

> Background levels=0.0700–0.1200
> Low levels=>0.1200–0.1900
> Moderate levels=>0.1900–0.2500
> Moderately high levels=>0.2500–0.3500
> High levels=>0.3500–0.4500
> Very high levels=>0.4500

This, quantitative ecological analogue biomonitoring method allows stakeholders and regulators to identify significant changes in sulfur loading over time, which can then facilitate mitigation measures if the data suggest the potential for adverse environmental effects and such actions are necessary. Another significant finding of the power of the ecological analogue approach to biomonitoring of air pollutants is shown from the biomass characteristics of jack pine needles. Analyses of the jack pine needles found at each ecological analogue site in 2009 showed significant differences in the ratios of needle biomass between first-, second-, and third-year needles, and these differences corresponded with foliar sulfur loadings as follows:

		Needle biomass ratio		
% S ratio	% S loading level	Year 1 needles	Year 2 needles	Year 3 needles
0.2500–0.4500	High and moderately high	1.00	2.40	3.35
0.1900–0.2500	Moderate	1.00	1.34	1.67
0.0700–0.1900	Low and background	1.00	0.62	0.53

The branchlets at the high and moderately high sulfur-loading levels possess few needle pairs in the first year and progressively greater numbers of needle pairs in the second- and third years. In addition, all age classes at the high and moderately high sulfur-loading level exhibit a dull, gray–green color; are brittle; and frequently exhibit curled malformed needles on the branchlets. The color of the needles in other sulfur-loading levels is yellowish, light-green in moderate sites and dark-green in sites with low and background. The full significance of these needle biomass findings on growth and vigor of trees within the various sulfur-loading levels is yet to be determined.

10.5 CONCLUSIONS

Synecological studies in the AOSR of Alberta have been successful in locating 24 sites of EAT #3 useful for biomonitoring of the WBEA Forest Health Monitoring Program in this region. The verification of the suitability of these sites for use as short-term and long-term biomonitoring sites has been strongly indicated by data from

a. sulfur loading in jack pine needle tissues and
b. jack pine needle biomass and visual analysis of needle characteristics.

The ecological analogue approach has been shown to be markedly superior to more random sampling methods used in past studies in the AOSR and northeastern British Columbia. The past studies which did not use ecologically analogous sampling site selection methods were unable to show statistically significant differences in critical impact assessment metrics for a complete range of atmospheric pollutant concentrations. On the other hand, the ecological analogue approach used in the AOSR shows great promise as it has been able to document statistically significant differences in various critical pollutant metrics at different levels of inferred exposure to acidifying anthropogenic industrial air emissions.

ACKNOWLEDGMENT

The authors wish to thank the WBEA and its membership for the foresight to support this work.

REFERENCES

A.E.U.B., 2001. GIS Compilation of Structural Elements in Northern Alberta. Alberta Energy Utilities Board Release 1.0. 2001-01.

A.G.S./A.E.U.B., 1999. Geological Map of Alberta. Alberta Geological Survey & Alberta Energy Utilities Board Publication.

Andriashek, L.D., 2003. Quaternary geological setting of the Athabasca Oil Sands (in situ) area, Northeast Alberta. Alberta Energy Utilities Board/Alberta Geological Survey Earth Sciences Report 2002-03, 286p.

Bayrock, L., 1971. Surficial Geology, Bitumount, NTS 74L. Map 34. Research Council of Alberta, Edmonton, AB.

Bayrock, L., Reimchen, T., 1973. Surficial Geology of the Waterways Map Area, NTS74D. Alberta Research Council, Alberta Geological Survey: Alberta Scale 1:250 000.

Beckingham, J.D., Archibald, J.H., 1996. Field guide to ecosites of Northern Alberta. Canadian Forest Service Northern Forestry Centre. Special Report 5, 528p.

Beier, C., 1991. Atmospheric pollutants: separation of gaseous and particulate dry deposition of sulfur at a forest edge in Denmark. J. Environ. Qual. 20, 460–466.

Beier, C., Gunderson, P., 1989. Atmospheric deposition to the edge of a spruce forest in Denmark. Environ. Pollut. 60, 257–271.

Braun-Blanquet, J., 1928. Pflanzensoziologie, first ed. Springer-Verlag, Berlin, Germany, 865 pp.

Braun-Blanquet, J., 1932. Plant Sociology: The Study of Plant Communities. Hafner, London Transl. of first ed. 1928 by Fuller, G.D. and Conrad, H.S., 438 pp.

Braun-Blanquet, J., 1965. Plant Sociology: The Study of Plant Communities. Hafner, London, England 439 pp.

Cajander, A.K., 1909. Uber Waldtypen. Acta For. Fenn. 1, 1–175.

Canada Committee on Ecological Classification, 1999. National Atlas of Canada. 1:1,000,000 Map of Ecodistricts. Ottawa, ON, Canada.

Case, J.W., 1994. Pine River Monitoring Study of Sulfur Content and S^{34} in lichens, moss, conifers and soil. Report to Westcoast Energy Inc., Chetwynd, BC, Canada. Sept. 1994, 37 pp.

Clements, F.E., 1905. Research Methods in Ecology. Univ, Publ. Co, Lincoln, NE, 334 pp.

Clements, F.E., 1916. Plant Succession. An Analysis of the Development of Vegetation. Carnegie Institute, Washington, DC, 512 pp.

Cowles, H.C., 1899. The ecological relations of vegetation on the sand dunes of Lake Michigan. Bot. Gaz. 27, 95–117 167–202, 281–308, 361–391.

Daubenmire, R.F., 1952. Forest vegetation of northern Idaho and adjacent Washington, and its bearing on concepts of vegetation classification. Ecol. Monogr. 22, 301–330.

Daubenmire, R.F., 1966. Vegetation: identification of typal communities. Science 151, 291–298.

Daubenmire, R.F., 1968. Plant Communities: A Textbook of Plant Synecology. Harper & Row, 300 pp.

Downing, D.J., Pettapiece, W.W., 2006. Natural Regions and Subregions of Alberta. Government of Alberta Publication, Edmonton, AB No. T/852, 254 pp.

Draaijers, G.P.J., Ivens, W.P.M.F., Bleuten, W., 1988. Atmospheric deposition in forest edges measured by monitoring canopy throughfall. Water, Air, and Soil Pollut. 42, 129–136.

Draaijers, G.P.J., Van Ek, R., Bleuten, W., 1994. Atmospheric deposition in complex forest landscapes. Bound-Lay. Meteorol. 69, 343–366.

Duncan, D.B., 1955. Multiple range and multiple F tests. Biometrics 11, 1–42.

Gray, A., Hooker, J.D., 1880. Vegetation of the Rocky Mountain region. U.S. Geological Survey Territories. 6, No. 1. books.google.com/books/.../The_Vegetation_of_the_Rocky_Mounta.

Grisebach, A., 1866. Der gegenwartige Standpukt der Geographie der Pflanzen. Behm's. Geogr. Jahrb. 1, 373–402.

Grisebach, A., 1872. Die Vegetation der Erde nach ihrer klimatischen Anordnung. Wilhelm Engelmann, Leipzig, vol. I, 603 pp. vol. II, 635 pp.

Hager, H., Kazda, M., 1985. The influence of stand density & canopy position on sulfur content in needles of Norway Spruce. Water Air Soil Pollut. 25, 321–329.

Hasselrot, B., Grennfelt, P., 1987. Deposition of air pollutants in a wind-exposed forest edge. Water Air Soil Pollut. 34, 135–143.

Henderson, J.A., Davis, L.S., 1977. ECOSYM—an ecosystem classification and data storage system for natural resources management. U.S. Dept. of Agriculture, Forest Service. Report #9.

Jackson, L.L., Englemann, E.E., Peard, J.L., 1985. Determination of total sulphur in lichens and plants by combustion—infrared analysis. Environ. Sci. Technol. 19 (5), 437–441.

Jaques, D.R., 2002. Bio-monitoring the chronic effects of SO_2 on Lodgepole pine Ecosystems near the Pine River Gas Plant, NE B.C. Report to: Westcoast Energy Inc., Chetwynd, B.C. February, 12, 2002, 34p.

Jaques, D.R., 2007. Location and characterization of jack pine ecological analogue sitesin the oil sands region of Alberta. Report to: Biosphere Solutions, Calgary, Alberta, Canada. December 10, 2007. 18 pp.

Johnson, C.M., Nishita, H., 1952. Microestimation of sulfur in plant materials, soils and irrigation waters. Anal. Chem. 24, 736–742.

Jones, C.E., Associates, 2007. Terrestrial environmental effects monitoriong and acidification monitoring program. Report to: Wood Buffalo Environmental Association, Terrestrial Environmental Effects Monitoring committee, Fort McMurray, AB, Canada. February 2007, 198 pp.

Kerner, von M.A., 1863. Das Pflantzenleben der Donaulander, second ed. By Vierhapper, F., 1929. University-Verlag Wagner, Innsbruck, 452 pp.

Krajina, V.J., 1933. Die Pflanzengesellshaften des Mlynica-Tales in den Vysoke Tatry (Hoch Tatra). Mit besonderer Berucksichitigung der oekologischen Verhaltnisse. Bot. Centrslbl. Abt. 2 (50), 744–957 2 (51), 1–244.

Krajina, V.J., 1960. Can we find a common platform for the different schools of forest-type classification? Silva Fenn. 105, 50–59.

Krajina, V.J., 1965. Biogeoclimataic zones and biogeocoenoses of British Columbia. Ecology of Western North America, vol. 1. Ecological Society of America and Department of Botany, University of British Columbia, Vancouver, BC, 112 pp.

Krajina, V.J., 1969. Ecology of forest trees in British Columbia. Ecology of Western North America, vol. 2, Number 1. Department of Botany, University of British Columbia, Vancouver, BC, 146 pp.

Krouse, H.R., Legge, A.H., Brown, H.M., 1984. Sulphur gas emissions in the boreal forest: the West Whitecourt Case Study—V. Stable sulphur isotopes. Water Air Soil Pollut. 22, 321–347.

Krupa, S.V., Legge, A.H., 1998. Sulphur dioxide, particulate sulphur and its impacts on a boreal forest ecosystem. In: Ambasht, R.S. (Ed.), Modern Trends in Ecology and Environment. Backhuys Publishers, Leiden, The Netherlands, pp. 285–306.

Legge, A.H., Bogner, J., 1983. Ecological monitoring of sulphur in forests in Western Canada. Aquilo Ser. Botanica Tom 19, 119–139.

Legge, A.H., Jaques, D.R., Amundson, R.G., Walker, R.B., 1977. Field studies of pine periodically subjected to sulfur gas emissions. Water Air Soil Pollut. 8, 105–129.

Legge, A.H., Jaques, D.R., Harvey, G.W., Krouse, H.R., Brown, H.M., Rhodes, E.C., Nosal, M., Schellhase, H.U., Mayo, J., Hartgerink, A.P., Lester, P.F., Amundson, R.G., Walker, R.B., 1981. Sulphur gas emissions in the Boreal Forest: the West Whitecourt Case Study I: executive Summary. Water Air Soil Pollut. 15, 77–85.

Legge, A.H., Bogner, J.C., Krupa, S.V., 1988a. Foliar sulphur species in pine: a new indicator of a forest ecosystem under air pollution stress. Environ. Pollut. 55 (1), 15–27.

Legge, A.H., Corbin, J., Bogner, J., Strosher, M., Krouse, H.R., Laishley, E.J., Bryant, R.D., Cavey, M.J., Prescott, C.E., Nosal, M., Schellhase, H.U., Weidensaul, T.C., Mayo, J.M., 1988. Acidification in a temperate forest ecosystem: The role of sulphur gas emissions and sulphur dust. A final report submitted to the Whitecourt Environmental Study Group. University Calgary, Kananaskis Centre for Environmental Research, Calgary, AB, Canada.

Lester, P.F., Rhodes, E.C., Legge, A.H., 1987. Sulphur gas emissions in the boreal forest: the West Whitecourt Case Study—IV. Air quality and the meteorological environment. Water Air Soil Pollut. 27, 85–108.

Livingstone, B.E., Shreve, F., 1921. The Distribution of Vegetation in the United States as Related to Climatic Conditions. Carnegie Institute, Washington, DC Publ. 284:1–284.

Macoun, J., 1882. Manitoba & the Great Northwest. The World Publishing Co., Guelph, ON.

Merriam, C.H., 1890. Results of a biological survey of the San Francisco Mountain region and the desert of the Little Colorado, Arizona. N. Am. Fauna 3, 1–113.

Merriam, C.H., 1899. Results of a biological survey of Mount Shasta, California. N. Am. Fauna 1901, 257–270.

Moss, C.E., 1910. The fundamental units of vegetation. New Phytol. 9, 18–53.

Natural Resources Canada, 2009. Digital Elevation Model Data of Alberta. Ottawa, ON, Canada.

Nichols, G.E., 1917. The interpretation and application of certain terms and concepts of the ecological classification of plant communities. Plant World 20 (305–329), 341–353.

Prietzel, J., Mayer, B., Legge, A.H., 2004. Cumulative impact of 40 years of industrial sulphur emissions on a forest soil in west-central Alberta (Canada). Environ. Pollut. 132, 129–144.

Raunkiaer, C., 1913. Formationsstatistiske Undersogelser paa Skagens Odde. Bot. Tids. 33, 197–228.

Schimper, A.F.W., 1903. Plant Geography on a Physiological Basis. Clarendon Press, Oxford, 839 pp.

Schouw, J.F., 1823. Grundzuge Einer Allegmeinen Pflanzengeographie. Reimer, Berlin, Germany, 524 pp.

Sorenson, T., 1948. A method of establishing groups of equal amplitude in plant sociology based on similarity of speies content. Det Kong. Danske Vidensk. Biol. Skr. Copenhagen 5 (4), 1–34.

Sukachev, V., 1928. Principles of classification of the spruce communities of European Russia. J. Ecol. 16, 1–18.

Sukachev, V., 1932. Die Untersuchchung der Waldtypen des osteuropaischen Flachlandes, XI:191-250. In: Abderhalden, E. (Ed.), Handbuch der biologischen Arbeitsmethoden. Urban & Schwarzenberg, Berlin, Germany.

Sukachev, V., 1945. Biogeocoenology and phytocoenology. Sov. Acad. Sci., U.S.S.R., Siberian Branch 47, 429–431.

Turchenek, L., Lindsey, D., 1982. Soils inventory of the Alberta Oil Sands Environmental Research Program (AOSERP) Study Area. Alberta Environment, Edmonton, AB, Canada. AOSERP Report 122, 221 pp.

Vestal, A.G., 1913. An associational study of Illinois sand prairie. Bull. Ill. Nat. Hist. Surv. 10, 1–94.

Visher, S.S., 1916. The biogeography of the northern Great Plains. Geogr. Rev. 2, 89–115.

von Humboldt, A., 1805. Essay sur la Geographie des Plantes. Par Al. de Humboldt et A. Bonpland, regige par Al. De Humboldt. Levrault, Schoell et Cie, Paris, 155 pp.

von Humboldt, A., 1806. Ibeen zu einer Physiognomik der Gewachse. Cotta, Stuttgart, 28 pp.

Warming, E., 1909. Oecology of Plants. An Introduction to the Study of Plant Communities. Oxford Univ. Press, London, 422 pp. (English ed. of Danish publication, Platesumfund. 1895.).

Weathers, K.C., Cadenasso, M.L., Pickett, S.T.A., 2001. Forest edges as nutrient and pollutant concentrators: potential synergisms between fragmentation, forest canopies, and the atmosphere. Conserv. Biol. 15 (6), 1506–1514.

Weaver, J.E., 1917. A Study of the Vegetation of Southeastern Washington and Adjacent Idaho. Univ. of Nebraska Studies, Lincoln, NE 17 (1):1–133.

Chapter 11

Tracing Industrial Nitrogen and Sulfur Emissions in the Athabasca Oil Sands Region Using Stable Isotopes

B.C. Proemse[1] and B. Mayer

Department of Geoscience, University of Calgary, Calgary, Alberta, Canada
[1]Corresponding author: e-mail: bcpromse@ucalgary.ca

ABSTRACT

The rapid development in the Athabasca Oil Sands Region (AOSR) in northeastern Alberta, Canada, has raised concerns about the impact of the industrial emissions on the surrounding terrestrial and aquatic ecosystems. Stable isotope techniques may help to trace the transport and fate of industrial emissions provided that they are isotopically distinct from background isotope ratios in environmental receptors. In order to trace nitrogen (N) and sulfur (S) emissions released by the oil sands industry, chemical and isotopic compositions of various N and S compounds in emissions, in atmospheric deposition, and in several environmental receptors were determined. It was found that $\delta^{18}O$ values of nitrate and sulfate and $\Delta^{17}O$ values of nitrate are indicators that constitute excellent new monitoring tools for tracing industrial N and S emissions in the surrounding environment. Application of quantitative and qualitative stable isotope tracers revealed that industrial N and S emissions were observable in the surrounding environment within ca. 30 km distance to the major emission sources.

11.1 INTRODUCTION

Fossil fuels will continue to dominate the global energy supply for several decades (Sims et al., 2007). While conventional energy resources continue to decline, unconventional resources such as heavy oil will become

Disclaimer: The content and opinions expressed by the authors in this chapter do not necessarily reflect the views of the Wood Buffalo Environmental Association (WBEA) or of the WBEA membership.

increasingly important (Sims et al., 2007). Canada's oil resources, 98% of which is heavy oil, are currently the third largest in the world (Chapter 1).

The Athabasca oil sand deposits located in northeastern Alberta, Canada, constitute an enormous heavy oil reservoir that has been estimated to hold 2.36×10^{11} m^3 crude bitumen (ERCB, 2010). Due to the expanding oil sands industry, there has been a rapid development in the exploitation of this unconventional oil resource over the past decade. There are currently six open pit mines and more than 13 *in situ* projects operational (ERCB, 2010), and the town of Fort McMurray has been growing to >61,370 inhabitants (Statistics Canada, 2011). This has increased urban and industrial emissions in the region and is of concern regarding their potential effects on terrestrial and aquatic ecosystems.

The mining, extraction, and upgrading of the bitumen result in air emissions from stacks, heavy haulers, and other equipment operating in the open pit mines, on-road vehicles, tailing ponds, outgassing from mine windblown dust from, for example, coke or dust from dry tailings and roads (Timoney and Lee, 2009). Mining and upgrading of the oil sands require energy that is derived from fossil fuel combustion accompanied by nitrogen oxide (NO_x) and sulfur oxide (SO_x) emissions that are of environmental concern (Allen, 2004; Environment Canada, 2011; Galloway et al., 2008; Hazewinkel et al., 2008; Schindler et al., 2006; Sims et al., 2007; Whitfield et al., 2009). Elevated atmospheric deposition may be followed by an increase in biomass production over the short term (Jäger and Krupa, 2009), but elevated N and S may also affect plant productivity, diversity, and community at a later stage and may potentially lead to eutrophication and acidification (Allen, 2004).

Previously, monitoring efforts of atmospheric pollutants in the AOSR were solely based on concentration measurements. Stable isotope techniques may help to trace the transport and fate of industrial emissions provided such emissions are isotopically distinct from background components. Depending on the distinctiveness of the isotopic signature of the industrial emissions, their isotopic fingerprints may be useable as either qualitative or quantitative tracers. If N and S isotope ratios of industrial emissions in the AOSR differ significantly from those of other sources, receptor models (Watson et al., 1984, 2008; Chapter 18), and multi-end-member mixing analyses can be applied to estimate industrial contributions to nearby and distant sites (quantitative tracer). Therefore, chemical and isotopic compositions ($\delta^{15}N$, $\delta^{34}S$, $\delta^{18}O$, $\Delta^{17}O$) of various N and S compounds in emissions and environmental receptors were determined to identify quantitative and qualitative tracers that are suitable for tracking the transport and fate of industrial emissions in surrounding terrestrial and aquatic ecosystems in the AOSR.

11.2 STUDY AREA AND SAMPLING

The Athabasca Oil Sands Region (AOSR) in northeastern Alberta is located in the Western Canadian Sedimentary Basin and is one of three oil sands developments in Alberta (Athabasca, Peace River, and Cold Lake). The largest

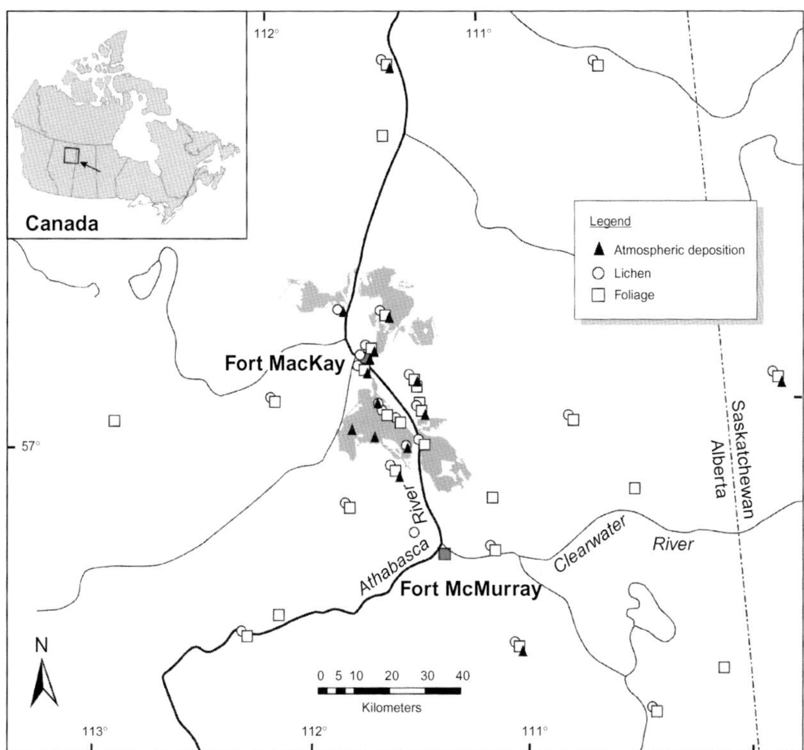

FIGURE 11.1   Map of study area showing sampling sites for atmospheric deposition (triangle), lichens (circle), and conifer needles (rectangle). The gray shaded area represents the open pit mining sites as of 2008.

town near the oil sands development is Fort McMurray. Located approximately 65 km north of Fort McMurray is the First Nation community of Fort McKay (Figure 11.1). The undisturbed landscape of the AOSR is characterized by a mixture of peat lands and upland forests composed of jack pine, spruce, and aspen (Whitfield et al., 2009) often growing on an eluviated Dystric Brunisol consisting mostly of sand (Soil Classification Working Group, Canada, 1998). Some of the surrounding forested soils are known to be acid sensitive (Whitfield et al., 2010), particularly in regions of the Precambrian Shield and the Athabasca Sedimentary Basin (Shewchuk, 1982). More than 65,000 ha of boreal landscape have been disturbed due to mining operations and tailing ponds (Timoney and Lee, 2009).

We determined chemical and isotopic compositions ($\delta^{15}N$, $\delta^{34}S$, $\delta^{18}O$, $\Delta^{17}O$) of various N and S compounds in stack emitted fine particulate matter ($PM_{2.5}$), in atmospheric bulk and throughfall deposition, and in bioindicators such as lichens and conifer needles sampled at various distances to one of the major emission stacks (Figure 11.1). Note, however, that there are numerous other stationary, mobile, and fugitive emission sources spread throughout the AOSR (Chapter 12). Eluants from $PM_{2.5}$ filters (Chapter 8) were analyzed for $\delta^{15}N$ ($n=7$), $\delta^{18}O$ ($n=7$), $\Delta^{17}O$ ($n=5$) of nitrate, $\delta^{15}N$ of ammonium ($n=5$), $\delta^{34}S$ ($n=10$), and $\delta^{18}O$ ($n=9$) of sulfate. Atmospheric nitrate ($n=455$ for $\delta^{15}N$ and $\delta^{18}O$, $n=30$ for $\Delta^{17}O$), ammonium ($\delta^{15}N$, $n=134$), and sulfate ($n=294$ and $n=328$ for $\delta^{34}S$ and $\delta^{18}O$, respectively) samples from bulk deposition and throughfall collectors at 16 sites located at distances ranging between 3 to 113 km from the oil sands operations (upgrading and mining) were also analyzed. Deposition collectors were exposed for approximately 6-month periods between May 2008 and May 2009 using ion exchange resins for the collection of atmospheric nitrate, ammonium, and sulfate (Fenn and Poth, 2004). Thirty-two lichen samples from 24 sites in 3–113 km distance from the operations collected in summer 2008 were analyzed for total N contents, $\delta^{15}N$ of total N, total S contents and $\delta^{34}S$ of total S to investigate whether isotopically distinct industrial N and S emissions impact lichen N and S. For conifer needles (pine and spruce) collected from 25 sites at distances between 3 and 119 km from the operations, total N contents, $\delta^{15}N$ of total N, $\delta^{34}S$ of total S, and total S and sulfate-S in foliage were determined ($n=143$).

11.3 METHODS

Concentration analyses of nitrate (NO_3^-), ammonium (NH_4^+) and sulfate (SO_4^{2-}) were conducted by ion chromatography or automated colorimetry. Isotope abundance ratios were measured in the Isotope Science Laboratory at the University of Calgary by isotope ratio mass spectrometry (IRMS), and results are reported in the internationally accepted δ notation defined as

$$\delta_{sample}(‰) = \left[\left(R_{sample}/R_{standard}\right) - 1\right] \times 1000 \tag{11.1}$$

where, R is the $^{15}N/^{14}N$, $^{34}S/^{32}S$, or $^{18}O/^{16}O$ ratio of the sample and a standard, respectively. $\delta^{15}N$ values are reported relative to atmospheric N_2 (AIR), $\delta^{34}S$ values relative to Vienna Canyon Diablo Troilite, and $\delta^{18}O$ relative to Vienna Standard Mean Ocean Water. Nitrogen isotope ratios ($\delta^{15}N$) and oxygen isotope ratios ($\delta^{18}O$) of nitrate were analyzed by IRMS using N_2O gas generated via the bacterial denitrification method (Casciotti et al., 2002; Sigman et al., 2001) with long-term external precisions of $\pm0.5‰$ and $\pm1.0‰$, respectively (Proemse et al., 2012a). $\Delta^{17}O–NO_3$ values were determined at IsoLab at the University of Washington using a modified bacterial denitrification method (Kaiser et al., 2007). $\Delta^{17}O–NO_3$ values are reported as the deviation of $\delta^{17}O$ from the mass dependent relationship ($\delta^{17}O=0.52 \times \delta^{18}O$) and expressed as $\Delta^{17}O$ ($\Delta^{17}O=\delta^{17}O - 0.52 \times \delta^{18}O$). Long-term external precision is 0.6‰. $\delta^{15}N$ of ammonium was measured using the ammonium diffusion method (Sebilo et al., 2004) with an external precision of $\pm0.6‰$ (Proemse et al., 2012a).

Total nitrogen content and nitrogen isotope ratios of solids were determined using an elemental analyzer (EA) interfaced with an IRMS (Preston and Owens, 1983). The precision for $\delta^{15}N$ values is $\pm0.15‰$, and for total N contents typically less than $\pm5\%$ of the measured value.

Sulfur and oxygen isotope ratios ($\delta^{34}S$, $\delta^{18}O$) of sulfate were determined on $BaSO_4$ precipitated from the samples (Proemse et al., 2012b). $\delta^{34}S$ values were analyzed using SO_2 gas via continuous flow-isotope ratio mass spectrometry (EA-CF-IRMS) (Giesemann et al., 1994). $\delta^{18}O$ values of sulfate were determined on CO generated by pyrolysis of $BaSO_4$ in the presence of graphite at a temperature of 1450 °C followed by IRMS (Kornexl et al., 1999).

For determination of the total sulfur contents and $\delta^{34}S$ values in solid materials, total sulfur of the sample material was converted to $BaSO_4$ using the Parr-bomb technique (Siegfriedt et al., 1951) followed by EA-CF-IRMS as described above. The external precision for $\delta^{34}S$ values is $\pm0.5‰$. Total S contents of lichen samples were determined gravimetrically by weighing the sample and the resulting $BaSO_4$ precipitate. Measurement precision is typically better than $\pm10\%$ of the determined value, except for samples with very low S contents.

The total sulfur content in ground foliage samples was determined using a Leco elemental sulfur analyzer at Pacific Soil Analysis Inc. (Vancouver, British Columbia). Foliar inorganic sulfur content (SO_4–S) was determined according to the method of Johnson and Nishita (1952). The foliar organic sulfur content was calculated as the difference between the foliar total S content and the foliar inorganic S content.

11.4   RESULTS AND DISCUSSION

11.4.1   Stack Emitted Particulate Matter ($PM_{2.5}$)

We determined the isotopic compositions of nitrate, ammonium, and sulfate in particulate matter <2.5 μm in diameter ($PM_{2.5}$) emitted from two stacks (Stack A and Stack B) from one oil sands operation and compared them to the nitrogen and sulfur isotopic compositions of source materials and upgrading products (Proemse et al., 2012a). Particles were collected on filter packs from diluted stack emissions using real-world PM measurement instruments (Chapter 8), and filter extracts were analyzed for the isotopic composition of nitrate (NO_3), ammonium (NH_4), and sulfate (SO_4). The results show that the isotopic composition of nitrate in $PM_{2.5}$ is distinct compared to those reported for atmospheric nitrate in the literature, where $\delta^{15}N$ values of atmospheric nitrate are reported to typically vary from -15 to $+15‰$, and $\delta^{18}O$–NO_3 values range from $\sim30‰$ to $95‰$ (Kendall, 1998; Kendall et al., 2007). Nitrate in stack emitted $PM_{2.5}$ had $\delta^{15}N$ values of $9.4‰$ (Stack A) and $16.1\pm1.2‰$ (Stack B). Stack emitted nitrate was significantly enriched in ^{15}N compared to total N in oil sand (δ^{15}N $= 2.5$‰), by-products of upgrading (δ^{15}N between -0.3‰ and 1.3‰), and atmospheric N$_2$ (δ^{15}N $= 0$‰). This suggests that elevated ^{15}N contents in particle-associated nitrate released from tested stacks constitute a potential tracer for some industrial nitrogen emissions. Nitrate in PM$_{2.5}$ samples from Stack B had an average Δ^{17}O value of 0.5 ± 0.9‰ and was therefore not mass-independently enriched in ^{17}O. This is distinct from Δ^{17}O values typically observed for atmospheric nitrate, ranging from 20‰ to 31‰ (Michalski et al., 2003). Therefore, Δ^{17}O values of particle-associated nitrate released from the two tested stacks constitute potentially an excellent tracer for some industrial nitrate emissions. δ^{18}O values of nitrate were 36.0‰ for Stack A and 17.6 ± 1.8‰ for Stack B and therefore also significantly lower than δ^{18}O values of atmospheric nitrates (Kendall et al., 2007). δ^{15}N values of ammonium in PM$_{2.5}$ from Stack B spanned a large range with values between -4.5‰ to $+20.1$‰. Ammonium concentrations in Stack A were too low for isotope analyses.

δ^{34}S values of sulfate in PM$_{2.5}$ were 7.3 ± 0.3‰ for Stack A and 9.4 ± 2.0‰ for Stack B, and therefore slightly enriched in ^{34}S compared to δ^{34}S in bitumen (4.3 ± 0.3‰) and coke (3.9 ± 0.2‰). δ^{18}O values of sulfate in PM$_{2.5}$ were 18.9 ± 2.9‰ and 14.2 ± 2.8‰ for Stack A and Stack B, respectively. Emissions of particulate matter account for less than 1% of total N and S emissions in Stack A and typically less than 10% in Stack B. Therefore, data on gaseous stack emissions as well as mobile emission sources (e.g., heavy hauler exhaust) and their isotopic compositions are needed to fully characterize industrial N and S emissions (Chapters 7 and 8).

11.4.2 Atmospheric Deposition

Anthropogenic N and S emissions in the AOSR contribute to nitrate (NO$_3$), ammonium (NH$_4$), and sulfate (SO$_4$) deposition in the surrounding environment. We have investigated the isotopic composition of atmospheric NO$_3$, NH$_4$, and SO$_4$ in open field bulk and throughfall deposition (Proemse et al., 2012b). Between May 2008 and May 2009, NO$_3$–N and NH$_4$–N deposition was monitored at 16 sites, and SO$_4$–S at 15 sites between 3 and 113 km from the oil sands operations.

Industrial sites within 12 km distance of main operations showed considerably elevated nitrate and ammonium deposition rates in throughfall of up to 11.0 kg NO$_3$–N ha^{-1} yr^{-1} (at 12 km) and 18.3 kg NH$_4$–N ha^{-1} yr^{-1} (at 3 km) compared to distant forested plots (>90 km). Nitrate and ammonium deposition rates in open field bulk deposition samplers at industrial sites were lower than 2.0 kg NO$_3$–N ha^{-1} yr^{-1} and 4.7 kg NH$_4$–N ha^{-1} yr^{-1}, indicating that dry deposition must be responsible for the high-throughfall deposition rates on industrial sites. Both bulk and throughfall sulfate deposition rates were also highest on industrial sites within 12 km distance to the operations with deposition rates as high as 11.7 kg SO$_4$–S ha^{-1} yr^{-1} for bulk deposition and up to 39.2 kg SO_4–S ha^{-1} yr^{-1} for throughfall. At forested plots more than 90 km distant to the oil sands operations, nitrate, ammonium, and sulfate bulk deposition rates were low (0.5 kg NO_3–N ha^{-1} yr^{-1}, 1 kg NH_4–N ha^{-1} yr^{-1}, and 1.4 kg SO_4–S ha^{-1} yr^{-1}, respectively). Annual ammonium deposition rates exceeded nitrate deposition rates at 15 out of 16 sites, indicating that ammonium is an important component of nitrogen input (Fenn and Ross, 2010). Chapter 12 reports modeled deposition predicted to be >10 kg S ha^{-1} yr^{-1}, 10 kg N ha^{-1} yr^{-1} within the immediate main operations (mine/upgrader) area, and 1.6–2.2 kg S ha^{-1} yr^{-1}, 1.1–2.0 kg N ha^{-1} yr^{-1} within the model domain (northeastern Alberta), but distant from major sources.

We found $\Delta^{17}O$ values of $PM_{2.5}$ associated nitrate in one industrial stack near 0‰, whereas atmospheric nitrate at low nitrate deposition sites (=high $1/(NO_3$–N)) had $\Delta^{17}O$ values between 22.8‰ (summer) and 32.0‰ (winter) (Figure 11.2). $\Delta^{17}O$ values of atmospheric nitrate in the vicinity of the operations (<12 km) were found to be lower (as low as 15‰) than reported in other atmospheric nitrate studies as a result of nitrate contributions from local industrial emission sources. $\delta^{18}O$ values of atmospheric nitrate showed a similar trend with lowest values typically associated with high NO_3–N deposition rates (Figure 11.2). We applied a two-end member mixing analysis and calculated minimum industrial contributions to atmospheric nitrate deposition in the AOSR based on $\delta^{18}O$ and $\Delta^{17}O$ values measured for atmospheric nitrate deposition and for stack emitted $PM_{2.5}$, respectively. Results showed that up to 68% of the deposited nitrate, particularly at sites within 29 km distance to the main operations appeared to be derived from industrially emitted nitrate. This preliminary calculation requires, however, further refinement once additional industrial emission sources (e.g., vehicle fleet) are isotopically characterized.

Industrial sources of reduced forms of nitrogen with $\delta^{15}N$–NH_4 values as high as 20.1‰ appeared to also affect $\delta^{15}N$–NH_4 values of deposited ammonium at selected sites. The site averaged $\delta^{15}N$–NH_4 values in bulk and throughfall atmospheric deposition varied between -6.3‰ and 11.3‰ with highest $\delta^{15}N$–NH_4 values found close to the operations (<12 km). A similar trend of ^{15}N enrichment closer to stacks tested was observed for $\delta^{15}N$–NO_3 values in bulk and throughfall deposition for winter periods, however, not during summer periods. This suggests that elevated ^{15}N contents may serve as a qualitative tracer of industrial derived ammonium and nitrate. Due to the large variability of $\delta^{15}N$ values of particle-associated ammonium in the tested stack a quantitative assessment of industrial contributions to atmospheric ammonium deposition is currently not possible.

Sulfur isotope ratio measurements of atmospheric sulfate deposited within the AOSR revealed that at few selected locations sulfate depleted in ^{34}S likely derived from H_2S emissions from tailing ponds contributed to local atmospheric sulfate deposition (Proemse et al., 2012b). In general, however, $\delta^{34}S$ values of sulfate in atmospheric bulk and throughfall deposition at distant sites

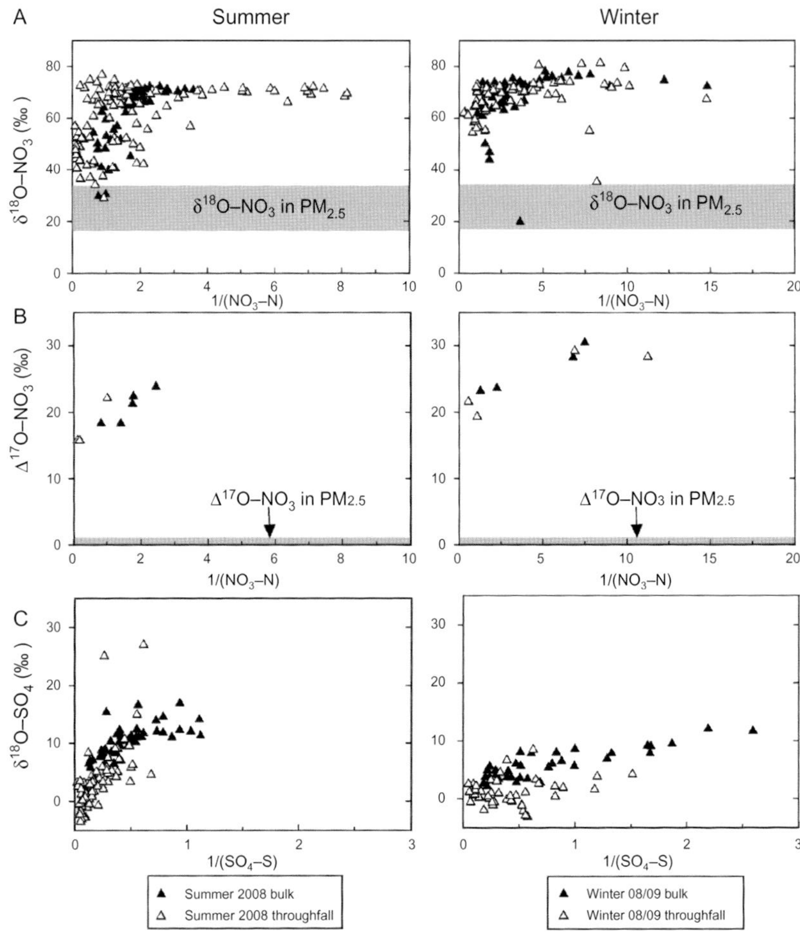

FIGURE 11.2 $\delta^{18}O$ values (A) and $\Delta^{17}O$ values (B) of atmospheric nitrate in bulk and through-fall deposition during summer (left) and winter (right) plotted versus the inverse deposition rates show that both low $\delta^{18}O$ and $\Delta^{17}O$ values occur in samples associated with high NO_3–N deposition rates (kg/ha), indicating industrial derived NO_3. The distinct trend of $\delta^{18}O$–SO_4 values in atmospheric sulfate deposition (C) with sulfate deposition rate allowed for the quantification of industrial derived SO_4 (Proemse et al., 2012b).

($>$90 km from main operations) was not isotopically different from $\delta^{34}S$ values of sulfate at high deposition sites in close proximity to the operations ($<$30 km). Also, the $\delta^{34}S$ values of sulfate in PM$_{2.5}$ from the two tested stacks were not sufficiently different from $\delta^{34}S$ of sulfate in atmospheric bulk and throughfall deposition at sites close and distant to the industrial emissions in the AOSR. Therefore, sulfur isotope ratios appear to be of limited use for tracing industrial contributions to atmospheric sulfate deposition in the AOSR.

In contrast, $\delta^{18}O$ values of atmospheric sulfate in bulk and throughfall showed a distinct trend in deposition rates at various sites in the AOSR. $\delta^{18}O$ values of sulfate at high deposition rates on industrial sites within 12 km distance to the operations were more than 10‰ lower than $\delta^{18}O$ values of sulfate at sites with low sulfate deposition typically at further distant forested plots (Figure 11.2). This suggests that $\delta^{18}O$ values may constitute an excellent tracer for assessing industrial contributions to atmospheric sulfate deposition. Two-end-member mixing calculations revealed that bulk sulfate deposition at sites <27 km distance to center of the operations and industrial emissions was characterized by elevated (>50%) industrial sulfate contribution (Proemse et al., 2012b). This preliminary calculation requires, however, further refinement once additional industrial emission sources (e.g., vehicle fleet) are isotopically characterized.

11.4.3   Lichens as Bioindicators

Epiphytic lichens take up atmospheric compounds from gases as well as from wet and dry particulate matter, providing the potential to monitor atmospheric S and N pollution (Balaguer and Manrique, 1991; Dahlman et al., 2004; Gries et al., 1997; Wadleigh, 2003). Element uptake may occur as particle adsorption, extra- or intracellular ion exchange and intracellular uptake (Bargagli and Mikhailova, 2002). Some compounds such as sulfur dioxide (SO_2), hydrogen sulfide (H_2S), ammonia (NH_3), ammonium (NH_4), and NO_x were reported to have a profound effect on lichens (Dahlman et al., 2004; Hawksworth and Rose, 1970; Krupa, 2003; Loppi and Frati, 2006; Loppi and Nascimbene, 2010; Tozer et al., 2005; Van Dobben and Ter Braak, 1999) but particulate matter (e.g., $PM_{2.5}$) may also influence the element composition (Carvalho and Freitas, 2011). It appears that regardless of the phase of a pollutant or nutrient (gaseous, aqueous, solid), its content in lichen may change, but environmental stress may also be reflected in diversity and growth (Loppi and Nascimbene, 2010; Rheault et al., 2003). Therefore, lichens have been used extensively as bioindicators for anthropogenic pollution (Nimis and Purvis, 2002; Nimis et al., 2002) and may be an integral part of air pollution monitoring programs, in particular in areas of difficult accessibility.

Previous studies of sulfur and nitrogen isotope ratios on lichens suggest that this environmental receptor may archive information about the sources of S and N if the isotopic composition of the pollution source differs from the isotopic composition of background in the environment (Case and Krouse, 1980; Takala et al., 1991; Tozer et al., 2005; Wadleigh, 2003; Wadleigh and Blake, 1999; Wiseman and Wadleigh, 2002). For instance, Case and Krouse (1980) reported that S isotope ratios in lichens closer to a point source were similar to those of the S emissions. Nitrogen isotope ratios have also been used as an indicator for pollution sources in lichens (Tozer et al., 2005) and mosses (Pearson et al., 2000), and to examine N uptake by lichens

(Dahlman et al., 2004). Therefore, we investigated whether epiphytic lichens in the AOSR are suitable for monitoring N and S emissions from the oil sand operations and to evaluate whether stable isotope techniques can provide further insights into the sources of N and S accumulated by the lichens.

In August and September 2008, 32 epiphytic lichen samples of *Evernia mesomorpha* were collected from trees at 24 sites at distances between 3 and 113 km from the operations. In addition, lichen samples from the edge of the forest stand and the interior of the forest were obtained at one site at 14 km distance. Total nitrogen contents in lichen samples ranged from 0.6% to 2.7%. Nitrogen contents were about three times higher with an average N content of $2.2 \pm 0.3\%$ in samples from sites within 7 km distance to the operations compared to lichens from sites more than 111 km distant that had an average N content of $0.7 \pm 0.1\%$. Lichen samples from sites further than 7 km distance from main operation emissions showed little variability in total N content with an average of $1.0 \pm 0.3\%$ (Figure 11.3).

$\delta^{15}N$ values of total nitrogen in lichens ranged from $-7.4\permil$ to $+11.6\permil$. Only duplicate lichen samples from the nearest site at 3 km distance were highly enriched in ^{15}N with $\delta^{15}N$ values of $+11.6\permil$ and $+9.8\permil$, respectively (Figure 11.3). Excluding the $\delta^{15}N$ values at this site, the mean $\delta^{15}N$ value for all other sites was $-5.3 \pm 1.0\permil$. The total N content of lichens collected at the forest edge of the site at 14 km distance was higher (1.2%) compared to samples from the interior of the forest stand (0.8%). However, there was no difference in the N isotopic composition of the lichen samples from the forest edge ($-4.8\permil$) compared to samples from the interior of the stand ($-4.7\permil$).

Plotting isotopic compositions versus inverse concentration data reveals whether data patterns can be explained by mixing of N from two or more sources (Krouse, 1980). To investigate the sources of nitrogen in *E. mesomorpha*, the $\delta^{15}N$ values were plotted versus the inverse N content. Figure 11.4 shows that there are at least two sources of nitrogen. Lichen samples were typically depleted in ^{15}N compared to atmospheric N_2 ($\delta^{15}N = 0\permil$) with a

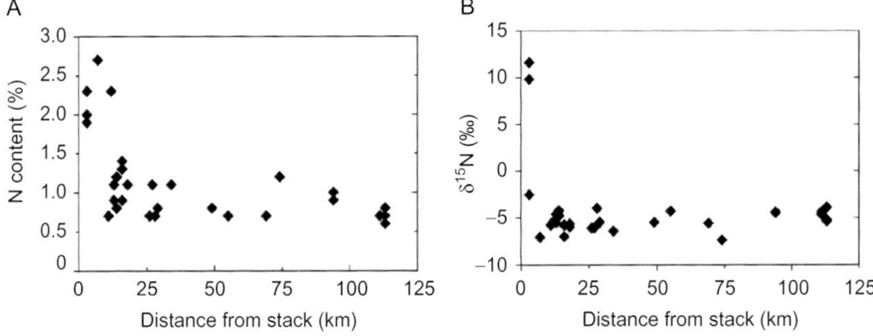

FIGURE 11.3 Nitrogen contents (A) and $\delta^{15}N$ values (B) of the lichen samples plotted versus distance from the oil sands operations. Both N content and N isotope ratios are highest close to the operations.

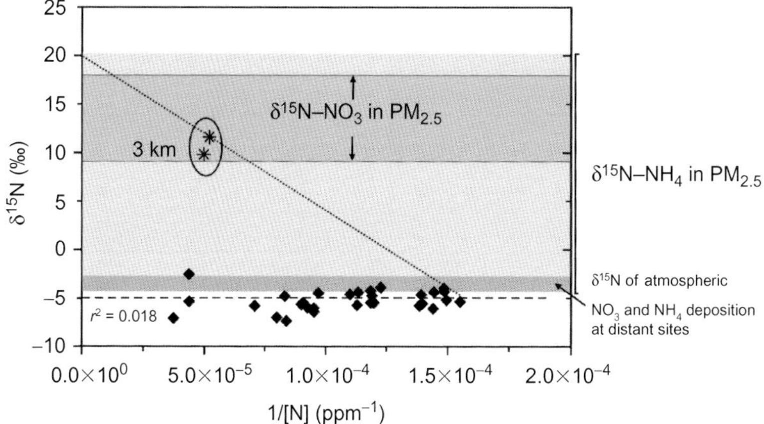

FIGURE 11.4 δ^{15}N values of total N in lichen samples (*Evernia mesomorpha*) versus 1/[N]. Except for the δ^{15}N values of lichens from the site at 3 km distance to the oil sands operations, δ^{15}N values are rather constant around -5.3 ± 1.0‰.

mean δ^{15}N value of -5.3 ± 1.0‰, independent of their nitrogen content. Only the two lichen samples obtained from the site at 3 km distance, the site closest to the main emissions, were markedly enriched in ^{15}N and were also associated with high N contents. Extrapolating a trendline between background N, indicated by lowest N content, and nitrogen with elevated δ^{15}N values intersects the *y*-axis at $+20$‰ (Figure 11.4). This suggests that lichens at the site very close the operations (3 km) are partly affected by a N source with a δ^{15}N value around $+20$‰. Ammonium in PM$_{2.5}$ emitted had δ^{15}N values of up to $+20.1$‰. Ammonium is preferentially taken up by some lichens compared to uptake of other N compounds such as amino acids, nitrate, and particles may be adsorbed onto the lichen surface (Bargagli and Mikhailova, 2002; Carvalho and Freitas, 2011; Dahlman et al., 2004; Miller and Brown, 1999). This suggests that industrial emitted particulate NH$_4$ with high δ^{15}N values has likely caused the elevated δ^{15}N values of total N in lichens at this site ($+11.6$‰ and 9.8‰). Nitrate in stack emitted PM$_{2.5}$ was also enriched in ^{15}N compared to atmospheric N$_2$ (δ^{15}N between 9.4‰ and 17.9‰) and may also be a potential source of nitrogen at the site at 3 km distance.

All other lichen samples displayed little variability of δ^{15}N values (-5.3 ± 1.0‰). Bulk and throughfall samples at low deposition rates averaged -3.6 ± 0.9‰ for δ^{15}N–NH$_4$ and -3.2 ± 1.5‰ for δ^{15}N–NO$_3$ during summer. Tozer et al. (2005) suggested that diffusive assimilation of gaseous NH$_3$ into lichens is associated with isotope fractionation and may lead to ^{15}N depleted nitrogen isotope ratios in lichens. This nitrogen isotope fractionation step may be responsible for lichen δ^{15}N values being slightly lower than δ^{15}N–NH$_4$ and δ^{15}N–NO$_3$ values at sites with low atmospheric N deposition.

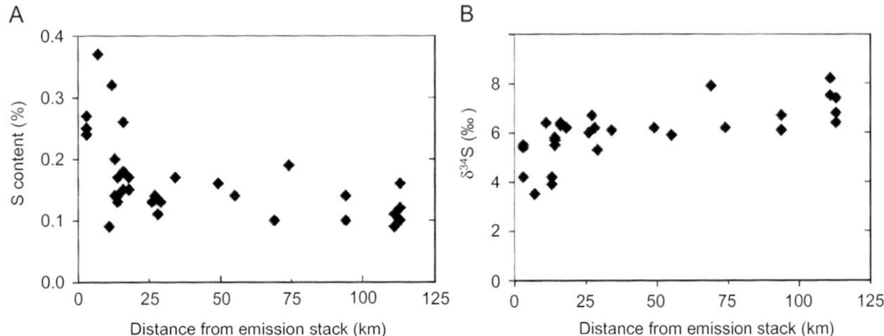

FIGURE 11.5   Sulfur contents (A) and $\delta^{34}S$ values (B) of lichen samples (*Evernia mesomorpha*) with distance from the oil sands operations. The S contents are higher close to the major operations and are associated with lower $\delta^{34}S$ values compared to S contents and isotope ratios of lichens than from sites 94 and 113 km distant to the operations.

This suggests that nitrogen in lichen samples at all sites except the site at 3 km distance is likely from atmospheric ammonium or nitrate, and additional isotope fractionation during the uptake of N appears to be responsible for slightly lower $\delta^{15}N$ values of total N in lichens compared to those of atmospheric ammonium and nitrate.

Total sulfur contents in lichen samples ranged from 0.09% to 0.37% (Figure 11.5A). The S content was up to three times higher within 7 km distance averaging $0.3 \pm 0.1\%$ and about two times higher within 16 km distance (average $0.21 \pm 0.08\%$) to the operations compared to more distant sites at 111 and 113 km where average total S contents were $0.12 \pm 0.03\%$. This range of S contents is similar to sulfur concentrations in lichens reported for the AOSR in 1980 (Addison and Puckett, 1980). They observed highest S concentrations ($>0.3\%$) in lichens within 10 km of the oil sand operations, similar to the distance range observed in this study. The average $\delta^{34}S$ value of total S in lichen samples was $+5.7 \pm 1.5\%$ (Figure 11.5B). Similar $\delta^{34}S$ values were determined for bitumen ($+4.3 \pm 0.3\%$), oil sand ($+6.5 \pm 0.4\%$), and for elemental sulfur blocks ($+5.3 \pm 0.5\%$) in the AOSR. Atmospheric sulfate in summer samples with low deposition rates had an average $\delta^{34}S$ value of $4.3 \pm 0.3\%$, similar to $\delta^{34}S$ values in atmospheric sulfate at high deposition rates ($\sim 5\%$). The total S content in lichens collected at the forest edge at the site at 14 km distance was slightly higher (0.17%) compared to that of lichens from the interior of that stand (0.14%). There was no difference in the S isotopic composition of lichens collected at the edge of this site ($+5.8\%$) compared to samples from the interior of the forest stand ($+5.7\%$). $\delta^{34}S$ values of different sources of sulfur, atmospheric sulfate deposition at sites closer to and more distant to operations, and total sulfur in lichen samples are therefore not distinct and hence are in principle not suitable to identify different sources of total sulfur in lichen samples. However, at two sites within 12–16 km of oil sands operations, which are also close to highway 63, open pit mines, tailing ponds and upgraders, elevated total

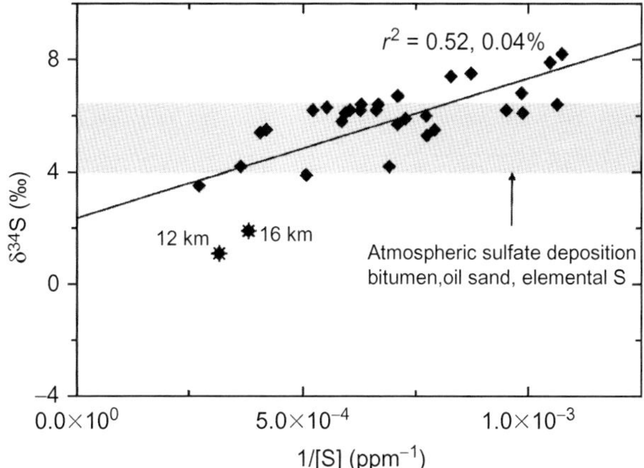

FIGURE 11.6 $\delta^{34}S$ values of total S in lichens versus the inverse sulfur content suggest a source with a low $\delta^{34}S$ value (<2‰) that results in increasing S contents in lichens. The gray bar indicates $\delta^{34}S$ values of atmospheric sulfate deposition, bitumen, oil sand, and elemental S from a sulfur storage block.

S contents(0.32% and 0.26%, higher than the average S content of $0.17\pm0.07\%$) were observed associated with the lowest $\delta^{34}S$ values (1.1‰ and 1.9‰). Plotting the $\delta^{34}S$ values versus the inverse S content of lichens (Figure 11.6) indicates that there is an additional S source associated with low $\delta^{34}S$ values (<2‰). Atmospheric deposition sites close to tailing ponds collected sulfate with low $\delta^{34}S$ values in summer 2008 (Proemse et al., 2012b), likely due to H_2S emissions from the tailing ponds. Bacterial sulfate reduction (BSR) occurs in tailing ponds in the AOSR (Ramos-Padrón et al., 2011), and BSR is associated with significant sulfur isotope fractionation resulting in low $\delta^{34}S$ values for H_2S produced in sediments or reducing aquatic systems (Canfield, 2001) that may subsequently outgas in part to the atmosphere. Sulfur from H_2S emissions may then be either directly incorporated into lichens or taken up after its oxidation via SO_2 to SO_4. It remains unclear to what extent gas phase emissions (SO_2, H_2S) from stationary sources, mobile sources, and tailing ponds contribute to the sulfur content in lichens. More work is needed to evaluate whether gaseous stack, vehicle, and pond emissions are isotopically distinct and if those emissions are responsible for decreasing $\delta^{34}S$ values in lichen samples with elevated total S contents. Similar to N, the total S content in *E. mesomorpha* sampled at the edge of the stand at 14 km distance was higher compared to the sample obtained in the interior of the stand, suggesting that lichen samples collected at the edge of stands provide an early warning parameter for elevated atmospheric deposition.

11.4.4   Conifer Needles

Forest ecosystems are sensitive to anthropogenic air pollution, with sulfur dioxide (SO_2) and ozone (O_3) causing more damage to plants than all other air pollutants combined (Kozlowski, 1980). The gases are absorbed by trees through the stomata and may cause visible and/or nonvisible injuries (Kozlowski, 1980). Total sulfur (S) concentration in pine needles were found to be a good indicator of SO_2 concentrations in ambient air (Manninen et al., 1991). In contrast to lichens, trees take up sulfur from both the atmosphere and soils (Winner et al., 1978). Hence, both atmospheric and pedospheric sources have to be considered, although sulfur concentrations in leaves tend to be more influenced by direct uptake of ambient SO_2 as opposed to sulfate (SO_4) uptake by roots (Cape, 2009). Conifers that absorb gaseous SO_2 store the sulfur in the form of sulfate rather than organic S compounds (S_{org}) (Kaiser et al., 1993). During root uptake, S is mostly taken up from the soil as SO_4 and is subsequently reduced within the tree and incorporated into S_{org} (Tausz, 2007). However, in the case of an oversupply of SO_4, for example due to high atmospheric sulfate deposition, the tree may not be able to convert all SO_4 into S_{org}, resulting in an accumulation of inorganic sulfur in the plant tissue. Hence, both S uptake pathways, via leaf absorption or root uptake, appear to result in an accumulation of SO_4 within the foliage. The ratio of total inorganic sulfur (SO_4–S) to S_{org} in pine needles has been suggested as an effective indicator for S stress (Legge et al., 1988). Legge et al. (1988) found ratios of 0.29 at background sites to ratios of 0.88 at a site with high sulfur loading in the vicinity of a sour gas plant in Alberta, Canada.

S metabolism is different for different plant species and the uptake of sulfur by conifers may also depend on the nitrogen (N) balance of the tree (Manninen and Huttunen, 2000). For example, in nutrient poor northern forest ecosystems, comparatively low atmospheric nitrogen dioxide (NO_2) and ammonia (NH_3) concentrations can contribute significantly to the nitrogen content of the plant and/or its growth potential (Cape, 2009; Manninen and Huttunen, 2000). Nitrogen uptake strongly depends on the species, for example, pine trees and spruce trees were found to respond differently to elevated atmospheric NO_2 concentrations (Manninen and Huttunen, 2000). Increasing biomass growth due to a higher availability of N may also cause a decrease of the N content within the plant (Cape, 2009 and references therein). Pedospheric N sources are organic N (e.g., amino acids), ammonium (NH_4), and nitrate (NO_3).

In addition to the determination of S and N contents, the isotopic composition of total sulfur and total nitrogen in foliage may provide information about the sources of S and N in trees (Ammann et al., 1999; Krouse, 1977; Winner et al., 1978). $\delta^{34}S$ of total S in the youngest needles had the same sulfur isotope ratios as SO_2 emissions near a sour gas processing plant in Alberta, Canada, whereas older needles showed lower $\delta^{34}S$ values (Winner et al., 1978). Winner et al. (1978) also found differences in $\delta^{34}S$ values with tree height. Ammann et al. (1999) were able to estimate plant uptake of NO_2 derived from vehicle emissions by determining $\delta^{15}N$ values of nitrogen in pine needles. They observed a trend toward higher $\delta^{15}N$ values closer to a highway in Switzerland due to nitrogen emissions from vehicle exhaust enriched in ^{15}N.

We determined the total S content and $[SO_4-S]/[S_{org}]$ ratios of jack pine and white and black spruce needles sampled in 2008, and investigated whether $[SO_4-S]/[S_{org}]$ ratios combined with $\delta^{34}S$ values are suitable early warning parameters indicating S stress in forest ecosystems. In addition, total nitrogen and $\delta^{15}N$ values of pine and spruce needles were determined to investigate whether the chemical and isotopic compositions of nitrogen elucidate environmental effects of elevated nitrogen deposition in the AOSR. Jack pine needles were sampled at 24 sites, black spruce at 5 and white spruce at 2 sites located at various distances (3–119 km) to the main operations. Needles were sampled from the most recent growth year (2008) from five different trees per site. At one site at 14 km distance, needles from the edge of the forest as well as from the interior of the stand were sampled.

The total sulfur contents in jack pine needles were similar to those of spruce needles (on average 924 ± 105 and 936 ± 110 ppm, respectively). Black and white spruce needles showed a trend of decreasing total S contents with increasing distance to the oil sand operations ($r^2=0.87$), whereas jack pine needles appeared to have rather constant total S contents with distance (Figure 11.7A). $[SO_4-S]/[S_{org}]$ ratios in jack pine needles were generally <0.4, whereas site averaged $[SO_4-S]/[S_{org}]$ ratios in spruce needles were as high as 0.76 ± 0.33 (at 13 km), with ratios >0.4 observed at sites within 27 km from the marker point (Figure 11.7B). This indicates that spruce trees may be more sensitive to sulfur loading resulting in increasing total S contents in recent growth pine needles within 27 km distance of the main operations.

In a study in Finland, pine needles were found to be more indicative of atmospheric SO_2 changes compared to spruce because stomatal SO_2 uptake by pine needles is more abundant compared to spruce (Manninen and Huttunen, 2000). However, Manninen and Huttunen (2000) also found lower $[SO_4-S]/[S_{org}]$ ratios for pine needles compared to spruce under NO_2+SO_2 fumigation experiments because the mature pine trees were very efficient in assimilating sulfate from SO_2 into organic sulfur. Total S contents and $[SO_4-S]/[S_{org}]$ ratios of pine needles from the edge of the site at 14 km distance were identical to those of pine needles collected within the stand at this site, although this site is only 14 km from the oil sands operations. This seems to confirm that pine trees are very effective in coping with elevated SO_2 concentrations. Total S content and $[SO_4-S]/[S_{org}]$ ratios in spruce needles may therefore be a better indicator of elevated ambient SO_2 concentrations than these parameters measured in pine needles. $\delta^{34}S$ values of total sulfur in conifer needles differed significantly between species (Figure 11.7C). $\delta^{34}S$ values of spruce needles varied between $-5‰$ and $+5‰$ with the lower values occurring in white spruce needles, whereas pine needle $\delta^{34}S$ values typically varied between $+5‰$ and $+15‰$. Highest $\delta^{34}S$ values were observed in jack pine needles at the edge of the site at 14 km distance. Sulfate in $PM_{2.5}$ emissions from two stacks tested was characterized by $\delta^{34}S$ values of $7.3\pm0.3‰$ (Stack A) and $9.4\pm2.0‰$ (Stack B). $\delta^{34}S$ values for atmospheric sulfate

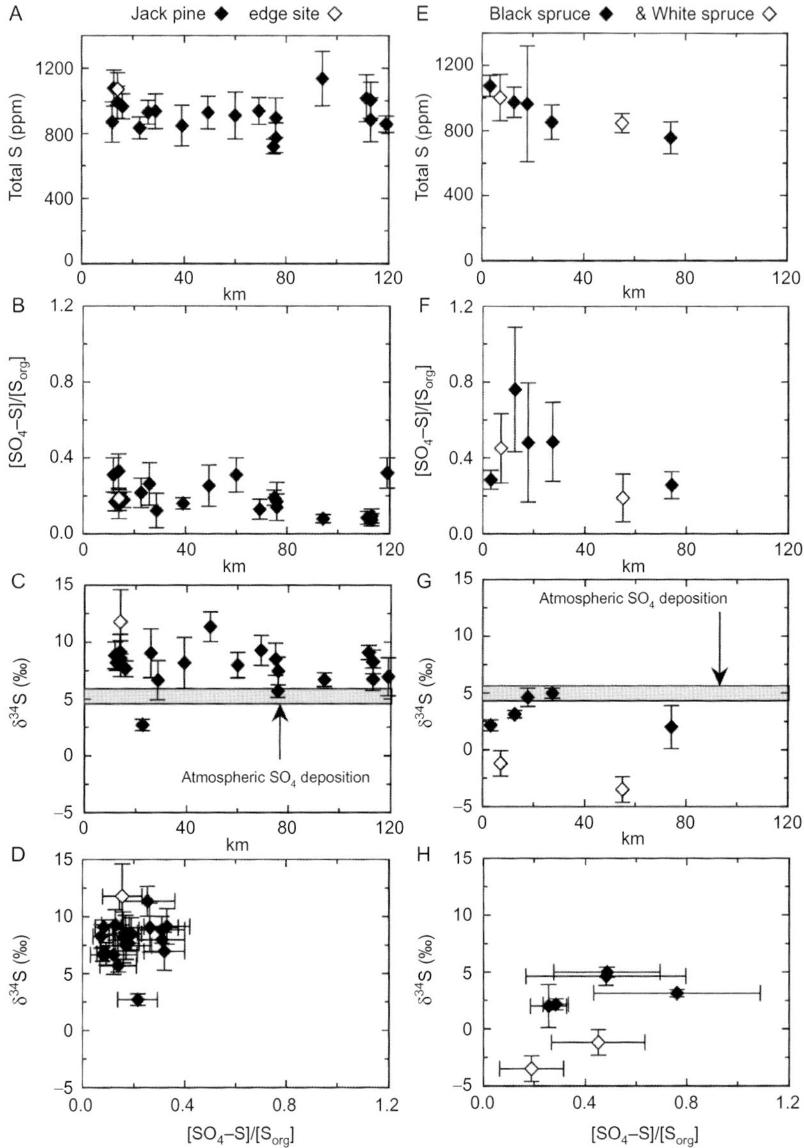

FIGURE 11.7 Site averaged (mean ± SD) total S content (A, E), [SO$_4$–S]/[Sorg] ratios (B, F) and δ^{34}S values of total S (C, G) in jack pine needles (left) and spruce needles (right) versus distance. There was a slight trend toward higher δ^{34}S values with increasing [SO$_4$–S]/[Sorg] ratios (H) for spruce needles, but not for jack pine needles (D). The gray bar represents δ^{34}S values of atmospheric sulfate deposition from distant deposition sampling sites.

deposition sites with low rates of <1.4 kg S ha^{-1} (around $+5‰$) were not isotopically different compared to δ^{34}S values of sulfate at sites with high deposition rates of >10 kg S ha^{-1} in the AOSR. Hence, industrial PM$_{2.5}$ sulfate stack emissions or atmospheric sulfate deposition alone cannot be responsible for δ^{34}S values higher than $+10‰$ in foliage of trees.

Stomatal SO$_2$ uptake is more abundant by pine needles compared to spruce needles and pine trees are effective in reducing foliar S via reemission of H$_2$S (Manninen and Huttunen, 2000). H$_2$S emissions measured for Scotch pine foliage in a study in Germany were 18 times higher compared to H$_2$S emissions by Blue and Norway spruce twigs (Kindermann et al., 1995). H$_2$S emissions are typically depleted in ^{34}S resulting in the accumulation of ^{34}S in the remaining S in the pine needles, thus in elevated δ^{34}S values of total S in foliage as observed in jack pine needles in the AOSR. Sulfur stress of pine trees that leads to increasing emissions of reduced S from the foliage is therefore reflected by δ^{34}S values that can be several per mil higher than those measured for industrial PM$_{2.5}$ emissions from two stacks in the AOSR (Proemse et al., 2012a) or for atmospheric sulfate deposition (Proemse et al., 2012b). This makes δ^{34}S values of total S in foliage an excellent indicator of S stress in pine trees resulting from elevated atmospheric S inputs. This hypothesis is supported by the fact that pine needles from the edge of the site at 14 km distance likely more exposed to ambient SO$_2$ or atmospheric sulfate deposition compared to needles within the stand, had the highest δ^{34}S values of total S (Figure 11.7C).

The ranges of total N contents in pine ($0.7\pm0.2\%$ to $1.1\pm0.1\%$) and spruce needles ($0.8\pm0.1\%$ to $1.5\pm0.1\%$) were comparable, with slightly higher values observed for pine needles (Figure 11.8A). Pine needle N contents appear rather constant with respect to distance to the oil sand operations, whereas total nitrogen contents in spruce increased slightly within 18 km of the operations. Site averaged δ^{15}N values of total nitrogen in pine and spruce needles varied between $-5.6‰$ and $-0.1‰$ and δ^{15}N values did not show a trend with distance or total N content (Figure 11.8B and C). δ^{15}N values of nitrate in stack emitted PM$_{2.5}$ was characterized by elevated δ^{15}N values of $9.4‰$ (Stack A) and $16.1\pm1.2‰$ (Stack B), indicating that these emissions were significantly enriched in ^{15}N compared to oil sand (δ^{15}N $\sim2.5‰$), by-products of upgrading (δ^{15}N from $-0.3‰$ to $1.3‰$), and atmospheric N$_2$ (δ^{15}N $=0‰$). Ammonium in PM$_{2.5}$ was characterized by a large δ^{15}N range with values between $-4.5‰$ to $+20.1‰$. δ^{15}N values of atmospheric nitrate and ammonium deposition were elevated within 30 km distance to one of the major emission stacks, up to $6.3‰$ for δ^{15}N–NO$_3$ and up to $11.3‰$ for δ^{15}N–NH$_4$. Although there was a trend toward higher δ^{15}N values in spruce needles between 27 and 3 km distance from the operations, high δ^{15}N values as observed for nitrate or ammonium of PM$_{2.5}$ from two stacks and atmospheric nitrate and ammonium deposition at sites within 30 km distance to the operations were not observed in total nitrogen

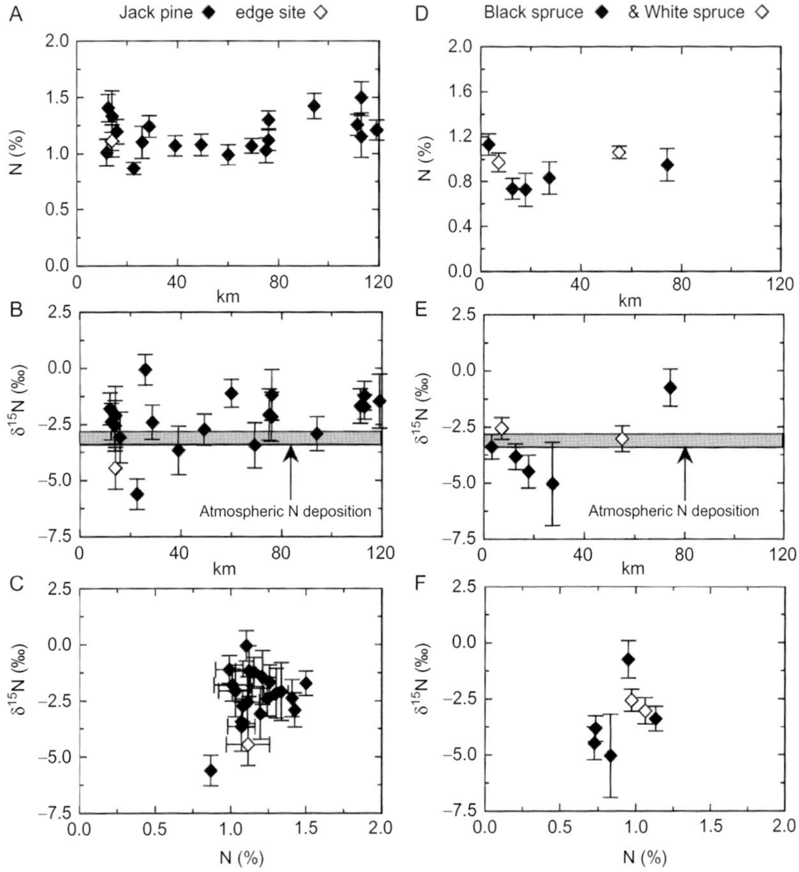

FIGURE 11.8 Site averaged (mean ± SD) total N content (A, D) and $\delta^{15}N$ values of total nitrogen (B, E) for jack pine needles (left) and spruce needles (right) versus distance. There was no correlation between the two parameters (C, F), neither for spruce nor for pine needles. The gray bar in (B) and (E) represents $\delta^{15}N$ values of summer 2008 atmospheric nitrate and ammonium deposition at distant deposition sampling sites.

of conifer foliage. $\delta^{15}N$ values of atmospheric nitrate and ammonium deposition in summer at distant sites (>90 km) were −3.2 ± 1.5‰ and −3.6 ± 0.9‰, respectively, similar to $\delta^{15}N$ values observed in total nitrogen of conifer needles. Similar $\delta^{15}N$ values in atmospheric N inputs to forest ecosystems and in total N of conifer foliage typically occurs when all available N is taken up by vegetation, providing little or no opportunity for nitrogen isotope fractionation. This seems to be the case in the AOSR. However, $\delta^{15}N$ values of gaseous industrial and natural (e.g., from forest fires) NO_x emissions were not determined in the AOSR but may constitute a major source of nitrogen for conifer needles (Manninen and Huttunen, 2000).

11.5   SUMMARY

We investigated whether stable isotope techniques constitute quantitative and qualitative tracers of industrial nitrogen (N) and sulfur (S) emissions from the oil sands operations. Eluants from $PM_{2.5}$ samples from two stacks, and atmospheric bulk/throughfall deposition samples from various sites located at distances between 3 and 113 km to the main mining and upgrading operations were analyzed for the nitrogen and triple oxygen isotope ratios ($\delta^{15}N$, $\delta^{18}O$, and $\Delta^{17}O$) of nitrate, the $\delta^{15}N$ values of ammonium as well as the sulfur and oxygen isotope ratios ($\delta^{34}S$, $\delta^{18}O$) of sulfate. Industrial $PM_{2.5}$ emissions from the tested stacks have distinct ^{18}O and ^{17}O contents of nitrates and unique ^{18}O contents of sulfate that appear suitable as quantitative tracers for nitrate and sulfate derived from industrial emissions.

The analyses of the isotopic compositions of atmospheric nitrate and sulfate in bulk deposition and throughfall at 16 sites revealed the presence of isotopically distinct nitrate and sulfate from industrial emissions within 30 km distance to mining/upgrading operations associated with elevated nitrate and sulfate deposition rates. Oxygen isotope ratios of atmospheric nitrate ($\delta^{18}O$ and $\Delta^{17}O$) and of atmospheric sulfate ($\delta^{18}O$) in bulk deposition and throughfall were used to quantify industrial contributions to atmospheric nitrate and sulfate deposition. Results revealed that industrial N and S contributions were significant ($>50\%$) at sites <30 km to the oil sands operations and to the mining sites. Isotopic compositions of atmospheric nitrate and sulfate deposition suggest negligible industrial contributions at low deposition rates at sites >90 km from the oil sands operations.

$\delta^{15}N$ values of nitrate and ammonium in $PM_{2.5}$ emissions from the tested stacks were variable, but typically enriched in ^{15}N compared to oil sand material and by-products from the upgrading. Nitrogen isotope ratios therefore provide a qualitative tracer of N emissions with unusually elevated $\delta^{15}N$ values. High $\delta^{15}N$ values of atmospheric ammonium and nitrate deposition only occur within 30 km distance to the operations.

$\delta^{34}S$ values of oil sand material, by-products from the upgrading, $PM_{2.5}$ emissions from two stacks and atmospheric sulfate deposition are not isotopically distinct and are less indicative of sulfur sources. However, $\delta^{34}S$ in atmospheric sulfate deposition at some sites close to tailing ponds revealed reduced sulfur gas emissions constituting a qualitative tracer for this additional S source.

We also analyzed environmental receptor samples to investigate whether isotopically distinct industrial N or S compounds (qualitative tracers as mentioned above) were detectable in the surrounding ecosystems. Elevated N and S contents in lichen samples were observed within 16 km distance from the operations. $\delta^{15}N$ values of total N in lichens indicate industrial derived N at the site closest (3 km) to a major emission source. $\delta^{15}N$ values of total N in lichen samples from further distant sites (>3 km) and $\delta^{34}S$ values of total S in lichen samples from all sites did not reveal clear evidence of industrially derived N and S contributions. However, low $\delta^{34}S$ values ($<2‰$) in lichen samples from sites close to tailing ponds were also indicative of reduced S emissions from the ponds. $\delta^{15}N$ and $\delta^{34}S$ values of conifer needles from samples taken within the forest stands did not reveal clear evidence of industrial N and S contributions based on stable isotope tracer techniques. Total S contents and $[SO_4–S]/[S_{org}]$ ratios of pine needles currently show little effect of industrial S loading.

We conclude that stable isotope techniques combined with concentration and flux measurements can provide unique insights into the transport and the fate of industrial N and S emissions in surrounding terrestrial environments in the AOSR.

11.6 ACKNOWLEDGMENTS

This study was funded by the Wood Buffalo Environmental Association (WBEA), the Natural Sciences and Engineering Research Council of Canada (NSERC CRDPJ 372151-08), and the Canada School for Energy and the Environment (CSEE). We thank the staff of WBEA for their assistance in project coordination and for obtaining samples. Resin extracts and concentration data of atmospheric deposition samplers were provided by Dr. Mark Fenn and Christopher Ross from the USDA Forest Service Laboratory in Riverside (California, USA). Lichen samples were provided by Dr. Shanti Berryman and Justin Straker (Integral Ecology Group Ltd., Victoria, British Columbia, Canada). The conifer needles were collected by Dr. Dennis Jaques (Ecosat Geobotanical Surveys Inc., Vancouver, British Columbia, Canada). Dr. Judith Chow and Dr. John Watson (Desert Research Institute, Reno, Nevada, USA) provided stack emitted $PM_{2.5}$ samples.

REFERENCES

Addison, P., Puckett, K., 1980. Deposition of atmospheric pollutants as measured by lichen element content in the Athabasca oil sands area. Can. J. Bot. 58, 2323–2334.

Allen, E., 2004. Effects of Nitrogen Deposition on Forests and Peatlands: A Literature Review and Discussion of the Potential Impacts of Nitrogen Deposition in the Alberta Oils Sands Region. Wood Buffalo Environmental Association, Fort McMurray, AB, 114 pp.

Ammann, M., Siegwolf, R., Pichlmayer, F., Suter, M., Saurer, M., Brunold, C., 1999. Estimating the uptake of traffic-derived NO_2 from ^{15}N abundance in Norway spruce needles. Oecologia 118, 124–131.

Balaguer, L., Manrique, E., 1991. Interaction between sulfur dioxide and nitrate in some lichens. Environ. Exp. Bot. 31, 223–227.

Bargagli, R., Mikhailova, I., 2002. Accumulation of inorganic contaminants. In: Nimis, P., Scheidegger, C., Wolseley, P. (Eds.), Monitoring with lichens—Monitoring lichens. Proceedings of the NATO Advanced Research Workshop on Lichen Monitoring, Wales, United Kingdom, 16-23 August 2000. Kluwer Academic Publishers, Dordrecht, The Netherlands, pp. 65–84.

Canfield, D., 2001. Isotope fractionation by natural populations of sulfate-reducing bacteria. Geochim. Cosmochim. Acta 65, 1117–1124.

Cape, J., 2009. Plants as accumulators of atmospheric emissions. In: Legge, A. (Ed.), Air Quality and Ecological Impacts: Relating Sources to Effects. Developments in Environmental Science, Vol. 9, Elsevier, Amsterdam, The Netherlands, pp. 61–98.

Carvalho, A., Freitas, M., 2011. Airborne chemical elements: Correlation between atmospheric levels and lichen content. Int. J. Environ. Health 5, 134–147.

Casciotti, K.L., Sigman, D.M., Hastings, M.G., Böhlke, J.K., Hilkert, A., 2002. Measurement of the oxygen isotopic composition of nitrate in seawater and freshwater using the denitrifier method. Anal. Chem. 74, 4905–4912.

Case, J., Krouse, H., 1980. Variations in sulphur content and stable sulphur isotope composition of vegetation near a SO_2 source at Fox Creek, Alberta, Canada. Oecologia 44, 248–257.

Dahlman, L., Persson, J., Palmqvist, K., Näsholm, T., 2004. Organic and inorganic nitrogen uptake in lichens. Planta 219, 459–467.

Environment Canada, 2011. Integrated monitoring plan for the oil sands: Air Quality Component. No. En14-45/2011E-PDF, Minister of the Environment, 72 pp.

ERCB, 2010. Alberta's energy reserves 2009 and supply/demand outlook 2010-2019. Energy Resources Conservation Board ST98-2010, pp. 1–232.

Fenn, M., Poth, M., 2004. Monitoring nitrogen deposition in throughfall using ion exchange resin columns: A field test in the San Bernardino Mountains. J. Environ. Qual. 33, 2007–2014.

Fenn, M., Ross, C., 2010. Sulfur and nitrogen deposition in the Athabasca oil sands region. In: Proceedings of The Air and Waste Management Association 103rd Annual Conference and Exhibition. Calgary, Canada, pp. 6 2010-A-664-AWMA.

Galloway, J., Townsend, A., Erisman, J., Bekunda, M., Cai, Z., Freney, J., Martinelli, L., Seitzinger, S., Sutton, M., 2008. Transformation of the nitrogen cycle: Recent trends, questions, and potential solutions. Science 320, 889–892.

Giesemann, A., Jaeger, H.J., Norman, A.-L., Krouse, H.R., Brand, W.A., 1994. Online sulfur-isotope determination using an elemental analyzer coupled to a mass spectrometer. Anal. Chem. 66, 2816–2819.

Gries, C., Sanz, M.-J., Romagni, J., Goldsmith, S., Kuhn, U., Kesselmeier, J., Nash, T., 1997. The uptake of gaseous sulphur dioxide by non-gelatinous lichens. New Phytol. 135, 595–602.

Hawksworth, D., Rose, F., 1970. Qualitative scale for estimating sulphur dioxide air pollution in England and Wales using epiphytic lichens. Nature 227, 145–148.

Hazewinkel, R.R.O., Wolfe, A.P., Pla, S., Curtis, C., Hadley, K., 2008. Have atmospheric emissions from the Athabasca Oil Sands impacted lakes in northeastern Alberta, Canada? Can. J. Fish. Aquat. Sci. 65, 1554–1567.

Jäger, H.-J., Krupa, S.V., 2009. Hormesis—Its relevance in phytotoxicology. In: Legge, A.H. (Ed.), Air Quality and Ecological Impacts: Relating Source to Effects. Developments in Environmental Science, Vol., 9, Elsevier, Amsterdam, The Netherlands, pp. 137–152.

Johnson, C.M., Nishita, H., 1952. Microestimation of sulfur in plant materials, soils, and irrigation waters. Anal. Chem. 24, 736–742.

Kaiser, W., Dittrich, A., Heber, U., 1993. Sulfate concentrations in Norway spruce needles in relation to atmospheric SO_2: A comparison of trees from various forests in Germany with trees fumigated with SO_2 in growth chambers. Tree Physiol. 12, 1–13.

Kaiser, J., Hastings, M.G., Houlton, B.Z., Röckmann, T., Sigman, D.M., 2007. Triple oxygen analysis of nitrate using the denitrifier method and thermal decomposition of N2O. Anal. Chem. 79, 599–607.

Kendall, C., 1998. Tracing nitrogen sources and cycling in catchments. In: Kendall, C., McDonnell, J. (Eds.), Isotope Tracers in Catchment Hydrology. Elsevier, Amsterdam, The Netherlands, pp. 519–576.

Kendall, C., Elliott, E., Wankel, S., 2007. Tracing anthropogenic inputs of nitrogen to ecosystems. In: Michener, R., Lajtha, K. (Eds.), Stable Isotopes in Ecology and Environmental Science. Blackwell Publishing, Oxford, United Kingdom, pp. 375–449.

Kindermann, G., Hüve, K., Slovik, S., Lux, H., Rennenberg, H., 1995. Emission of hydrogen sulfide by twigs of conifers—A comparison of Norway spruce (*Picea abies* (L.) Karst.), Scotch pine (*Pinus sylvestris* L.) and Blue spruce (*Picea pungens* Engelm.). Plant Soil 168, 421–423.

Kornexl, B.E., Gehre, M., Höfling, R., Werner, R.A., 1999. On-line $\delta^{18}O$ measurement of organic and inorganic substances. Rapid Commun. Mass Spectrom. 13, 1685–1693.

Kozlowski, T., 1980. Impacts of air pollution on forest ecosystems. Bioscience 30, 88–93.

Krouse, H., 1977. Sulphur isotope abundance elucidate uptake of atmospheric sulphur emissions by vegetation. Nature 265, 45–46.

Krouse, H., 1980. Sulphur isotopes in our environment. In: Fritz, P., Fontes, J. (Eds.), Handbook of Environmental Isotope Geochemistry. Elsevier, New York, USA, pp. 435–471.

Krupa, S., 2003. Effects of atmospheric ammonia (NH_3) on terrestrial vegetation: A review. Environ. Pollut. 124, 179–221.

Legge, A., Bogner, J., Krupa, S., 1988. Foliar sulphur species in pine: A new indicator of a forest ecosystem under air pollution stress. Environ. Pollut. 55, 15–27.

Loppi, S., Frati, L., 2006. Lichen diversity and lichen transplants as monitors of air pollution in a rural area of central Italy. Environ. Monit. Assess. 114, 361–375.

Loppi, S., Nascimbene, J., 2010. Monitoring H_2S air pollution caused by the industrial exploitation of geothermal energy: The pitfall of using lichens as bioindicators. Environ. Pollut. 158, 2635–2639.

Manninen, S., Huttunen, S., 2000. Response of needle sulphur and nitrogen concentrations of Scots pine versus Norway spruce to SO_2 and NO_2. Environ. Pollut. 107, 421–436.

Manninen, S., Huttunen, S., Torvela, H., 1991. Needle and lichen sulphur analyses on two industrial gradients: Water. Air Soil Pollut. 59, 153–163.

Michalski, G., Scott, Z., Kabiling, M., Thiemens, M., 2003. First measurements and modelling of $\Delta^{17}O$ in atmospheric nitrate. Geophys. Res. Lett. 30 (16), 1870.1–1870.4.

Miller, J., Brown, D., 1999. Studies of ammonia uptake and loss by lichens. Lichenologist 31, 85–93.

Nimis, P., Purvis, O., 2002. Monitoring lichens as indicators of pollution. In: Nimis, P., Scheidegger, C., Wolseley, P. (Eds.), Monitoring with Lichens—Monitoring Lichens. Proceedings of the NATO Advanced Research Workshop on Lichen Monitoring in Wales, United Kingdom, 16–23 August 2000. Kluwer Academic Publishers, Dordrecht, The Netherlands, pp. 7–10.

Nimis, P., Scheidegger, C., Wolseley, P., 2002. Monitoring with Lichens—Monitoring Lichens. Proceedings of the NATO Advanced Research Workshop on Lichen Monitoring in Wales, United Kingdom, 16-23 August 2000. Kluwer Academic Publishers, Dordrecht, The Netherlands, 408 pp.

Pearson, J., Wells, D., Seller, K., Bennett, A., Soares, A., Woodall, J., Ingrouille, M., 2000. Traffic exposure increases natural ^{15}N and heavy metal concentrations in mosses. New Phytol. 147, 317–326.

Preston, T., Owens, N.J.P., 1983. Interfacing an automatic elemental analyser with an isotope ratio mass spectrometer: The potential for fully automated total nitrogen and nitrogen-15 analysis. Analyst 108, 971–977.

Proemse, B.C., Mayer, B., Chow, J.C., Watson, J.G., 2012a. Isotopic characterization of nitrate, ammonium and sulfate in stack $PM_{2.5}$ emissions in the Athabasca Oil Sands Region, Alberta, Canada. Atmos. Environ. 60, 555–563

Proemse, B.C., Mayer, B., Fenn, M.E., 2012b. Tracing industrial sulfur contributions to atmospheric sulfate deposition in the Athabasca Oil Sands Region, Alberta, Canada. Appl. Geochem. http://dx.doi.org/10.1016/j.apgeochem.2012.08.006.

Ramos-Padrón, E., Bordenave, S., Lin, S., Bhaskar, I.M., Dong, X., Sensen, C.W., Fournier, J., Voordouw, G., Gieg, L.M., 2011. Carbon and sulfur cycling by microbial communities in a gypsum-treated oil sands tailings pond. Environ. Sci. Technol. 45, 439–446.

Rheault, H., Drapeau, P., Bergeron, Y., Esseen, P.-A., 2003. Edge effects of epiphytic lichens in managed black spruce forests of eastern North America. Can. J. Forest Res. 33, 23–32.

Schindler, D.W., Dillon, P.J., Schreier, H., 2006. A review of anthropogenic sources of nitrogen and their effects on Canadian aquatic ecosystems. Biogeochemistry 79, 25–44.

Sebilo, M., Mayer, B., Grably, M., Biliou, D., Mariotti, A., 2004. The use of the "ammonium diffusion" method for $\delta^{15}N - NH_4^+$ and $\delta^{15}N - NO_3^-$ measurements: Comparison with other techniques. Environ. Chem. 1, 99–103.

Shewchuk, S., 1982. An acid deposition perspective for northeastern Alberta and northern Saskatchewan. Water Air Soil Pollut. 18, 413–419.

Siegfriedt, R., Wiberley, S., Moore, R., 1951. Determination of sulfur after combustion in small oxygen bomb. Rapid Trimetric Method. Anal. Chem. 23, 1008–1011.

Sigman, D.M., Casciotti, K.L., Andreani, M., Barford, C., Galanter, M., Böhlke, J.K., 2001. A bacterial method for the nitrogen isotopic analysis of nitrate in seawater and freshwater. Anal. Chem. 73, 4145–4153.

Sims, R., Schock, R., Adegbululgbe, A., Fenhann, J., Konstantinaviciute, I., Moomaw, W., Nimir, H., Schlamadinger, B., Torres-Martinez, J., Turner, C., Uchiyama, Y., Vuori, S., Wamukonya, N., Zhang, X., 2007. Energy supply. In: Metz, B., Davidson, O., Bosch, P., Dave, R., Meyer, L. (Eds.), Climate Change 2007: Mitigation. Contribution of Working Group III to the Fourth Assessment Report of the Intergovernmental Panel on Climate Change. Cambridge University Press, Cambridge, United Kingdom and New York, USA, pp. 72.

Soil Classification Working Group, 1998. The Canadian System of Soil Classification, third ed. Agriculture and Agri-Food Canada Publication. http://sis.agr.gc.ca/cansis/taxa/cssc3/intro.html

Statistics Canada, 2011. Urban areas. Statistics Canada 2011 Census. http://www12.statcan.gc.ca/census-recensement/2011/dp-pd/hlt-fst/pd-pl/Table-Tableau.cfm?LANG=Eng&T=209&CMA=860&S=0&O=D&RPP=25.

Takala, K., Olkkonen, H., Krouse, H., 1991. Sulphur isotope composition of epiphytic and terricolous lichens and pine bark in Finland. Environ. Pollut. 69, 337–348.

Tausz, M., 2007. Sulfur in forest ecosystems. In: Hawkesford, M., De Kok, L. (Eds.), Sulfur in Plants. An Ecological Perspective. Springer, Verlag, Dordrecht, Boston, London, pp. 59–75.

Timoney, K., Lee, P., 2009. Does the Alberta tar sands industry pollute? The scientific evidence. Open Conserv. Biol. J. 3, 65–81.

Tozer, W.C., Hackell, D., Miers, D.B., Silvester, W.B., 2005. Extreme isotopic depletion of nitrogen in New Zealand lithophytes and epiphytes: The result of diffusive uptake of atmospheric ammonia?. Oecologia 144, 628–635.

Van Dobben, H., Ter Braak, C., 1999. Ranking of epiphytic lichen sensitivity to air pollution using survey data: A comparison of indicator scales. Lichenologist 31, 27–39.

Wadleigh, M., 2003. Lichens and atmospheric sulphur: What stable isotopes reveal. Environ. Pollut. 126, 345–351.

Wadleigh, M., Blake, D., 1999. Tracing sources of atmospheric sulphur using epiphytic lichens. Environ. Pollut. 106, 265–271.

Watson, J., Cooper, J., Huntzicker, J., 1984. The effective variance weighting for least squares calculations applied to the mass balance receptor model. Atmos. Environ. 18, 1347–1355.

Watson, J., Chen, L.-W.A., Chow, J., Doraiswamy, P., Lowenthal, D., 2008. Source apportionment: Findings from the U.S. Supersites Program. J. Air Waste Manage. Assoc. 58, 265–288.

Whitfield, C., Aherne, J., Watmough, S., 2009. Modeling soil acidification in the Athabasca oil sands region, Alberta, Canada. Environ. Sci. Technol. 43, 5844–5850.

Whitfield, C., Aherne, J., Watmough, S., McDonald, M., 2010. Estimating the sensitivity of forest soils to acid deposition in the Athabasca oil sands region, Albetra. J. Limnol. 69, 201–208.

Winner, W., Bewley, J., Krouse, H., Brown, H., 1978. Stable sulfur isotope analysis of SO_2 pollution impact on vegetation. Oecologia 36, 351–361.

Wiseman, R., Wadleigh, M., 2002. Lichen response to changes in atmospheric sulphur: Isotopic evidence. Environ. Pollut. 116, 235–241.

Chapter 12

Air Quality Modeling in the Athabasca Oil Sands Region

M.J.E. Davies[1]

Stantec Consulting, Calgary, Alberta, Canada

[1]*Corresponding author: e-mail: mervyn.davies@stantec.com*

ABSTRACT

The Athabasca Oil Sands Region of northeastern Alberta has been, and will continue to be, a significant source of SO_2 and NO_X emissions. Ambient air quality models that simulate transport, dispersion, chemical transformation, and deposition processes have been used in the region for the past 30 years to help manage ambient air quality due to these emissions. Model applications originally focused on evaluating SO_2 emissions and later shifted to include other chemical compounds such as NO_X. The early efforts focused on field studies to collect region-specific data to understand plume behavior and develop site-tuned models. As the regulatory models improved, the focus shifted from development of site-tuned models to adoption of the updated regulatory models. Over the last decade, the CALMET/CALPUFF model system has become the *de facto* standard for assessing air quality in the region by industry, regulatory, and multistakeholder organizations. A case study application of this model system applied to support a lichen bioindicator program is discussed. Specifically, the sulfur and nitrogen contents in lichen tissues collected at 359 sites located up to 150 km from the primary emission source region were compared to sulfur and nitrogen deposition predictions. Both the model predictions and the lichen measurements indicate that the main air quality footprint is within 20 km of the main emission sources.

12.1 INTRODUCTION

Northeastern Alberta has significant bitumen reserves, often referred to as oil sands. The first commercial scale bitumen extraction development commenced operation in 1967 with the Great Canadian Oil Sands (GCOS) operation (now Suncor) and was followed by the Syncrude Mildred Lake operation in 1978. Each of these operations comprised integrated mining, extraction,

Disclaimer: The content and opinions expressed by the author in this book chapter do not necessarily reflect the views of WBEA or of the WBEA membership.

Developments in Environmental Science, Vol. 11. http://dx.doi.org/10.1016/B978-0-08-099443-7.00012-0

and upgrading facilities producing synthetic crude oil as an end-product. Mining was primarily undertaken using electrically driven bucketwheel excavators, draglines, and conveyer belt systems. Both operations were substantive sources of sulfur dioxide (SO_2) emissions due to the use of a bitumen product (e.g., coke or coke burner off-gas) as a fuel source and the associated sulfur content of the bitumen. The Syncrude Aurora North and the Shell Muskeg River mine operations commenced production in 1998 and 2003, respectively. These two operations comprised shovel and truck mining and extraction facilities producing bitumen as an end-product, with upgrading to synthetic crude oil taking place off-site. In addition, more recent commercial operations use nonmining (i.e., *in situ*) methods that inject steam into the bitumen formation to facilitate extraction. Oxides of nitrogen (NO_X) emissions became more substantive in the region with diesel-fueled mine shovels and trucks replacing the older electrical-based mine equipment, and with natural gas being used to generate steam for the *in situ* operations.

Air quality modeling efforts in the region initially focused on SO_2 emissions due to direct (SO_2 fumigation) and indirect (acid deposition) effects of SO_2 emissions on terrestrial and aquatic ecosystems. With increasing NO_X emissions, the focus shifted to include the NO_X contribution to acid deposition, and a new concern associated with potential eutrophication effects. Parallel concerns in local communities focused on emissions that have a potential to produce odors or adverse health outcomes. These latter emissions include reduced sulfur compounds, volatile organic compounds, polycyclic aromatic compounds, and heavy metals.

As air quality simulation models provide a fundamental link between emissions discharged into the atmosphere and the air quality changes associated with these discharges, they have played a major role in the management of air quality in the oil sands region. Specifically, these models simulate transport, dispersion, chemical transformation, and deposition processes in the atmosphere. They have been used to predict the magnitude, and spatial concentration and deposition patterns associated with historical, current, and future development scenarios. This chapter summarizes the historical application of these models to the oil sands region and provides an overview of a recent air quality model application in the region.

12.2   HISTORICAL MODEL APPLICATIONS

The evolution and application of air quality models to the oil sands region has generally led or, in some cases, followed the development and application of the regulatory models in Alberta over the same period.

12.2.1   Regulatory Models

Alberta Environment (AENV) provides industry with air quality model guidelines and has provided air quality simulation models (e.g., Alberta

Environment, 1978, 1989, 1994). The historic AENV models focused on individual stacks (e.g., STACKS, FLARES), on individual facilities (e.g., SEEC, PLUMES), and on models to predict sulfur deposition (e.g., SULDEP3 and ADEPT2). The STACKS, FLARES, and SEEC models were screening models based on using predetermined wind speed, wind direction, and Pasquill–Gifford (PG) atmospheric stability class combinations; they focused on predicting maximum hourly average concentrations. The SULPDEP and ADEPT models were climatological dispersion models (CDMs) that focused on predicting seasonal and annual average sulfur compound deposition. The SULDEP model considers flat terrain, and the ADEPT model was enhanced to consider elevated terrain.

None of the AENV models had the ability to use sequential hourly time series meteorological data. The sequential time series approach allows for rigorous calculation of daily and annual average concentrations from the hourly values. This was one of the reasons why AENV transitioned to the use of U.S. EPA models, and subsequent model guidelines refer to application of these models (e.g., Alberta Environment, 1997, 2000, 2003). The most recent Alberta model guideline (Alberta Environment, 2009) identifies the SCREEN3, AERMOD-PRIME, and CALPUFF models. The SCREEN3 and AERMOD-PRIME models are viewed by AENV as being applicable for transport distances up to 25 km; the applicability of the CALPUFF model is up to 200 km.

12.2.2   The AOSERP/RMD and Industry Period

The Alberta Oil Sands Environmental Research Program (AOSERP) was established in 1975 to evaluate the environmental implications of oil sands development in the Athabasca oil sands region. AOSERP was initially sponsored by Alberta Environment and Environment Canada for a 10-year period, the initial period (1975–1980) focusing on accumulating baseline information and the follow-up period (1980–1985) improving the understanding of the atmosphere–biosphere interface. Smith (1981) summarized the research, conclusions, and recommendations based on the 1975–1980 period.

In 1979, the Alberta Government assumed the funding role for AOSERP through the Alberta Research Management Division (RMD). AOSERP, RMD, and industry recognized the value of air quality simulation models in assisting with the management of air emissions and commissioned a number of model studies that are described in the following sections.

12.2.2.1   The 1970s

Studies during this period focused on understanding plume behavior in the oil sands region and examining appropriate models that could be applied to evaluate the oil sands plant emissions. One of the first steps was to identify air quality model user needs (e.g., Angle, 1979).

Walmsley and Bagg (1977) from Environment Canada applied the CDM to predict seasonal and annual average SO_2 concentrations. One of the key challenges was to develop representative meteorological conditions for the two developments that are located within and adjacent to the Athabasca River valley, which has a north–south orientation. A statistical approach was used to generate synthetic STAR data based on correlations between the limited meteorological measurements near the developments and the longer term Fort McMurray Airport values. The airport is located in the Clearwater River valley that has a west–east orientation. Padro and Bagg (1978) applied the U.S. EPA single source CRSTER model to the GCOS plant. The model was used to predict maximum 1-, 3-, and 24-h, concentrations as well as annual average SO_2 concentrations. Again, one of the key challenges was the preparation of representative hourly meteorological data.

A number of field studies were conducted to collect plume rise and plume spread data from the GCOS powerhouse stack plume. Davison and Grandia (1979) systematically analyzed measurements from aircraft flights through the plume. The lateral plume spreads were found to be larger than those associated with the PG curves. The vertical plume spreads were found to be similar to those associated with the PG curves at 3.2 km downwind; the measured values, however, did not increase substantially at farther distances. Syncrude also examined the plume behavior of the GCOS plume based on photographic records, aircraft measurements, and local minisonde temperature and wind profiles (Slawson et al., 1980). The measurements were compared to three conceptual dispersion models in terms of plume spread, plume rise, and ambient SO_2 concentrations. The study examined the feasibility of developing a site-tuned plume rise and dispersion model for the Syncrude stack. Djurfors and Netterville (1978) developed plume rise algorithms for the strong wind shear conditions that were found in the region. Netterville (1979) conducted a study to examine concentration fluctuations to better understand the stochastic nature of the atmosphere.

Padro (1979) conducted a review of air quality models to determine their applicability to the AOSERP region. The appropriate models would have to account for elevated terrain features, and two models were considered for further consideration: a simple Gaussian model such as CRSTER that can be applied relatively quickly and a more complex Eulerian model such as LIRAQ or APDIC to deal with more complex considerations such as elevated terrain, atmospheric chemistry, and atmospheric deposition. Reid et al. (1979) further evaluated the LIRAQ and ADPIC models for application in the oil sands region. They concluded that the LIRAQ model, even though it had proven itself in an urban setting, did not handle stack sources well; limiting its application to the AOSERP area. The ADPIC model was viewed as having inadequate physics. As both models required significant computational resources, they were declared "computationally impractical." The review indicated that PATRIC, a simplified version of APDIC, had some potential for the region.

Other air quality (nonmodel)-related studies were undertaken over the same period. These included studies to examine the climatology of the area (Longley and Janz, 1978), the potential for enhanced fog formation from oil sands-related combustion sources (Croft et al., 1976; Murray and Kurtz, 1976; Murray and Low, 1979), the review of ambient air quality measurements (Strosher, 1978), and the review of snowpack and precipitation chemistry (Fanaki, 1978). Wilson (1979) conducted wind tunnel studies to determine the influence of tailings pond berms on the dispersion of plumes from nearby stacks.

12.2.2.2   The 1980s

The focus of the work during this period was on developing representative meteorological data and oil sands region-specific models. Additional work also focused on wind flow and chemistry model components.

Early efforts to develop a representative meteorological database in the region involved the installation of several meteorological monitoring stations and a 152-m tower in the floor of the Athabasca River valley as part of AOSERP. In the early 1980s, industry operated three Doppler acoustic sounders to examine vertical wind and turbulence profiles (two were associated with the existing Suncor and Syncrude operations and the third with a proposed operation; Davies, 1982).

While most of the wind data were collected from locations that represent the valley, some data were also collected for a few elevated sites. The AOSERP Birch Mountain Tower site is about 70 km north–northwest of the Mildred Lake monitoring station at an elevation of 850 m ASL. Data from 1976 to 1979 indicate that winds are primarily from the west to northwest sectors (Leahey and Hansen, 1981). There were 2344 minisonde releases (1975–1979) in the Mildred Lake area, and an analysis of the data indicated that winds at 1200 m above the ground were primarily from the west to northwest sectors, similar to those measured at Birch Mountain (Davison et al., 1981b). In summary, the wind data indicate that the upper level prevailing winds are from the west to northwest sectors and that the lower elevation sites are strongly influenced by local topography.

Strosher and Peters (1980) examined 2 years of ambient SO_2 concentration data measured at 10 monitoring stations in the AOSERP region to determine which meteorological conditions lead to high concentration events. While most of the high concentration events occurred during the day between 09:00 and 15:00, there was a range of conditions that produced the high events. They indicated that dispersion models for the region need to produce predictions that are consistent with the measurements.

Davison and Lantz (1980) conducted a review of air quality simulation requirements based on technical considerations, implementation considerations, and user needs. Based on this review, Davison et al. (1981a,b,c) developed a model to produce a concentration file (GLCGEN) and a postprocessor program to provide the desired output statistics (FRQDTN). The combined model system was also referred to as FREDIS. GLCGEN is a Gaussian model with dispersion based on similarity scaling and empirical regional plume measurements as summarized by Davison and Leavitt (1979). The hourly meteorological data used to evaluate the model performance were based on empirical algorithms derived from more than 2200 minisonde ascents in the region. A systematic sensitivity analysis of the model was undertaken.

Bottenheim and Strausz (1977, 1981) developed a chemically reactive plume model for the AOSERP study area. The interest in chemical modeling for the region was based on plume measurements conducted in 1977 (Fanaki, 1979; Fanaki et al., 1979). The model was found to be in reasonable agreement with the measurements. Cheng et al. (1987) summarized aircraft measurements conducted in March and December 1983. The SO_2 to particulate sulfate conversion varied from 0.02–2.81% per h in winter to 0.06–8.66% per h in summer. The NO to NO_2 conversion varied considerably from hour to hour, ranging from 0.2 to 21% per min.

Danard and Gray (1982) developed two mesoscale wind flow models: one centered over Mildred Lake and the other centered over Stoney Mountain. The input for the Mildred Lake application was from regional minisonde measurements and geostrophic winds from the Canadian Atmospheric Environment Service 700 and 850 mb wind analyses. The input for the Stoney Mountain application was based on sea level and 850 mb winds. The development of these models was in response to earlier studies that recognized the influence of complex terrain on air flow.

A number of model assessments focused on the deposition of acid-forming compounds. The Oil Sands Environmental Group developed and applied a long-range transport model to predict the deposition of acid-forming compounds (MEP, 1982). The model considered wet and dry deposition of sulfur and nitrogen compounds, and considered up to 14 existing and proposed oil sands developments. Fung and Davies (1987) applied a modified version of the MESOPUFF II model (referred to as LERTAD) to the oil sands area. This model focused on the wet and dry deposition of sulfur compounds, and the study was supported by the RMD (Research Management Division, Alberta Environment. Edmonton, AB). The RMD applied the MESOPUFF II, RELMAP, and SERTAD models to western Canada to determine an appropriate model to evaluate the deposition of acid-forming compounds (Cheng et al., 1990). A modified version of the RELMAP model (RELAD) was selected and has been used by Alberta Environment to examine deposition on a provincial scale.

The Fort McMurray Air Quality Task Force (AQTF), which comprised industry, community, and regulatory stakeholders, conducted a systematic review of ambient air quality in the region (Air Quality Task Force, 1987). The review focused on measurements and did not include model output. Dabbs (1985) conducted a systematic review of the atmosphere–biosphere interface by examining ambient air quality measurements, dispersion model predictions, and vegetation parameters. Multidisciplinary studies to examine the atmosphere–biosphere interface were proposed (Caton et al., 1987) to recognize the potential for acidification of terrestrial and aquatic systems. Other air quality (nonmodel)-related studies continued, including an updated study to examine the climatology of the area (Rudolph et al., 1984), the continuing review of ambient air quality measurements (Strosher, 1981), and the continuing review of snowpack and precipitation measurements (Davis et al., 1985; Murray, 1981; Olson et al., 1982).

12.2.2.3   The 1990s

This period saw the increasing use of models to support regulatory applications. A modified version of the U.S. EPA ISCST model and the Alberta Environment ADEPT models were applied in several impact assessment studies (e.g., Syncrude Canada, 1992, 1997, 1998). The plume rise and plume spread algorithms in the modified version of ISCST incorporate the findings from the previously indicated aircraft measurements (Davison and Grandia, 1979).

Suncor adapted the U.S. EPA RTDM model to assist with a supplemental emission control system as an interim management tool while the flue gas desulphurization unit was under construction. The model output was used to determine when to switch from burning a high sulfur coke product to a low sulfur content fuel oil.

Staff at Alberta Environment reviewed air quality studies conducted in, and relevant to, Alberta and the review also included much of the work conducted in the oil sands area (Angle and Sakiyama, 1991). The objective was to summarize current information about plume dispersion in Alberta and to provide professionals with a comprehensive, consistent reference document.

12.2.3   The RSDS and CEMA Period

In 1998, AENV created the Regional Sustainable Development Strategy (RSDS) for the Athabasca oil sands area. The RSDS led to the formation of the Cumulative Environmental Management Association (CEMA), a multistakeholder group representing all levels of government, industry, regulatory bodies, environmental groups, Aboriginal groups, and local health authorities. CEMA was responsible for addressing the following priority air quality issues: acidification of terrestrial and aquatic ecosystems, effects of air contaminant emissions on human health, bioaccumulation of trace metal emissions, and formation of ground-level ozone. CEMA recognized the need for and developed recommendations for evaluating cumulative environmental effects for the region.

The outcome of the CEMA initiatives was the development of several management frameworks. An Acid Deposition Management Framework (ADMF) addresses acid deposition from industrial activity to maintain the chemical characteristics (i.e., prevent excessive acidification) of soils and lakes in order to avoid adverse effects on ecosystems (Cumulative Environmental Management Association, 2004). Closely linked to the ADMF, an Interim Nitrogen (Eutrophication) Management Recommendations and Work Plan (NEP) addresses potential nitrogen eutrophication issues associated with the deposition of nitrogen compounds (Cumulative Environmental Management Association, 2008). An Ozone Management Framework (OMF) addresses issues and priorities related to ground-level O_3 and its precursors (Cumulative Environmental Management Association, 2006). This framework adopted the management responses outlined in the Alberta Particulate Matter and OMF (Alberta Environment, 2009). An Air Contaminants Management Framework (ACMF) addresses odors, human health effects, and ecological health risks due to contaminants (Cumulative Environmental Management Association, 2009).

By this time, the CALPUFF/CALMET model system has become the *de facto* model used to help develop the ADMF (Doram and Rawlings, 2003; Golder Associates, 2003, 2010), the NEP (Davies and Prasad, 2005), and the ACMF (Davies and Boulton, 2003; Picard, 2003), and for individual development Environmental Impact Assessments (EIAs). The CALPUFF/ CALMET model was appealing because it offers the ability to simulate emissions from multiple point sources (e.g., stacks) and area sources (e.g., mines), incorporates sulfur and nitrogen chemistry and deposition processes, uses varying meteorological fields across the model domain to reflect terrain influences, and addresses a model domain on the order of hundreds of kilometers to account for the overlap of multiple emission sources. The application of this model is supported in the Alberta Model Guideline, and the model system can be used to represent near-field and regional dispersion. The use of the model is not constrained by the U.S. EPA approach that only recognizes the model for source–receptor distances greater than 50 km. Studies have indicated that CALPUFF may overstate predictions for source–receptor distances greater than 200 km (Levya et al., 2002; Tonnesen et al., 2007).

Various photochemical models have been applied to the oil sands region, and the results have helped develop the OMF. The first photochemical modeling was based on the application of the SMOG model that was applied on a local basis to determine the potential for ozone formation. This was followed by the application of the CALGRID model to support upgrader expansion developments (Earthtech and Conor Pacific, 1998). Aircraft measurements were undertaken to determine the nature of ozone formation in the oil sands region (Rudolph et al., 2003, 2004). The measurements did not indicate substantive anthropogenic production. More recent ozone model studies used the Community Multi-scale Air Quality (CMAQ) model and the Sparse Matrix Operator Kernel Emissions (SMOKE) model to simulate emissions over a larger domain. These studies include an Environment Canada study (Fox and Kellerhals, 2007), a CEMA study (Morris et al.,

2010), and another Alberta Environment study (Mansell et al., 2010). The Environment Canada study focused on all of Alberta, while the CEMA and Alberta Environment studies focused on the Athabasca Oil Sands Region.

Two recent studies have reviewed the ambient air quality collected by Wood Buffalo Environmental Association (WBEA) and the CEMA air quality management frameworks. Kindzierski et al. (2009) reviewed a decade of WBEA ambient data to determine long-term trends in the region. For some compounds, there was a systematic increase with time (e.g., NO_2) and for others, no trends could be determined (e.g., SO_2, O_3). The findings are consistent with previous reviews (Wade et al., 2007). Foster (2010) conducted a technical review of the ADMF, NEP, and OMF that included a review of the application of the CALPUFF model for the ADMF and NEP. Recommendations for standardizing the source and emission inventory and the model parameters were made. These recommendations are consistent with a review of the application of the CALPUFF model to the oil sands area (Lundgren et al., 2008).

12.3   WBEA CASE STUDY: MODEL INPUT

The WBEA was created in 1997, and in 1998, WBEA assumed responsibility for ambient air quality monitoring in the oil sands region. WBEA is also involved with detecting, characterizing, and quantifying impacts of air emissions on terrestrial ecosystems and on traditional land resources through the Terrestrial Environmental Effects Monitoring (TEEM) Program.

In 2008, TEEM commissioned a study to collect and analyze lichen tissue samples at 359 locations within a nominal 150 km radius of existing oil sands mining and extraction operations. Two lichen species (*Hypogymnia physodes* and *Evernia mesomorpha*, when present at each site) were collected in August and September 2008. At the end of each day following collection, the samples were frozen. In the fall of 2008 and winter of 2009, the samples were cleaned, processed, and analyzed for total sulfur and total nitrogen (Berryman et al., 2010). A companion study, also commissioned by TEEM, applied the CALPUFF model to the region using representative emissions and meteorological conditions for the five-year period preceding the lichen collection program (i.e., October 2002 to September 2008). Specifically, the model was used to predict hourly, daily, seasonal, annual, and five-year average concentrations; and to predict seasonal, annual, and five-year average deposition (Davies et al., 2010). The findings of this model application are provided as a case study to illustrate the model input and setup, the model performance, and the model output. The predicted downwind sulfur and nitrogen deposition profiles are compared to downwind profiles of sulfur and nitrogen content in the lichen samples.

12.3.1 Spatial Boundaries

The input data requirements for the model depend on the size of the model domain and the geographic area where the model is applied. Model domains can range in extents from several tens of kilometers to evaluate a single facility up to several hundreds of kilometers to evaluate cumulative effects associated with the overlap of many sources. In the oil sands area of northeastern Alberta, existing and planned bitumen mining and extraction developments are located north of Fort McMurray, and *in situ* bitumen extraction operations primarily occur further south between Fort McMurray and the Cold Lake Air Weapons Range (CLAWR), and south of the CLAWR.

On this basis, a model domain that is 290 km in east to west extent and 700 km in north to south extent was selected (Figure 12.1). The model predictions, however, focus on a smaller 200 by 290 km area (referred to as the lichen domain) even though the simulation used data from the larger model domain. Two nested receptor grids were used to provide an understanding of the spatial concentration and deposition patterns in the lichen domain: a 2-km spacing within an area approximately 170 km by 105 km centered on the area near the main emission sources and a 5-km spacing for the remainder of the study area where the concentration and deposition pattern gradients were smaller. The two grids comprised 8657 receptor points. In addition, the 359 discrete locations corresponding to lichen sampling sites and 67 discrete locations corresponding to ambient air quality monitoring sites were included.

12.3.2 Source and Emission Inventory

As a model provides a linkage between the emissions and associated air quality changes, a source and emission inventory that provides the systematic identification and characterization of all substantive emission sources in the model domain is required. Combustion emission sources that produce sulfur and nitrogen compound emissions in the oil sands area include

- *Stacks*: Stacks vent products of combustion to the atmosphere and are associated with facilities that include gas-fired heaters and boilers that service oil sands extraction and upgrading facilities, gas-fired heaters that service conventional gas production facilities, gas-fired combustion turbine cogeneration facilities, coke-fired power plants that service bitumen upgrading facilities, and reciprocating engines that drive compressors at gas production facilities.
- *Mines*: Emissions associated with mine operations include combustion exhaust from diesel-powered shovel and truck fleets.
- *Nonindustrial sources*: These sources include community and highway traffic emissions, and domestic and commercial heating emissions from communities.

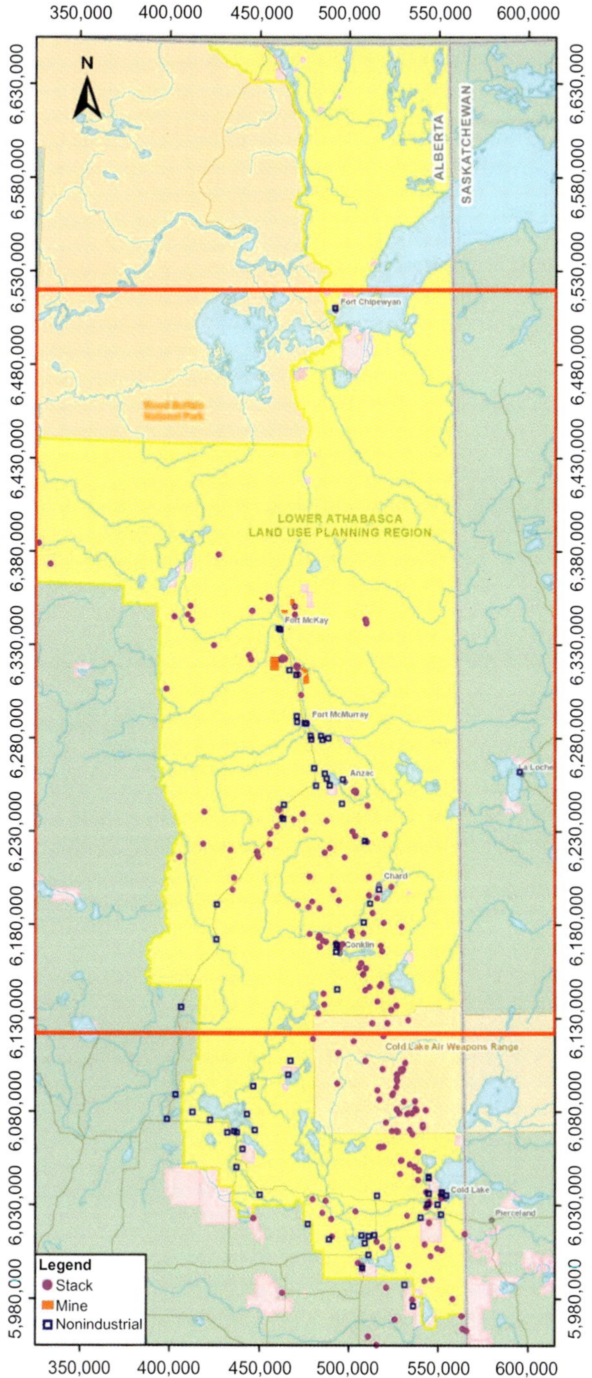

FIGURE 12.1 Emission sources located in the model domain. The lichen domain in indicated by the red boundary.

Noncombustion sources in the area include fugitive emissions:

- *Plant emissions*: Fugitive emissions from small leaks associated with valves, pipe fittings, seals, and vents. While these sources are individually small, collectively, they can be substantial for larger facilities such as upgraders.
- *Mine areas*: Exposed mine faces are also sources of fugitive hydrocarbon emissions, and windblown dust can originate from mine haul roads.
- *Tailings areas*: Tailings impoundments can be substantial sources of fugitive hydrocarbon emissions and also a source of windblown dust.

There is no "formal" source and emission inventory for the oil sands region where the term "formal" denotes an inventory that has an official regulatory or industry group status with respect to its preparation, maintenance, and distribution. The emission inventory information for the region tends to parallel an "open source" computer code concept with the continuing integration and evolution of the data from various information sources. Much source and emission information is in the public domain, from EIAs for projects located in this region and from industry group studies (e.g., CEMA). Each assessment tends to build on the information provided in previous assessments to achieve updated emission estimates. For this study, emissions for the main sources in the lichen domain were further confirmed through discussion with the respective operators.

Table 12.1 provides a summary of emissions in the model domain and compares, on a calendar year basis, the data from 2003 to 2008. For the purpose of presentation, the sources are grouped by location and type: stacks north of Fort McMurray, stacks between and including the CLAWR and Fort McMurray (i.e., South of Fort McMurray), and stacks in the Cold Lake area (i.e., South of the CLAWR). Mine fleet sources are classified as area sources, and all mines are located north of Fort McMurray. Virtually all SO_2 emissions in the model domain originate from stacks located north of Fort McMurray. Over the period 2003–2008, the SO_2 emissions have fluctuated from 274 to 314 t/d. The two main contributors to NO_X emissions in the model domain are the stacks and the mines north of Fort McMurray. Over the period 2003–2008, the NO_X emissions have increased steadily from 148 to 200 t/d.

Planned growth in the region is projected to further increase regional emissions. Assuming that all existing and planned developments operate simultaneously at their respective full capacities, the SO_2 emissions are projected to increase from 290 to 370 t/d, and the NO_X emissions are projected to increase from 200 to 738 t/d. The nominal full capacity projection based on disclosed developments corresponds to the 2030–2040 time frame. For purposes of interest, the highest historical SO_2 emissions in the region occurred during the 1980–1981 and the 1985–1996 periods when the combined average daily SO_2 emissions from the two then existing upgraders varied from 400 to 477 t/d, depending on the year. During that period, the NO_X emissions were typically in the 60 to 65 t/d range.

TABLE 12.1 Summary of Model Domain SO_2 and NO_X Emissions

	Source type/ location	2003 Year 1 (t/d)	2004 Year 1 (t/d)	2005 Year 2 (t/d)	2006 Year 3 (t/d)	2007 Year 4 (t/d)	2008 Year 5 (t/d)	Average (t/d)
SO_2	Stacks north of Fort McMurray	263.3	309.3	300.3	286.6	300.8	273.5	289.0
	Stacks south of Fort McMurray	3.3	3.9	4.1	4.7	3.8	8.6	4.7
	Stacks south of Air Weapons Range	5.8	6.8	7.3	9.5	8.2	5.8	7.2
	Area sources	1.2	1.2	1.1	1.4	1.4	1.9	1.4
	Nonindustry sources	0.3	0.3	0.3	0.3	0.3	0.3	0.3
	Sum	273.9	321.5	313.1	302.4	314.4	290.0	302.6
NO_X	Stacks north of Fort McMurray	60.2	67.1	68.9	74.5	79.2	69.3	69.9
	Stacks south of Fort McMurray	17.6	16.2	19.2	21.2	20.2	23.8	19.7
	Stacks south of Air Weapons Range	12.2	13.2	15.0	15.0	16.8	16.8	14.8
	Area sources	42.1	43.1	43.2	51.4	57.0	73.7	51.8
	Nonindustry sources	16.0	16.0	16.0	16.0	16.0	16.0	16.0
	Sum	148.2	155.5	162.2	178.1	189.3	199.7	172.2

Note: The designation Year 1, Year 2, and so on, refers to the 5-year period preceding the lichen sample collection (August and September 2008).

12.3.3   Topography

The valleys and elevated terrain features in the model domain are depicted in Figure 12.2, and they can affect surface wind flow patterns. The Athabasca River flows from the southwestern portion of the domain toward Fort McMurray and then flows north to Lake Athabasca. The Clearwater River flows from the east in Saskatchewan and merges with the Athabasca River at Fort McMurray.

Broadly speaking, the higher elevations are toward the south of the model domain and the lowest elevations are near the northeastern portion of the domain. Birch Mountain is near the central western boundary of the domain, west of the Athabasca River, and rises to an elevation of over 800 m above mean sea level (ASL). Muskeg Mountain is near the central eastern boundary

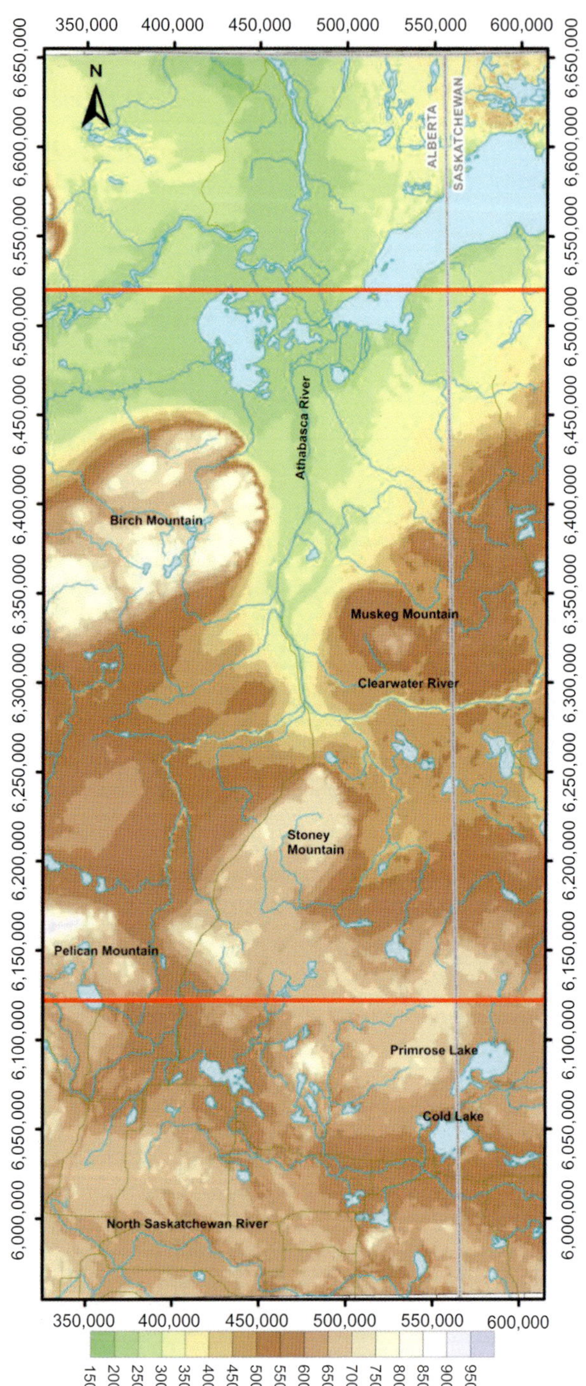

FIGURE 12.2 Terrain in the model domain. The lichen domain in indicated by the red boundary.

of the domain, east of the Athabasca River, and has elevations of over 650 m ASL. For comparison purposes, the existing operations are in valley locations at a nominal 300 m ASL. The terrain features in this region influence the transport and the dispersion of air emissions as they are transported by the winds from the source regions (e.g., Davison et al., 1981b; Leahey and Hansen, 1981).

12.3.4   Land Cover

Surface features (e.g., vegetation canopies) influence surface energy balances, atmospheric turbulence, and dry deposition processes. For these reasons, surface cover characteristics need to be determined for different land/vegetation types in the model domain. The model domain falls primarily in a boreal forest that is characterized by black spruce, white spruce, balsam fir, jack pine, tamarack, aspen, poplar, and white birch. Due to differing drainage patterns, soil types, and sun exposure, the vegetation cover throughout the region is nonuniform, and canopy heights may vary from less than 10 m to more than 30 m, based on these conditions and the tree types present.

Figure 12.3 shows the land cover types at a 5 km resolution based on the Earth Observation for Sustainable Development of Forests initiative by the Canadian Forest Service and the Canadian Space Agency. The model domain comprises 44% coniferous forest, 15% broadleaf forest, and 1% mixed forest. Wetlands comprise 19% of the domain, which is primarily trees and shrubs. The domain comprises 7% water, which includes Lake Claire and a portion of Lake Athabasca in the north; Gordon and Winifred lakes in the east-central portion; and Lac La Biche, Primrose Lake, and Cold Lake in the south. Other land covers include 7% herbs and 5% shrubs.

Each land cover type is characterized by a number of geophysical parameters that vary with season to reflect the presence or absence of snow cover, ice cover on water bodies, and leaf cover in broadleaf forests. These parameters include the following:

- The leaf area index (LAI) for deciduous forest canopies is the ratio of one side of the vegetation canopy surface area to the corresponding ground surface area. For coniferous canopies, the LAI is the ratio of both sides of the vegetation canopy to the corresponding ground surface areas. Higher dry deposition is associated with forest canopies that have higher LAI values. Monthly LAI indices for each vegetation type were based on satellite analyses by Bourque and Hassan (2008).
- The surface roughness length (Z_0) is a measure of surface features' influence on generating mechanical turbulence. The roughness length for a forest canopy is typically 1.0 m, whereas the roughness length for a flat grassy area is typically 0.01 m. Values of Z_0 vary with season, with smaller values occurring during the winter as a result of smoother surfaces associated with the loss of foliage and snow cover.

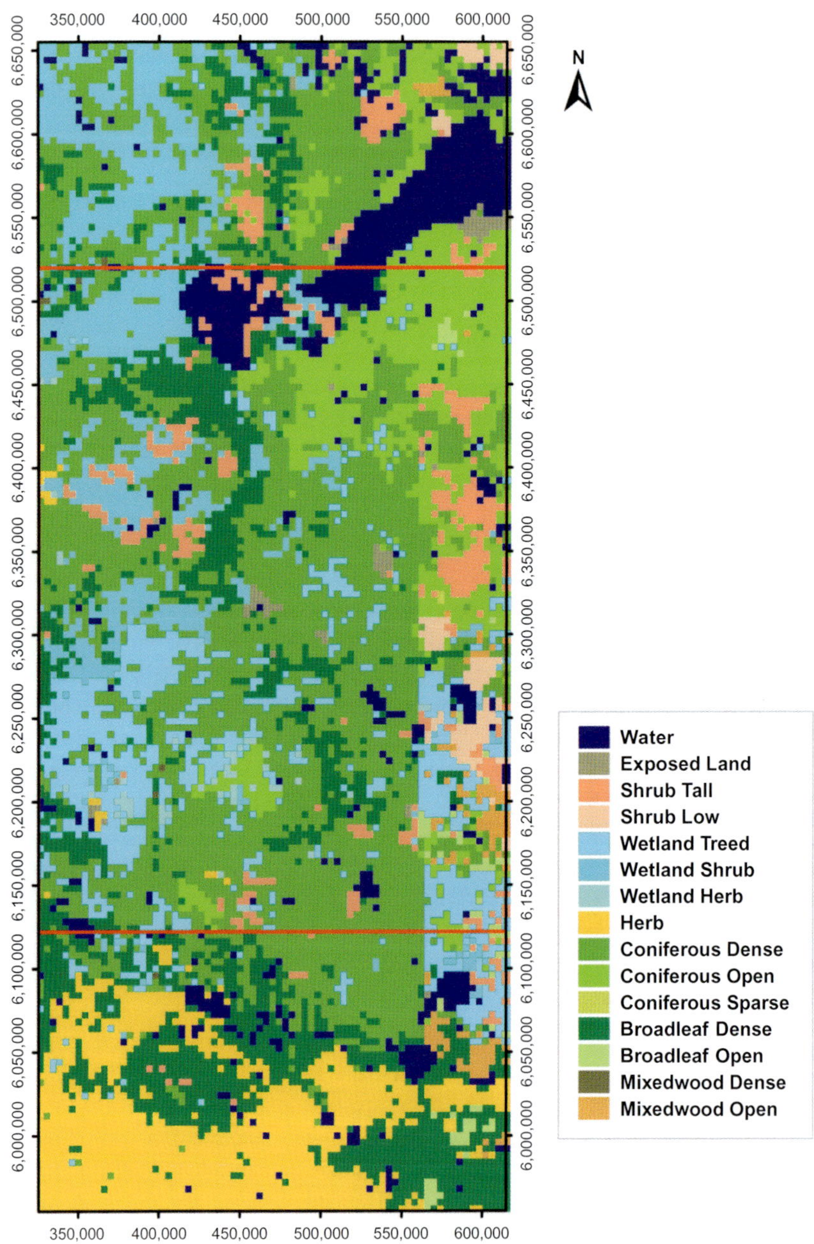

FIGURE 12.3 Land cover in the model domain. The lichen domain in indicated by the red boundary.

- Albedo is the fraction of sunlight reflected by the surface. Fresh snow may reflect 80% of the sunlight whereas a forest canopy may reflect only 10%. The values that were adopted are based on a blend of information from boreal forest measurements by Betts and Ball (1997) and on dispersion model generic recommendations (U.S. EPA, 2004).
- The Bowen ratio defines how energy received at the surface is distributed in the atmosphere, and is defined as the ratio of the sensible heat flux to the latent heat flux. Typical values range from 5 over semiarid regions to 0.5 over grasslands. The adopted values are based on a blend of boreal forest summer measurements (Barr et al., 1997), Saskatoon area values (Barr and Strong, 1996), and on dispersion model generic recommendations (Scire et al., 1999; U.S. EPA, 2004).
- The soil heat flux is the portion of the energy that is absorbed by the ground and is expressed as a fraction of the incoming heat flux. Typical values of 10–15% indicate that most of the incoming heat flux is transferred to the atmosphere. The values are based on generic recommendations (Scire et al., 1999) and are assumed to be constant with time.
- The anthropogenic heat flux accounts for energy input from urban development (e.g., residential heating). The anthropogenic contribution is negligible for rural areas.

These parameters were allowed to vary on a monthly basis for each land cover type and were assumed to be same for each simulation year.

12.3.5   WBEA Case Study: Meteorology

Meteorology determines the transport and dispersion of industrial emissions and, hence, plays a major role in determining air quality downwind of emission sources. Meteorological characteristics vary with time (e.g., season and time of day) and location (e.g., height, terrain, and land cover).

Wind direction and speed are measured by Environment Canada at airport locations and by WBEA at the ambient air quality monitoring sites. Figure 12.4 shows wind roses for various locations in the model domain. The Fort McMurray Airport winds are influenced by proximity to the Clearwater River valley that has an east–west orientation. Surface winds measured in Fort McMurray at the confluence of the Athabasca and Clearwater Rivers show the influence of both river valleys. Surface winds from the WBEA Fort McKay and Mildred Lake monitoring sites show a strong north–south bias that is associated with the north–south orientation of the Athabasca River valley in this area.

Given the paucity of meteorological data in the model domain and the limited representativeness of measured data to distant locations due to terrain influences, meteorological models are used to provide spatially and temporally varying wind and temperature fields across the model domain. These

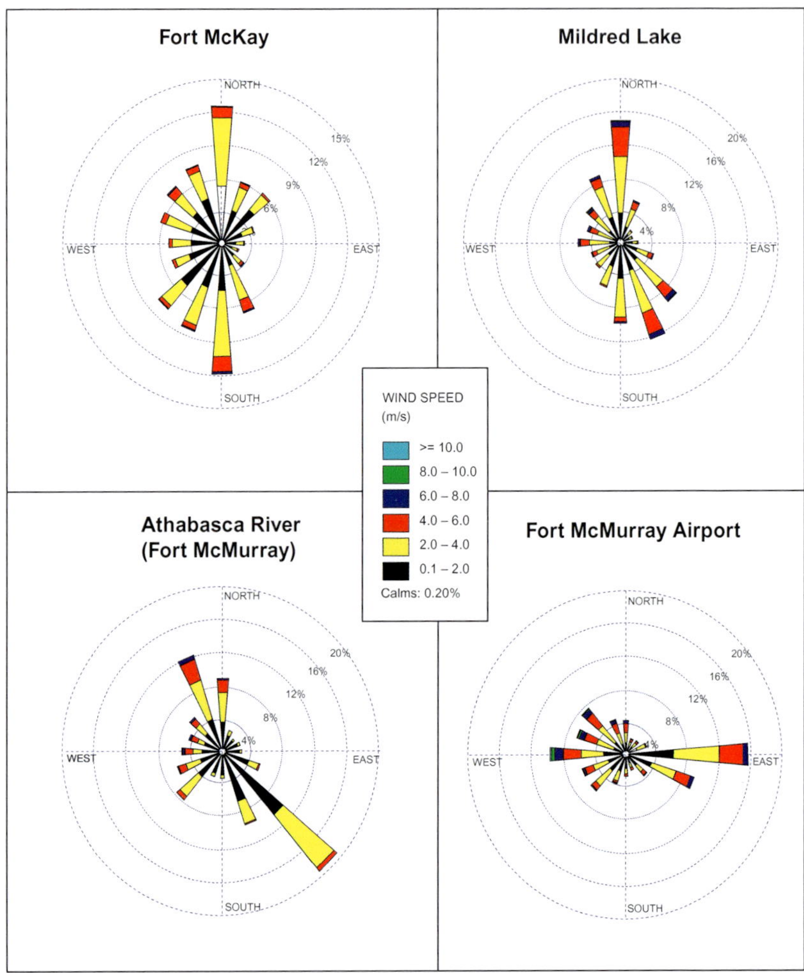

FIGURE 12.4   Wind roses at selected locations in the lichen domain.

models can be categorized as either prognostic or diagnostic. Prognostic models use meteorological measurements and fundamental equations of atmospheric motion to determine meteorological behavior between the observing stations. The MM5 model (a mesoscale meteorological model produced by Penn State University/NCAR) is an example of a prognostic model that has been applied to western Canada for air quality assessment purposes. Diagnostic models use interpolation schemes that rely on empirical relationships to account for topographical or other influences that can occur between the observing sites. The CALMET model is an example of a diagnostic model.

A combination of both models is often used for air quality assessments: the MM5 model output is used as an initial guess field (Scire et al., 2000) and the CALMET model to adjust the initial guess field for the kinematic effects of terrain, slope flows, and terrain blocking effects using finer scale terrain data. For this assessment, the meteorological data based on the MM5 model for 2003–2006 were obtained from Alberta Environment and were supplemented by Stantec Inc. (Calgary, AB) who applied the MM5 model for 2007 and 2008. The CALMET model used the MM5 and surface station meteorological data, and was run to provide three-dimensional wind and temperature profiles for a 5-km grid spacing across the model domain. While the model, as applied, will not resolve the micrometeorology that can occur within river valleys, it can resolve larger scale air flows within the model domain.

12.3.5.1   Predicted Winds

The predicted winds in the Fort McKay and Mildred Lake area using the MM5/CALMET model system show the north and south wind direction bias that reflects the orientation of the nearby Athabasca River valley. This finding is not surprising as the predicted value sites are strongly influenced by the measured wind data, and measured winds from those two sites were included as an input for the model.

The CALMET model winds at the 1000 m level were extracted for each simulation year. The 1000 m level was selected as it is the closest level in the CALMET run to the 1200 m winds derived from minisonde observations. While there are slight differences between the predicted and measured wind roses, the predictions indicated the most frequent winds are from the westerly and northwesterly sectors. This is consistent with the 1200 m wind measurements.

12.3.5.2   Predicted Atmospheric Stability Class

Atmospheric dispersion results from atmospheric turbulence, which can be related to atmospheric stability. Meteorologists define six stability classes (referred to as the PG classes):

- Unstable classes A, B, and C occurring during the day, when the earth is heated by solar radiation. The air next to the earth is heated and tends to rise, enhancing vertical motions.
- Stable classes E and F occur during the night, when the earth cools due to long-wave radiation losses. The air next to the earth cools, suppressing vertical motions.
- The neutral class D is associated with completely overcast conditions (day or night) when there is no net heating or cooling of the earth, and transitional periods between stable and unstable conditions, or during high-wind speed periods (winds greater than 6 m/s [or 22 km/h]).

The stability class frequency distributions based on the Fort McMurray Airport data using the Turner (1964) approach for estimating stability classes were compared with those estimated from the CALMET model for the WBEA Mildred Lake site. The stability class frequency distributions at the WBEA Mildred Lake site and the two airports are broadly similar. Unstable conditions are more frequent during the summer and during daytime periods. Stable conditions are more frequent during nighttime periods. This finding is consistent with the expected distribution of these classes.

12.3.5.3 Predicted Mixing Heights

The presence of an elevated inversion can trap effluents discharged into the atmosphere in the layer between the surface and the base of the inversion layer; this trapping can increase ground-level ambient concentrations relative to the absence of an inversion layer. Mixing heights are usually the highest (i.e., in the 1000–2000 m range) during daytime periods that are characterized by strong solar heating and the lowest (i.e., about 100 m) during the night. High wind speeds can also produce deep well-mixed layers.

Mixing heights are typically determined by analyzing vertical temperature profiles, which are not measured on a routine basis. Portelli (1977) reviewed temperature profiles from upper air stations across Canada and the northern United States, and calculated mean maximum afternoon mixing height values for Canada. Davison et al. (1981b) also determined seasonal mixing height statistics from the analysis of about 2200 minisonde temperature profiles collected in the Mildred Lake area for the period 1975–1978. The mixing heights obtained from the application of the MM5/CALMET system show diurnal and seasonal trends that are similar to those provided by the Portelli and the Davison et al. studies. The predicted maximum median afternoon value during the summer is about 1500 m, which is between the Portelli 1770 m and the Davison et al. 1000 m values. The predicted maximum median value during the winter is about 225 m, which is between the Portelli 180 m and the Davison et al. 270 m values.

The minimum values for each season are predicted to occur during the night when the mixing height tends to be determined from mechanical mixing. The CALMET model, as applied, sets the minimum mixing height to 50 m.

12.3.5.4 Predicted Precipitation

Most precipitation in the region occurs during the summer and is typically associated with convective activity. As convective activity is small scale, the extrapolation of airport measurements for a single event may not be representative of the larger region. Precipitation in the lichen domain has been found to be larger over elevated terrain (Rudolph et al., 1984).The CALMET model, as applied, uses the gridded precipitation data from the MM5 model.

For the CALMET performance evaluation, measured and predicted, precipitation data for Fort McMurray Airport, Cold Lake Airport, and Fort Chipewyan Airport are compared. Over the 2003–2008 period, the total annual precipitation at two of the airport sites are overpredicted by 58–88%. The largest overpredictions are at the Fort McMurray airport. On this basis, the CALPUFF model may overpredict wet deposition.

12.4   WBEA CASE STUDY: CALPUFF MODEL OPTIONS

The CALPUFF model provides the user with a number of assessment options and requires additional data. Some of the key options and required input data are discussed in this section.

The similarity scaling approach was used to calculate horizontal and vertical plume spreads as this approach reflects a more up-to-date understanding of dispersion in the boundary layer than the historical discrete PG dispersion approach. The similarity approach more realistically treats dispersion as a continuous function, whereas the PG approach only considers discrete classes. The probability distribution function (PDF) was used to account for downdrafts that occur under convective conditions. The PDF approach may increase the predicted concentrations resulting from stacks under these conditions. The Heffter (1965) adjustment was used to enhance horizontal dispersion for large distances.

As a plume/puff passes over complex terrain, it has the potential to move closer to the ground. The plume path coefficient (PPC) method can be used to account for the potential decrease in height above the ground. A PPC of 1.0 assumes that the plume trajectory is parallel to the terrain features. For this assessment, the following PPC values were adopted: 0.8, 0.7, 0.6, 0.5, 0.4, and 0.35 for PG categories A, B, C, D, E, and F, respectively (Davies and Prasad, 2005).

Most CALPUFF users tend to select one of the two traditional alternate chemical reaction schemes: the MESOPUFF II scheme or the RIVAD scheme. The RIVAD chemical scheme, however, is viewed as being more rigorous than the MESOPUFF II scheme (Morris et al., 2003) and has been commonly used for the region. The RIVAD chemical scheme treats the NO and NO_2 conversion process in addition to the NO_2 to NO_3^- and SO_2 to SO_4^{2-} conversions, with the equilibrium between gaseous HNO_3 and particulate NH_4NO_3 (Scire et al., 1999).

The chemical schemes in the CALPUFF model have recently been updated to add a new inorganic aerosol module (referred to as ISORROPIA) and an explicit aqueous-phase chemistry module to better represent gaseous to particle transformations. These modules are collectively referred to as the API chemistry scheme and are based on work by Karamchandani et al. (2008); a user's guide is available from TRC Environmental Corporation (2010). The updated API scheme is similar to that found in more sophisticated models such as CMAQ. While the RIVAD and ISORROPIA schemes predict similar

SO_4^{-2} concentrations, the two schemes differ in the partitioning of secondary HNO_3 and NO_3^{-} concentrations. The updated ISORROPIA scheme produces lower nitrate predictions (Wu and Scire, 2011). Limited comparisons, however, indicate that both schemes predict similar total sulfur deposition and total nitrogen deposition.

For this assessment, the more commonly used RIVAD chemical scheme was adopted. Future model applications for the region, however, will likely shift to adoption of the updated API scheme.

The conversion of NO to NO_2 was based on applying the ozone limiting method (OLM) to the NO_X concentrations predicted by CALPUFF. For the application of the OLM approach used in this assessment, hourly O_3 data from the WBEA Athabasca monitoring site were used for the simulation period. The OLM method was applied to the recombined NO and NO_2 predictions from the CALPUFF model.

The CALPUFF chemistry scheme requires ambient ammonia concentrations. Concentration data were obtained from regional WBEA measurements. The monthly average ammonia concentrations range from 1.0 $\mu g/m^3$ (or 1.4 ppb) in November to 2.6 $\mu g/m^3$ (or 3.8 ppb) in July.

12.5   WBEA CASE STUDY: MODEL PERFORMANCE

The determination of model performance relative to predicting deposition is a challenge to assess as there are virtually no deposition measurements in the lichen or the model domains. For this reason, intermediate SO_2 and NO_2 concentration predictions are compared to WBEA ambient measurements. The comparison, as applied, does not account for hour-to-hour and day-to-day variability associated with normal and abnormal emissions. For this reason, the average of the 25-top 1-h predicted and measured concentrations are calculated and compared to provide a more robust indicator of model performance. The ratio of the predicted over the observed concentration (P/O) was used as a performance measure. A P/O ratio of 1.0 indicates perfect agreement and P/O ratios between 0.5 and 2.0 indicate that the model is predicting within a factor of 2.

The CALPUFF model comparison was based on SO_2 and NO_2 concentration predictions. The measured values at the WBEA ambient air quality stations are compared to the model predictions for each year. Only complete years (i.e., 2004–2007) of data are compared. The comparisons between the maximum predicted and observed concentrations at each WBEA monitoring location are presented in tabular formats.

12.5.1 SO_2 Comparison

Table 12.2 compares the observed average of the top 25 1-h SO_2 concentrations with the predicted average of the top 25 1-h values. Table 12.3 compares the observed and the predicted annual SO_2 values. Note the year-to-year

TABLE 12.2 Comparison of the 25-Top 1-h Average SO_2 Concentrations at the 12 WBEA Ambient Monitoring Sites

25-Top 1-h	Observed ($\mu g/m^3$)					Predicted ($\mu g/m^3$)					Predicted/Observed				
	2004	2005	2006	2007	Mean	2004	2005	2006	2007	Mean	2004	2005	2006	2007	Mean
Fort McKay	138	182	149	161	158	101	118	97	81	99	0.73	0.65	0.65	0.50	0.63
Mildred Lake	197	211	242	254	226	204	201	187	166	190	1.04	0.95	0.77	0.65	0.85
Buffalo Viewpoint	230	172	160	190	188	159	134	105	157	139	0.69	0.78	0.66	0.83	0.74
Mannix	599	309	293	398	400	252	278	242	240	253	0.42	0.90	0.83	0.60	0.69
Patricia McInnes	130	103	65	99	99	67	67	67	84	71	0.52	0.65	1.03	0.85	0.76
Athabasca Valley	96	89	58	93	84	74	50	68	102	74	0.77	0.56	1.17	1.10	0.90
Fort Chipewyan	30	41	31	36	35	23	20	21	18	21	0.77	0.49	0.68	0.50	0.61
Albian Mine Site	146	198	124	119	147	84	72	55	60	68	0.58	0.36	0.44	0.50	0.47
Lower Camp	227	246	264	208	236	316	329	272	273	298	1.39	1.34	1.03	1.31	1.27
Millennium	201	183	151	205	185	111	94	111	112	107	0.55	0.51	0.74	0.55	0.59
Syncrude UE1	162	194	172	157	171	120	128	120	92	115	0.74	0.66	0.70	0.59	0.67
Anzac	–	–	39	47	43	62	46	52	97	64	–	–	1.33	2.06	1.70
Average	196	175	146	164	170	164	161	165	165	164	0.74	0.71	0.84	0.84	0.82

Note: Predicted/observed ratios <0.5 or >2 are shaded.

TABLE 12.3 Comparison of the Annual Average SO$_2$ Concentrations at the 12 WBEA Ambient Monitoring Sites

Annual	Observed (μg/m^3)					Predicted (μg/m^3)					Predicted/observed				
	2004	2005	2006	2007	Mean	2004	2005	2006	2007	Mean	2004	2005	2006	2007	Mean
Fort McKay	2.8	4.0	3.9	3.1	3.5	3.8	4.3	4.2	3.4	3.9	1.36	1.08	1.08	1.10	1.15
Mildred Lake	5.2	5.8	6.8	6.1	6.0	7.1	6.7	7.7	7.4	7.2	1.37	1.16	1.13	1.21	1.22
Buffalo Viewpoint	4.4	3.2	2.5	3.2	3.3	5.8	3.4	3.8	5.2	4.6	1.32	1.06	1.52	1.63	1.38
Mannix	11.2	7.8	7.3	7.2	8.4	9.2	9.3	7.7	7.9	8.5	0.82	1.19	1.05	1.10	1.04
Patricia McInnes	3.6	2.8	1.9	2.8	2.8	3.3	2.4	2.2	3.4	2.8	0.92	0.86	1.16	1.21	1.04
Athabasca Valley	2.8	2.6	2.0	2.3	2.4	3.0	2.0	2.3	3.6	2.7	1.07	0.77	1.15	1.57	1.14
Fort Chipewyan	0.9	0.9	0.8	0.9	0.9	0.4	0.6	0.5	0.5	0.5	0.44	0.67	0.63	0.56	0.57
Albian Mine Site	3.7	4.8	4.0	3.8	4.1	3.9	3.9	4.2	4.4	4.1	1.05	0.81	1.05	1.16	1.02
Lower Camp	5.6	5.7	6.6	5.6	5.9	7.5	9.8	9.3	7.9	8.6	1.34	1.72	1.41	1.41	1.47
Millennium	4.3	4.1	4.4	5.8	4.7	4.9	2.4	3.2	3.8	3.6	1.14	0.59	0.73	0.66	0.78
Syncrude UE1	2.8	3.5	3.4	2.7	3.1	4.7	5.3	5.2	3.9	4.8	1.68	1.51	1.53	1.44	1.54
Anzac	–	–	1.1	1.5	1.3	2.2	1.4	1.5	2.5	1.9	–	–	1.36	1.67	1.52
Average	4.3	4.1	3.7	3.8	4.0	3.9	3.8	3.9	3.9	3.9	1.14	1.04	1.15	1.23	1.16

Note: Predicted/observed ratios <0.5 or >2 are shaded.

variability associated with the observed SO_2 concentrations at some of the sites. While most of the year-to-year variability has been smoothed out by using the top 25 1-h measured values, some variability due to short-term, intermittent upset events is still evident at some of the sites.

High SO_2 concentrations are generally predicted where high concentrations are observed, and low SO_2 concentrations are generally predicted where low concentrations are observed. On average, the model is predicting the top 25 1-h values within a factor of 2 at 11 of the 12 monitoring sites, and the annual average concentrations within a factor of 2 at all the sites.

12.5.2 NO_2 Comparison

Table 12.4 compares the observed average of the top 25 1-h NO_2 concentrations with the predicted average of the top 25 1-h values. Table 12.5 compares the observed annual values with the predicted annual values. There is less year-to-year variability within the top 25 1-h measured NO_2 concentrations than in the corresponding SO_2 concentrations, inferring that short-term, intermittent upset events have a smaller influence on NO_2 concentrations.

High NO_2 concentrations are generally predicted where high concentrations are observed, and low NO_2 concentrations are generally predicted where low concentrations are observed. On average, the model is predicting the top 25 1-h values within a factor of 2 at 7 of the 8 monitoring sites, and the annual average within a factor of 2 at 6 of the 8 sites.

12.6   WBEA CASE STUDY: DEPOSITION

12.6.1   Calculation Approach

The CALPUFF model was used to predict SO_2, SO_4^{2-}, NO, NO_2, HNO_3, and NO_3^- deposition due to emission sources located in the model domain. The sulfur deposition represents the sum of the SO_2 and SO_4^{2-} deposition and is expressed in terms of sulfur as follows:

$$S = S_B + \frac{32}{64}\left[SO_2\right] + \frac{32}{96}\left[SO_4^{2-}\right]$$

$$S_B = \text{background sulphur deposition,}$$

where S is expressed in kg S/ha/a and the values in the brackets [ ] represent the sum of the predicted wet and dry deposition. Nitrogen deposition represents the sum of the NO, NO_2, HNO_3, and NO_3^- deposition and is expressed in terms of nitrogen (N) as follows:

$$N = N_B + \frac{14}{30}\left[NO\right] + \frac{14}{46}\left[NO_2\right] + \frac{14}{63}\left[HNO_3\right] + \frac{14}{62}\left[NO_3^-\right]$$

$$N_B = \text{background nitrogen deposition,}$$

TABLE 12.4 Comparison of the 25-Top 1-h Average NO_2 Concentrations at the WBEA Ambient Monitoring Sites

25 Top 1-h	Observed ($\mu g/m^3$)					Predicted ($\mu g/m^3$)					Predicted/observed				
	2004	2005	2006	2007	Mean	2004	2005	2006	2007	Mean	2004	2005	2006	2007	Mean
Fort McKay	67	80	68	76	73	87	86	85	76	84	1.30	1.08	1.25	1.00	1.16
Patricia McInnes	66	70	64	66	67	78	72	79	75	76	1.18	1.03	1.23	1.14	1.15
Athabasca Valley	99	84	108	86	94	74	75	77	72	75	0.75	0.89	0.71	0.84	0.80
Fort Chipewyan	37	31	37	48	38	11	11	14	13	12	0.30	0.35	0.38	0.27	0.33
Albian Mine Site	107	108	113	116	111	130	128	157	158	143	1.21	1.19	1.39	1.36	1.29
Millennium	92	101	99	112	101	89	88	101	102	95	0.97	0.87	1.02	0.91	0.94
Syncrude UE1	56	63	58	64	60	81	79	77	66	76	1.45	1.25	1.33	1.03	1.26
Anzac	–	–	51	77	64	44	36	36	47	41	–	–	0.71	0.61	0.66
Average	75	77	75	81	77	77	77	78	77	77	1.02	0.95	1.00	0.89	0.95

Notes: Predicted/observed ratios <0.5 or >2 are shaded.
NO_2 concentrations are only measured at eight WBEA sites.

TABLE 12.5 Comparison of the Annual Average NO_2 Concentrations at the WBEA Ambient Monitoring Sites

Annual	Observed ($\mu g/m^3$)					Predicted ($\mu g/m^3$)					Predicted/observed				
	2004	2005	2006	2007	Mean	2004	2005	2006	2007	Mean	2004	2005	2006	2007	Mean
Fort McKay	10.6	10.1	9.9	11.8	10.6	17.5	14.7	15.0	12.6	15.0	1.65	1.46	1.52	1.07	1.42
Patricia McInnes	10.4	10.9	9.5	11.3	10.5	19.5	18.6	19.4	21.0	19.6	1.88	1.71	2.04	1.86	1.87
Athabasca Valley	17.2	17.1	19.5	18.3	18.0	11.3	10.3	11.8	11.5	11.2	0.66	0.60	0.61	0.63	0.62
Fort Chipewyan	1.9	1.4	1.6	2.9	2.0	0.8	0.9	0.8	0.7	0.8	0.42	0.64	0.50	0.24	0.45
Albian Mine Site	18.8	17.2	17.1	21.7	18.7	24.4	23.2	26.8	26.1	25.1	1.30	1.35	1.57	1.20	1.35
Millennium	21.9	22.3	27.1	29.8	25.3	12.9	11.5	13.5	13.4	12.8	0.59	0.52	0.50	0.45	0.51
Syncrude UE1	8.3	8.4	8.0	11.2	9.0	15.8	13.9	14.2	10.7	13.7	1.90	1.65	1.78	0.96	1.57
Anzac	–	–	4.9	6.3	5.6	3.0	2.4	2.6	2.7	2.7	–	–	0.53	0.43	0.48
Average	12.7	12.5	12.2	14.2	12.9	12.9	13.0	13.3	13.0	13.1	1.20	1.13	1.13	0.85	1.04

Notes: Predicted/observed ratios <0.5 or >2 are shaded.
NO_2 concentrations are only measured at eight WBEA sites.

where N is expressed in kg N/ha/a and the values in the brackets [ ] represent the sum of the predicted wet and dry deposition. The multiplication coefficients account for molecular mass differences for the individual species.

12.6.2 Background Deposition

The CALPUFF model predictions are based on emissions located in the 290 by 700 km model domain and do not include contributions from sources located outside the model domain. Alberta Environment has applied the RELAD mesoscale model using western Canadian nonindustry sources and all industry sources excluding the WBEA oil sands sources. The model therefore provides an indication of background sulfur and nitrogen deposition values for the WBEA region. The background deposition values were obtained from Alberta Environment on a 1° latitude by a 1° longitude grid cell basis (Cheng, 2009). Figure 12.5 shows the background sulfur deposition (S_B) in units of kg S/ha/a and the background values range from 2.1 kg S/ha/a in the southwestern corner of the lichen domain to 1.4 kg S/ha/a in the northern portion of the lichen domain. Figure 12.6 shows the background nitrogen deposition (N_B) in units of kg N/ha/a and the background values range from 1.8 kg N/ha/a in the southwest corner of the lichen domain to 1.0 kg N/ha/a in the northeast corner of the lichen domain.

12.6.3 Predicted Sulfur Deposition (With Background)

The total sulfur equivalent deposition with the background contribution is shown in Figure 12.7. For locations distant from the main SO_2 emission sources, the sulfur deposition values are in the 1.6–2.2 kg S/ha/a range, and these values are dominated by the background contribution. For locations near the main SO_2 emission sources, values greater than 5 kg S/ha/a are predicted. Values greater than 10 kg S/ha/a are predicted within the upgrader/mine development boundaries. Near the primary SO_2 emission sources, wet deposition dominates. Further downwind, the wet and dry depositions contributions are similar.

12.6.4 Predicted Nitrogen Deposition (With Background)

The total nitrogen equivalent deposition with the background contribution is shown in Figure 12.8. For locations distant from the main NO_X emission sources, the nitrogen deposition values are in the 1.1–2.0 kg N/ha/a range, and these values are dominated by the background contribution. For locations near the main NO_X emission sources, values greater than 5 kg N/ha/a are predicted. Values greater than 10 kg N/ha/a are predicted within the mine development boundaries. Near the primary NO_X emission sources, dry deposition dominates. Further downwind, the wet and dry deposition contributions are similar.

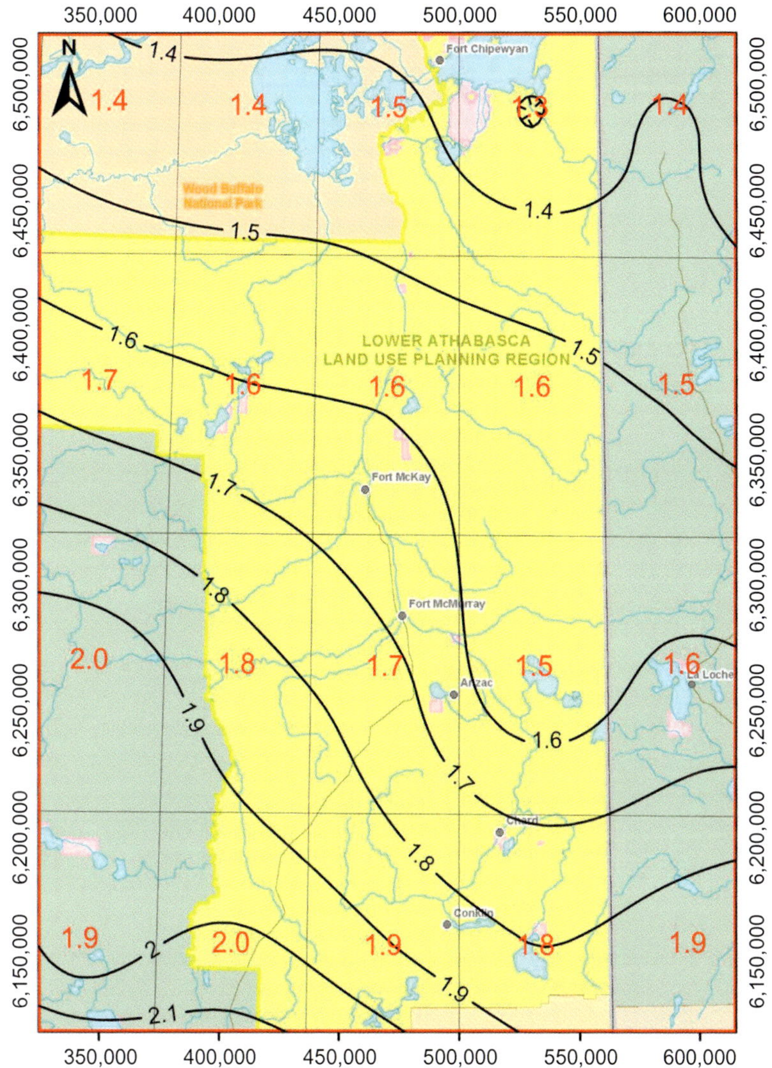

FIGURE 12.5   Background sulfur deposition (kg S/ha/a) in the lichen domain.

12.6.5 | Correlations Between Model Predictions

The predicted five-year average Total SO_2 Deposition, Total SO_4^{-2} Deposition, SO_2 Concentration, SO_4^{-2} Concentration, Total S Deposition, Total S Dry Deposition, and Total S Wet Deposition are strongly correlated. That is, the high concentrations and depositions tend to occur near the existing sources and decrease with increasing distance from the source areas. The year-to-year patterns are also similar.

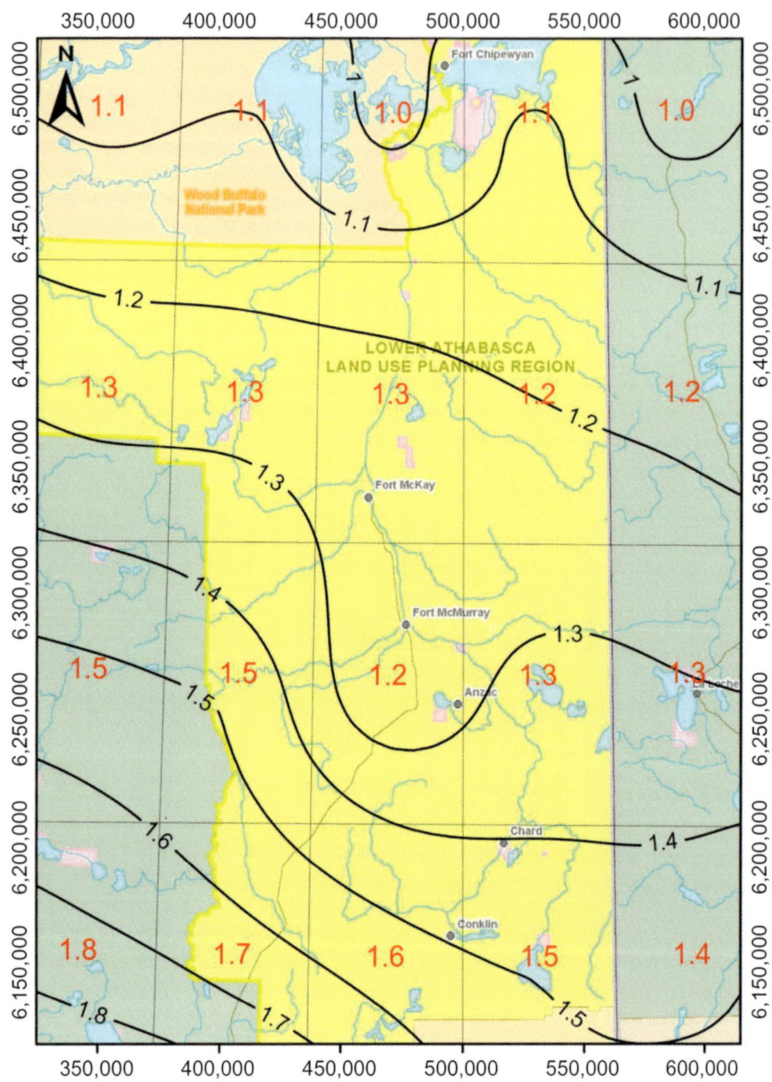

FIGURE 12.6   Background nitrogen deposition (kg N/ha/a) in the lichen domain.

Similarly, the nitrogen compound concentration and deposition patterns also appear to be highly correlated. That is, the high concentrations and depositions tend to occur near the existing sources and decrease with increasing distance from the source areas. The year-to-year patterns also appear to be similar.

The high degree of autocorrelation among the predicted concentration and deposition parameters indicates that the corresponding patterns are similar. The predicted five-year S and N deposition (kg S or N/ha/a) were selected for comparison with the corresponding lichen tissue S or N content.

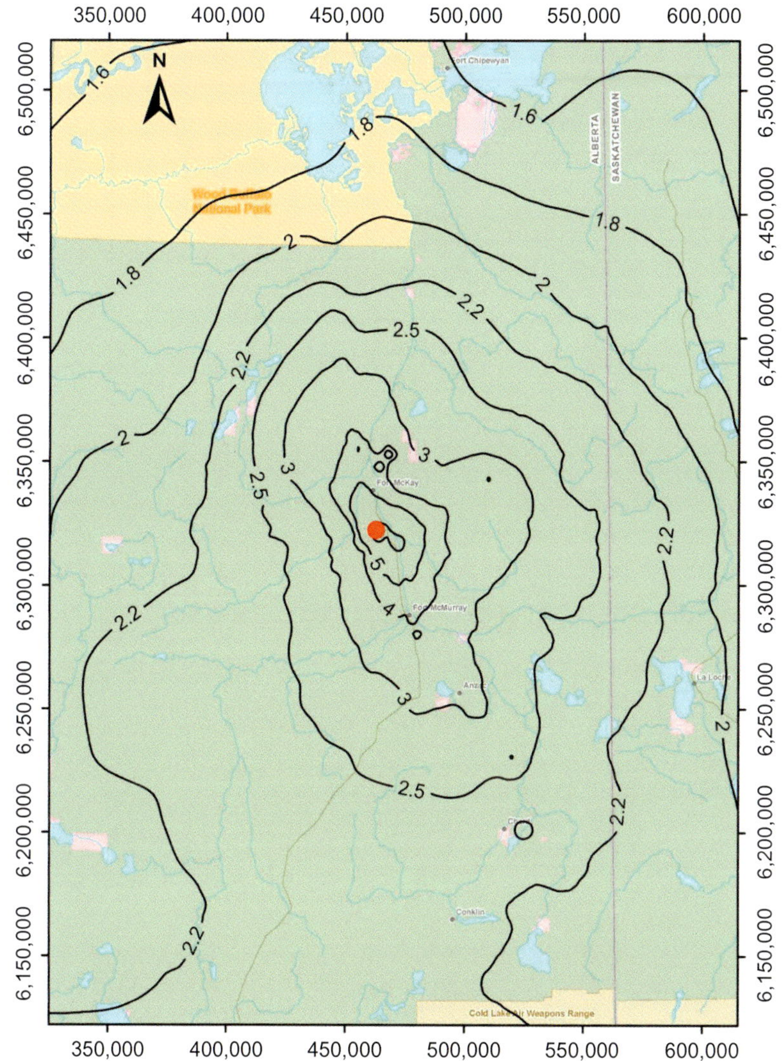

FIGURE 12.7   Sulfur deposition (kg S/ha/a) in the lichen domain (background + oil sands).

12.7   WBEA CASE STUDY: LICHEN COMPARISON

Two-dimensional plots of model predictions and lichen tissue measurements are provided as first order comparison tools. Specifically, the downwind variation of the lichen sulfur concentrations and the predicted total sulfur deposition are compared. For the sulfur comparison, the distance is calculated as that from each lichen site to the nearest of the two substantive SO_2 sources in the lichen domain. Similarly, the downwind variation of the lichen nitrogen concentration and the predicted total nitrogen deposition are also compared.

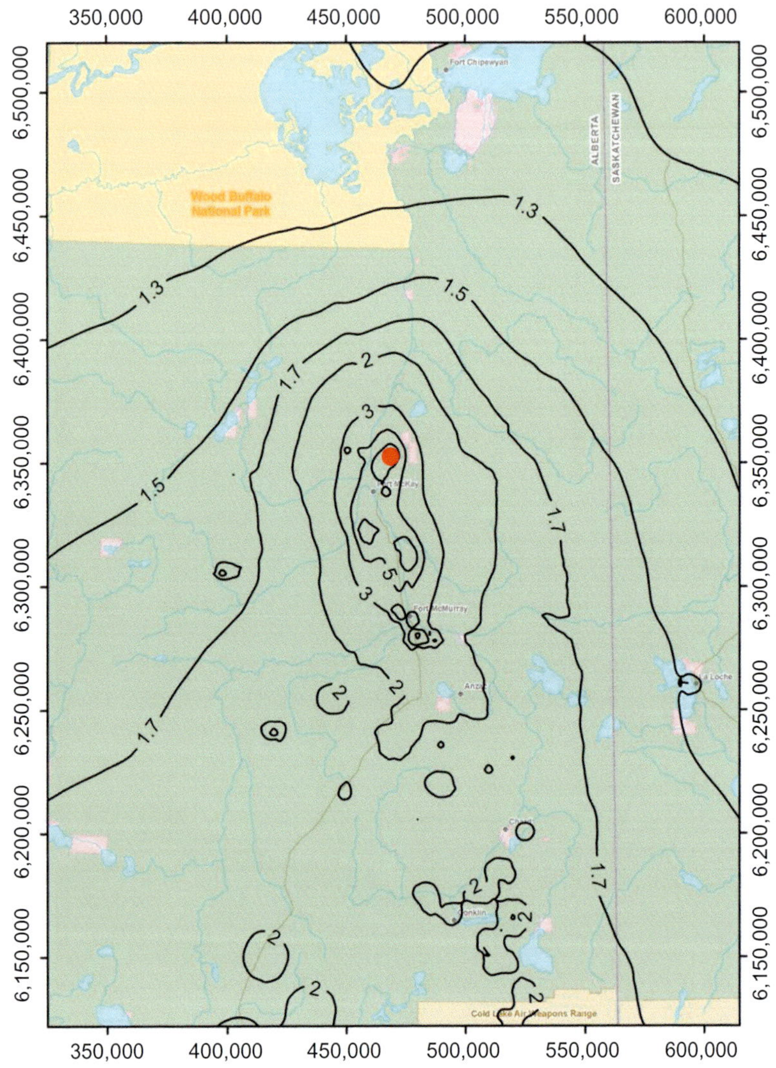

FIGURE 12.8   Nitrogen deposition (kg N/ha/a) in the lichen domain (background + oil sands).

For the nitrogen comparison, the distances are calculated as being the distance from each lichen site to the nearest of the six substantive NO_x sources in the region.

The lichen data included samples collected within and at the edge of the tree canopies, included duplicates, and replicate laboratory analyses. The model predictions for the 359 lichen sites are compared to the lichen tissue measurements.

12.7.1   Sulfur Compounds

Figure 12.9 shows the total sulfur deposition plot predicted for the lichen sample collection sites. There is a clear trend toward decreasing deposition with increasing distance from the main SO_2 emission sources. The primary influence of the SO_2 sources tends to be within a nominal 20 km radius where the deposition is greater than 5 kg S/ha/a. The influence of the sources decreases from 5 kg/ha/a at 20 km to 3 kg S/ha/a at 50 km. At a distance of 100 km, the deposition converges to the background value of about 2.2 kg S/ha/a. The scatter represents contributions from other smaller SO_2 emitting sources, upwind/downwind influences, meteorological influences, and terrain influences.

Figure 12.10 shows the sulfur content variation with distance for both lichen species. There is also a trend toward decreasing sulfur content with increasing distance from the main SO_2 emission sources. The primary influence of the SO_2 sources tends to be within a nominal 20 km radius where the sulfur content is greater than 0.15%. For distances greater than 20 km, the average sulfur content is $0.077 \pm 0.020\%$. The relative scatter is greater than that associated with the predicted sulfur deposition. This additional scatter can include additional influences due to one or more of the following: site elevation, terrain slope, terrain aspect, vegetation type, canopy edge location, canopy closure, and natural variability.

12.7.2   Nitrogen Compounds

Figure 12.11 shows the total nitrogen deposition plot predicted for the lichen sample collection sites. There is a clear trend for decreasing deposition with increasing distance from the main NO_X emission sources. The primary influence of the NO_X sources tends to be within a nominal 20-km radius where the deposition is greater than 4 kg N/ha/a. The influence of the sources decreases from 4 kg/ha/a at 20 km to 2 kg N/ha/a at 50 km. At a distance of 100 km, the deposition converges to the background value of about 1.8 kg N/ha/a. The scatter represents contributions from other smaller NO_X emitting sources, upwind/downwind influences, meteorological influences, and terrain influences

Figure 12.12 shows the nitrogen content variation with distance for both lichen species. There is also a trend for decreasing nitrogen content with increasing distance from the main NO_X emission sources. The primary influence of the NO_X sources tends to be within a nominal 15-km radius where the nitrogen content is greater than 1.5%. For distances greater than 20 km, the average nitrogen content is $0.85 \pm 0.16\%$. As with the sulfur content, the relative scatter of the nitrogen content is greater than that associated with the predicted nitrogen deposition. This additional scatter is attributable to the same factors as for the sulfur content.

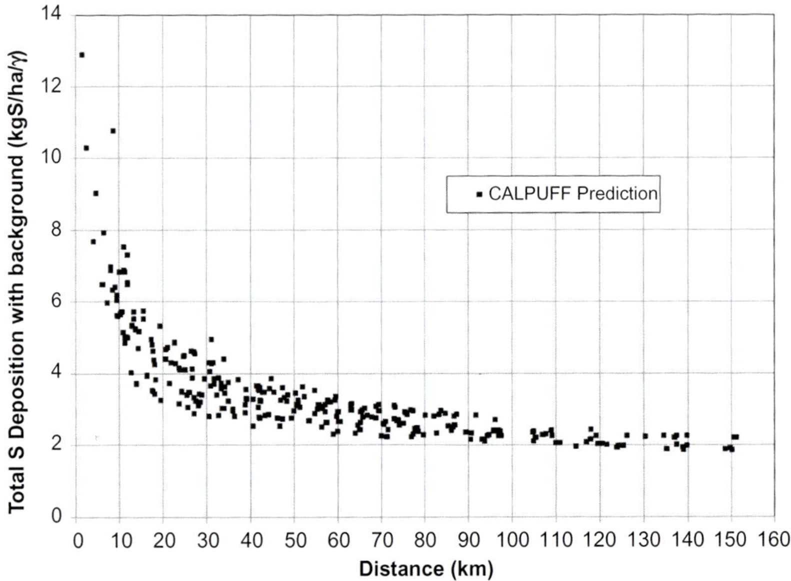

FIGURE 12.9   Predicted (CALPUFF) sulfur deposition (kg S/ha/a).

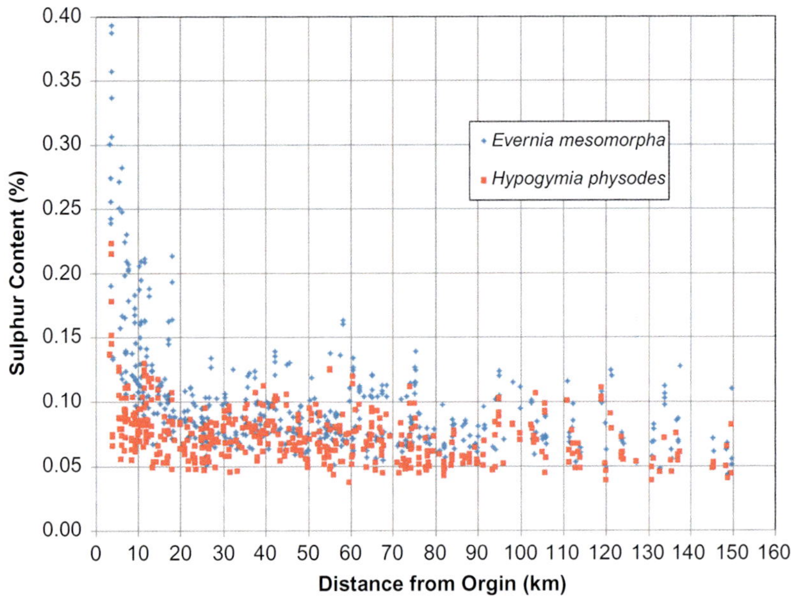

FIGURE 12.10   Measured lichen sulfur content (% S).

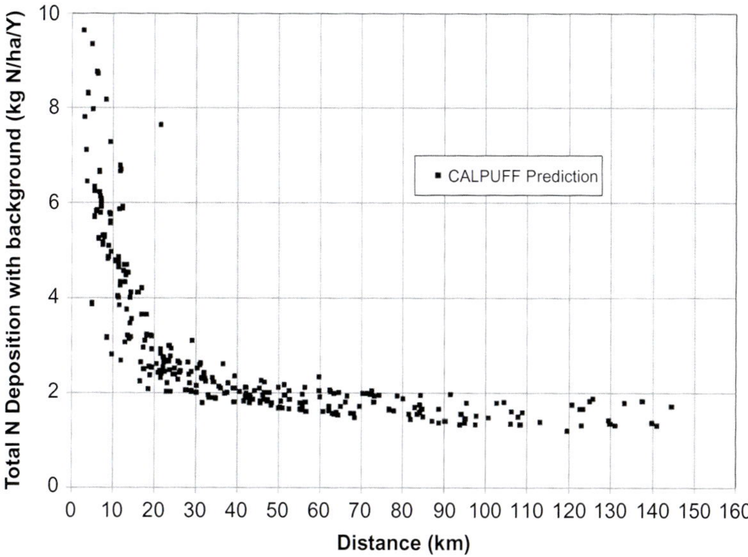

FIGURE 12.11   Predicted (CALPUFF) nitrogen deposition (kg N/ha/a).

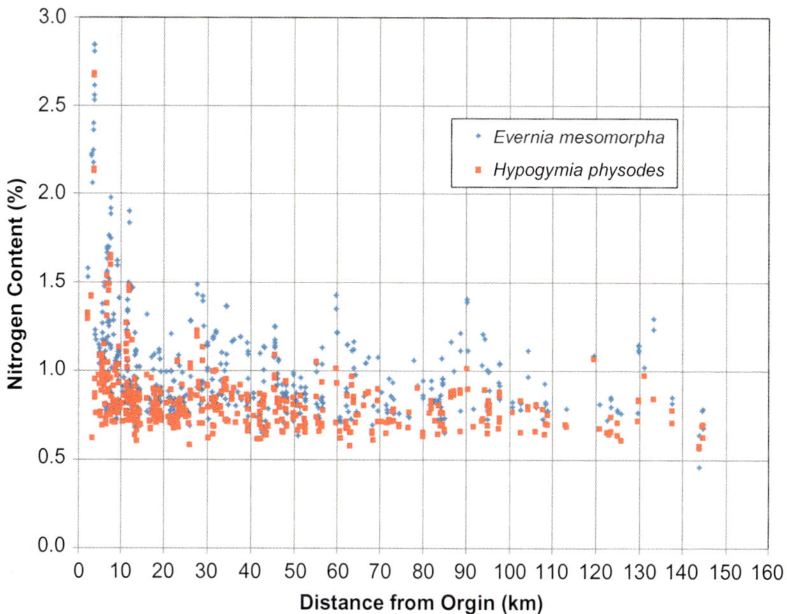

FIGURE 12.12   Measured lichen nitrogen content (% N).

12.8 CONCLUSIONS

12.8.1 Model History

The application of air quality simulation models over the last 30 years to the oil sands area of northeastern Alberta has sometimes led the evolution of regulatory models over the same period. The first models focused on predicting maximum 1-h SO_2 concentrations that could occur on a local scale (i.e., within 10 km) and then on the examination of sulfur deposition in response to the interest in acid rain. At that time, SO_2 emissions from the upgraders were the primary pollutant of interest. The focus expanded to evaluate NO_2 concentrations and nitrogen deposition due to increased NO_X emissions associated with the shift to shovel and truck mining, and *in situ* bitumen extraction; and also expanded to evaluate cumulative effects (i.e., the overlap from multiple sources). Considerable modeling efforts were undertaken by regulators, industry, and multistakeholder organizations during this period.

The number of bitumen extraction facilities in the region is projected to increase regional emissions, indicating the continuing need for these and improved models. While the model performance has been evaluated numerous times, improvements to model inputs and the model physics and chemistry will further increase confidence in model predictions. Potential improvements include the use of hourly emission data instead of longer term averages, improved chemistry algorithms, and improved meteorological representation. While some improvements may not necessarily improve the accuracy of the model predictions, it may improve the performance statistics. Enhanced source and ambient monitoring programs in the region also provide additional information that can further increase our understanding of the atmosphere–biosphere interface, and lead to more refined air quality management tools.

12.8.2 Case Study Findings

During the last 10 years, the CALMET/CALPUFF model system has become the *de facto* modeling approach for the region. The CALMET component offers the advantage of providing three-dimensional wind and temperature fields in response to terrain and ground-cover influences with minimal local data input. The CALPUFF component accounts for sulfur and nitrogen chemistry, and associated wet and dry deposition process. The model has been used to predict 1-h SO_2 and NO_2 concentrations, which are compared to the ambient regulatory objectives. A comparison of model predictions and ambient air quality measurements indicates that the model is predicting within a factor of 2 at most stations. The model has also been used to predict long-term spatial sulfur and nitrogen deposition patterns. These predictions and lichen sample measurements indicate that the main air quality footprint is typically within 20 km from the emission sources.

ACKNOWLEDGMENTS

The author would like to thank fellow Stantec coworkers Reid Person, Kanwardeep Bajwa, and Yan Shen, who provided support for the CALMET/CALPUFF modeling presented in this case study, and Shanti Berryman and Justin Straker (both now with Integral Ecology Group Ltd.), who were responsible for the collection and analysis of the lichen data. In addition, the author would like to thank industry regulators (e.g., Alberta Environment) and industry associations (e.g., WBEA and CEMA) for their confidence and support over the last 30 years in the oil sands region.

REFERENCES

Air Quality Task Force, 1987. Fort McMurray regional air quality, assessment 1975–1986. https://ercb.andornot.com/Record/ERCB3200.

Alberta Environment, 1978. Guidelines for Dispersion Calculations. Standards & Approvals Division. Alberta Environment, Edmonton, AB, 108 pp.

Alberta Environment, 1989. Guidelines for Atmospheric Dispersion Modelling in Alberta. Standards & Approvals Division. Alberta Environment, Edmonton, AB Draft, 87 pp.

Alberta Environment, 1994. Air Quality Model Guidelines. Air and Water Approvals Division. Alberta Environment, Edmonton, AB Draft, 15 pp plus appendices.

Alberta Environment, 1997. Air Quality Model Guidelines. Draft. Air and Water Approvals Division. Alberta Environment, Edmonton, AB.

Alberta Environment, 2000. Air Quality Model Guideline. Environmental Sciences Division. Alberta Environment, Edmonton, AB, 44 pp.

Alberta Environment, 2003. Air Quality Model Guideline. Science and Standards Branch. Alberta Environment, Edmonton, AB, 36 pp.

AlbertaEnvironment, 2009. Air Quality Model Guideline. Climate Change, Air and Land Policy Branch. Alberta Environment, Edmonton, AB, 44 pp.

Angle, R.P. 1979. Air quality modelling and user needs. Prepared for the Alberta Oil Sands Environmental Research Program by Alberta Environment, Air Quality Control Branch, Alberta Environment, Edmonton, AB, Canada. AOSERP report 74, 34 pp.

Angle, R.P., Sakiyama, S.K., 1991. Plume Dispersion in Alberta. Standards and Approvals Division. Alberta Environment, Edmonton, AB, 428 pp.

Barr, A.G., Strong, G.S., 1996. Estimating regional surface heat and moisture fluxes above prairie cropland from surface and upper-air measurements. J. Appl. Meteor. 35, 1716–1735.

Barr, A.G., Betts, A.K., Desjardins, R.L., MacPherson, J.I., 1997. Comparison of regional surface fluxes from boundary-layer budgets and aircraft measurements above boreal forest. J. Geophys. Res. 102 (D24), 29, 213–29, 218.

Berryman, S., Straker, J., Krupa, S., Davies, M., Ver Hoef, J., Brenner, G., 2010. TEEM Lichen Project Interim Report: mapping the characteristics of air pollutant deposition patterns in the Athabasca Oil Sands Region using epiphytic lichens as bio-indicators. Prepared for the Wood Buffalo Environmental Association by Stantec Consulting Ltd. Calgary, AB, Canada.

Betts, A.K., Ball, J.H., 1997. Albedo over the Boreal Forest. J. Geophys. Res. 102 (D24), 28901–28910.

Bottenheim, J.W., Strausz, O.P., 1977. Review of pollutant transformation processes relevant to the Alberta oil sands area. Prepared by the University of Alberta, Hydrocarbon Research

Centre for the Alberta Oil Sands Environmental Research Program Edmonton, AB, Canada. AOSERP report 25, 166 pp.

Bottenheim, J.W., Strausz, O.P., 1981. Development of a chemically reactive plume model for application in the AOSERP study area. Prepared by Department of Chemistry, University of Alberta for the Alberta Oil Sands Environmental Research Program, Edmonton, AB, Canada. AOSERP report 117, 159 pp.

Bourque, C., Hassan, Q., 2008. Leaf area index review and determination for the greater Athabasca oil sands region of northern Alberta, Canada. Prepared by Faculty of Forestry and Environmental Management, University of New Brunswick and Department of Geomatics Engineering, Schulich School of Engineering, University of Calgary for the Cumulative Environmental Management Association (CEMA), Fort McMurray, AB, Canada.

Caton, R.B., Davis, C.S., Davies, M.J.E., Stevens, D.L.M., Wallace, R.R., Yarranton, G.A., Crowther, R.A., 1987. An aerometric program for effects studies in the Athabasca oil sands region. Prepared by Concord Scientific Corporation, Dominion Ecological Consulting Ltd., Yarranton Holdings Ltd. and Aquatic Resource Management Ltd. for Research Management Division, Alberta Environment, AB, Canada.

Cheng, L., 2009. Personal communication via email with Lawrence Cheng, Senior Air Quality Scientist, Air Policy Branch, Alberta Environment, Edmonton, AB, Canada. November 3, 2009.

Cheng, L., Davis, A., Peake, E., Rogers, D., 1987. The use of aircraft measurements to determine transport, dispersion and transformation rates of pollutants emitted from oil sands extraction plants in Alberta. Prepared by Atmospheric Sciences Department, Alberta Research Council and Kananaskis Centre for Environmental Research, University of Calgary for Alberta Environment, Research Management Division Edmonton, AB, Canada. AOSERP report 133, 117 pp.

Cheng, L., Wong, R., Thomson, B., 1990. A comparison of mesoscale acid deposition models as applied to northern Alberta and Saskatchewan. Prepared by Alberta Research Council and Environment Canada for Alberta Environment, Edmonton, AB, Canada, 191 pp.

Croft, B.R., Lamb, A., Dawson, R.N., 1976. A preliminary investigation into the magnitude of fog occurrence and associated problems in the oil sands area. Prepared by Stanley Associates Engineering Ltd. for Alberta Oil Sands Environmental Research Program, Edmonton, AB, Canada. Project ME 3.3, 87 pp.

Cumulative Environmental Management Association, 2004. Recommendations for the Acid Deposition Management Framework for the Oil Sands Region of North-Eastern Alberta. CEMA, Fort McMurray, AB, 39 pp.

Cumulative Environmental Management Association, 2006. Ozone Management Framework for the Regional Municipality of Wood Buffalo Area. CEMA, Fort McMurray, AB, 19 pp.

Cumulative Environmental Management Association, 2008. Proposed Interim Nitrogen (Eutrophication) Management Recommendations and Work Plan for the Regional Municipality of Wood Buffalo Area. CEMA, Fort McMurray, AB, 64 pp.

Cumulative Environmental Management Association, 2009. Recommendations for the Management of Air Contaminant Emissions in the Regional Municipality of Wood Buffalo. CEMA, Fort McMurray, AB, 38 pp.

Dabbs, D. L. (Ed.), 1985. Atmospheric emissions monitoring and vegetation effects in the Athabasca oil sands region. Syncrude Canada Ltd. Environmental Research Monograph 1985-5. Prepared by Dabbs Environmental Services with contributions From Western Research, Concord Scientific Corporation, Hardy Associates (1978) Ltd. and CASE Bio-management for Syncrude Canada Ltd, Edmonton, AB, Canada, 127 pp.

Danard, M., Gray, M., 1982. Modelling topographic effects on winds in the Alberta oil sands area. Prepared by Atmospheric Dynamics Corporation for the Research Management Division, Alberta Environment, Edmonton, AB, Canada. AOSERP report 126, 89 pp.

Davies, M.J.E., 1982. An inter-comparison of wind profiles obtained from a Doppler acoustic wind sensor with those obtained from tracking helium filled balloons. Presented at the Pacific Northwest International Section of the Air Pollution Control Association Meeting. Vancouver, BC, Canada.

Davies, M.J.E, Boulton, W., 2003. Predicted ambient concentrations and deposition of priority substances released to the air in the oil sands region. CEMA Contract Number 2005-0009. Prepared by RWDI AIR Inc. for the Cumulative Environmental Management Association, Fort McMurray, AB, Canada, 422 pp.

Davies, M.J.E., Prasad, S., 2005. NO_X dispersion and chemistry assumptions in the CALPUFF model. CEMA Contract 2003-0034. Prepared by RWDI AIR Inc. for the Cumulative Environmental Management Association, Fort McMurray, AB, Canada, 233 pp.

Davies, M.J.E., Person, R., Bajwa, K., Shen, Y., 2010. CALPUFF modelling in support of TEEM's lichen bioindicator project. Prepared by Stantec Consulting Ltd. for the Wood Buffalo Environmental Association, Fort McMurray, AB, Canada.

Davis, C.S., Fellin, P., Stevens, D., Todd, S., Heidorn, K., 1985. Evaluation, analysis, and assessment of snowpack and precipitation data from a monitoring network in the AOSERP study area. Prepared by Concord Scientific Corporation for Alberta Environment, Research Management Division, Alberta Environment, Edmonton, AB, Canada. RMD report 86/36, 341 pp.

Davison, D.S., Grandia, K.L., 1979. Plume dispersion measurements from an oil sands extraction plant, June 1977. Prepared by INTERA Environmental Consultants Ltd. for the Alberta Oil Sands Environmental Research Program, Alberta Environment, Edmonton, AB, Canada. AOSERP report 52, 209 pp.

Davison, D.S., Lantz, R.B., 1980. Review for requirements for air quality simulation models. Prepared by INTERA Environmental Consultants Ltd. for the Alberta Oil Sands Environmental Research Program, Alberta Environment, Edmonton, AB, Canada. AOSERP report 104, 86 pp.

Davison, D.S., Leavitt, E.D., 1979. Analysis of AOSERP plume sigma data. Prepared by INTERA Environmental Consultants Ltd. for the Alberta Oil Sands Environmental Research Program, Alberta Environment, Edmonton, AB, Canada. AOSERP report 63, 251 pp.

Davison, D.S., Leavitt, E.D., McKenna, R.R., Rudolph, R.C., Davies M.J.E., 1981a. Airshed management system for the Alberta oil sands. Volume I: a Gaussian frequency distribution model. Prepared by Intera Environmental Consultants Ltd. and Western Research and Development for the Research Management Division, Alberta Environment, Edmonton, AB, Canada. AOSERP report 119, 149 pp.

Davison, D.S., Hansen, M.C., Rudolph, R.C., Davies M.J.E., 1981b. Airshed management system for the Alberta oil sands. Volume II: meteorological data. Prepared by Intera Environmental Consultants Ltd. and Western Research and Development for the Research Management Division, Alberta Environment, Edmonton, AB, Canada. AOSERP report 120, 89 pp.

Davison, D.S., Davies, M.J.E., Rudolph, R.C., Hansen, M.C., 1981c. Airshed management system for the Alberta oil sands. Volume III: verification and sensitivity studies. Prepared by Intera Environmental Consultants Ltd. and Western Research and Development for the Research Management Division, Alberta Environment, Edmonton, AB, Canada. AOSERP report 124, 129 pp.

Djurfors, S., Netterville, D., 1978. Buoyant plume rise in nonuniform wind conditions. J. Air Pollut. Control Assoc. 28 (8), 780–784.

Doram, D., Rawlings, M.A., 2003. Evaluation of possible management frameworks for acid deposition in the Athabasca Oil Sands region. Prepared by Golder Associates for the NO_x–SO_x Management Working Group, Cumulative Environmental Management Association, Fort McMurray, AB, Canada, 113 pp.

Earthtech and Conor Pacific, 1998. Initial CALGRID Ozone Modelling in the Athabasca Oil Sands Region. Syncrude Canada Ltd, Fort McMurray, AB.

Fanaki, F. (1978), Meteorology and air quality winter field study in the AOSERP study area, March 1976. Prepared by Fisheries and Environment Canada, Atmospheric Environment Service for the Alberta Oil Sands Environmental Research Program Alberta Environment, Edmonton, AB, Canada. AOSERP report 27, 249 pp.

Fanaki, F., 1979. Air System summer field study in the AOSERP study area, June 1977. Prepared by Atmospheric Environment Service for the Alberta Oil Sands Environmental Research Program, Alberta Environment, Edmonton, AB, Canada. AOSERP report 68, 248 pp.

Fanaki, F., Mickle, R., Lusis, M., Kovalick, J., Markes, J., Froude, F., Arnold, J., Gallant, A., Melnychuk, S., Brymer, D., Gaudenzi, A., Moser, A., Bagg, D., 1979. Air System winter field study in the AOSERP study area, February 1977. Prepared by Fisheries and Environment Canada, Atmospheric Environment Service for the Alberta Oil Sands Environmental Research Program, Alberta Environment, Edmonton, AB, Canada. AOSERP report 24, 182 pp.

Foster, K., 2010. Technical review of acid deposition management framework, eutrophication work plan and ozone management framework. Prepared by Owl Moon Environmental for the NO_x/SO_2 Management Working Group, Cumulative Environmental Management Association, Fort McMurray, AB, Canada, 172 pp.

Fox, D., Kellerhals, M., 2007. Modelling of Ozone Levels in Alberta: Base Case, Sectoral Contributions and a Future Scenario. Prairie and Northern Region. Environment Canada, Edmonton, AB.

Fung, C.S., Davies, M.J.E., 1987. Prepared by Concord Environmental, for Research Management Division, Alberta Environment, Edmonton, AB, Canada, 67 pp.

Golder Associates, 2003. Evaluation of historic and future acid deposition effects on soils in the Athabasca oil sands region. Final report submitted to NO_x–SO_x Management Working Group, Cumulative Environmental Management Association, Fort McMurray, AB, Canada. September 2003.

Golder Associates, 2010. The assessment of acid deposition on Alberta oil sands region—phase 2 of stage 2 implementation of the CEMA Acid Deposition Management Framework. Prepared for Cumulative Environment Management Association. Fort McMurray, AB, Canada.

Heffter, J.L., 1965. The variation of horizontal diffusion parameters with time for travel periods of one hour or longer. J. Appl. Meteor. 4, 153–156.

Karamchandani, P., Chen, S.Y., Seigneur, C., 2008. CALPUFF chemistry upgrade. AER final report CP277-07-01 prepared for API (American Petroleum Institute), Washington, DC, USA. February 2008.

Kindzierski, W.B., Chelme-Ayala, P., EI-Din, M.G., 2009. Ambient air quality data summary and trend analysis: part i main report. Prepared by University of Alberta for the Wood Buffalo Environmental Association, Fort McMurray, AB, Canada.

Leahey, D.M., Hansen, M., 1981. Analysis of wind data for Mildred Lake, Birch Mountain Bitumount Towers. Prepared by Western Research for Canstar Oil Sands, Fort McMurray, AB, Canada. December 1981.

Levya, J., Spengler, J., Hlinka, D., Sullivan, D., Moon, D., 2002. Using CALPUFF to evaluate the impacts of power plant emissions in Illinois: model sensitivity and implications. Atmos. Environ. 36, 1063–1075.

Longley, R.W., Janz, B., 1978. The climatology of the Alberta Oil Sands Environmental Research Program study area. Prepared by Fisheries and Environment Canada, Atmospheric Environment Service for the Alberta Oil Sands Environmental Research program, Alberta Environment, Edmonton, AB, Canada. AOSERP report 39, 102 pp.

Lundgren, J., Roberts, S., Chadder, D.S., 2008. Development of a work plan for CALPUFF Modelling in the Regional Municipality of Wood Buffalo. Prepared by RWDI Air for Alberta Environment, Edmonton, AB, Canada, 68 pp.

Mansell, G., Shah, T., Johnson, J., Nopmongcol, U., Piyachaturawat, P., Jung, J., Morris, R., Pollock, T., Shi, J., Staniaszek, P., 2010. Modelling particulate matter and ground level ozone in north eastern Alberta. Prepared by Environ (EC) Canada Inc. and Millennium EMS Solutions Ltd. for Alberta Environment. Edmonton, AB, Canada.

MEP, 1982. Long range transformation and deposition of air pollution from the Alberta oil sands plants. Prepared by Meteorological and Environmental Planning Limited for the Oil Sands Environmental Study Group. Fort McMurray, AB, Canada, 77 pp + appendices.

Morris, R.E., Tana, C., Yarwood, G., 2003. Evaluation of the sulphate and nitrate formation mechanism in the CALPUFF Modelling System. Presented at A&WMA Conference. Guideline on Air Quality Models: The Path Forward. October 22 to 24 2003. Mystic, CT.

Morris, R., Shah, T., Nopmongcol, U., Johnson, J., Jung, J., Piyachaturawat, P., Pollock, T., Singh, R., Thompson, M., Shi, J., Staniaszek, P., 2010. PM and ozone chemistry modelling in the Alberta oil sands area using the Community Multi-scale Air Quality (CMAQ) Model. Prepared by ENVIRON (EC) Canada Inc. and Millennium EMS Solutions Ltd. for Cumulative Environment Management Association. Fort McMurray, AB, Canada, 359 pp.

Murray, W.A., 1981. The 1981 snowpack survey in the AOSERP study area. Prepared by Promet Environmental Group Ltd. for the Alberta Oil Sands Environmental Research Program, Alberta Environment, Edmonton, AB, Canada. AOSERP report 125, 137 pp.

Murray, W.A., Kurtz, J., 1976. A study of potential ice fog and low temperature water fog occurrence at Mildred Lake, Alberta. Syncrude Canada Ltd. Environmental Research Monograph 1976-4. Prepared by the MEP Company for Syncrude Canada Ltd. Fort McMurray, AB, Canada, 63 pp.

Murray, W., Low, T., 1979. An observational study of fog in the AOSERP study area. Prepared by Promet Environmental Group Ltd. for the Alberta Oil Sands Environmental Research Program, Alberta Environment, Edmonton, AB, Canada. AOSERP report 86, 54 pp.

Netterville, D.D.J., 1979. Concentration fluctuations in plumes. Syncrude Canada Ltd. Environmental Research Monograph 1979-4. Fort McMurray, AB, Canada, 288 pp.

Olson, R., Thomson, B., Bertram, H., R. Peters., 1982. Athabasca oil sands precipitation chemistry studies: 1976-1979 and 1981. Prepared by Research Management Division, Alberta Environment and Alberta Environmental Centre for Research Management Division, Alberta Environment, Edmonton, AB, Canada. AOSERP Report No. 129, 59 pp.

Padro, J., 1979. Review of dispersion models and possible applications in the AOSERP study area. Prepared by Atmospheric Environment Services, Environment Canada for the Alberta Oil Sands Environmental Research Program, Alberta Environment, Edmonton, AB, Canada. AOSERP Project ME 4.2.1, 64 pp.

Padro, J., Bagg, D., 1978. Application of the CRSTER model to the Alberta oil sands region. Prepared by Atmospheric Environment Services, Environment Canada for the Alberta Oil

Sands Environmental Research Program, Alberta Environment, Edmonton, AB, Canada. AOSERP report 21, 35 pp.

Picard, D., 2003. A priority ranking of air emissions in the oil sands region. Prepared by Clearstone Engineering Ltd and Golder Associates Ltd. for the Trace Metal and Air Contaminant Working Group, Cumulative Environmental Management Association, Fort McMurray, AB, Canada, 155 pp.

Portelli, R.V., 1977. Mixing Heights, Wind Speeds and Ventilation Coefficients for Canada. Fisheries and Environment Canada and Atmospheric Environment, Toronto, ON Climatological Studies, Number 31.

Reid, J.D., Walmsley, J.L., Findleton, L.B., Christie, A.D., 1979. An assessment of the models LIRAQ and ADPIC for application to the Alberta oil sands area. Prepared by Atmospheric Environment Service, Environment Canada for the Alberta Oil Sands Environmental Research Program, Alberta Environment, AB, Canada. AOSERP report 66, 95 pp.

Rudolph, R.C., Oleskiw, M.M., Stuart, R.A., 1984. Climatological analysis of recent data from the Athabasca Oil Sands area. Prepared by Intera Environmental Consultants Ltd. for the Alberta Oil Sands Environmental Research Program, Alberta Environment, Edmonton, AB, Canada. AOSERP report 130, 199 pp.

Rudolph, R., Shauck, M., Zanin, G., Alvarez, S., Owen, E., Buhr, M., 2003. Analysis of 2001-2002 airborne ozone and ozone precursor measurements in the oil sands. Prepared by AMEC Earth & Environmental Limited for the Cumulative Effects Management Association Fort McMurray, AB, Canada, 229 pp.

Rudolph, R., Shauck, M., Zanin, G., Alvarez, S., Owen, E., Buhr, M., 2004. Analysis of airborne ozone and ozone precursor measurements in the oil sands, summer 2001 and 2002. Prepared by AMEC Earth & Environmental, Baylor University and Sonoma Technologies for the Cumulative Effects Management Association Fort McMurray, AB, Canada. 536 pp.

Scire, J.S., Strimaitis, D.G., Yamartino, R.J., 1999. A User's Guide for the CALPUFF Model (Version5.0). Earth Technologies Inc., Concord, MA.

Scire, J.S., Robe, F.R., Ferneau, M.E., Yamartino, R.J., 2000. A User's Guide for the CALMET Meteorological Model. Earth Technologies Inc, Concord, MA, 332 pp.

Slawson, P.R., Davidson, G.A., Maddukuri, C.S., 1980. Environmental research report 1980-1. Prepared by Envirodyne Ltd. for Syncrude Canada, Dispersion Modeling of a Plume in the Tar Sands Area Fort McMurray, AB, Canada, 316 pp.

Smith, S.B., 1981. Alberta oil sands environmental research program 1975-1980: summary report. Prepared by S.B. Smith Environmental Consultants Limited for the Alberta Oil Sands Environmental Research Program, Alberta Environment, Edmonton, AB, Canada. AOSERP report 118, 170 pp.

Strosher, M.M., 1978. Ambient air quality in the AOSERP study area, 1977. Prepared for the Alberta Oil Sands Environmental Research Program, Alberta Environment, Edmonton, AB, Canada. AOSERP report 30, 74 pp.

Strosher, M.M., 1981. Background air quality in the AOSERP study area, March 1977 to March 1980. Prepared for Pollution Control Division, Alberta Environment, Edmonton, AB, Canada. Alberta Oil Sands Environmental Research Program. AOSERP Report L 73, 84 pp.

Strosher, M.M., Peters, R.R., 1980. Meteorological factors affecting ambient SO_2 concentrations near an oil sands extraction plant. Prepared for Alberta Environment, Pollution Control Division, Edmonton, AB, Canada. Alberta Oil Sands Environmental Research Program. AOSERP report 106, 87 pp.

Syncrude Canada, 1992. Air quality Assessment for the Continued Improvement and Development of the Syncrude Mildred Lake Operation. Concord Environmental Corporation for Syncrude Canada Ltd, Fort McMurray, AB.

Syncrude Canada, 1997. Air Quality Implications of NO_X Emissions from the Proposed Syncrude Aurora Mine. BOVAR Environmental for Syncrude Canada Ltd., Fort McMurray, AB.

Syncrude Canada, 1998. Mildred lake upgrader expansion application and environmental impact assessment. Submitted to Alberta Energy and Utilities Board and Alberta Environmental Protection.

Tonnesen, G., Wang, Z., Omary, M., Chen, C.J., 2007. Assessment of nitrogen deposition: modeling and habitat assessment. CEC-500-2005-032. Prepared by University of California, Riverside for California Energy Commission Public Interest Energy Research (PIER) Program. Riverside, CA, USA.

TRC Environmental Corporation, 2010. CALPUFF chemistry updates: user's instructions for API chemistry options. Prepared for Western Energy Supply and Transmission Associates (WEST Associates, Tucson, AZ) by TRC. Lowell. MA, USA, 21 pp.

Turner, D.B., 1964. A diffusion model for an urban area. J. Appl. Meteor. 3 (1), 83–91.

U.S. EPA, 2004. User Guide for the AERMOD Meteorological Processor (AERMET). United States Environmental Protection Agency, Research Triangle Park, NC EPA-454/B-03-002.

Wade, K.S, Brown, S.G., Roberts, P.T., 2007. Characterization of ambient air quality in the oil sands area of northern Alberta. Prepared by Sonoma Technology, Inc. for Trace Metals and Air Contaminants Working Group, Cumulative Environmental Management Association, Fort McMurray, AB, USA, 80 pp.

Walmsley, J.L., Bagg, D.L., 1977. Calculations of annual averaged sulphur dioxide concentrations at ground level in the AOSERP study area. Prepared by Fisheries and Environment Canada, Atmospheric Environment Service for the Alberta Oil Sands Environmental Research Program, Alberta Environment, Edmonton, AB, Canada. AOSERP report 19, 40 pp.

Wilson, D.J., 1979. Wind tunnel simulation of plume dispersion at Syncrude. Prepared by University of Alberta, Department of Mechanical Engineering for Syncrude Canada Ltd., Fort McMurray, AB, Canada. Environmental Research Monograph 1979-1, 198 pp.

Wu, Z.X., Scire, J.S., 2011. Revisiting the 1995 SWWYTAF Air Quality Modeling Study using the new API chemistry options in CALPUFF. Air & Waste Management Association 104th Annual Conference and Exhibition, Orlando, FL, USA. June 21–24, 2011.

Chapter 13

WBEA Receptor Modeling Study in the Athabasca Oil Sands: An Introduction

S. Krupa[1]

Plant Pathology, University of Minnesota-Twin Cities, St. Paul, Minnesota, USA
[1]*Corresponding author: e-mail: krupa001@umn.edu*

The following five chapters in this volume (Chapters 14–18) represent the receptor modeling or source apportionment studies involving multiple air pollutant emission sources in the Athabasca Oil Sands Region (AOSR).

In that context, the terrestrial environment or the boreal forests in the Athabasca Oil Sands are exposed to air emissions (human made and natural) from both stationary and mobile sources as appropriate. Of particular concern is the atmospheric deposition of sulfur (S) and nitrogen (N) and their direct (through shoot uptake) and indirect (through root uptake) effects or both on the forests and native vegetation in the region. The oxides of S and N (criteria or regulated pollutants in Canada and the United States) are well-known phytotoxic agents. They are also known to adversely acidify soils and surface waters. Here, specific cause and effect relationships needed to be established spatially, particularly given the contributions of diverse air emission sources in the Athabasca Oil Sands development (Chapters 7 and 8) and the influence of regional scale transport and mixing. Oil Sands source emissions include both at the surface (e.g., mining) and at elevations above the ground from various smoke stacks, resulting in complex plume dispersion and their temporally variable deposition (wet and dry; Chapter 12).

In addressing the aforementioned issues, a significant limitation is logistics. Continuous measurements of criteria or regulated and other relevant air pollutant measurements are being made at 15 locations close to the sources where ground level access and electricity are available (Chapter 4; also

---

Developments in Environmental Science, Vol. 11. http://dx.doi.org/10.1016/B978-0-08-099443-7.00013-2

http://www.wbea.org/air-monitoring/ambient-air-monitoring). However, on a regional scale, neither ready access by road nor electricity is available at potential air quality monitoring and vegetation effect assessment sites (up to 150 km in radius from the center of the emissions area). Access to such sites is only by a helicopter. Therefore, reliance had to be placed on the use of passive samplers to assess the relative air quality.

Passive absorbent and adsorbent techniques are available for collecting time-integrated air samples and their subsequent chemical analysis for the quantification of average or total concentrations of atmospheric constituents. In that context, appropriate species of both lower (mosses, lichens) and higher (herbs, shrubs, and trees) plants can also be used as passive samplers.

Under appropriate growth conditions and exposures, many plant species accumulate S and N in their tissues and the associated spatial differences in such a property can be used to map plume impacts. In addition, accumulation of S and N is required and essential for the growth and maintenance of a given species and any excesses above those values can be related to negative changes in their growth and productivity.

Thus, in the early 2000s a bioindicator program was started in the Oil Sands using the epiphytic lichen *Hypogymnia physodes* (Figure 13.1) growing on tree bark. Epiphytic lichens obtain their nutrients and other chemical constituents primarily from the atmosphere as gases and particles, since the outer tree bark on which they grow, are essentially composed of dead cells. *Hypogymnia* was found to be the most frequently occurring epiphytic lichen in the study area and according to the world literature it is the most studied epiphytic lichen for S, N, and other elemental accumulation and impacts (http://dx.doi.org/10.1016/S0269-7491(02)00133-1).

A preliminary study conducted during 2002 using some 20 sampling sites in the Oil Sands showed high S and N levels in *Hypogymnia* at certain locations

FIGURE 13.1 *Hypogymnia physodes* on a pine branch (modified from and by the courtesy of J. Bohdal, NaturFoto.cz.2006).

compared to others with low concentrations. Likely reasons are spatial and temporal difference in plume dispersion, deposition, and uptake. A highly expanded study with some 359 sites was conducted during 2008. The results also showed sites with high S and N concentrations compared to the low, but in addition, sites with high versus low S:N ratios. As noted previously, in explaining these results the spatial influences of specific emission source types (anthropogenic versus natural, stationary versus mobile) will need to be quantified. That is also necessary for explaining any environmental effects.

Addressing the aforementioned issues requires source apportionment or identification of the contributions of the individual source types to the total spatially variable atmospheric deposition. Source apportionment can be achieved by the application of receptor models. Receptor models are focused on the behavior of the ambient environment at the point of impact as opposed to dispersion models that focus on the emissions, transport, dilution, and physicochemical transformation that occur starting at the source and following their transport and also the products of the transformation to the sampling site. In that context, using plume dispersion models alone at a regional scale can be satisfactory; however, using physicochemical models can lead to measurable uncertainties due to variable chemical transformation rates and deposition velocities specific to a region. There is an increasing need for the two approaches to be combined.

Virtually all of our knowledge of source apportionment through receptor modeling is based on separate collection and elemental analysis of coarse and fine particles through the use of a dichotomous sampler or other similar methods. From a regulatory viewpoint, such an approach is extremely valuable, however, with regard to a regional scale environmental impact assessment (e.g., 150 km in radius from the diverse sources, as in the present case), caution is warranted. Dichotomous type samplers are expensive, require power to operate and elemental analysis of sufficient number of sampler filters can be financially restrictive. Equally importantly data from a few sites likely will not be representative of the region where air pollutant exposure and higher plant effects are stochastic by their very nature varying significantly in time and space.

More recently, epiphytic lichens as in the present case have been used in source apportionment studies and in the descriptions of geographic differences in atmospheric deposition or exposure patterns as in the entire Netherlands. There are several approaches to receptor modeling, specific approach(s) being determined by the objective and the type of available data. Among these approaches are the Chemical Mass Balance (CMB), multifactorial analysis such as Target Transformation Factorial Analysis (TTFA), Positive Matrix Factorization (PMF), and U.S. EPAs Unmix. Each of the methods has their advantages and limitations. In the present study, comparisons were made of the results from the application of PMF, Unmix, and CMB to the same set of samples from some 100+ sampling sites. These sites were selected based on the content of S and N in lichen tissues as described previously. Chapter 18 provides a detailed description of the receptor modeling methods used in the present study and the associated results.

As a prelude to the receptor modeling, the required inorganic elemental concentrations of lichen tissues were quantified (some 43 elements) using two different analytical instruments: Inductively Coupled Plasma (ICP)-Mass Spectroscopy (MS) and high-resolution ICP-MS and the associated discrepancies in the results were identified between the two and correction measures were taken at the outset (Chapter 14).

In addition to these analyses, stable isotopes of Led, ^{206}Pb, ^{207}Pb, ^{208}Pb, and their ratios were used to identify the influences of distal transport versus proximal source emissions at the lichen sites and such results were integrated with the data output from the receptor modeling (Chapter 15). Here, attempt was made to identify the contributions of local, specific types of emission sources by using stable isotopes of Mercury, ^{199}Hg, ^{202}Hg, and ^{204}Hg in the lichen tissue (Chapter 16).

Further, in addition to the inorganic elements, a preliminary study was conducted on the concentrations of poly-aromatic hydrocarbons (PAHs) in *Hypogymnia* tissue collected during 2008 at 20 of 300+ same sampling sites used for inorganic (S and N) elemental analysis (Chapter 17). Although significant correlations were observed between crustal materials (in the receptor modeling) and the PAHs data, sampling and analytical variability and bias, indicated that it constituted a significant gap in many previous lichen-PAH biomonitoring studies. Nevertheless, additional work on PAHs is under way for future data inclusion in receptor modeling.

In the end, several source types or factors and their relative contributions to the lichen elemental, S and N concentrations (including select trace elements and stable isotope tracers) were identified by location and the overall implications of those results are discussed, with reference to future needs and directions (Chapter 18).

Chapter 14

Method for Extraction and Multielement Analysis of *Hypogymnia physodes* samples from the Athabasca Oil Sands Region

E.S. Edgerton*[,1], J.M. Fort*, K. Baumann*, J.R. Graney[†], M.S. Landis[‡], S. Berryman[§] and S. Krupa[¶]

Atmospheric Research & Analysis, Inc., Cary, North Carolina, USA

[†]*Geological Sciences and Environmental Studies, Binghamton University, Binghamton, New York, USA*

[‡]*US EPA, Office of Research and Development, Research Triangle Park, Durham, North Carolina, USA*

[§]*Integral Ecology Group Ltd., P.O. Box 23012, Cook St. RPO, Victoria, British Columbia, Canada*

[¶]*Plant Pathology, University of Minnesota-Twin Cities, St. Paul, Minnesota, USA*

[1]*Corresponding author: e-mail: eedgerton@atmospheric-research.com*

ABSTRACT

A microwave-assisted digestion technique followed by ICPMS (inductively coupled plasma-mass spectrometry) analysis was used to measure concentrations of 43 elements in *Hypogymnia physodes* samples collected in the Athabasca Oil Sands Region (AOSR) of northern Alberta, Canada. Analysis of multiple standard reference materials, replicate samples, and digestion blanks indicates that 34 elements were routinely quantifiable in small samples of lichens (25–30 mg) from the AOSR. Analysis of As and Se was performed by dynamic reaction cell ICPMS techniques to minimize polyatomic interferences and improve detection limits. Data from 121 sampling locations show that concentrations of many elements are higher (factors of 1.5–3) near the oil sands

Developments in Environmental Science, Vol. 11. http://dx.doi.org/10.1016/B978-0-08-099443-7.00014-4

operations (i.e., within 50 km) than further away (i.e., beyond 50 km). Statistical analysis shows that many of the 34 elements (particularly the rare earth elements) are highly correlated, but others only have weak correlations with a few other elements. Linear regression of element concentrations versus Al indicates a range of behavior across elements that likely reflect multiple accumulation processes. Comparison with previous studies of *H. physodes* shows that elemental concentrations within the AOSR are generally comparable but often toward the lower end of those reported for remote and background areas of the northern hemisphere. Sb and Pb concentrations, in particular, are among the lowest reported for *H. physodes*.

14.1 INTRODUCTION

Epiphytic lichens have been used for more than half a century to assess environmental effects of air pollution (Hawksworth and Rose, 1970; Tuominen, 1967). Numerous studies have used native or relocated lichen specimens to determine global distribution of radionuclides from thermonuclear weapons tests (Ferry et al., 1967); heavy metal concentrations in remote, high elevation regions (Bergamaschi et al., 2002); depleted uranium in war zones (Di Lella et al., 2003); heavy metal gradients in and around industrial facilities (Gailey et al., 1985; Nieboer et al., 1972); and to recolonize areas with low or absent populations of once native species (Bennett et al., 1996). This chapter describes the digestion and analytical method used to quantify elemental concentrations in *Hypogymnia physodes* samples collected in the Athabasca Oil Sands Region (AOSR) of northern Alberta, Canada. As described in other chapters of this book, the objective of this work is to quantify spatial gradients and atmospheric sources of elements in *H. physodes*, and, ultimately, assess potential effects of emissions from extraction and processing operations within the AOSR. *H. physodes* was chosen for this work because it is relatively pollution tolerant and found at most of the forested sites in the AOSR; hence, it is a good candidate species for defining deposition gradients across the region.

Several factors were considered in the selection of digestion and analytical techniques. First, we wanted to perform a complete digestion in order to measure both mineral-bound and organically bound elements. Given the variety of oil production operations in the AOSR from mining to refining, we would expect particulate emissions in a broad range of sizes (ultrafine to supercoarse), chemical states, and chemical matrices. Complete digestion of the lichen samples was therefore necessary to bring all size and source categories into solution. Second, we needed a large number of elements (major, minor, and trace) to account for the potentially large number of sources and processes (natural and anthropogenic) that could contribute to lichen concentrations. Third, we wanted a technique that could be used on relatively small sample sizes (i.e., 25–50 mg). This was necessary to ensure that there would be sufficient sample mass for replicate analyses, independent analyses of Hg isotopes, polyaromatic hydrocarbons, and archives. Finally, we needed a solution that was amenable to Pb isotope analysis. Based on these considerations, we selected microwave-assisted digestion with a sequential peroxide and acid addition procedure, followed by analysis via inductively coupled plasma-mass spectrometry (ICPMS). Details of the lichen digestion and analytical procedures, statistical analysis, and comparison to results from other multielement lichen studies are the major components covered in this chapter.

14.2   MATERIALS AND METHODS

14.2.1   Lichen Samples

Samples of *H. physodes* were collected by the WBEA Terrestrial Environmental Effects Monitoring (TEEM) program in 2008, during which more than 800 lichen samples were collected from 369 circular plots roughly 0.4 ha in size (see Figure 14.1). The lichen sampling protocol followed methods developed by the U.S. Forest Service Air Quality Monitoring Program in the Pacific Northwest, with a few modifications (Berryman et al., 2010). Lichens were collected from tree branches at a height of at least 1.5 m from the forest floor and placed in clean, unused metalized polyester Kapak bags. Samples were collected from a minimum of six overstory trees, primarily jack pine (*Pinus banksiana*) or black spruce (*Picea mariana*), in or near each study plot. Duplicate samples (field replicates) were collected when sufficient material was available. The ecosite phase or wetland type was also classified for each stand, and the plant community type was described (following classifications in Beckingham and Archibald, 1996 and Vitt et al., 1996).

Following collection, samples were air dried the same day and cleaned to remove host plant material and other debris. Samples were then dried at 65 °C for 2 h and sealed in clean envelopes. Aliquots of each sample were then analyzed for total sulfur (total S) and total nitrogen (total N) concentration at the University of Minnesota Research Analytical Laboratory (UMRAL; Berryman et al., 2010). Total S concentration was determined with a LECO Corp. Sulfur Determinator (Model No. S144-DR) by dry combustion of a 100–150 mg aliquot covered by tungsten oxide in an oxygen atmosphere at 1350 °C. The sulfur detector was calibrated with three LECO plant reference materials (LECO 1026, orchard leaves; LECO 1025, orchard leaves; LECO 1010, tobacco leaves). Total N concentration was determined with a LECO Corp. FP-528 Nitrogen Analyzer by dry combustion of a 150–500 mg aliquot in a gel capsule. The nitrogen analyzer was calibrated using three LECO plant reference materials (LECO 1006, rice flour; LECO 1026, orchard leaves; LECO 1052, ethylenediaminetetraacetic acid). Total S and total N concentrations from the 369 sample sites were 378–1783 and 5645–26,799 µg/g, respectively.

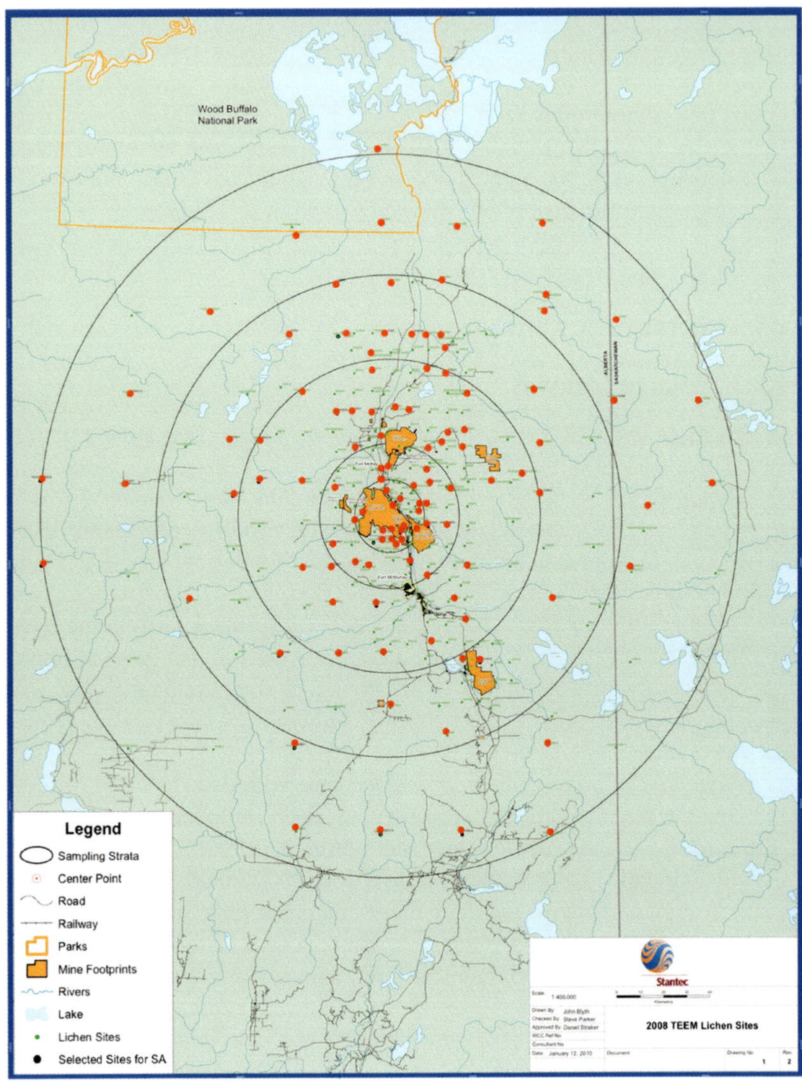

FIGURE 14.1 Lichen sampling locations. Red circles indicate sites used in this study. Note that <121 circles are shown due to proximity of site.

Samples used for the multielement study were obtained from the subset of 121 circled sites in Figure 14.1.

The sample selection strategy for multielement analysis involved a staged approach as follows:

The first group of 50 targeted lichen samples with
- high N and S concentration,
- low N and S concentration,

- high N:S ratio, or
- low N:S ratio.

The next group of 20 samples were selected to focus on

- high N and S samples near the main oil sands production operations,
- low N and S (background) sites to the far north of the study area, and
- low N and S (background) sites to the east near the Saskatchewan/ Alberta border.

The last group of 51 samples were used to fill in and around sites of interest from the first 70 samples and to ensure that roughly equal numbers of samples had been selected from proximal sites (i.e., <50 km from the midpoint between the two main oil processing facilities, $n=61$) and distal sites ($n=60$). As discussed in Chapter 15, sample selection was also designed to select an equal number of sites from locations with similar soil moisture conditions based on the ecosite phase classification noted during lichen collection (specifically wet vs. dry sites).

14.2.2   Sample Digestion

Prior to digestion for multielement analysis, lichen samples were ground and homogenized in a SPEX (Metuchen, NJ) 8000 M ball mill equipped with zirconium balls. Grinding was performed at the University of Michigan laboratory of Dr. Joel D. Blum.

Aliquots of homogenized lichen samples (25–35 mg) were digested in batches of 30–35 in a sequential peroxide and acid addition procedure similar to that developed by Jalkanen and Hasanen (1996) using a MARS Express (CEM Corp., Matthews, NC) microwave digestion system. The first digestion step included addition of 2.0 mL of 30% ultrapure H_2O_2 (J.T. Baker Ultrex-II) with heating to 100 °C for 10 min to break down organic matrices. After cooling, 1.0 mL of ultrapure HNO_3 (J.T. Baker Ultrex-II) and 0.25 mL of ultrapure HF (J.T. Baker Ultrex-II) were added to each vessel and heated to 180 °C for 10 min to dissolve the mineral matrices. After cooling, 5 mL of ASTM Type II ultrapure (18.2 MΩ cm) water was added to each vessel and heated to 180 °C for 20 min. Digestates were then vacuum filtered through a Whatman #541 ashless paper filter and brought up to a final volume of 25 mL with ultrapure water. Each batch included two to three digestion blanks, two to three replicate samples, and one to two aliquots each of two standard reference materials (SRMs). The SRMs were (i) Community Bureau of Reference (BCR) 482 (elements in lichens) and (ii) National Institute of Standards and Technology (NIST) 1648a (urban particulate matter). Digestion aliquots for BCR 482 and NIST 1648a were in the range of 25–30 and 1–2 mg, respectively.

14.2.3   Multielement Analysis via ICPMS

Sample extracts were analyzed with a Perkin Elmer/Sciex dynamic reaction cell (DRC) II ICPMS for 43 elements. Operating conditions for the ICPMS are shown in Table 14.1. Two independent reference solutions and one

TABLE 14.1 ICPMS Operating Conditions

Sample throughput	
Nebulizer	Elemental Scientific PFA-ST
Spray chamber	Elemental Scientific PFA Pure Chamber Spray Chamber and PFA PureCap Endcap
Injector	Elemental Scientific PFA/Sapphire
Torch	Quartz
Plasma conditions	
Nebulizer flow	0.92 L/min
Auxiliary gas flow	1.2 L/min
Plasma gas flow	15 L/min
Lens voltage	6.4 V
RF power	1.2 kW
Ion focusing	
Detector mode	Dual
Analog stage voltage	-1750 V
Pulse stage voltage	1250 V
Quadrupole rod offset (std) (QRO)	-4.50 V
Cell rod offset (std) (CRO)	-16.00 V
Discriminator threshold	25.00 V
Cell path voltage (std) (CPV)	-11.00 V
Dynamic reaction cell gas A	Ammonia
Dynamic reaction cell gas B	Oxygen
Cell gas A flow	0.6 sccm
Cell gas B flow	0.7 sccm
DRC Neb flow	0.92 L/min
DRC mode (QRO)	-8.50 V
DRC mode (CRO)	-1.00 V
DRC mode (CPV)	-16.00 V
Sampling parameters	
Mass range	5–270 amu

TABLE 14.1 ICPMS Operating Conditions—Cont'd

Resolution	0.7 amu
Uptake	~700 µL/min
Number of replicates	3
Total acquisition time per sample	~20 min

SRM (NIST 1643e, trace elements in water) were analyzed to confirm the ICPMS calibration, and 10 digestion spikes (0.1–1.0 ppb) were analyzed to assess matrix interferences and estimate method detection limits (MDLs). Arsenic (As) and selenium (Se) were measured using the DRC function of the ICPMS to reduce polyatomic interferences and improve detection limits. The DRC is located before the quadrupole chamber and uses a reagent or collision gas to reduce interfering ions using charge transfer/collision-induced and chemical dissociation. In this case, oxygen was used as the reagent gas for As (quantified as $^{75}As^{16}O$) and ammonia was used as the reagent gas for Se (quantified as ^{78}Se) in the DRC. Multiple dilutions ($2\times$, $5\times$, and $10\times$) were performed on each digestate to optimize quantification. Data were then reviewed for SRM recoveries, replicate precision, and stability of concentrations across dilutions. Following data review, the $5\times$ dilution was used for purposes of reporting and analysis. Several samples were also analyzed at the USEPA-NERL using a ThermoFinnigan Element2 high-resolution magnetic sector field ICPMS (HR-ICPMS) to estimate potential polyatomic interferences around 75 amu that hinder quantification of ^{75}As (i.e., $^{40}Ar^{35}Cl$, $^{39}K^{36}Ar$) and to confirm that $^{75}As^{16}O$ measured by ARA accurately represented As concentrations.

14.3   RESULTS AND DISCUSSION

14.3.1   Detection Limits

MDLs for each element are shown in Table 14.2. MDLs were calculated using a slight modification of EPA method 200.1 (40 CFR Appendix B, part 136) as three times the standard deviation (SD) of the digestion spike concentrations times the Student's t value for a 99% confidence interval. Also shown in Table 14.2 are the minimum observed concentrations for each element in the 121 sample data set. For most elements, the minimum observed concentration was one or two orders of magnitude greater than the MDL, suggesting that the method used for lichen sample digestion and analysis in this study

TABLE 14.2 Detection Limits and Minimum Observed Concentrations for Lichen Digestates

Element	Analytical isotope/ mass	LOD (µg/g)	Min. observed (µg/g)	Element	Analytical isotope/ mass	LOD (µg/g)	Min. observed (µg/g)
Al	27	25.3	579	Ni	60	0.063	1.22
As	75	0.389	na	P	31	8.57	365
As (DRC)	91	0.018	0.108	Pb	208	0.007	1.12
Ba	137	1.53	15.0	Pd	108	0.011	0.019
Be	*9*	*0.018*	*0.006*	Pr	141	0.019	0.063
Bi	209	0.001	0.008	*Pt*	*195*	*0.004*	*<0.001*
Ca	44	216	2035	Rb	85	0.051	2.08
Cd	114	0.006	0.060	Sb	123	0.003	0.014
Ce	140	0.001	0.576	Se	77	0.133	na
Co	59	0.006	0.189	Se (DRC)	78	0.014	0.232
Cr	52	0.360	0.625	Si	28	61.6	1413
Cs	133	0.009	0.053	Sm	147	0.002	0.049
Cu	63	0.249	1.78	*Sn*	*118*	*0.052*	*0.006*
Fe	56	15.3	405	Sr	88	0.038	5.23
K	39	1.21	1830	*Ta*	*181*	*0.014*	*<0.001*
La	139	0.005	0.300	Th	232	0.001	0.052
Li	7	0.077	0.720	Ti	49	0.142	14.1
Mg	26	2.11	382	Tl	205	0.001	0.003
Mn	55	0.132	50.3	U	238	0.001	0.024
Mo	98	0.040	0.068	V	51	0.045	1.28
Na	23	1.03	22.1	*W*	*182*	*0.022*	*<0.001*
Nb	*93*	*0.030*	*0.028*	Zn	68	0.962	26.8
Nd	143	0.002	0.272				

is well suited for trace element quantification. However, this was not the case for Be, Nb, Pt, Sn, Ta, and W (italicized in Table 14.2). For these elements, concentrations were below the MDL anywhere from 1% of the time (Nb) to 50% of the time (Pt) and were rarely present at concentrations more than two to five times the detection limit.

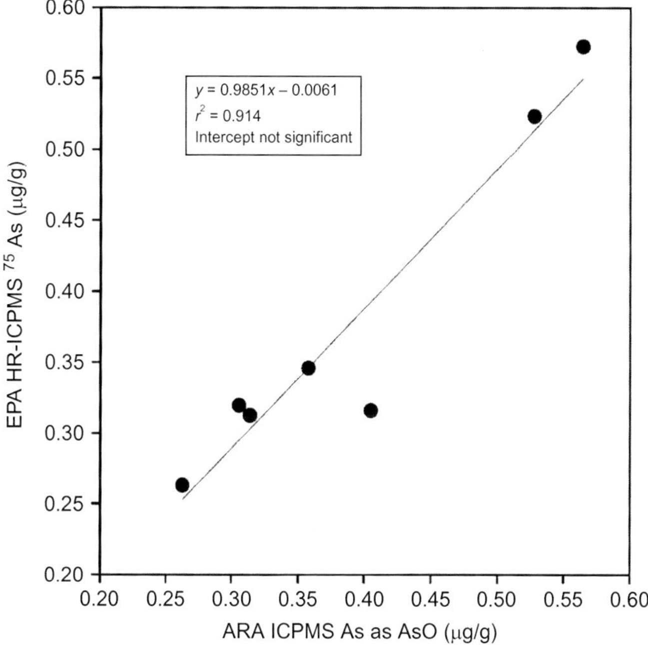

FIGURE 14.2   Scatter plot of EPA ^{75}As versus ARA ^{75}As^{16}O.

Table 14.2 shows two sets of results for As and Se, one for analysis in normal mode (^{75}As, ^{77}Se) and the other for analysis via DRC (^{75}As^{16}O, ^{78}Se). The DRC-mode MDLs are roughly an order of magnitude lower than the normal-mode MDLs. Subsequent discussion of As and Se in this chapter therefore refers specifically to DRC-mode concentrations. Figure 14.2 shows a scatter plot of the USEPA HR-ICPMS analysis of ^{75}As versus the DRC-mode ICPMS analysis of ^{75}As^{16}O. Results show a strong correlation ($r^2 = 0.914$) with a slope near unity (0.985), suggesting that the DRC-ICPMS method is appropriate for measurement of sub-part per billion (in solution) concentrations of As. In contrast, results for normal-mode ^{75}As indicate poor comparability with the USEPA HR-ICPMS. This is indicated by a weak correlation ($r^2 = 0.379$), a slope of 0.66, and a significantly negative intercept (-0.51). These results suggest that historical As data for lichens should be viewed with caution, unless the analytical technique in question has been demonstrated to be largely artifact free.

14.3.2   SRM Recoveries

Recovery data for certified elements in solid-phase SRMs are shown in Table 14.3. Results show good recovery ($\pm 10\%$) and high reproducibility ($\pm 5\%$) for most elements. The two exceptions are Cr and Ti, with recoveries of 35% and 76%, respectively, in urban particulate matter (NIST

TABLE 14.3 Recoveries from Digested Standard Reference Materials

	NIST 1648a ($n = 10$)			BRC-482 ($n = 12$)		
Element	Conc. ($\mu g/g$)	Mean recovery (%)	SD (%)	Conc. ($\mu g/g$)	Mean recovery (%)	SD (%)
Al	34,300	100.8	5.7	1103	101.3	5.7
As	115.50	104.8	5.8	0.85	98.1	8.5
Ca	58,400	98.3	5.5	*2624*	*90.0*	*7.1*
Cd	73.70	101.4	5.9	0.56	88.8	3.7
Co	17.93	98.6	4.5	*0.32*	*97.1*	*5.7*
Cr	402	35.6	8.4	4.12	78.8	9.8
Cu	610	96.4	2.7	7.03	97.2	3.9
Fe	39,200	97.5	2.9	*804*	*94.2*	*4.0*
Mg	8130	97.8	5.7	*578*	*93.4*	*6.3*
Mn	790	101.2	1.8	*33.0*	*92.1*	*4.0*
Ni	81.1	103.0	4.7	2.47	96.6	4.4
Pb	6550	101.5	5.0	40.9	88.0	2.2
Rb	51.0	92.8	4.2	*7.12*	*110.3*	*2.2*
Sb	45.4	94.0	11.7	*0.35*	*87.4*	*9.8*
Sr	215	101.0	5.7	*10.35*	*91.7*	*1.8*
Ti	4021	73.6	13.8	*34.2*	*129.1*	*30.2*
V	127	99.5	6.6	*3.74*	*98.3*	*3.0*
Zn	4800	98.0	7.1	100.6	94.9	5.2

Bold elements are certified; italicized elements are for reference purposes only.

1648a) and 80% and 129%, respectively, in lichen (BCR 482). Several previous studies have also reported low recoveries for Cr in NIST 1648a when using similar digestion techniques (Jalkanen and Hasanen, 1996; Rasmussen et al., 2007).

14.3.3 Lab Replicates

Results of laboratory replicate extractions are summarized in Table 14.4 as the median absolute percentage difference (MAPD) and the SD of percentage differences between replicates. MAPD is a useful statistic because it conveys the midpoint of the absolute deviations from exact agreement. For example,

TABLE 14.4 Summary Statistics for Duplicate Extractions ($n = 19$)

Element	MAPD (%)	SD (%)	Element	MAPD (%)	SD (%)
Al	3.2	4.6	Nd	3.1	4.8
As	1.6	4.9	Ni	2.0	5.7
Ba	2.9	3.9	P	3.7	5.3
Be	12.6	26.1	Pb	2.1	3.1
Bi	3.3	6.6	Pd	8.9	13.4
Ca	1.2	3.8	Pr	5.1	6.7
Cd	3.6	7.2	Pt	18.5	73.5
Ce	2.1	4.7	Rb	1.7	5.7
Co	3.5	6.8	Sb	5.5	52.8
Cr	2.2	7.3	Se	4.3	5.7
Cs	2.4	4.2	Si	3.3	7.4
Cu	3.9	10.6	Sm	2.4	6.7
Fe	1.9	3.0	Sn	6.4	61.8
K	2.4	3.7	Sr	2.2	4.4
La	3.4	5.5	Ta	5.5	45.0
Li	2.9	4.3	Th	4.2	6.4
Mg	2.3	5.3	Ti	7.9	16.5
Mn	1.4	3.8	Tl	5.5	5.3
Mo	5.2	6.8	U	4.2	6.1
Na	4.5	7.0	V	2.8	5.4
Nb	10.9	29.3	W	22.3	45.5
			Zn	2.6	4.6

the MAPD for Al in Table 14.4 indicates that half of the replicate samples agreed within ±3.2%. MAPD ranges from 1.2% (Ca) to 22.3% (W) and SD ranges from 3.0% (Fe) to 73.5% (Pt). In general, MAPD is <5% and SD is <10% for the majority of elements. These findings indicate that the samples were well homogenized in the grinding process and that measurement precision for the most part is adequate to detect relatively small (10–20%) differences between sampling locations. Exceptions include Be, Bi, Nb, Pd, Pt, Sn, Ta, Tl, and W, elements with concentrations typically only a factor of

2–3 above the MDL. Given the relatively low concentrations and poor reproducibility of these nine elements, they will be excluded from further discussion in this chapter.

14.3.4 Data Overview

Statistical summaries for each of the 34 quantifiable elements are shown in
Table 14.5 for two subsets of sampling sites. The distal group ($n=60$)
includes sites >50 km from the center of oil sand production operations,

TABLE 14.5 Element Concentrations (µg/g) in Distal (D) and Proximal (P) Lichen Samples

Element	Distal ($n=60$)	Proximal ($n=61$))	P/D ratio	Element	Distal ($n=60$)	Proximal ($n=61$)	P/D ratio
Al	986	2324	2.4	P	675	722	1.1
As	0.251	0.454	1.8	Pb	2.45	2.92	1.2
Ba	28.2	40.4	1.4	Pr	0.124	0.336	2.7
Ca	10,105	18,594	1.8	Rb	6.42	8.38	1.3
Cd	0.168	0.152	0.9	Sb	0.022	0.034	1.6
Ce	1.09	2.92	2.7	Se	0.430	0.899	2.1
Co	0.292	0.792	2.7	Si	2615	5765	2.2
Cr	0.978	2.63	2.7	Sm	0.095	0.256	2.7
Cs	0.106	0.175	1.7	Sr	9.9	24.3	2.5
Cu	2.93	3.79	1.3	Th	0.111	0.307	2.8
Fe	711	1932	2.7	Ti	25.6	53.1	2.1
K	2737	3496	1.3	U	0.042	0.091	2.2
La	0.519	1.39	2.7	V	2.94	9.53	3.2
Li	0.95	1.96	2.1	Zn	51.6	46.6	0.9
Mg	570	827	1.5				
Mn	182	116	0.6	Total N	7653	8426	1.1
Mo	0.113	0.332	2.9	Total S	591	813	1.4
Na	39.8	63.0	1.6				
Nd	0.51	1.42	2.8				
Ni	1.93	5.15	2.7				

and the proximal group ($n=61$) includes sites <50 km from the center of oil sand operations. The proximal versus distal site groupings allow us to determine the influence of near-field sources in this study and to compare our results to those from other studies from a local source versus background source contribution perspective. Results show a broad range of concentrations across elements, but a fairly narrow range of proximal:distal ratios (P/D ratios). Calcium (Ca) is the most abundant element and, on average, represents 1–2% of lichen mass. Aluminum (Al), iron (Fe), potassium (K), phosphorus (P), and silicon (Si) are also major elements and can represent 0.1–0.6% of lichen mass. If we assume that Ca, Al, Fe, K, and Si are present as oxides, then, taken together, they account for about 2.5% and 4.8% of sample mass at distal and proximal sites, respectively. P/D ratios range from 0.6 (Mn) to 3.2 (V) and appear to cluster for the most part into three groups: 0.9 ± 0.3 (4 elements), 1.5 ± 0.3 (10 elements), and 2.5 ± 0.3 (20 elements). Alkali and alkaline earth metals compose the group with P/D ratios around 1.5, while Al, Si, most transition metals, and rare earth elements (REE) are in the group with P/D ratios around 2.5.

Pearson correlation coefficients (r) for the 121 sample data set are shown in Table 14.6. In general, there are a large number of significant and highly significant correlations among elements. The highest correlations ($r \sim 0.98$) are among the REEs (La, Ce, Pr, Nd, and Sm) and Al. Interestingly, the non-REEs As and Se are also very highly correlated with the REEs, and with Al and Fe. Several elements, including Cd, P, and Zn, exhibit very few significant correlations, while others, including Ca, K, Mg, and Pb, exhibit significant correlations with many elements but with r values in a fairly narrow range (i.e., 0.4–0.7). Uniquely, Mn is negatively correlated with most other elements.

Figure 14.3A through F consists of scatter plots of selected elements versus Al for all lichen samples. Elements were selected to represent major components, REEs, transition metals, trace elements, and four elements (P, Mo, Mn, and Zn) that exhibit interesting behavior with respect to Al. Al is often used as a normalizing element for comparing lichen concentrations, but very similar relationships would be observed for any of the REEs, given the high correlations between Al and these elements. Each scatter plot contains a best fit regression line (solid) and also a line corresponding to the upper crust element:Al ratio from Taylor and McLennnan (1995). The latter is included simply as a point of reference to illustrate relative behavior across groups of elements. As discussed by Reimann and DeCaritat (2000, 2005), a host of geochemical and biochemical processes, in addition to anthropogenic activities, can cause element/Al ratios to depart from the global average. Thus, elemental ratios should be used only as another tool for exploring complex multielement data sets, rather than as a diagnostic of anthropogenic versus geogenic sources. As detailed in Chapter 18, location-specific differences in the overburden and the geologic formations that are mined during the

TABLE 14.6 Pearson Correlation Coefficients for Elements in *Hypogymnia physodes* ($n = 121$)

(1) Li–Cr

	Li	Na	Mg	Al	Si	P	K	Ca	Ti	V	Cr
Li	1.00	0.909**	0.674**	0.984**	0.947**	0.047	0.489**	0.671**	0.965**	0.907**	0.835**
Na	0.909**	1.00	0.686**	0.934**	0.940**	0.066	0.465**	0.652**	0.930**	0.837**	0.844**
Mg	0.674**	0.686**	1.00	0.682**	0.730**	0.390**	0.718**	0.698**	0.706**	0.551**	0.582**
Al	0.984**	0.934**	0.682**	1.00	0.952**	0.023	0.474**	0.649**	0.969**	0.911**	0.835**
Si	0.947**	0.940**	0.730**	0.952**	1.00	0.045	0.496**	0.661**	0.980**	0.867**	0.850**
P	0.047	0.066	0.390**	0.023	0.045	1.00	0.810**	0.292**	0.011	−0.018	0.003
K	0.489**	0.465**	0.718**	0.474**	0.496**	0.810**	1.00	0.491**	0.468**	0.409**	0.373**
Ca	0.671**	0.652**	0.698**	0.649**	0.661**	0.292**	0.491**	1.00	0.654**	0.512**	0.629**
Ti	0.965**	0.930**	0.706**	0.969**	0.980**	0.011	0.468**	0.654**	1.00	0.907**	0.842**
V	0.907**	0.837**	0.551**	0.911**	0.867**	−0.018	0.409**	0.512**	0.907**	1.00	0.775**
Cr	0.835**	0.844**	0.582**	0.835**	0.850**	0.003	0.373**	0.629**	0.842**	0.775**	1.00
Mn	−0.341**	−0.354**	−0.197*	−0.353**	−0.352**	−0.087	−0.308**	−0.174	−0.358**	−0.387**	−0.349**
Fe	0.949**	0.908**	0.656**	0.962**	0.916**	−0.053	0.402**	0.668**	0.947**	0.896**	0.811**
Co	0.933**	0.878**	0.787**	0.934**	0.945**	0.095	0.516**	0.742**	0.936**	0.825**	0.830**
Ni	0.921**	0.844**	0.652**	0.922**	0.897**	0.056	0.480**	0.642**	0.929**	0.971**	0.811**
Cu	0.776**	0.776**	0.577**	0.781**	0.800**	0.081	0.412**	0.504**	0.804**	0.732**	0.687**

Zn	−0.102	−0.034	0.095	−0.108	−0.036	0.160	0.026	0.201[*]	−0.083	−0.105	−0.059
Se	0.965[**]	0.888[**]	0.647[**]	0.979[**]	0.917[**]	−0.017	0.443[**]	0.631[**]	0.958[**]	0.938[**]	0.818[**]
Rb	0.438[**]	0.443[**]	0.247[**]	0.469[**]	0.422[**]	0.100	0.337[**]	0.137	0.431[**]	0.380[**]	0.416[**]
Sr	0.435[**]	0.408[**]	0.702[**]	0.406[**]	0.458[**]	0.404[**]	0.544[**]	0.681[**]	0.431[**]	0.293[**]	0.395[**]
As	0.957[**]	0.943[**]	0.694[**]	0.970[**]	0.954[**]	0.029	0.461[**]	0.660[**]	0.960[**]	0.874[**]	0.826[**]
Mo	0.909[**]	0.848[**]	0.530[**]	0.902[**]	0.850[**]	0.026	0.407[**]	0.560[**]	0.889[**]	0.964[**]	0.786[**]
Cd	−0.177	−0.092	0.018	−0.168	−0.136	0.254[**]	0.049	0.033	−0.182[*]	−0.274[**]	−0.125
Sb	0.317[**]	0.360[**]	0.232[*]	0.308[**]	0.337[**]	0.011	0.107	0.274[**]	0.337[**]	0.311[**]	0.479[**]
Cs	0.907[**]	0.853[**]	0.550[**]	0.913[**]	0.873[**]	−0.032	0.375[**]	0.545[**]	0.887[**]	0.818[**]	0.786[**]
Ba	0.468[**]	0.514[**]	0.620[**]	0.441[**]	0.529[**]	0.391[**]	0.496[**]	0.721[**]	0.487[**]	0.387[**]	0.556[**]
La	0.975[**]	0.906[**]	0.641[**]	0.981[**]	0.942[**]	−0.056	0.392[**]	0.667[**]	0.968[**]	0.916[**]	0.831[**]
Ce	0.975[**]	0.899[**]	0.636[**]	0.982[**]	0.926[**]	−0.051	0.395[**]	0.661[**]	0.958[**]	0.914[**]	0.825[**]
Pr	0.972[**]	0.896[**]	0.635[**]	0.982[**]	0.925[**]	−0.058	0.391[**]	0.656[**]	0.957[**]	0.910[**]	0.823[**]
Nd	0.970[**]	0.895[**]	0.647[**]	0.981[**]	0.928[**]	−0.062	0.392[**]	0.655[**]	0.958[**]	0.907[**]	0.823[**]
Sm	0.966[**]	0.888[**]	0.647[**]	0.978[**]	0.928[**]	−0.074	0.384[**]	0.646[**]	0.955[**]	0.900[**]	0.820[**]
Pb	0.527[**]	0.641[**]	0.420[**]	0.567[**]	0.608[**]	−0.136	0.070	0.511[**]	0.627[**]	0.542[**]	0.496[**]
Th	0.973[**]	0.899[**]	0.618[**]	0.983[**]	0.919[**]	−0.069	0.377[**]	0.634[**]	0.953[**]	0.911[**]	0.821[**]
U	0.984[**]	0.908[**]	0.656[**]	0.982[**]	0.951[**]	−0.041	0.413[**]	0.640[**]	0.974[**]	0.909[**]	0.836[**]

Continued

TABLE 14.6 Pearson Correlation Coefficients for Elements in *Hypogymnia physodes* ($n=121$)—Cont'd

(2) Mn–Mo

	Mn	Fe	Co	Ni	Cu	Zn	Se	Rb	Sr	As	Mo
Li	−0.341[**]	0.949[**]	0.933[**]	0.921[**]	0.776[**]	−0.102	0.965[**]	0.438[**]	0.435[**]	0.957[**]	0.909[**]
Na	−0.354[**]	0.908[**]	0.878[**]	0.844[**]	0.776[**]	−0.034	0.888[**]	0.443[**]	0.408[**]	0.943[**]	0.848[**]
Mg	−0.197[*]	0.656[**]	0.787[**]	0.652[**]	0.577[**]	0.095	0.647[**]	0.247[**]	0.702[**]	0.694[**]	0.530[**]
Al	−0.353[**]	0.962[**]	0.934[**]	0.922[**]	0.781[**]	−0.108	0.979[**]	0.469[**]	0.406[**]	0.970[**]	0.902[**]
Si	−0.352[**]	0.916[**]	0.945[**]	0.897[**]	0.800[**]	−0.036	0.917[**]	0.422[**]	0.458[**]	0.954[**]	0.850[**]
P	−0.087	−0.053	0.095	0.056	0.081	0.160	−0.017	0.100	0.404[**]	0.029	0.026
K	−0.308[**]	0.402[**]	0.516[**]	0.480[**]	0.412[**]	0.026	0.443[**]	0.337[**]	0.544[**]	0.461[**]	0.407[**]
Ca	−0.174	0.668[**]	0.742[**]	0.642[**]	0.504[**]	0.201[*]	0.631[**]	0.137	0.681[**]	0.660[**]	0.560[**]
Ti	−0.358[**]	0.947[**]	0.936[**]	0.929[**]	0.804[**]	−0.083	0.958[**]	0.431[**]	0.431[**]	0.960[**]	0.889[**]
V	−0.387[**]	0.896[**]	0.825[**]	0.971[**]	0.732[**]	−0.105	0.938[**]	0.380[**]	0.293[**]	0.874[**]	0.964[**]
Cr	−0.349[**]	0.811[**]	0.830[**]	0.811[**]	0.687[**]	−0.059	0.818[**]	0.416[**]	0.395[**]	0.826[**]	0.786[**]
Mn	1.00	−0.326[**]	−0.261[**]	−0.358[**]	−0.244[**]	0.173	−0.356[**]	−0.168	−0.246[**]	−0.310[**]	−0.354[**]
Fe	−0.326[**]	1.00	0.888[**]	0.904[**]	0.744[**]	−0.160	0.964[**]	0.425[**]	0.359[**]	0.945[**]	0.898[**]
Co	−0.261[**]	0.888[**]	1.00	0.898[**]	0.765[**]	0.020	0.909[**]	0.409[**]	0.558[**]	0.935[**]	0.802[**]
Ni	−0.358[**]	0.904[**]	0.898[**]	1.00	0.742[**]	−0.060	0.952[**]	0.383[**]	0.406[**]	0.894[**]	0.931[**]
Cu	−0.244[**]	0.744[**]	0.765[**]	0.742[**]	1.00	0.004	0.737[**]	0.444[**]	0.349[**]	0.803[**]	0.778[**]

Zn	0.173	−0.160	0.020	−0.060	0.004	1.00	−0.138	−0.306[**]	0.288[**]	−0.088	−0.100
Se	−0.356[**]	0.964[**]	0.909[**]	0.952[**]	0.737[**]	−0.138	1.00	0.443[**]	0.362[**]	0.944[**]	0.914[**]
Rb	−0.168	0.425[**]	0.409[**]	0.383[**]	0.444[**]	−0.306[**]	0.443[**]	1.00	−0.026	0.447[**]	0.381[**]
Sr	−0.246[**]	0.359[**]	0.558[**]	0.406[**]	0.349[**]	0.288[**]	0.362[**]	−0.026	1.00	0.437[**]	0.296[**]
As	−0.310[**]	0.945[**]	0.935[**]	0.894[**]	0.803[**]	−0.088	0.944[**]	0.447[**]	0.437[**]	1.00	0.885[**]
Mo	−0.354[**]	0.898[**]	0.802[**]	0.931[**]	0.778[**]	−0.100	0.914[**]	0.381[**]	0.296[**]	0.885[**]	1.00
Cd	0.213[*]	−0.219[*]	−0.042	−0.200[*]	0.099	0.152	−0.229[*]	0.172	0.029	−0.106	−0.220[*]
Sb	−0.246[**]	0.317[**]	0.308[**]	0.325[**]	0.364[**]	0.075	0.296[**]	0.094	0.178	0.336[**]	0.352[**]
Cs	−0.303[**]	0.854[**]	0.857[**]	0.823[**]	0.771[**]	−0.116	0.882[**]	0.634[**]	0.320[**]	0.895[**]	0.827[**]
Ba	−0.156	0.411[**]	0.579[**]	0.477[**]	0.439[**]	0.409[**]	0.394[**]	0.056	0.657[**]	0.463[**]	0.416[**]
La	−0.339[**]	0.963[**]	0.932[**]	0.935[**]	0.787[**]	−0.106	0.978[**]	0.431[**]	0.395[**]	0.970[**]	0.917[**]
Ce	−0.339[**]	0.964[**]	0.925[**]	0.930[**]	0.773[**]	−0.119	0.983[**]	0.438[**]	0.388[**]	0.966[**]	0.915[**]
Pr	−0.339[**]	0.961[**]	0.926[**]	0.928[**]	0.769[**]	−0.116	0.983[**]	0.437[**]	0.387[**]	0.964[**]	0.910[**]
Nd	−0.335[**]	0.959[**]	0.933[**]	0.929[**]	0.766[**]	−0.120	0.982[**]	0.439[**]	0.392[**]	0.966[**]	0.902[**]
Sm	−0.334[**]	0.953[**]	0.936[**]	0.924[**]	0.760[**]	−0.122	0.978[**]	0.437[**]	0.393[**]	0.963[**]	0.890[**]
Pb	−0.141	0.585[**]	0.524[**]	0.545[**]	0.522[**]	0.168	0.566[**]	0.039	0.249[**]	0.601[**]	0.569[**]
Th	−0.344[**]	0.960[**]	0.916[**]	0.921[**]	0.767[**]	−0.120	0.981[**]	0.440[**]	0.372[**]	0.961[**]	0.912[**]
U	−0.344[**]	0.949[**]	0.940[**]	0.925[**]	0.786[**]	−0.085	0.974[**]	0.428[**]	0.406[**]	0.967[**]	0.902[**]

Continued

TABLE 14.6 Pearson Correlation Coefficients for Elements in *Hypogymnia physodes* ($n=121$)—Cont'd

(3) Cd–U

	Cd	Sb	Cs	Ba	La	Ce	Pr	Nd	Sm	Pb	Th	U
Li	−0.177	0.317[**]	0.907[**]	0.468[**]	0.975[**]	0.975[**]	0.972[**]	0.970[**]	0.966[**]	0.527[**]	0.973[**]	0.984[**]
Na	−0.092	0.360[**]	0.853[**]	0.514[**]	0.906[**]	0.899[**]	0.896[**]	0.895[**]	0.888[**]	0.641[**]	0.899[**]	0.908[**]
Mg	0.018	0.232[*]	0.550[**]	0.620[**]	0.641[**]	0.636[**]	0.635[**]	0.647[**]	0.647[**]	0.420[**]	0.618[**]	0.656[**]
Al	−0.168	0.308[**]	0.913[**]	0.441[**]	0.981[**]	0.982[**]	0.982[**]	0.981[**]	0.978[**]	0.567[**]	0.983[**]	0.982[**]
Si	−0.136	0.337[**]	0.873[**]	0.529[**]	0.942[**]	0.926[**]	0.925[**]	0.928[**]	0.928[**]	0.608[**]	0.919[**]	0.951[**]
P	0.254[**]	0.011	−0.032	0.391[**]	−0.056	−0.051	−0.058	−0.062	−0.074	−0.136	−0.069	−0.041
K	0.049	0.107	0.375[**]	0.496[**]	0.392[**]	0.395[**]	0.391[**]	0.392[**]	0.384[**]	0.070	0.377[**]	0.413[**]
Ca	0.033	0.274[**]	0.545[**]	0.721[**]	0.667[**]	0.661[**]	0.656[**]	0.655[**]	0.646[**]	0.511[**]	0.634[**]	0.640[**]
Ti	−0.182[*]	0.337[**]	0.887[**]	0.487[**]	0.968[**]	0.958[**]	0.957[**]	0.958[**]	0.955[**]	0.627[**]	0.953[**]	0.974[**]
V	−0.274[**]	0.311[**]	0.818[**]	0.387[**]	0.916[**]	0.914[**]	0.910[**]	0.907[**]	0.900[**]	0.542[**]	0.911[**]	0.909[**]
Cr	−0.125	0.479[**]	0.786[**]	0.556[**]	0.831[**]	0.825[**]	0.823[**]	0.823[**]	0.820[**]	0.496[**]	0.821[**]	0.836[**]
Mn	0.213[*]	−0.246[**]	−0.303[**]	−0.156	−0.339[**]	−0.339[**]	−0.339[**]	−0.335[**]	−0.334[**]	−0.141	−0.344[**]	−0.344[**]
Fe	−0.219[*]	0.317[**]	0.854[**]	0.411[**]	0.963[**]	0.964[**]	0.961[**]	0.959[**]	0.953[**]	0.585[**]	0.960[**]	0.949[**]
Co	−0.042	0.308[**]	0.857[**]	0.579[**]	0.932[**]	0.925[**]	0.926[**]	0.933[**]	0.936[**]	0.524[**]	0.916[**]	0.940[**]
Ni	−0.200[*]	0.325[**]	0.823[**]	0.477[**]	0.935[**]	0.930[**]	0.928[**]	0.929[**]	0.924[**]	0.545[**]	0.921[**]	0.925[**]
Cu	0.099	0.364[**]	0.771[**]	0.439[**]	0.787[**]	0.773[**]	0.769[**]	0.766[**]	0.760[**]	0.522[**]	0.767[**]	0.786[**]

Zn	0.152	0.075	−0.116	0.409**	−0.106	−0.119	−0.116	−0.120	−0.122	0.168	−0.120	−0.085
Se	−0.229*	0.296**	0.882**	0.394**	0.978**	0.983**	0.983**	0.982**	0.978**	0.566**	0.981**	0.974**
Rb	0.172	0.094	0.634**	0.056	0.431**	0.438**	0.437**	0.439**	0.437**	0.039	0.440**	0.428**
Sr	0.029	0.178	0.320**	0.657**	0.395**	0.388**	0.387**	0.392**	0.393**	0.249**	0.372**	0.406**
As	−0.106	0.336**	0.895**	0.463**	0.970**	0.966**	0.964**	0.966**	0.963**	0.601**	0.961**	0.967**
Mo	−0.220*	0.352**	0.827**	0.416**	0.917**	0.915**	0.910**	0.902**	0.890**	0.569**	0.912**	0.902**
Cd	1.00	−0.102	−0.051	−0.002	−0.175	−0.182*	−0.182*	−0.177	−0.174	−0.213*	−0.183*	−0.180*
Sb	−0.102	1.00	0.280**	0.351**	0.321**	0.313**	0.311**	0.308**	0.302**	0.270**	0.308**	0.318**
Cs	−0.051	0.280**	1.00	0.367**	0.910**	0.909**	0.909**	0.907**	0.905**	0.470**	0.914**	0.916**
Ba	−0.002	0.351**	0.367**	1.00	0.434**	0.416**	0.412**	0.409**	0.402**	0.394**	0.399**	0.441**
La	−0.175	0.321**	0.910**	0.434**	1.00	0.998**	0.997**	0.997**	0.994**	0.608**	0.993**	0.989**
Ce	−0.182*	0.313**	0.909**	0.416**	0.998**	1.00	0.999**	0.998**	0.996**	0.591**	0.997**	0.988**
Pr	−0.182*	0.311**	0.909**	0.412**	0.997**	0.999**	1.00	0.999**	0.997**	0.592**	0.997**	0.987**
Nd	−0.177	0.308**	0.907**	0.409**	0.997**	0.998**	0.999**	1.00	0.999**	0.587**	0.995**	0.987**
Sm	−0.174	0.302**	0.905**	0.402**	0.994**	0.996**	0.997**	0.999**	1.00	0.579**	0.993**	0.986**
Pb	−0.213*	0.270**	0.470**	0.394**	0.608**	0.591**	0.592**	0.587**	0.579**	1.00	0.579**	0.590**
Th	−0.183*	0.308**	0.914**	0.399**	0.993**	0.997**	0.997**	0.995**	0.993**	0.579**	1.00	0.988**
U	−0.180*	0.318**	0.916**	0.441**	0.989**	0.988**	0.987**	0.987**	0.986**	0.590**	0.988**	1.00

Asterisks indicate significance at $p < 0.05$ (*) or $p < 0.01$ (**).

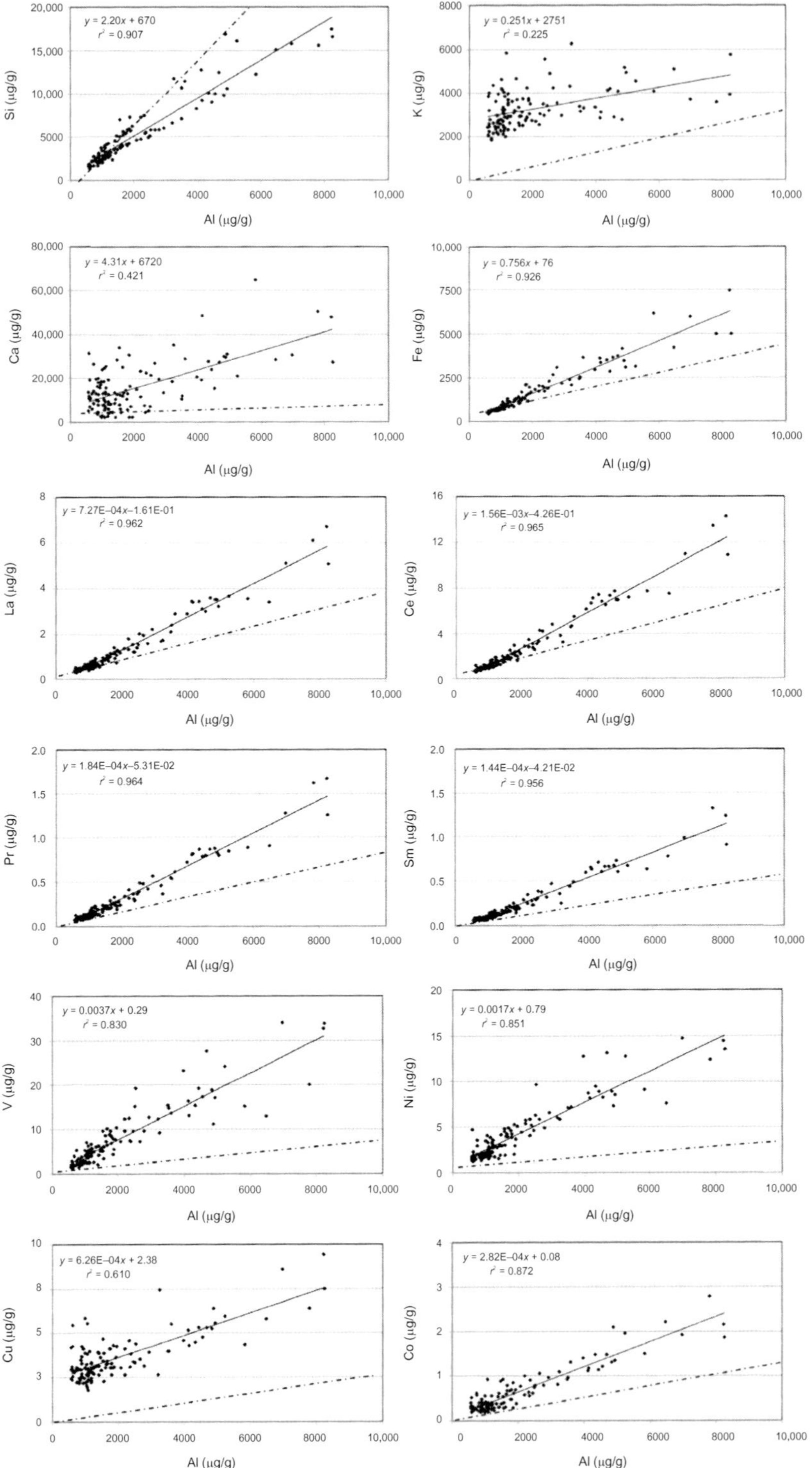

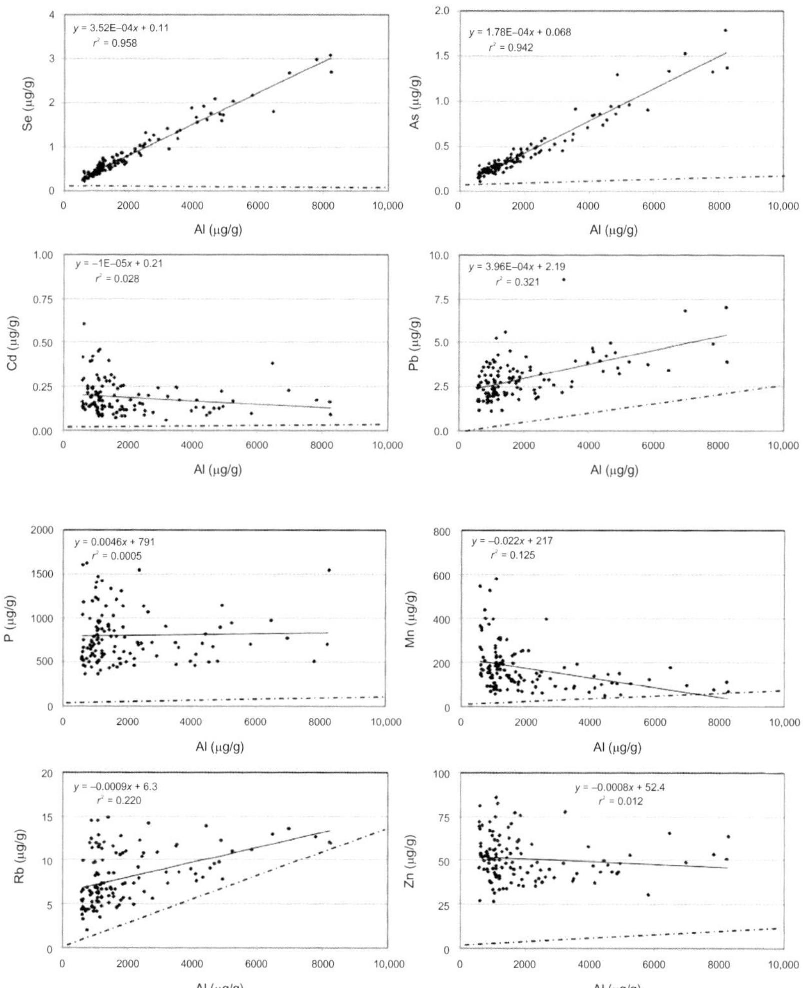

FIGURE 14.3   (A) Scatter plots of major elements (Si, K, Ca, and Fe) versus Al. (B) Scatter plots of selected REEs (references, La, Ce, Pr, and Sm) versus Al. (C) Scatter plots of selected transition metals (V, Ni, Cu, and Co) versus Al. (D) Scatter plots of selected trace elements (Se, As, Cd, and Pb) versus Al. (E) Scatter plots of P, Mn, Rb, and Zn versus Al.

extraction of the oil sands are a major reason why element:Al ratios should be expected to depart from the global average in the AOSR.

Si and Fe in Figure 14.3A exhibit strong correlations with Al and regression line intercepts that are small in relation to observed concentrations. The regression line for Si indicates Si/Al ratios that are lower than the crustal average; however, close inspection suggests there may be two distinct relationships, one of which is near the crustal average and the other about a factor of 2 lower. The Fe/Al slope, on the other hand, is almost a factor of 2 higher than the crustal average. Ca and K differ from Si and Fe in several respects. First, the regression line intercepts are relatively high, suggestive of a process for accumulating Ca and K which is independent of Al. The regression slope for Ca is at least a factor of 10 higher than the crustal ratio, while that for K is not statistically different from the crustal value.

The REEs La, Ce, Pr, and Sm exhibit very strong correlations with Al ($r^2 > 0.95$) and virtually negligible intercepts (see Figure 14.3B). Regression line slopes in all cases are roughly 1.7 times the crustal average. The transition elements V, Ni, Cu, and Co exhibit lower correlations with Al than the REEs (see Figure 14.3C). Regression intercepts are near zero for V, Ni, and Co, while that for Cu is quite large. Regression slopes for V and Ni are five to eight times the crustal ratio, while those for Cu and Co are two to three times the crustal ratio.

The relatively volatile trace elements Se, As, Cd, and Pb exhibit interesting and diverse relationships to Al (Figure 14.3D). Se and As are both highly correlated with Al ($r^2 \sim 0.95$) and exhibit slopes at least a factor of 100 higher than the crustal average. Cd exhibits a very large intercept and negative slope, while Pb exhibits a large intercept and positive slope. Despite the negative slope, observed Cd/Al ratios are at least a factor of 10 higher than crustal ratios across the range of observed concentrations. For Pb, the slope is roughly a factor of 1.5 higher than the crustal value. The elements P, Mn, Rb, and Zn show weak to nonsignificant relationships to Al and dominant intercepts relative to observed concentrations (Figure 14.3E). Data for Mn, and to a lesser extent Zn and P, are suggestive of an inverse relationship to Al. P/Al and Zn/Al ratios are invariably much higher than crustal values, while Mn/Al and Rb/Al often approach crustal values. The diversity of Al relationships shown in Figure 14.3A–E suggests that a range of processes contribute to the accumulation of minor and trace elements in *H. physodes* and underscore the need for multivariate analyses, such as principal component analysis, to elucidate source contributions (Chapter 18).

14.3.5   Comparison with *H. physodes* Data from Previous Studies

Data from previous studies are compared with the distal and proximal subsets of AOSR samples in Table 14.7. Given the large number of published studies, only a limited cross section could be included in this discussion. In general, studies were excluded from comparison if they (i) reported less than four elements, (ii) did not present data from a "control" or "background" site, or (iii) employed sample digestion and analytical techniques that were deemed incommensurable with the current study. Data in Table 14.7 must be approached with caution because they span broad

TABLE 14.7 Element Concentrations (μg/g) in *Hypogymnia physodes* from the AOSR and Other Locations

Reference	This study	This study	USDA (1999)	Pfeiffer and Barclay-Estrup (1992)	USDA (1999)	Bennett et al. (1996)	USDA (1999)	USDA (2012)	Valeeva and Moskovchenko (2002)	Pilegaard (1979)	USDA (1999)	Gailey et al. (1985)	Herzig et al. (1989)	Olszowski et al. (2011)	Cloquet et al. (2006)	Guttova et al. (2011)	Jeran et al. (1996)	Bergamaschi et al. (2002)
Country			CA	CA	USA	USA	USA	USA	Russia	Denmark	Finland	Scotland	Switzerland	Poland	France	Slovakia	Slovenia	Sagarmatha NP, Nepal
Location	AOSR distal	AOSR proximal	Salt Spr. Is., BC	Thunder Bay, ON	Olympic NP, WA	Northern WI	Acadia NP, ME	Nat'l Forests	Western Siberia	"bckgnd"	Rural sites	Near Tayside	First Neuenegg	Rural	Metz	Rural	Rural	(3200 m) 1999
Element/ year	2008	2008	Pre-1986	1990	Pre-1988	1992	Pre-1988	1986–1997	1993–2002	1977	1985–1986	1981	1985	2009	2002–2003	2006	2005	1999	1999
Al	986	2324	1948	185	743	580	163	191–880	n.d.	n.d.	n.d.	n.d.	363	n.d.	838	n.d.	n.d.	n.d.	n.d.
As	0.25	0.45	n.d.	0.90	n.d.	2.6	n.d.	n.d.	n.d.	n.d.	n.d.	n.d.	n.d.	n.d.	n.d.	n.d.	1.24	0.37	0.43
Ca	10,105	18,594	13,513	n.d.	14,500	12,593	6795	882–9800	n.d.	n.d.	n.d.	n.d.	25,500	n.d.	n.d.	n.d.	n.d.	n.d.	n.d.
Cd	0.17	0.15	n.d.	0.20	0.22	n.d.	0.4	0.4–2.0	0.5	0.58	0.7	n.d.	0.22	0.54	0.31	0.46	1.1	n.d.	0.04
Co	0.29	0.79	n.d.	n.d.	n.d.	0.3	n.d.	n.d.	0.87	n.d.	n.d.	n.d.	0.43	n.d.	n.d.	n.d.	0.55	0.81	1.87
Cr	0.98	2.63	n.d.	n.d.	1.87	1.2	0.4	0.5–6.7	4.6	5	2.1	1.4	1.8	0.22	n.d.	1.8	n.d.	3.70	2.60
Cu	2.93	3.79	2.60	0.80	4.70	5.3	2.8	0.8–9.3	6	8.4	7.3	18.5	4.3	3.7	6.7	4.2	n.d.	n.d.	5.00
Fe	711	1932	1549	114	754	622	151	188–954	1013	1100	540	871	480	350	n.d.	n.d.	1253	1659	1843
K	2737	3496	2472	n.d.	2810	2944	2496	n.d.	n.d.	n.d.	n.d.	n.d.	3095	n.d.	n.d.	n.d.	4094	9053	4228
La	0.52	1.40	n.d.	n.d.	n.d.	n.d.	n.d.	n.d.	n.d.	n.d.	n.d.	n.d.	n.d.	n.d.	n.d.	n.d.	1.2	1.60	3.50

TABLE 14.7 Element Concentrations (μg/g) in *Hypogymnia physodes* from the AOSR and Other Locations—Cont'd

Mg	570	827	680	n.d.	834	745	695	188–954	n.d.	n.d.	n.d.	n.d.	510	n.d.	n.d.	n.d.	n.d.	3038	3899
Mn	182	116	162	n.d.	255	32.2	205	77–665	270	24	131	43	n.d.	n.d.	n.d.	524	n.d.	86.1	79.1
Ni	1.93	5.15	4.10	n.d.	4.00	1.8	1.1	0.7–3.6	3	0.11	2.6	1.8	n.d.	1.1	n.d.	1.8	n.d.	n.d.	n.d.
Pb	2.45	2.92	60	3.9	38.0	7.8	31	5.3–9.9	17.7	44	18	11	28	9	14.7	8	n.d.	n.d.	14.10
Rb	6.42	8.38	n.d.	n.d.	n.d.	n.d.	n.d.	n.d.	n.d.	n.d.	n.d.	n.d.	n.d.	n.d.	n.d.	n.d.	14	27.6	15.5
Sb	0.02	0.03	n.d.	n.d.	n.d.	n.d.	n.d.	n.d.	n.d.	n.d.	n.d.	n.d.	n.d.	n.d.	n.d.	n.d.	0.35	0.60	1.40
Se	0.43	0.90	n.d.	n.d.	n.d.	1.7	n.d.	n.d.	n.d.	n.d.	n.d.	n.d.	n.d.	n.d.	n.d.	n.d.	0.27	0.85	0.25
V	2.94	9.53	n.d.	n.d.	n.d.	n.d.	n.d.	n.d.	n.d.	5.6	n.d.	n.d.	n.d.	n.d.	n.d.	n.d.	n.d.	2.80	2.20
Zn	51.6	46.6	38.0	49.0	53.0	78.4	45	32–125	80	85	86	57	44	55	72	48	90	45.2	31.9

reaches of time and space and because details of sample collection, site ecology, and local sources are unknown or uncontrolled across studies. Graney et al. (Chapter 15), for example, discuss the nontrivial effects of wet versus dry soil conditions that translate to ecology-based differences in canopy composition that in turn affect element concentrations in lichens. The studies in Table 14.7 are listed in order of approximate distance from the AOSR, ranging from about 1200 km for Salt Spring Island, BC, Canada to over 10,000 km for Sagarmatha NP, Nepal. This clearly exemplifies the large geophysical range of *H. physodes*, which is one of its preeminent characteristics for biomonitoring studies.

To place the results from this study into context with other work, the following synopsis is provided for several of the studies listed in Table 14.7.

- Bennett et al. (1996) conducted a transplant study in 1992 to determine whether *H. physodes* could be recolonized in the Indiana Dunes National Lakeshore on the south shore of Lake Michigan. Control samples for the experiment were collected from Toft Point in extreme northeast WI and were analyzed for 20 elements via ICPMS.
- Bergamaschi et al. (2002) performed a study under the auspices of the EV-K2-CNR program to quantify trace element concentrations in lichen samples from remote, high elevation areas. The authors reported data for 20 elements in samples collected from the Sagarmatha National Park at elevations of 3200 and 5090 m.
- Cloquet et al. (2006) collected samples of several lichen species in and around the city of Metz, France (2006 population ca. 300,000). Samples of *H. physodes* were collected at 6 of the 36 study sites and analyzed for Al, Cd, Cu, Pb, and Zn via ICPMS.
- Gailey et al. (1985) conducted a transplant study in 1981 to evaluate accumulation of heavy metals in and around the town of Armadale, a small industrial town in central Scotland. Transplant samples were collected near Tayside, Scotland (80 km northeast of Armadale), in an area distant from heavy metal pollution and analyzed for Fe, Mn, Zn, Pb, Cu, Cr, and Ni via AAS.
- Olszowski et al. (2011) conducted a translocation study in rural southwest Poland in 2009–2010 to obtain heavy metal data in a region with little historical monitoring. Control samples for this study were collected in the Bory Niemendenski Forest and analyzed for Mn, Zn, Pb, Ni, Cd, and Cu via atomic absorption spectrometry.
- Pfeiffer and Barclay-Estrup (1992) conducted a survey of metal concentration in lichen samples collected in and around the city of Thunder Bay, Ontario, Canada.
- Valeeva and Moskovchenko (2002) collected samples across the Tyumen Oblast of northwestern Siberia to determine the effect of oil exploration and other human activities on trace element concentrations. A total of 42

Hypogymnia samples were collected and analyzed from 10 sites, including a background site on the Yamal Peninsula.

- The U.S. Department of Agriculture (USDA) published a literature review of nutrient and metal concentrations in 1999 and maintains the National Lichens and Air Quality Database and Clearinghouse (USDA, 1999, 2012). Together, these resources contain a wealth of data on numerous lichen species, including *H. physodes*, from global studies conducted between 1977 and 1997 and from otherwise unpublished studies in U.S. National Forests conducted between 1986 and 1997.

Inspection of concentration data across sites in Table 14.7 shows relatively-large ranges for all elements. AOSR samples tend to exhibit some of the highest concentrations for the major elements, such as Al, Ca, Fe, and K. Ni and V concentrations in the AOSR proximal group also appear somewhat elevated compared to the results from other studies. At the same time, AOSR results include some of the lowest reported concentrations for a number of trace elements, including As, Sb, Cd, and Pb. In fact, the Sb and Pb concentrations may be the lowest ever reported for *H. physodes*.

14.4 CONCLUSIONS

The methods used in this work are appropriate for multielement analysis of small (25–30 mg) lichen samples. The microwave-assisted procedure using sequential additions of H_2O_2 and $HNO_3 + HF$ effectively digested the organic and mineral portions of the samples and liberated most elements quantitatively into solution for subsequent analysis. Possible exceptions include Cr and Ti, which showed poor recoveries for NIST 1643a (urban particulate matter) but higher recoveries for BCR 482 (lichen). Matrix-matched ICPMS detection limits were found to be suitable for quantification of 34 of the 43 target elements. Multiple analyses of SRMs and duplicate extractions show that accuracy and precision are on the order of 5–15% for many elements. For future studies, larger sample sizes or preconcentration might be needed to reliably quantify the platinum group elements, plus Be, Bi, Nb, Sn Ta, Tl, and W. As and Se should be analyzed by high-resolution ICPMS or DRC ICPMS to achieve optimum detection limits and to minimize polyatomic interferences. The large data set produced in this effort shows considerable texture both in terms of interelement relationships and as a function of distance to the AOSR oil production operations (distal vs. proximal subsets). In general, the concentrations reported in this work are well in line with those reported for *H. physodes* samples collected from other parts of Canada, the USA, and remote areas of the northern hemisphere. The major elements tend to be at or near the upper end of reported concentrations, while a number of trace elements (As, Sb, Pb, Cd) are at or below the lower end of concentrations reported in other studies.

ACKNOWLEDGMENTS

We thank Justin Straker (Stantec) for his efforts in lichen sample collection and cleaning and Joel Blum (University of Michigan) for lichen sample grinding. This work was funded by WBEA.

REFERENCES

Beckingham, J.D., Archibald, J.H., 1996. Field Guide to Ecosites of Northern Alberta. Northern Forestry Centre, Forestry Canada, Northwest Region, Edmonton, Alberta.

Bennett, J.P., Dibben, M.J., Lyman, K.J., 1996. Element concentrations in the lichen *Hypogymnia physodes* after 3 years of transplanting along Lake Michigan. Environ. Exp. Bot. 36 (3), 255–270.

Bergamaschi, L., Rizzio, E., Valcuvia, M.G., Verza, G., Profumo, A., Gallorini, M., 2002. Determination of trace elements and evaluation of their enrichment factors in Himalayan lichens. Environ. Pollut. 120, 137–144.

Berryman, S., Straker, J., Krupa, S., Davies, M., Ver Hoef, J., Brenner, G., 2010a. Mapping the characteristics of air pollutant deposition patterns in the Athabasca Oil Sands Region using epiphytic lichens as bioindicators. In: Interim Report Submitted to the Wood Buffalo Environmental Association, Fort McMurray, AB, Canada.

Cloquet, C., Carignan, J., Libourel, G., 2006. Atmospheric pollution dispersion around an urban area using trace metal concentrations and Pb isotope compositions in epiphytic lichens. Atmos. Environ. 40, 574–587.

Di Lella, L.A., Frati, L., Loppi, S., Protano, G., Riccobono, F., 2003. Lichens as biomonitors of uranium and other trace elements in an area of Kosovo heavily shelled with depleted uranium rounds. Atmos. Environ. 37, 5445–5449.

Ferry, B.W., Baddeley, M.S., Hawksworth, D.L., Tuominen, Y., 1967. Studies on the strontium uptake of the *Cladonio alpestris* thallus. Ann. Bot. Fenn. 4, 1–28.

Gailey, F.A.Y., Smith, G.H., Rintoul, L.J., Lloyd, O.L., 1985. Metal deposition in central Scotland, as determined by lichen transplants. Environ. Monit. Assess. 5, 291–309.

Guttova, A., Lackovicova, A., Pisut, I., Pisut, P., 2011. Decrease in air pollution load in urban environment of Bratislava (Slovakia) inferred from accumulation of metal elements in lichens. Environ. Monit. Assess. 182, 361–373.

Hawksworth, D.L., Rose, F., 1970. Qualitative scale for estimating sulphur dioxide air pollution in England and Wales using epiphytic lichens. Nature 227, 145–148.

Herzig, R., Liebendörfer, L., Urech, M., Ammann, K., Cuecheva, M., Landolt, W., 1989. Passive biomonitoring with lichens as a part of an integrated biological measuring system for monitoring air pollution in Switzerland. Int. J. Environ. Anal. Chem. 35 (1), 43–57.

Jalkanen, L.M., Hasanen, E.K., 1996. Simple method for the dissolution of atmospheric aerosol samples by inductively coupled plasma mass spectrometry. J. Anal. At. Spectrom. 11, 365–369.

Jeran, Z., Jacimovic, R., Batic, F., Smodis, B., Wolterbeek, H.T., 1996. Atmospheric heavy metal pollution in Slovenia derived from results of epiphytic lichens. Fresenius J. Anal. Chem. 354, 681–687.

Nieboer, E., Ahmed, H.M., Puckett, K.J., Richardson, D.H.S., 1972. Heavy metal content of lichens in relation to distance from a nickel smelter in Sudbury, Ontario. Lichenologist 5, 291–304.

Olszowski, T., Tomaszewska, B., Goraina-Wloodarczyk, K., 2011. Air quality in non-industrialized area in the typical Polish countryside based on measurements of selected pollutants in immission and deposition phase. Atmos. Environ. 50, 139–147. http://dx.doi.org/10.1016/j.atmosenv. 2011.12.049.

Pfeiffer, H.N., Barclay-Estrup, P., 1992. The use of a single lichen species, *Hypogymnia physodes*, as an indicator of air quality in Northwestern Ontario. The Bryologist 95 (1), 38–41.

Pilegaard, K., 1979. Heavy metals in bulk precipitation and transplanted *Hypogymnia physodes* and *Dicranoweisia cirrata* in the vicinity of a Danish steelworks. Water Air Soil Pollut. 11, 77–91.

Rasmussen, P.E., Wheeler, A., Hassan, N., Filiatreault, A., Lanouette, M., 2007. Monitoring personal, indoor and outdoor exposures to metals in airborne particulate matter: risk of contamination during sampling, handling and analysis. Atmos. Environ. 41, 5897–5907.

Reimann, C., DeCaritat, P., 2000. Intrinsic flaws of element enrichment factor (EFs) in environmental geochemistry. Environ. Sci. Technol. 34, 5084–5091.

Reimann, C., DeCaritat, P., 2005. Distinguishing between natural and anthropogenic sources for elements in the environment: regional geochemical surveys versus enrichment factors. Sci. Total Environ. 337, 91–107.

Taylor, S.R., McLennnan, S.M., 1995. The geochemical evolution of the continental crust. Rev. Geophys. 33, 241–265.

Tuominen, Y., 1967. Studies on the strontium uptake of the *Cladonio alpestris* thallus. Ann. Bot. Fenn. 4, 1–28.

USDA Forest Service, 1999. A Review of Lichen and Bryophyte Elemental Content Literature with Reference to Pacific Northwest Species. March 1999.

USDA Forest Service, 2012. National Lichens and Air Quality Database and Clearinghouse. http://gis.nacse/lichenair. Accessed February 21, 2012.

Valeeva, E.I., Moskovchenko, D.V., 2002. Trace-element composition of lichens as an indicator of atmospheric pollution in northern West Siberia. Polar Geogr. 26 (4), 249–269.

Vitt, D.H., Halsey, L.A., Thormann, M.N., Martin, T., 1996. Peatland Inventory of Alberta Phase 1: Overview of Peatland Resources of the Natural Regions and Subregions of the Province. National Centres of Excellence. Sustainable Forest Management Centre, University of Alberta, Edmonton, Alberta.

Chapter 15

Coupling Lead Isotopes and Element Concentrations in Epiphytic Lichens to Track Sources of Air Emissions in the Athabasca Oil Sands Region

J.R. Graney[*,1], M.S. Landis[†] and S. Krupa[‡]

[*]*Geological Sciences and Environmental Studies, Binghamton University, Binghamton, New York, USA*

[†]*US EPA, Office of Research and Development, Research Triangle Park, North Carolina, USA*

[‡]*Plant Pathology, University of Minnesota-Twin Cities, St. Paul, Minnesota, USA*

[1]*Corresponding author: e-mail: jgraney@binghamton.edu*

ABSTRACT

A study was conducted that coupled use of element concentrations and lead (Pb) isotope ratios in the lichen *Hypogymnia physodes* collected during 2002 and 2008, to assess the impacts of air emissions from the Athabasca Oil Sands Region (AOSR, Canada) mining and processing operations. The lichens selected from the 2002 data set were from 15 samples sites collected on an N–S and E–W grid centered between the oil sands processing sites. The lichens selected for analysis in 2008 were collected using a stratified, nested circular grid approach radiating away from the oil sands processing sites, and included 121 sampling sites as far as 150 km from the mining and processing operations. Spatial analysis indicates three main element groupings including a geogenic source (aluminum and others) related to oil sands mining, an oil processing source (vanadium and others), and a grouping that is likely related to biogeochemical processes (manganese and others). An exponential decrease in concentration of the geogenic grouping of elements versus distance from the mining sites was found, whereas near source concentrations of elements typically associated with oil processing are more homogeneous spatially than the geogenic elements. The mining and oil processing

Disclaimer: The content and opinions expressed by the authors do not necessarily reflect the views of the WBEA or of the WBEA membership. This chapter has been subjected to U.S. EPA Agency review and approved for publication.

Developments in Environmental Science, Vol. 11. http://dx.doi.org/10.1016/B978-0-08-099443-7.00015-6

related element groupings are superimposed over the elemental signature that reflects lichen biogeochemical processes. The ranges in Pb isotope ratios were similar in 2002 and 2008, suggesting that sources of Pb accumulated by the lichens did not change substantially between 2002 and 2008. The Pb isotope ratios from lichens collected beyond 50 km from the mining and processing sites cluster into a grouping with a $^{207}Pb/^{206}Pb$ ratio of 0.8650 and a $^{208}Pb/^{206}Pb$ ratio near 2.095. This grouping likely reflects the regional background Pb isotope ratio signature. The lowering of the $^{207}Pb/^{206}Pb$ and $^{208}Pb/^{206}Pb$ ratios near the mining and processing operations indicates that other Pb sources, likely related to the oil sands mining and processing, are contributing to the Pb source signature. This assessment was confirmed through the analysis of source and stack samples. The Pb isotope ratios were a better predictor of the extent of the source contribution than the element concentrations because the Pb isotope ratios are not affected by either the metabolic processing of elements by the lichens or by moisture related controls on atmospheric deposition processes at the collection sites.

The main goals for this project to determine (i) the efficacy of using Pb isotopes to assist in identifying the sources of atmospherically deposited air pollutants in the AOSR and (ii) whether coupling Pb isotopes with elemental concentrations can help to elucidate the causes for spatial differences in the accumulation of elements by epiphytic lichens in relation to emission sources were successful. An approach that couples Pb isotopes, spatial analysis, and element concentrations is recommended for future source attribution studies in the AOSR.

15.1   INTRODUCTION

In the Athabasca Oil Sands Region (AOSR), there are both natural and anthropogenic air emission sources. The anthropogenic sources include oil sand mining, oil extraction and refining facilities, and transportation sources. Natural sources include forest fires and wind-blown dust from soils. Assessing regional scale effects of air pollution sources in the AOSR is difficult due to logistical constraints. Roadway infrastructure and electrical services are lacking outside the mining and processing facilities. As an alternative to direct atmospheric measurements, plant species can be used as surrogates in air quality studies. Lower order (e.g., mosses, lichens) and higher order (e.g., jack pine, aspen) plant species are possibilities for use as biological indicators of air quality over large spatial scales in the AOSR (Addison and Puckett, 1980; Nash and Gries, 1995; Puckett, 1988).

Lichens are commonly used in bio-monitoring studies to determine spatial and temporal gradients in air pollution (Conti and Cecchetti, 2001; Purvis et al., 2004). Epiphytic lichens obtain most of their nutrients from the atmosphere in the form of wet and dry deposition of aerosols and gases (Nash, 1989). Lichens do not contain waxy cuticles or root systems like vascular plants, and likely exchange materials across their entire surface (Nash, 1996). They can be sensitive to air pollutants, particularly SO_2 and NO_x, and have served as an indicator of adverse environmental conditions on local, regional and global spatial scales (Bergamaschi et al., 2002; Garty, 2001;

Getty et al., 1999; Nash, 1996) as well as temporal scales in transplantation experiments (Ayrault et al., 2007; Bergamaschi et al., 2007; Rusu et al., 2006; Spiro et al., 2004). Effects of air pollution on lichen communities (shifts in populations and spatial diversity) and the elemental composition of their tissue can be measured to determine pollutant deposition patterns from emission sources (Garty, 2000a,b; Nash, 1989). These spatial features in lichen elemental concentrations offer opportunities to apply receptor modeling in examining the basis for such differences in regard to pollutant emissions and their accumulation in the lichens (Chapter 18). In ideal cases, mining and processing facilities and their predominant metal emissions provide unique elemental fingerprints for source characterization.

In addition to elemental composition, the distributions of Pb isotopes can be used to determine the relative contribution of sources of air pollution. Pb has four major isotopes, 204, 206, 207, and 208. ^{208}Pb is formed from the radioactive decay of ^{232}Th, ^{207}Pb from ^{235}U, and ^{206}Pb from ^{238}U. ^{204}Pb is referred to as common Pb (has no radioactive parent and is much less abundant than the other isotopes). The uranium and thorium parents have differing decay rates resulting in predictable changes in Pb isotope ratios (Faure, 1986). The Pb isotope ratios from source materials reflect the age of the Pb incorporated into the parent material (e.g., ore deposit, coal, sediments, oil, oil sands), which is preserved during the subsequent industrial process that emitted the Pb into the environment (Ault et al., 1970; Graney et al., 1995). Following emission from high-temperature processes such as smelting, coal combustion, or oil refining, Pb species typically nucleate or condense into atmospheric aerosols very quickly. Pb isotope ratios are not believed to undergo significant fractionation during the industrial processes or subsequent biogeochemical processes resulting in Pb accumulation by the lichens.

In North America Pb isotope ratios from lichens have been used to track metal source, transport, and deposition processes in southern Quebec (Carignan and Gariepy, 1995), and other parts of northeastern North America (Carignan et al., 2002) and western Canada (Simonetti et al., 2003). The results from these studies suggest that in consort with metal concentrations, the use of Pb isotope ratios from lichens can be used to differentiate between sources of Pb bioaccumulated within the lichens on local, regional, and global scales.

The main goals for this project were to (i) determine the efficacy of using Pb isotopes to assist in identifying the sources of atmospherically deposited air pollutants in the AOSR and (ii) whether coupling Pb isotopes with elemental concentrations can elucidate the causes for spatial differences in the accumulation of elements in epiphytic lichens in relation to emission sources. An integrated analysis combining Pb isotopes, distance from known sources, and selected element concentrations demonstrates that both goals for this project were met and this technique may be successfully applied in future source attribution studies.

15.2 MATERIALS AND METHODS

The sensitivity of lichens to pollutants may be species specific, in part influenced by morphology and other factors (Nieboer et al., 1978). Metal accumulation by lichens can also be species dependent (Garty et al., 1979; Nash, 1989) in part because of physiological differences in lichen species (Bergamaschi et al., 2007; Cercasov et al., 2002). For purposes of this hybrid source attribution effort, it was decided to focus on analysis of one species of lichens from the AOSR rather than combining results from several species. In the AOSR, *H. physodes* was an ideal choice because it is an epiphytic lichen that is commonly used in air quality monitoring due to its widespread distribution and tolerance to SO_2 (Garty, 2001). Previous studies in the AOSR (Berryman et al., 2004, 2010) have shown that *H. physodes* exhibited distinct spatial variations in their total sulfur (S) and total nitrogen (N) elemental composition in relation to distance from known emission sources.

The methods used to collect the lichens are described in detail in Berryman et al. (2004, 2010). Briefly, all lichen samples were collected during the summer in 2002 and 2008 from the tips of tree branches. GPS was used to obtain the sample location coordinates and elevation at each collection site. The field personnel that harvested the lichens classified the sample location stand characteristics into ecosite based designations. The ecosite classification that was used followed an edaphic grid of moisture and nutrient regimes suitable for northeast Alberta (Beckingham and Archibald, 1996), supplemented by the addition of fens and bogs (Vitt et al., 1996). Samples from 15 locations and 5 duplicate samples (from a total of 44 samples) from the 2002 collection campaign were selected for inclusion in this study. The 2002 samples were collected on an N–S and E–W grid centered between the two main oil sands processing sites. The 2008 data set included lichens collected from 369 sites using a stratified, nested circular grid approach radiating away from the oil sands processing sites. Samples from 121 sites were selected for elemental and Pb isotope ratio determinations in this effort. This included lichens from sites as far as 150 km from the mining and processing locations. Comparisons of results from 2002 to 2008 will determine if temporal changes are apparent between the data sets. Bulk source and stack samples were also collected and analyzed for source apportionment purposes (Chapter 18 for details on the bulk source samples; Chapter 7 for details on stack samples).

Chapter 14 provides a detailed description of the lichen and source sample digestion and elemental analysis methods. Briefly, the elements were extracted from the lichens in an H_2O_2–HNO_3–trace HF mixture using microwave digestion. The concentrations of the elements in the extracts were analyzed using a Perkin Elmer Elan 9000 DRC ICP-MS at ARA Inc. Aliquots of the lichen and source sample digests that had been used for the multi-element determinations were subsequently measured for stable Pb isotope ratios at the U.S. Environmental Protection Agency National Exposure

Research Laboratory (EPA NERL) using a Thermo Scientific (Franklin, MA) Element2 inductively coupled plasma high-resolution magnetic sector field mass spectrometer (ICP-SFMS). The method that was developed to measure the Pb isotope ratios included: (i) self-aspiration of sample through a cyclonic spray chamber with an uptake rate of 200 μl min^{-1} to maximize signal stability, (ii) optimization of the detector dead time in the scanning speed operation mode (the dead time calculation is used to optimize the accuracy of the Pb isotope ratio measurements), (iii) utilization of low-resolution detection mode to produce flat-topped peak shapes, and (iv) use of a narrow mass width window (10% of the peak top-width) scanned at a high sweep rate. The concentration levels determined to minimize detector signal processing inefficiencies was assessed by analysis of National Institute of Standards and Technology Standard Reference Material 981 (NIST SRM 981, Pb Isotope Standard) at varying concentrations levels (from 0.10 to 5 ppb). Optimal isotope ratio results were found when Pb concentrations were lower than 2.0 ppb.

Lichen and source sample digests typically required a fivefold or greater dilution to achieve the less than 2 ppb sample concentration needed for Pb isotope ratio determination. A bracketing technique was then used to correct for mass bias during the Pb isotope ratios measurements (Krachler et al., 2004; Yip et al., 2008). This mass bias correction is instrumentation dependent and is now recognized as a conventional procedure needed during the measurement of Pb isotope ratios to optimize the accuracy of the results (Yip et al., 2008). In this study, correcting for mass bias was achieved by analyzing an aliquot of NIST SRM 981, before and after every lichen or source sample extract. The average of the Pb isotope ratios from the bracketing samples was used to correct the results from the lichen sample for ICP-SFMS mass bias (Krachler et al., 2004). Each of the diluted lichen and source sample extracts was sequentially analyzed four times between the NIST SRM 981 runs. The relative standard error from the four sets of sequential measurements was averaged to determine the relative precision of the isotope ratio measurements. All Pb isotope ratio results using the ICP-SFMS are reported using two-sigma relative standard error notation. In this study, it was found that high precision isotope ratios could be obtained on samples with Pb concentrations as low as 0.10 ppb. Using an uptake rate of 200 μl min^{-1} and 10 min per set of four analyses resulted in a sample consumption of 2 ml of sample per Pb isotope ratio determination.

15.3   RESULTS

15.3.1   Theory for Using Pb Isotopes to Help Identify Sources

There are several methods to depict results from Pb isotope ratio measurements. Plotting results as ^{208}Pb/^{206}Pb isotope ratios (y-axis) versus ^{207}Pb/^{206}Pb isotope ratios (x-axis) often yields either linear arrays or triangular fields of data points

(Sondergaard et al., 2010). Individual datum on such plots typically reflects the age of the Pb incorporated into the parent material (e.g., ore, coal, sediments, oil, oil sands), which is preserved during the subsequent process that emitted the Pb into the environment. Results in linear arrays or triangular fields indicate homogenization processes that mix Pb from several sources (Faure, 1986). In Fig. 15.1 the variance in ^{208}Pb/^{206}Pb versus ^{207}Pb/^{206}Pb isotope ratios from several of the major ore and coal deposits from North America has been plotted. On ^{208}Pb/^{206}Pb versus ^{207}Pb/^{206}Pb plots such as Fig. 15.1, older Pb will be found in the upper right quadrant of the diagram, and younger Pb in the lower left quadrant due to the differences in the age of the Pb incorporated into the samples of interest. The decrease in ^{207}Pb/^{206}Pb ratios from old to young Pb reflects the difference in decay rate of the parent ^{235}U and ^{238}U isotopes. The difference in ^{208}Pb/^{206}Pb can reflect differences in the amount of parent ^{232}Th and ^{238}U. Source materials with higher thorium concentrations generate higher ^{208}Pb/^{206}Pb ratios. Coupling elemental concentrations on the y-axis versus either distance or Pb isotope ratios on the x-axis can yield additional insights related to the source of the Pb and other elements.

15.3.2 Comparing Results from 2002 to 2008

The ^{208}Pb/^{206}Pb versus ^{207}Pb/^{206}Pb isotope ratios measured in this study (Fig. 15.2A and B) define a field in which mixing from several Pb sources is suggested. The differences in the Pb isotopes ratios are significant (based on two-sigma relative error as noted in Fig. 15.2A) and several end member compositions and groupings are apparent. The end members (samples with the lowest and highest isotope ratios) may represent a fingerprint for specific air emission sources. This possibility will be discussed later. Note that there is a similar range

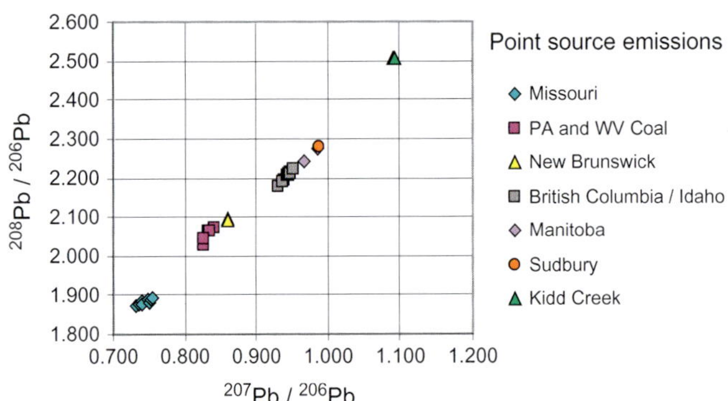

FIGURE 15.1   Variance in ^{208}Pb/^{206}Pb versus ^{207}Pb/^{206}Pb isotope ratios from several major ore and coal deposits from North America. The sources of data used to construct this plot were compiled in Graney et al. (1995).

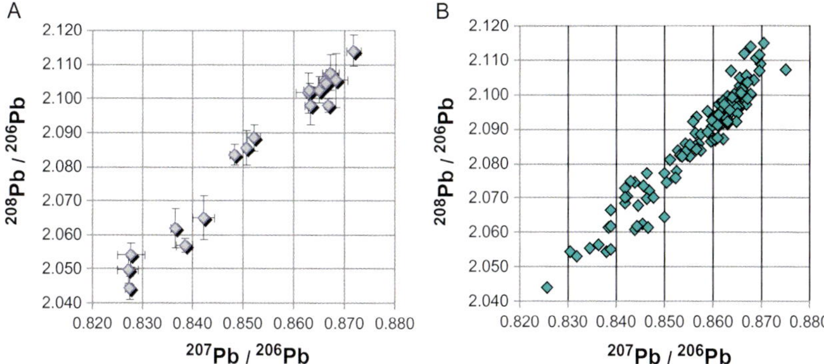

FIGURE 15.2   The ^{208}Pb/^{206}Pb versus ^{207}Pb/^{206}Pb isotope ratios measured in lichens collected in (A) 2002 and (B) 2008. Two sigma error bars are included within (A). Two sigma errors bars are not shown in (B) for sake of clarity but are comparable to those displayed in (A).

in isotope ratio fields in 2002 and 2008, suggesting sources of Pb incorporated into the lichens have not changed substantially between 2002 and 2008.

When examining the relation between the Pb isotope ratios versus distance from the oil sands processing facilities (Fig. 15.3A and B), the lowest ^{207}Pb/^{206}Pb (and ^{208}Pb/^{206}Pb ratios, not shown here) are found closest to the processing sites, and the highest ^{207}Pb/^{206}Pb and ^{208}Pb/^{206}Pb ratios are found in the samples 50 km or more from the processing sites. Based on source sample analysis (to be discussed in detail later), lower ^{207}Pb/^{206}Pb and ^{208}Pb/^{206}Pb ratios found near the processing sites reflect contributions from the Pb in the bitumen from the oil sands superimposed over the Pb sequestered in the matrix of the parent materials. The samples collected beyond 50 km from the processing site cluster into an isotope ratio grouping with a ^{207}Pb/^{206}Pb ratio near 0.8650 and a ^{208}Pb/^{206}Pb ratio near 2.095. This grouping could reflect a regional background Pb isotope ratio signature because the matrix in the oil sands deposits (the sand and clay components) is the main component in the soil at the distal sites. However global scale transport processes may also contribute to the Pb isotope signature at the distal sites, this possibility will be explored in Section 15.4. The lowering of the ^{207}Pb/^{206}Pb and ^{208}Pb/^{206}Pb ratios as one approaches the processing sites indicates other Pb sources (related to the oil sands processing) are influencing the measured Pb isotope ratios. The $r^2 > 0.70$ power law function fits for the isotope ratio versus distance relation suggests source proximity is the major determinant of isotope composition in the AOSR lichens.

Combining Pb concentration in lichens and distance from the oil processing facilities revealed a similar concentration distance relation in 2002 and 2008 (Fig. 15.3C and D). Of note is the limited difference in Pb concentrations between the proximal and distal sites. The r^2 value for the power law function fits for the Pb concentration versus distance is <0.30. This suggests

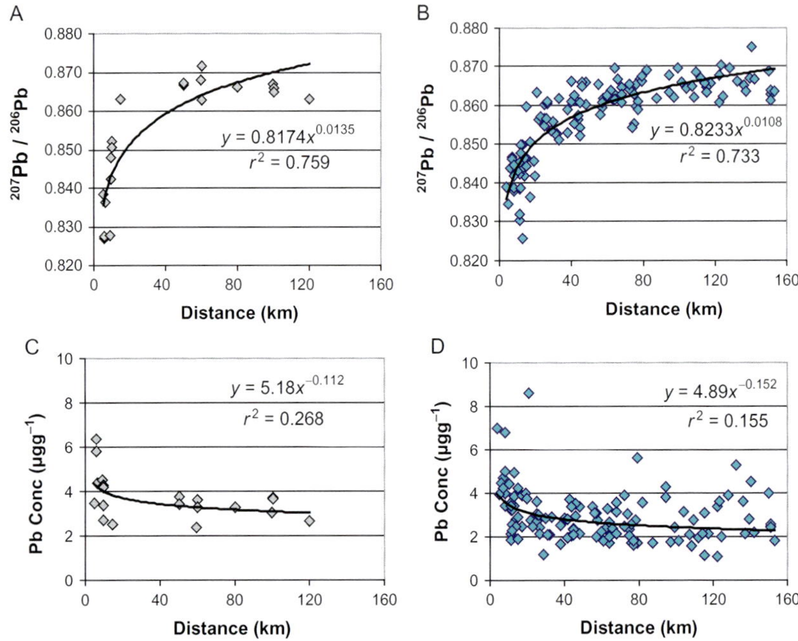

FIGURE 15.3 ^{207}Pb/^{206}Pb isotope ratios versus distance from the midpoint between the main oil sands processing facilities in (A) 2002 and (B) 2008; and Pb concentration as a function of distance in (C) 2002 and (D) 2008. Results from regression analysis using a power law relation displayed on (A)–(D).

that proximity to sources is not the only factor that influences Pb concentrations in lichens. There seems to be an exponential decrease in Pb concentrations between 0 and 50 km from the oil processing sites (proximal sites), versus greater variability in the concentrations at the more distal sites (>50 km). The spatial distribution could reflect the size and morphology of aerosols that transport Pb to the lichens. The deposition velocity, or the speed at which an aerosol will deposit to lichens, is inversely proportional to the mass median aerodynamic diameter (Landis and Keeler, 2002). Therefore, the observed exponential decreases in element concentrations near mining sources are consistent with the emission of large coarse mode particles during the oil sands mining process (abrasion followed by suspension). During the processing of the oil sands, the Pb source incorporated into the lichens may be different, and reflect Pb transformation during processing to a volatile phase, with subsequent incorporation into secondary fine mode aerosols. Secondary aerosols smaller size distribution would result in lower deposition velocities and a smaller near field deposition gradient than the larger primary particles. Another possibility is that the variability in the Pb concentrations at the distal sites may also be related to differences in the host vegetation at the sampling sites, which affected how Pb was accumulated by the lichens. To address these possibilities, the relations between other elements and Pb isotope ratios needs to be examined.

15.3.3   Coupling Pb Isotopes and Metal Concentrations in Lichens

To help explain the significance of the Pb concentration and isotopic data, we need to consider that concentrations of metals in lichens in the AOSR probably reflect source differences, as well as source proximity, and the biogeochemistry of lichens (Addison and Puckett, 1980). Comparing other elements to Pb from a source and process perspective would likely involve assessing differences between (i) elements from geogenic wind-blown dust sources versus (ii) those from oil sands processing versus (iii) those that may reflect the nutrient and metabolic needs of lichens which could be related to ecological variables as well as air quality factors. To assess these possibilities, we examine the spatial distribution of a likely representative element from the geogenic group (Aluminum, Al), the oil sands processing group (Vanadium, V), and an element from the nutrient group (Manganese, Mn) coupled with Pb isotope ratio findings.

An examination of the distance from the source versus concentration results suggests metals such as Al (and several others) are likely emitted as coarse mode particles near the processing site (Fig. 15.4). This is reflected in an exponential decrease in Al concentration versus distance in 2002 and 2008 (Fig. 15.4A and B) and the resulting $r^2 > 0.75$ for the power law best fits. Al concentration versus Pb isotope ratio plots in 2002 and 2008 (Fig. 15.4C and D) indicate a cluster of samples with high $^{207}Pb/^{206}Pb$ isotope ratios when Al concentrations were low, and a greater range in, and lower $^{207}Pb/^{206}Pb$ isotope ratios, when Al concentrations were enhanced. In 2002 and 2008, Al concentrations and Pb isotope ratios displayed similarities from a distance from source perspective. The larger 2008 data set allowed the Al concentrations versus Pb isotope ratios to better define groups of samples from a distance from source perspective, and confirmed major enhancement of Al concentrations within 20 km from mining sites suggested by the 2002 samples.

Combining V concentration, Pb isotope ratios, and distance using the 2008 data set results in groupings of samples (Fig. 15.5A and B) similar to the Al results. The exponential decay in V concentrations away from the processing sites is also expressed by a good fit using a power law regression.

Combining Mn concentration, Pb isotope ratios, and distance from the 2008 data set indicate a different type of behavior for this micronutrient than found for the Al and V (Fig. 15.5C and D). Mn concentrations in lichens seem to increase away from the mining and processing sites, and the power law fit for this relation is not as well defined. Unlike Al and V, and Pb; the lowest

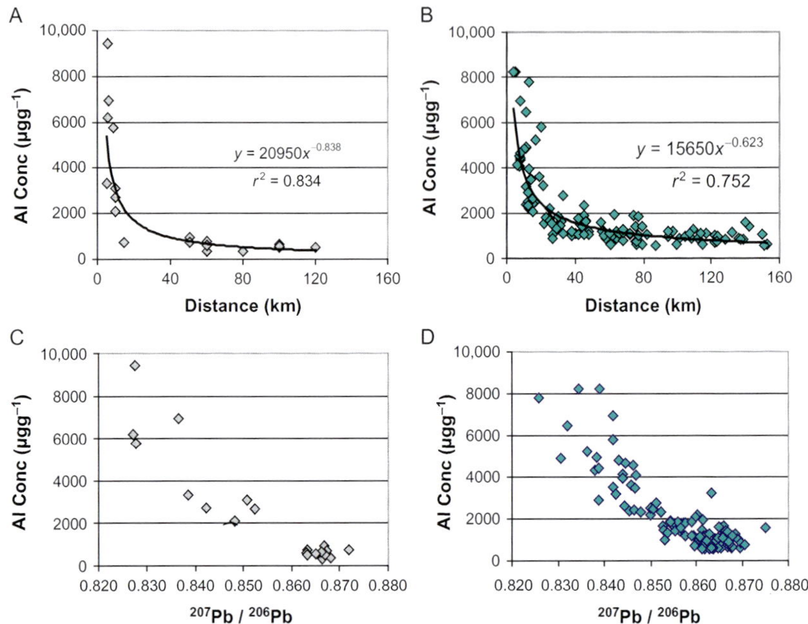

FIGURE 15.4   Aluminum (Al) concentrations as a function of distance and from oil sands processing facilities in (A) 2002 and (B) 2008, and Al concentration versus Pb isotope ratios in (C) 2002 and (D) 2008. Results from regression analysis using a power law relation displayed on (A) and (B).

concentrations of Mn are typically found at locations that are closest to the mining and processing sites. Comparing Mn concentration versus Pb isotopes yields a J shaped pattern, suggesting that the source of the Pb (and other elements), as reflected in $^{207}Pb/^{206}Pb$ ratios, has some relation to Mn accumulation (or lack thereof) by the lichens (Fig. 15.5C and D).

15.3.4   Contouring Lichen Concentrations to Aid in Spatial Assessments

Another way to express the relation between Pb isotopes and elemental concentration is through use of contoured results of spatial distributions. A graphical contouring program (SURFER v10, Golden Software, Boulder, CO) was used to visualize such spatial relations. In Fig. 15.6A, the distribution of the 2008 lichen sample sites is plotted on a UTM E and UTM N grid system. Note, the grid ends at the sample collection points to minimize extrapolation during the contouring process. The generation of the concentration contours involved kriging of the concentration data without smoothing the results. An example of the contouring results is presented for Al in Fig. 15.6B. The highest Al concentrations in the lichens are centered over the three (in 2008) main

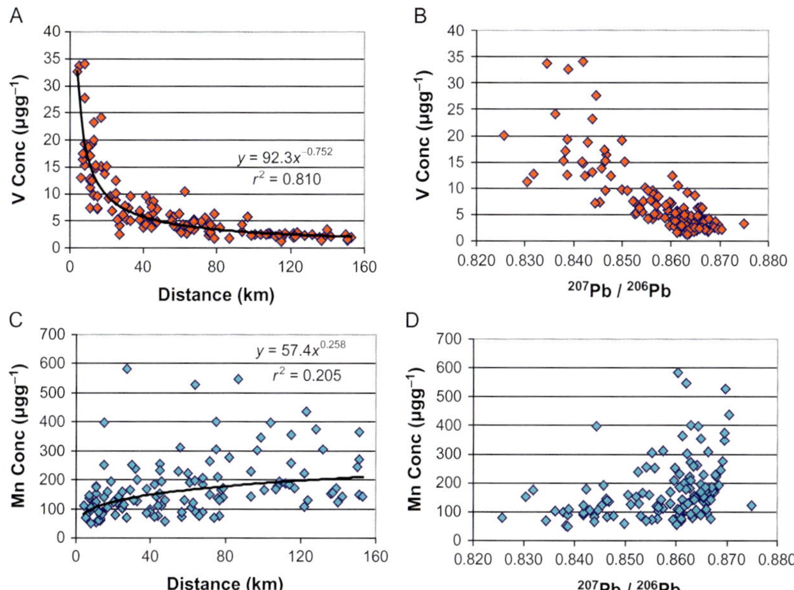

FIGURE 15.5   (A) Vanadium (V) concentrations as a function of distance from oil sands production facilities, (B) vanadium concentration versus $^{207}Pb/^{206}Pb$ isotope ratios (C) manganese concentrations as a function of distance from oil sands production facilities, (D) Manganese (Mn) concentration versus $^{207}Pb/^{206}Pb$ isotope ratios. All from the 2008 data set. Results from regression analysis using a power law relation displayed on (A) and (C).

open pit mining operations which are located within the 50 km on a side grid cell square bounded by UTM 450,000–500,000 East and 630,000–635,000 North. As will be discussed in the source apportionment chapter in this volume (Chapter 18), other elements including Fe, La and Ce have a spatial distribution similar to that of the Al, and reflect similarities in sources and subsequent deposition. Elements such as V (as well as Ni, Mo) reveal a pattern that is more homogeneous (uniform) than the geogenic elements (Fig. 15.6C). This type of spatial distribution for V and related elements would be expected from emission of aerosols with subsequent fine particle deposition to the lichens. This spatial distribution of measured concentrations of V related elements in lichens is similar to the modeled pattern for the deposition of S expected from aerosol emissions from the oil processing facilities (Chapter 12). The lowest $^{207}Pb/^{206}Pb$ and $^{208}Pb/^{206}Pb$ isotope ratios are centered over the mining and processing sites, in consort with the highest Al and V concentrations (Fig. 15.6C and D).

The topography generated by Surfer for the sampling domain (Fig. 15.6F) was based on the GPS measured elevation at all 369 of the lichen sample collection sites. The gently northward dipping topographic relief of the

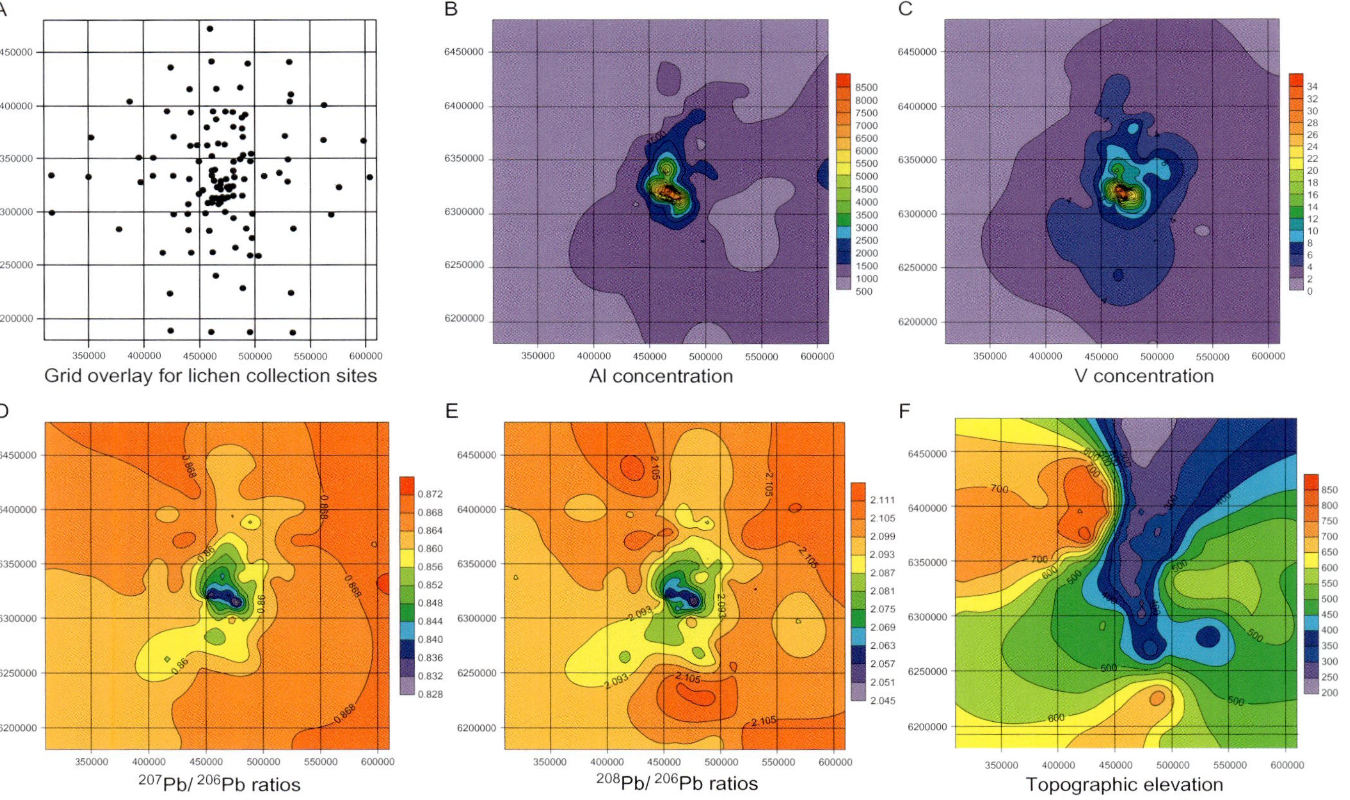

FIGURE 15.6 SURFER generated spatial distributions plots from the 121 lichen selection sites from the 2008 dataset. (A) Location of the 121 lichen selection sites on an UTM E and UTM N x–y metric grid. Contoured (B) Al concentrations, (C) V concentrations, (D) $^{207}Pb/^{206}Pb$ ratios, (E) $^{208}Pb/^{206}Pb$ ratios, and (F) topographic elevation (in meters). Contours grouped and color coded as noted on the bar scales to the right of the figures. All concentrations are in $\mu g\,g^{-1}$. Lower concentrations, elevations, and isotope ratios in purple hues, and higher concentrations, elevations, and isotope ratios in redder hues.

Athabasca River drainage basin is well expressed, as are the higher elevations on the plateau located to the northwest of Ft. McMurray. This regional difference in elevation exerts some control on the resulting element concentrations and Pb isotope ratios in the lichens. The elongation of the low $^{207}Pb/^{206}Pb$ and $^{208}Pb/^{206}Pb$ ratios in the north-south direction may directly reflect the topographic expression of the Athabasca River Valley (Fig. 15.6F). And the coupled Pb isotope and V concentration pattern in lichens may reflect near surface enhanced atmospheric deposition processes in the river valley. $^{207}Pb/^{206}Pb$ ratios greater than 0.8600 and $^{208}Pb/^{206}Pb$ ratios greater than 2.090 dominate outside of the topographic lows, except for an area just southwest of Ft. McMurray which may represent the impacts from traffic along the main roadway corridor.

The concentrations of elements in lichens that may reflect metabolic needs differ from those of the geogenic and oil processing groups (Fig. 15.7). Zn (Fig. 15.7A) is depleted in the lichens near the mining and processing sites. The depletion trend seems to follow the surface topography to some extent, following preferential airflow along the topographic expression of the river valleys (to the north). The baseline value for Zn concentrations is 25 ppm, this may reflect minimum nutrient/metabolic needs of the lichens. Mn depletion in the lichens is greatest near the mining and processing sites, and this inverse concentration trend is more pronounced than the Zn well beyond the visible footprint of the oils sands operations (Fig. 15.7B). The Mn depletion trend also seems to follow the surface topography, to some extent mirroring the Zn. The baseline value for Mn concentrations seems to be 40 ppm, possibly reflecting minimum nutrient/metabolic needs for the lichens. The highest Mn concentrations occur at the highest topographic elevations (on the plateaus to the northwest of Ft. McMurray). The distribution of the Pb concentrations in the lichens is unlike that of Zn and Mn (Fig. 15.7C). There seems to be high concentrations of Pb near the mining operations as well as the city of Ft. McMurray. In contrast, locations with higher concentrations of K away from the mining and processing sites (Fig. 15.7D) appear to correspond to locations with lower Pb concentrations.

Some of differences in the spatial variation of elements such as Zn, Mn, Pb, and K may be related to differences in the vegetation that host the lichens, which in turn reflect variability in soil moisture and nutrient regimes at the collection sites. Hydrology, canopy, open water, and edge fetch, as well as vegetation type are important factors in aerosol capture and deposition dynamics. This study was not designed to directly assess the synergy between these parameters. However, as a first approximation for grouping synergistic effects, we used the ecosite classifications from Berryman et al. (2010) to group the lichens from the 121 sample locations into wet and dry site designations based on soil moisture regime. The dry sites ($n = 61$) were jack pine and mixed hardwood dominant, and the wet sites ($n = 60$) included bogs and fens where Black Spruce and Tamarack were common.

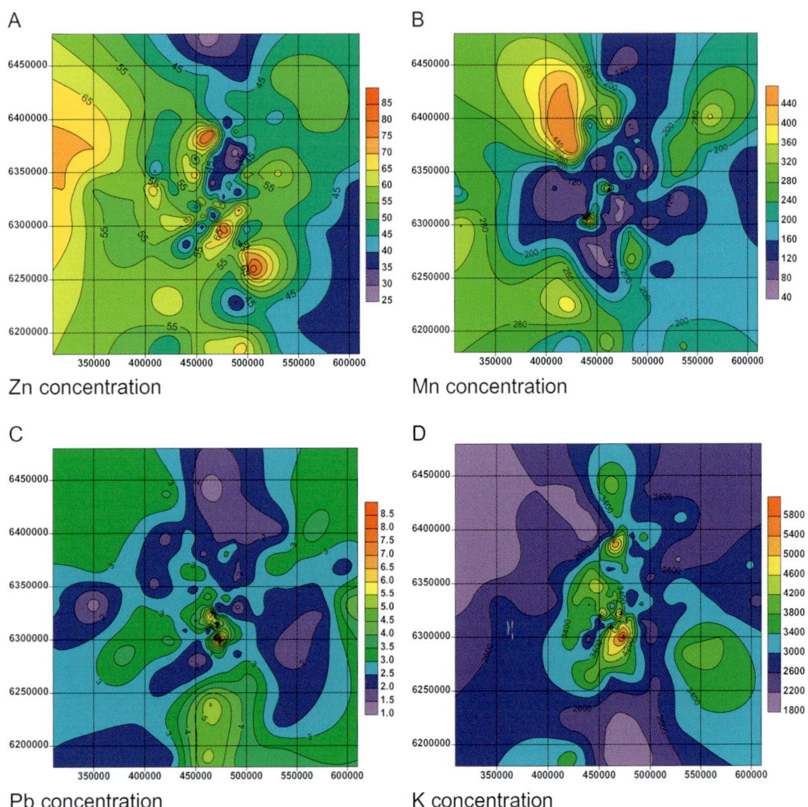

FIGURE 15.7 Spatial distributions of metal concentrations in lichens collected in 2008 at 121 sampling sites. (A) Zn concentrations, (B) Mn concentrations, (C) Pb concentrations, and (D) K concentrations. See Figure 15.6 for locations of the sampling sites. All concentrations are in $\mu g\ g^{-1}$. Contours grouped and color coded using notation on the bar scales to the right of the figures. Lower concentrations in purple hues, and higher concentrations in redder hues.

To assess whether this soil moisture regime parameter could account for variability observed in Pb concentrations away from the mining and processing facilities, Pb concentrations and isotope ratios versus distance were plotted using the wet versus dry site designation (Fig. 15.8A and B). In addition, results from paired two sample t-tests were computed to aid in this graphical assessment (Table 15.1). The paired two sample t-tests were chosen so a distance variable could be included in the analysis. Only the distal samples, samples located greater than 50 km from the processing sites, were included in the statistical analysis to avoid complexities from near source contributions. The concentrations and Pb isotope ratios from samples at similar distances from the processing sites, but differing soil moisture regimes, could thereby be compared to one another. The $P(T \leq t)$ for the matched sample distance

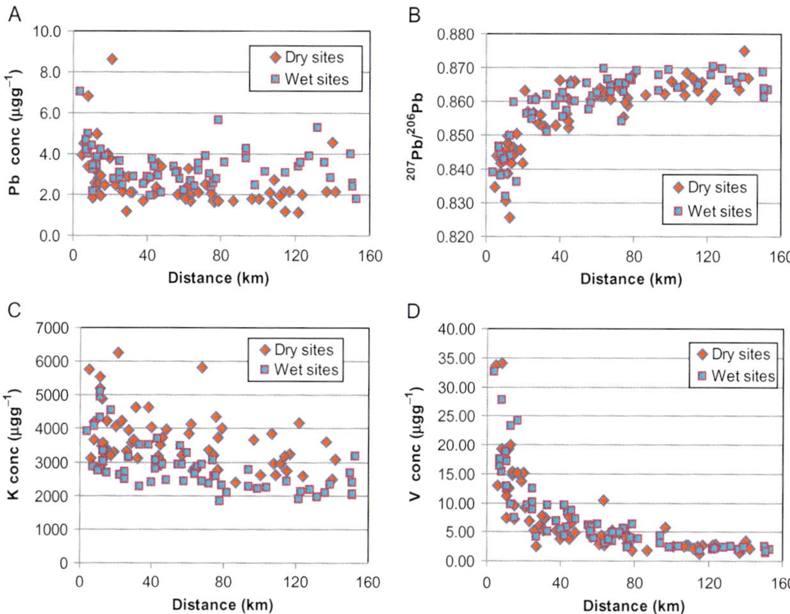

FIGURE 15.8   (A) Pb concentration versus distance, (B) ^{207}Pb/^{206}Pb isotope ratios versus distance, (C) K concentration versus distance, and (D) V concentration versus distance. Distance refers to the midpoint between the two oil sands processing facilities. Ecosites grouped into wet versus dry sites based on hydrologic characteristics of the soils at the lichen sampling locations.

TABLE 15.1 Results from *t*-Tests of Paired Two Sample for Means (from Samples >50 km from Processing Sites)

Mean	Dry sites	Wet sites	$P(T \leq t)$ one-tail
Distance (km)	93.82	90.14	0.005
^{207}Pb/^{206}Pb	0.8636	0.8653	0.022
Aluminum	1068	1004	0.188
Vanadium	3.37	3.72	0.149
Molybdenum	0.130	0.130	0.495
Lead	2.06	3.34	1.48E^{-06}
Zinc	47.38	56.59	0.004
Potassium	3316	2493	5.96E^{-07}
Sulfur	752	520	6.26E^{-07}

Distance is in km, element concentrations in microgram/gram.
Dry and wet sites refer to Ecosite grouping (see text for explanation).

variable was 0.005 (Table 15.1), and comparison to this value will be used as the significance determinant for this study. Dry versus wet site was found to have a significant impact on Pb concentrations in lichens away from the mining and processing sources ($P \ll 0.005$), the lichens from the wet sites contained higher Pb concentrations than the dry sites. However, the variance in Pb isotopes does not seem to be soil moisture regime dependent ($P > 0.005$). To further test this wet-dry dependence hypothesis, a spatial distribution plot for K (Fig. 15.8C) and V (Fig. 15.8D) is provided. Outside of the mining areas, highest K and S concentrations correspond to dry sites ($P \ll 0.005$ in Table 15.1), especially where jack pine is prevalent. The behavior of V, Mo, and Al was different from that of Pb and K, concentrations for these elements do not seem to be soil moisture regime dependent ($P > 0.005$ in Table 15.1).

The relation of element concentrations in lichens to soil moisture regime may reflect processes directly associated with the host vegetation including stem flow and canopy density or architecture interactions, in addition to humidity gradients and source proximity. The specific reasons for our wet–dry findings will require further study, of more importance for this work was that ecosite type does not seem to influence the Pb isotope ratios (Fig. 15.8B). It appears that the Pb isotope ratios are a more robust indicator of source impacts than Pb concentrations. This finding allows us to explore the possibilities of using source samples in conjunction with lichen samples to apportion sources and their impacts on spatial scales.

15.3.5 Source Samples

Source samples from the oil sands mining and processing operations, including raw oil sand from mining operations and exposures along the Athabasca River, mine haul road materials, coke, fly ash, stack samples, tailings sand, and vehicle emissions were collected from the AOSR (Chapter 18). ARA Inc. provided the extracts from the digested source samples that were diluted at EPA NERL and then measured for Pb isotope ratio determinations using ICP-SFMS. The results from the samples were grouped into several categories for ease of interpretation (Fig. 15.9A). The processed oil sands materials cluster into a group of points with the lowest ^{207}Pb/^{206}Pb and ^{208}Pb/^{206}Pb ratios. The Pb isotope ratios from the stack samples overlap those from the processed oil sands materials. The tailings sand clusters into a group of points with the highest ^{207}Pb/^{206}Pb and ^{208}Pb/^{206}Pb ratios. The raw oil sands have mid-range ^{207}Pb/^{206}Pb and ^{208}Pb/^{206}Pb ratios.

Based on the results displayed in Fig. 15.9A, the isotopic composition of the source materials in the AOSR region includes an end member (source) that is related to the oil processing operations, as well as a tailings sand end member. The raw oil sands would include a mixture of the Pb within the bitumen and

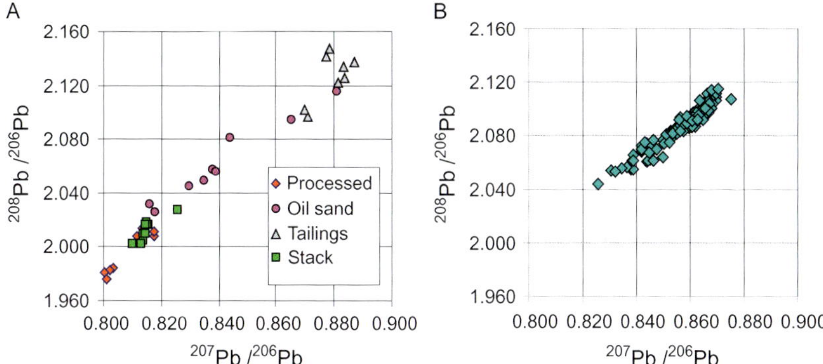

FIGURE 15.9   The ^{208}Pb/^{206}Pb versus ^{207}Pb/^{206}Pb isotope ratios measured in (A) source and stack samples and (B) 2008 lichen samples. The similarities in types of source samples have been grouped by color codes.

additional Pb in the earth materials that compose the matrix of the oil sand reservoir. In simplest terms from a regional context, the background (distal samples) should include Pb from the sand and clay matrix of the oil sand reservoir over which the Pb isotope signature of the bitumen has been superimposed. When the isotope ratio results from the source samples are compared to those from the lichens (Fig. 15.9B), the lichens proximal to the mining and processing sites were found to overlap the raw oil sands isotope ratio field, whereas the more distal lichens overlap the tailings sands fields. None of the lichen samples have an isotopic signature that is dominated by the processed oil sand signature. The end member contributions to the lichen samples can be calculated using mixing equations to approximate the source of the Pb to individual lichen samples. For example, if it is assumed that only (i) oil processing and (ii) tailings sand end members contributed Pb to the lichens, at most 50% and as little as 1% (or less) of the Pb in the lichens could come from the oil processing end member (from the Pb in the bitumen). However, the two end member solution likely overestimates the oil processing end member contribution because a third end member, fugitive dust from mining oil sands, is likely also a significant source of Pb. Other forms of source apportionment are needed to better constrain contributions, particularly the importance of fugitive dust emissions from the mining operations. Source apportionment will be the focus in a subsequent chapter (Chapter 18).

15.4   DISCUSSION

15.4.1   Other Studies That Document Emissions of Metals from Point Sources of Pollution in Canada

The emissions of trace metals from point sources in Canada have typically involved characterizing impacts from smelters (Aznar et al., 2008; Gallon et al., 2006; Simonetti et al., 2004; Telmer et al., 2004) rather than oil sands processing facilities. Telmer et al. (2004) indicated exponential decreases in metal concentrations with distance from the Horne smelter (Rouyn, Quebec) based on results from snow pack sampling. They suggested metals in the snowpack represented emissions from the smelter (Cu, Pb and others) as well as contributions from background sources and smelter emissions (Sr, Al and others) versus metals from only background sources (K, Rb, Mn, and Cs). Their sampling extended to distances as far as 50 km from the smelter and they suggested that wet and dry deposition occurred within 15 km from the smelter, but wet deposition predominated beyond 15 km from the stack. Gallon et al. (2006) examined variation in Pb isotope ratios in lake sediments to suggest evidence of emissions from the Horne smelter could be found in lake sediment 10, 25, and 150 km from the smelter, but not in lake sediment collected 300 km away from the smelter. Simonetti et al. (2004) used metal concentrations and Pb isotope ratios in conjunction with sampling within plumes from aircraft and on the ground sampling to identify impacts from the Horne smelter. They suggested that the Pb isotope ratios in ground level pollution reflected contributions from the local (smelter) as well as global (Eurasian) sources, but that apportioning contributions from the sources was difficult because the Pb isotope and metal contribution signal from the smelter was not homogeneous.

All of these studies suggested exponential decreases in metal concentrations with distance from point sources of emissions similar to the spatial gradients found in this study by using atmospheric deposition and lake sediment records, rather than the biomonitors used in this study. In contrast, Aznar et al. (2008) is an example of a study that merged deposition patterns from smelter emissions through use of Pb isotope ratios in lichens. Their results indicated contamination declined exponentially with distance from a copper smelter in Quebec but was only detectable at distances 30 km or less from the smelter. Further, examples of results from coupling Pb isotopes with lichens as biomonitors in other types of atmospheric deposition studies are summarized in Section 15.4.3.

15.4.2 Controls on Metal Concentrations and Accumulation by Lichens in the AOSR

As documented in Chapter 14, the metal concentrations in *H. physodes* from the AOSR are lower than those in lichens from urban areas and locations near smelters (Bergamaschi et al;, 2007; Doucet and Carignan, 2001; Spiro et al., 2004). And the concentrations for many metals in *H. physodes* from distal collection sites in the AOSR are comparable to those lichens from remote locations such as the Himalayas (Bergamaschi et al., 2002). This suggests that from a metal accumulation standpoint, lichens at distal locations in the AOSR (greater than 50 km from processing and mining sites) do not have elevated concentrations, and concentrations in lichens within 50 km from the major sources in the AOSR are not as elevated as at other industrial and urban sites.

However, several caveats when comparing these *H. physodes* concentration results to those from other lichen studies should be noted. Although there have been numerous studies in which metal concentrations have been measured in lichens (e.g., see references from Conti and Cecchetti, 2001), how metals are accumulated by epiphytic lichens is a source of ongoing debate. Physiological differences in lichen species may make comparisons between studies difficult, as studies that have compared species have noted (Bergamaschi et al., 2007; Cercasov et al., 2002). To avoid need for interspecies comparisons was a major reason why only *H. physodes* was chosen for this study. *H. physodes* is also known to be tolerant of acidic environments that include high SO_2 concentrations as well as low pH conditions (Hauck, 2011) likely present near oil processing facilities.

Some authors suggest metal accumulation in lichens may be element specific reflecting metabolic needs. Chiarenzelli et al. (2001) found that major and micronutrient needs as reflected in Zn, K, and Ca concentrations in lichen tissue had relatively consistent values regardless of lichen species. And they suggested that particulate deposition rather than substrate was the likely source for much of the variation in other metals incorporated into the lichen tissue. Others suggest canopy foliage and stem flow contributions would also affect metal deposition rates to lichen surfaces. The metals within these particles on the thalli may be solubilized by acid deposition or organic acids generated by the lichens (Nash, 1989). Solubilized metals can bind to ion exchange sites on cell walls or pass through cell membranes (Nash and Gries, 1995). However, lichens also have methods to concentrate metals on exterior surfaces to avoid passage through the cell membrane. Williamson et al. (2004) observed that lichens sequester Ca in the form of the Ca-oxalate weddellite on *H. physodes* surfaces. This metal fixation on external surfaces could be an avoidance mechanism to limit possible toxic effects. In other cases, neutralization of acidic compounds may occur in areas with high aerosol concentrations of carbonate minerals producing products such as gypsum on lichen surfaces (Garty and Garty-Spitz, 2011).

Some authors suggest that the trapping of relatively large particles on the lichen thalli is a main cause of elevated concentrations of metals in lichens (Garty et al., 1979). Irregularities in surfaces of lichens as well as in the shapes of particles may affect attachment to thalli. Williamson et al. (2004) used SEM-EDX to examine particulate matter on the surfaces of transplanted lichen thalli near a copper smelter in Karabash Russia. They suggested particles <2.5 μm in diameter were less efficiently captured by *H. physodes* than larger particles. This may suggest that coarser particles from abrasion from the mining operations in AOSR would be more effectively captured by *H. physodes* in comparison to finer particles generated from the oil processing facilities. Some authors suggest that lichens do not capture and accumulate metals in a manner resulting in increasing concentrations within the lichens. Rather, metal accumulation and removal may reflect equilibrium, steady-state processes (Spiro et al., 2004; Williamson et al., 2004) in which metal incorporation and metal loss both take place. The Spiro et al. (2004) study was conducted near smelting operations in Russia, and documented through the use of the Pb isotope ratios, that Pb within transported lichen tissue quickly undergoes exchange processes. Their results indicate that lichens produce a transient record resulting from the accumulation and loss of Pb in lichen tissues. Such steady-state processes suggest that lichens may not act as a passive bioaccumulator of pollutants, but continually recycle some of the metals, perhaps as a detoxification mechanism. This recycling process may explain why the Pb concentrations in the lichens from the AOSR are not elevated as might be expected and why Pb isotope ratios rather than concentrations are a more robust method for source attributions. So although metal concentrations in lichens will reflect proximity to metal sources in the form of spatial gradients, they may not quantitatively reflect metal emission and deposition rates. Pb isotopic ratios would therefore be a more robust way to monitor transport and deposition processes because Pb isotopic ratios would reflect source profiles regardless of concentration gradients affected by metabolic processes.

To our knowledge, the increase in Mn, and possibly Zn, concentrations with distance away from mining and processing operations in the AOSR has not been noted in other studies in urban or industrial areas. Some authors suggest that the Mn (and perhaps Zn) is incorporated into the lichens by leaching from the canopy, after translocation to the canopy through root system induced processes, rather than from atmospheric deposition (Ceburnis and Steinnes, 2000). This suggests that Mn and Zn accumulation in lichens may be influenced by canopy interactions reflecting ecological variables. Another possibility for near source depletion in Mn concentrations suggested by Hauck (2008) is that *H. physodes* is Mn sensitive and may use the production of depsidone physodalic acid to reduce Mn uptake while not inhibiting the uptake of Fe and Zn.

In summary, the physiological processes on and within the lichens that control metal accumulation are complex (Hauck and Huneck, 2007; Pawlik-Skowronska et al., 2002). And the methods by which particles and gases are dry- and wet-deposited onto lichens reflect a complex interplay between site factors such as microclimate, nutrient supply, air pollution and forest fires in northern coniferous forests Hauck (2011). These site factors might all affect metal accumulation rates in lichens in the AOSR. In the case of AOSR, the concentrations of elements in *H. physodes* are within the range found in numerous studies and may be on the lower end of the range typical for urban areas (Chapter 14). Further study is needed to determine the reasons for the spatial gradients in Mn and other elemental concentrations in lichens in the AOSR including (i) characterizing the size and chemical composition of aerosols, specifically the importance of fugitive dust on controlling deposition fluxes, (ii) the importance of S species and their interaction with fugitive dust emissions on controlling pH on lichen surfaces, and (iii) the influence of relative humidity gradients between wet and dry ecosites. This moisture gradient may affect dry versus wet deposition and transformation processes, perhaps on an element by element basis.

15.4.3   Other Work That Examined Pb Isotopes in Lichens

Studies in which metal concentrations and Pb isotopes in lichens have been measured for source attribution are listed in Table 15.2. The studies in Table 15.2 included both determinations of anthropogenic versus natural impacts, as well as local versus regional source contributions. A summary of findings from studies in Table 15.2 that are most pertinent to AOSR source attribution issues follow.

The work of Carignan and Gariepy (1995) was probably the first study that tried to use Pb isotope ratios from lichens to constrain impacts from a major point source of emissions, in this case, the Noranda Smelter in Quebec. The Noranda smelter often used massive sulfide ore from the Kidd Creek deposit as smelter feed. As noted from Fig. 15.1, Kidd Creek ore has very high $^{207}Pb/^{206}Pb$ and $^{208}Pb/^{206}Pb$ ratios, and the lichens collected closest to the Noranda smelter had the highest $^{207}Pb/^{206}Pb$ and $^{208}Pb/^{206}Pb$ ratios (Fig. 15.10). A spatial gradient was also apparent in the other samples, with lichens from south Quebec having lower ratios than those from the mid Quebec and Hudson Bay regions. Although the Pb isotopes from lichen samples collected near the smelter did not overlap directly with those of ore deposits from Kidd Creek, the lichens near the smelter contained evidence of smelter air emission impacts from a Pb isotope ratio perspective. This study found that, at least on a regional scale, point source emission impacts can be found in the Pb isotope ratios from lichens. It should be noted that the source signature from this smelter is very different in Pb isotope space from the processed oil sands signature from the AOSR. The Kidd Creek ores have very high $^{207}Pb/^{206}Pb$ and $^{208}Pb/^{206}Pb$ ratios in comparison to source materials from the AOSR.

In a follow-up 2002 study by Carignan et al., Pb isotope ratios from lichens in eastern North America indicated several source signatures (Fig. 15.11A). Samples from northern Quebec contained high $^{207}Pb/^{206}Pb$ and $^{208}Pb/^{206}Pb$ once again suggesting some contributions from the Kidd Creek ore deposits were being released to the atmosphere in the stack emissions from the smelter near Noranda. In contrast, samples from the northeast United States had the lowest $^{207}Pb/^{206}Pb$ and $^{208}Pb/^{206}Pb$ ratios, which could indicate contributions from coal- and oil-fired power plants as well as other anthropogenic sources that contained Pb from ore deposits from Missouri (see Fig. 15.1). Samples from eastern Canada had higher $^{207}Pb/^{206}Pb$ and $^{208}Pb/^{206}Pb$ ratios than those from the Northeast United States suggesting differences in anthropogenic emissions from Canadian sources. Samples from the St. Lawrence River valley are likely mixtures from Canadian and United States

TABLE 15.2 Synopsis of Previous Studies That Used Pb Isotopes in Lichens to Aid in Source Attribution

Authors	Date	Location	Study goals
Carignan and Gariepy	1995	Quebec, Canada	Determining impacts near Noranda smelter
Carignan et al.	2002	Eastern Canada	Follow up from 1995 smelter study, potential for determining Canadian versus United States source contributions
Cloquet et al.	2006a	France	Distinguishing between urban anthropogenic sources
Cloquet et al.	2006b	France	Used Zn and Pb isotopes, follow up from 2006a study
Cloquet et al.	2009	France	Follow up from 2006 studies, included lichen transplants
Dolgopolova et al.	2006	Russia	Used Zn and Pb isotopes to assess impacts near mining operations
Monna et al.	1999	Sicily	Impact of natural emissions from volcanoes
Monna et al.	2006	South Africa	Distinguishing between urban anthropogenic sources
Simonetti et al.	2003	Western Canada	Latitudinal transects, potential for determining Arctic influences, and Canadian versus United States source contributions
Spiro et al.	2004	Russia	Determine smelter impacts on transplanted lichens

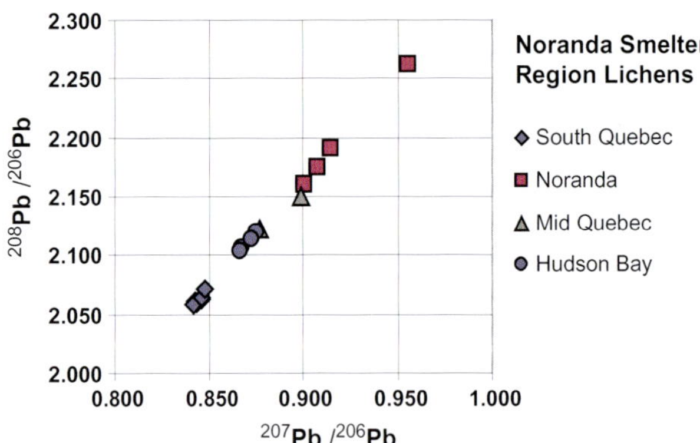

FIGURE 15.10 Pb isotope ratios from lichens collected in the Province of Quebec near the Noranda smelter as reported in Carignan and Gariepy (1995).

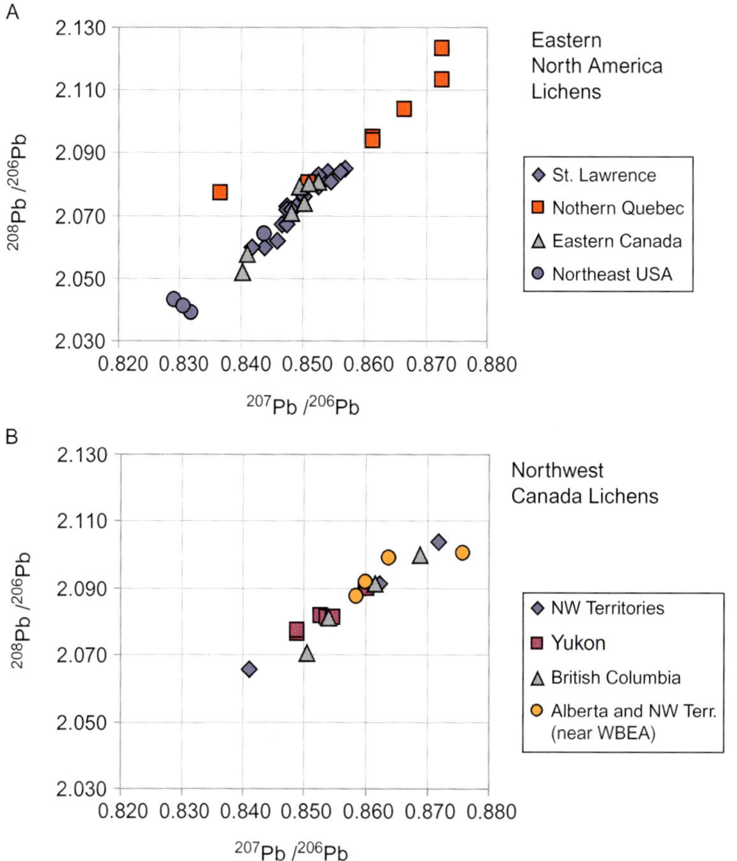

FIGURE 15.11   Pb isotope ratios from lichens collected in (A) eastern North America as reported in Carignan et al. (2002) and (B) northwestern Canada as reported in Simonetti et al. (2003).

sources. The Pb isotope ratios from the AOSR lichens in this study cover the entire range found in the Carignan et al. (2002) study (Fig. 15.11A), although the Pb sources in the AOSR are very different from those in eastern North America. Reasons for these similarities and differences will be explored below.

In a study by Simonetti et al. (2003), Pb isotope ratios from lichens in western North America were collected. The results suggested north–south spatial gradients in Pb isotope ratios were present within the data set. Only the samples from the northern portion of their sampling domain are presented here (Fig. 15.11B), these would be the samples closest to the AOSR and might provide values for background conditions in the AOSR away from the oil sands processing sources. In Fig. 15.11B, the samples from the Northwest (NW) territories and British Columbia exhibited the widest range in Pb isotope ratios, whereas the samples from the Yukon were more restricted in Pb isotope space. Several samples from northern Alberta were included in this study, and the results overlapped with those from the other sample sites. Although the Pb isotope ratios from the Simonetti et al. (2003) study directly overlap with those from the AOSR from this investigation, no cluster of data points that might be used to constrain a regional background value for the AOSR area are apparent. Results from aerosol studies may be needed to provide better insights for defining the regional background in the AOSR.

15.4.4 Comparing Pb Isotope Ratios in Aerosols to those in Lichens from the AOSR

The results from the global scale Bollhoefer and Rosman aerosol studies (2001, 2002) indicate a wide range of values on Pb isotope ratio plots (Fig. 15.12). Samples of aerosols from Canadian cities tend to have higher $^{207}Pb/^{206}Pb$ and $^{208}Pb/^{206}Pb$ ratios than those from cities in the United States. The difference in Pb isotope ratios in aerosols collected in Canada and the United States is believed to be due to isotopic differences in Pb ore sources used in industrial processes. If one mixes Pb from predominantly Canadian sources (Kidd Creek ore, Sudbury ore, British Columbia ore, and New Brunswick ore in Fig. 15.1) and compares that Pb isotope signature to predominantly United States sources (Idaho ore, Pennsylvania and West Virginia coal, and Missouri ore in Fig. 15.1) one would expect that Canadian cities would have higher $^{207}Pb/^{206}Pb$ and $^{208}Pb/^{206}Pb$ ratios in aerosols than those from the United States. The difference between the Canadian and United

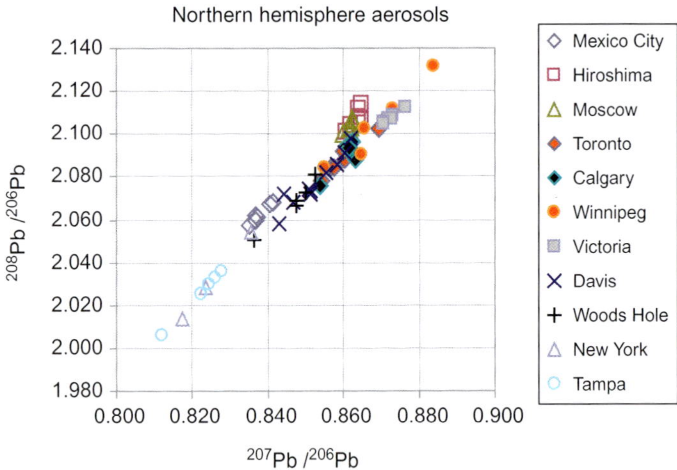

FIGURE 15.12 Pb isotope ratios from aerosols collected from several cities in Russia, Japan, Mexico, Canada, and the United States as reported in Bollhoefer and Rosman (2002).

States aerosols in Pb isotope space was noted in several studies by Sturges and Barrie (1987, 1989) and has been used by other investigators to help apportion emissions between United States and Canadian sources.

The samples from Mexico City, cities in Japan, and Moscow from the Bollhoefer and Rosman (2002) study tend to have higher $^{208}Pb/^{206}Pb$ ratios than sites from Canada and the United States (Fig. 15.12). This finding indicates a more thorogenic source of Pb is present in those samples. This thorogenic Pb signature is now commonly reported in studies of aerosols from China and indicates global scales differences in U–Th reservoirs. However, if one compares the results from the Bollhoefer and Rosman (2002) versus those of Simonetti et al. (2003), there is little evidence for global scale transport and deposition of thorogenic Pb to northwestern Canada. That is, the background Pb isotopic signature in the AOSR seems to be dominated by regional rather than global sources.

Also of note from the Bollhoefer and Rosman (2002) study are the low $^{207}Pb/^{206}Pb$ and $^{208}Pb/^{206}Pb$ ratios from Tampa and New York. Oil fired power plants are a major source of electricity in those regions of the United States. These low isotopic ratios are similar to the processed oil sands values from the AOSR and suggest that combustion of fuel oil may have an isotopic signature that is distinct from coal deposits and other Pb emission sources as previously suggested by Hurst (2002). With the continuing decrease in Pb concentrations in gasoline across the globe, further study is needed to determine if coal and oil sources have discernible differences in Pb isotopes that might be used in future source apportionment studies to assess impacts from fossil fuel use.

15.5   CONCLUSIONS

The concentrations of elements accumulated by lichens in the AOSR reflected proximity to oil sands mining and processing operations as well as ecosite variability at the lichen collection sites. Spatial analysis indicates three main element groupings including a geogenic source (Al and others) related to oil sands mining, an oil processing source (V and others), and biogeochemical processes (Mn and others) affected element accumulation in lichens.

- An exponential decrease in concentration of the geogenic grouping of metals versus distance from the mining sites was found, whereas near source concentrations of metals characteristic of oil processing have a more homogeneous spatial distribution than the geogenic materials. The mining and oil processing related elements are superimposed over the elemental signature that reflects the nutrient needs of the lichens.
- There is some enhancement in Pb concentrations within lichens at the proximal sites ($<$50 km from mining and processing operations), but Pb concentration enhancement is limited compared to other elements from the mining and processing groups. Elevated concentrations of Pb at the

distal sites reflects ecosite differences, wet sites have higher Pb concentrations than dry sites. The other elements from the mining and processing groups have concentrations that are less influenced by ecosite.

There are major differences in Pb isotope ratios at proximal versus distal locations. The ecosite type does not affect the Pb isotope ratio distribution at either the proximal or distal sites.

- The samples collected beyond 50 km from the mining and processing sites cluster into a Pb isotope ratio grouping with a $^{207}Pb/^{206}Pb$ ratio of 0.8650 and a $^{208}Pb/^{206}Pb$ ratio near 2.095. This grouping likely reflects the regional background Pb isotope ratio signature. The lowering of the $^{207}Pb/^{206}Pb$ and $^{208}Pb/^{206}Pb$ ratios near the mining and processing operations indicates other Pb sources, likely related to oil sands mining and processing, are contributing to the Pb source signature.
- The source attribution based on Pb isotope ratios in lichens was confirmed by source and stack sampling. The lowest $^{207}Pb/^{206}Pb$ and $^{208}Pb/^{206}Pb$ ratios were from the processed materials and stack samples, which reflect contributions from the Pb in the bitumen component of the raw oil sands. The highest $^{207}Pb/^{206}Pb$ and $^{208}Pb/^{206}Pb$ ratios were from the tailings materials, which includes the Pb in the sand and clay matrix from the raw oil sands. The $^{207}Pb/^{206}Pb$ and $^{208}Pb/^{206}Pb$ ratios from the lichens collected at the distal sites are similar to those from tailings materials. This result suggests Pb from the sand and clay fraction of the soils at distal locations is the major contributor to the regional background Pb isotope signature.

The Pb isotope ratios are a more robust indicator of source contributions than the Pb concentrations. The Pb isotope ratios are a better predictor of the extent of the source impacts on the lichens than Pb concentrations because the Pb isotope ratios are not affected by either metabolic processing or ecosite related controls on atmospheric deposition.

ACKNOWLEDGMENTS

We wish to thank the Members of the Wood Buffalo Environmental Association (WBEA) of the Athabasca Oil Sands Region (Canada) for generously supporting this scientific study. We are also grateful for their desire to bring our work into the open literature. We specifically appreciate the help and support of Drs. Kevin Percy and Allan Legge, and the many staff members of WBEA for their continued and dedicated help. We also thank Shanti Berryman and Justin Straker for their efforts in lichen collection and cleaning.

REFERENCES

Addison, P.A., Puckett, K.J., 1980. Deposition of atmospheric pollutants as measured by lichen content in the Athabasca oil sands area. Canadian Journal of Botany. 58, 2323–2334.

Ault, W.A., Senechal, R.G., Erlebach, W.E., 1970. Isotopic composition as a natural tracer of lead in the environment. Environ. Sci. Technol. 4, 305–311.

Ayrault, S., Clochiatti, R., Carrot, F., Daudin, L., Bennett, J.P., 2007. Factors to consider for trace element deposition biomonitoring surveys with lichen transplants. Sci. Total Environ. 372, 717–727.

Aznar, J.C., Richer-Lafleche, M., Cluis, D., 2008. Metal contamination in the lichen Alectoria sarmentosa near the copper smelter of Murdochville, Quebec. Environ. Pollut. 156, 76–81.

Beckingham, J.D., Archibald, J.H., 1996. Field Guide to Ecosites of Northern Alberta. Northern Forestry Centre, Forestry Canada. Northwest Region, Edmonton, AB, Canada.

Bergamaschi, L., Rizzio, E., Valcuvia, M.G., Verza, G., Profumo, A., Gallorini, M., 2002. Determination of trace elements and evaluation of their enrichment factors in Himalayan lichens. Environ. Pollut. 120, 137–144.

Bergamaschi, L., Rizzio, E., Giaveri, G., Loppi, S., Gallorini, M., 2007. Comparison between the accumulation capacity of four lichen species transplanted to a urban site. Environ. Pollut. 148, 468–476.

Berryman, S., Geiser, L., Brenner, G., 2004. Depositional gradients of atmospheric pollutants in the Athabasca Oil Sands region, Alberta, Canada: an analysis of lichen tissue and lichen communities. Lichen Indicator Pilot Program 2002-2003. Final Report Submitted to the Terrestrial Environmental Effects Monitoring (TEEM) Science Sub-committee of the Wood Buffalo Environmental Association (WBEA). Fort McMurray, AB, Canada.

Berryman, S., Straker, J., Krupa, S., Davies, M., Ver Hoef, J., Brenner, G., 2010. Mapping the characteristics of air pollutant deposition patterns in the Athabasca Oil Sands Region using epiphytic lichens as bioindicators. Interim Report Submitted to the Terrestrial Environmental Effects Monitoring (TEEM) Science Sub-committee of the Wood Buffalo Environmental Association (WBEA). Fort McMurray, AB, Canada.

Bollhoefer, A., Rosman, K.J.R., 2001. Isotopic source signatures for atmospheric lead: the Northern Hemisphere. Geochim. Cosmochim. Acta 65 (11), 1727–1740.

Bollhoefer, A., Rosman, K.J.R., 2002. The temporal stability in lead isotopic signatures at selected sites in the Southern and Northern Hemispheres. Geochim. Cosmochim. Acta 66 (8), 1375–1386.

Carignan, J., Gariepy, C., 1995. Isotopic composition of epiphytic lichens as a tracer of the sources of atmospheric lead emissions in southern Quebec, Canada. Geochim. Cosmochim. Acta 59, 4427–4433.

Carignan, J., Simonetti, A., Gariepy, C., 2002. Dispersal of atmospheric lead in northeastern North America as recorded by epiphytic lichens. Atmos. Environ. 36, 3759–3766.

Ceburnis, D., Steinnes, E., 2000. Conifer needles as biomonitors of atmospheric heavy metal deposition: comparison with mosses and precipitation, role of the canopy. Atmos. Environ. 34, 4265–4271.

Cercasov, V., Pantelica, A., Salagean, M., Caniglia, G., Scarlat, A., 2002. Comparative study of the suitability of three lichen species to trace-element air monitoring. Environ. Pollut. 119, 129–139.

Chiarenzelli, J., Aspler, L., Dunn, C., Cousens, B., Ozarko, D., Powis, K., 2001. Multi-element and rare earth element composition of lichens, mosses, and vascular plants from the Central Barrenlands, Nunavut, Canada. Appl. Geochem. 16, 245–270.

Cloquet, C., Carignan, J., Libourel, G., 2006a. Atmospheric pollutant dispersal around an urban area using trace metal concentrations and Pb isotopic compositions in epiphytic lichens. Atmos. Environ. 40, 574–587.

Cloquet, C., Carignan, J., Libourel, G., 2006b. Isotopic composition of Zn and Pb deposition in an urban/periurban area of northeastern France. Environ. Sci. Technol. 40, 6594–6600.

Cloquet, C., De Muynck, D., Sinoret, J., Vanhaecke, F., 2009. Urban/peri-urban aerosol survey by determination of the concentration and isotopic composition of Pb collected by transplanted Hypogymina physodes. Environ. Sci. Technol. 43, 623–629.

Conti, M.E., Cecchetti, G., 2001. Biological monitoring: lichens as bioindicators of air pollution assessment a review. Environ. Pollut. 114, 471–492.

Dolgopolova, A., Weiss, D.J., Seltmann, R., Kober, B., Mason, T.F.D., Coles, B., Stanley, C.J., 2006. Use of isotope ratios to assess sources of Pb and Zn dispersed in the environment during mining and ore processing within the Orlovka-Spokoinoe mining site (Russia). Appl. Geochem. 21, 563–579.

Doucet, F.J., Carignan, J., 2001. Atmospheric Pb isotopic composition and trace metal concentration as revealed by epiphytic lichens: an investigation related to two altitudinal sections in eastern France. Atmos. Environ. 35, 3681–3690.

Faure, G., 1986. Principles of Isotope Geology. John Wiley & Sons, New York.

Gallon, C., Tessier, A., Gobeil, C., Carignan, R., 2006. Historical perspective of industrial lead emissions to the atmosphere from a Canadian smelter. Environ. Sci. Technol. 40, 741–747.

Garty, J., 2000a. Environment and elemental content of lichens. In: Markert, B., Friese, K. (Eds.), Trace Elements—Their Distribution and Effects in the Environment. Elsevier Science B.V., Amsterdam, pp. 245–276.

Garty, J., 2000b. Trace metals, other chemical elements and lichen physiology: research in the nineties. In: Markert, B., Friese, K. (Eds.), Trace Elements—Their Distribution and Effects in the Environment. Elsevier Science B.V., Amsterdam, pp. 277–322.

Garty, J., 2001. Biomonitoring atmospheric heavy metals with lichens: theory and application. Crit. Rev. Plant Sci. 20 (4), 309–371.

Garty, J., Garty-Spitz, R.L., 2011. Neutralization and neoformation: analogous processes in the atmosphere and in lichen thalli—a review. Environ. Exp. Bot. 70, 67–79.

Garty, J., Galun, M., Kessel, M., 1979. Localization of heavy metals and other elements accumulated in the lichen thallus. New Phytol. 82, 159–168.

Getty, S.R., Gutzler, D.S., Asmerom, Y., Shearer, C.K., Free, S.J., 1999. Chemical signals of epiphytic lichens in southwestern North America: natural versus man-made sources for airborne particulates. Atmos. Environ. 33, 5095–5104.

Graney, J.R., Halliday, A.N., Keeler, G.J., Nriagu, J.O., Robbins, J.A., Norton, S.A., 1995. Isotopic record of lead pollution in lake sediments from the northeastern United States. Geochim. Cosmochim. Acta 59, 1715–1728.

Hauck, M., 2008. Metal homeostasis in Hypogymnia physodes is controlled by lichen substances. Environ. Pollut. 153, 304–308.

Hauck, M., 2011. Site factors controlling epiphytic lichen abundance in northern coniferous forests. Flora 206, 81–90.

Hauck, M., Huneck, S., 2007. Lichen substances affect metal adsorption in Hypogymnia physodes. J. Chem. Ecol. 33, 219–223.

Hurst, R.W., 2002. Lead isotopes as age-sensitive, genetic markers in hydrocarbons: 2. kerogens, crude oils, and unleaded gasoline. Environ. Geosci. 9, 1–7.

Krachler, M., Le Roux, G., Kober, B., Shotyk, W., 2004. Optimising accuracy and precision of lead isotope measurement (^{206}Pb, ^{207}Pb, ^{208}Pb) in acid digests of peat with ICP-SMS using individual mass discrimination correction. J. Anal. At. Spectrom. 19, 354–361.

Landis, M.S., Keeler, G.J., 2002. Atmospheric mercury deposition to Lake Michigan during the Lake Michigan mass balance study. Environ. Sci. Technol. 36, 4518–4524.

Monna, F., Aiuppa, A., Varrica, D., Dongarra, G., 1999. Pb isotope composition in lichens and aerosols from eastern Sicily: insights into the regional impact of volcanoes on the environment. Environ. Sci. Technol. 33, 2517–2523.

Monna, F., Poujol, M., Losno, R., Dominik, J., Annegarn, H., Coetzee, H., 2006. Origin of atmospheric lead in Johannesburg, South Africa. Atmos. Environ. 40 (34), 6554–6566.

Nash III, T.H., 1989. Metal tolerance in lichens. In: Shaw, A.J. (Ed.), Heavy Metal Tolerance in Plants: Evolutionary Aspect. CRC Press, Boca Raton, FL, pp. 119–131.

Nash III, T.H., 1996. Nutrient, elemental accumulation and mineral cycling. In: Nash III, T.H. (Ed.), Lichen Biology. Cambridge University Press, Cambridge, pp. 136–153.

Nash, T.H., Gries, C., 1995. The use of lichens in atmospheric deposition studies with an emphasis on the Arctic. Sci. Total Environ. 160, 729–736.

Nieboer, E.A., Richardson, D.H.S., Tomassini, F.D., 1978. Mineral uptake and release by lichens: an overview. Bryologist 81 (2), 226–246.

Pawlik-Skowronska, B., Sanita di Toppi, L., Favali, M.A., Fossati, F., Pirszel, J., Skowronski, T., 2002. Lichens respond to heavy metals by phytochelatin synthesis. New Phytol. 156, 95–102.

Puckett, K.J., 1988. Bryophytes and lichens as monitors of metal deposition. In: Nash III, T.H., Wirth, V. (Eds.), Lichens, Bryophytes and Air Quality. Cramer, Berlin- Stuttgart, pp. 231–267.

Purvis, O.W., Chimonides, P.J., Jones, G.C., Mikhailova, I.N., Spiro, B., Weiss, D., Williamson, B.J., 2004. Lichen biomonitoring in one of the most polluted areas in the world. Proc. R Soc. Lond. 272, 277–285.

Rusu, A.-M., Jones, G.C., Chimonides, P.D.J., Purvis, O.W., 2006. Biomonitoring using the lichen Hypogymnia physodes and bark samples near Zlatna, Romania immediately following closure of a copper ore processing plant. Environ. Pollut. 143, 81–88.

Simonetti, A., Gariépy, C., Carignan, J., 2003. Tracing sources of atmospheric pollution in western Canada using Pb isotopic composition and heavy metal abundances in epiphytic lichens. Atmos. Environ. 37, 2853–2865.

Simonetti, A., Gariepy, C., Banic, C.M., Tanabe, R., Wong, H.K., 2004. Pb isotopic investigation of aircraft-sampled emissions from the Horne smelter (Rouyn, Quebec): implications for atmospheric pollution in northeastern North America. Geochim. Cosmochim. Acta 68, 3285–3294.

Sondergaard, J., Asmund, G., Johansen, P., Elberling, B., 2010. Pb isotopes as tracers of mining-related Pb in lichens, seaweed and mussels near a former Pb-Zn mine in West Greenland. Environ. Pollut. 158, 1319–1326.

Spiro, B., Weiss, D.J., Purvis, O.W., Mikhailova, I., Williamson, B.J., Coles, B.J., Udachin, V., 2004. Lead Isotopes in Lichen Transplants around a Cu Smelter in Russia Determined by MC-ICP-MS Reveal Transient Records of Multiple Sources. Environ. Sci. Technol. 38, 6522–6528.

Sturges, W.T., Barrie, L.A., 1987. Lead 206/207 isotope ratios in the atmosphere of North America as tracers of US and Canadian emissions. Nature 329, 144–146.

Sturges, W.T., Barrie, L.A., 1989. Stable lead isotope ratios in arctic aerosols: evidence for the origin of arctic air pollution. Atmos. Environ. 23, 2513–2519.

Telmer, K., Bonham-Carter, G.F., Kliza, D.A., Hall, G.E.M., 2004. The atmospheric transport and deposition of smelter emissions: evidence from the multi-element geochemistry of snow, Quebec, Canada. Geochim. Cosmochim. Acta 68, 2961–2980.

Vitt, D.H., Halsey, L.A., Thormann, M.N., Martin, T., 1996. Peatland inventory of Alberta Phase 1: overview of peatland resources of the natural regions and subregions of the province.

National Centres of Excellence, Sustainable Forest Management Centre, University of Alberta, Edmonton, AB, Canada.

Williamson, B.J., Mikhailova, I., Purvis, O.W., Udachin, V., 2004. SEM-EDX analysis in the source apportionment of particulate matter on Hypogymnia physodes lichen transplants around the Cu smelter and former mining town of Karabash, South Urals, Russia. Sci. Total Environ. 322, 139–154.

Yip, Y., Chung-wah Lamb, J., Tong, W., 2008. Applications of lead isotope ratio measurements. Trends Anal. Chem. 27, 460–480.

Chapter 16

Mercury Concentration and Isotopic Composition of Epiphytic Tree Lichens in the Athabasca Oil Sands Region

J.D. Blum[*,1], M.W. Johnson[*], J.D. Gleason[*], J.D. Demers[*], M.S. Landis[†] and S. Krupa[‡]

[*]*Earth and Environmental Sciences, University of Michigan, Ann Arbor, Michigan, USA*
[†]*Office of Research and Development, US EPA, Mail Drop E205-03, Research Triangle Park, North Carolina, USA*
[‡]*Plant Pathology, University of Minnesota-Twin Cities, St. Paul, Minnesota, USA*
[1]*Corresponding author: e-mail: jdblum@umich.edu*

ABSTRACT

Mercury (Hg) is a toxic heavy metal that is found associated with fossil fuel deposits and that can be released to the atmosphere during fossil fuel combustion and/or processing. Hg emitted to the atmosphere can be deposited to aquatic and terrestrial ecosystems where it can be methylated by bacteria. Methylmercury is strongly biomagnified in food webs and this leads to toxic levels in high trophic level fish, the consumption of which is a major human exposure pathway. Epiphytic tree lichens have been widely used to investigate the relationship between atmospheric point sources of Hg and regional Hg deposition patterns. An intensive study of Hg concentration and stable isotopic composition of the epiphytic tree lichen *Hypogymnia physodes* was carried out in the area within 150 km of the Athabasca oil sands region (AOSR) industrial developments. Concentrations of Hg were comparable to background values measured in previous studies from remote areas and were far below the values observed near significant atmospheric industrial sources of Hg. Spatial patterns provide no evidence for a significant atmospheric point source of Hg from the oil sands developments, and Hg accumulation actually decreases in lichens within 25 km of the northern AOSR development, presumably due to physiological responses of *H. physodes* to enhanced SO_2 deposition.

Disclaimer: This chapter has been subjected to U.S. EPA review and approved for publication. The content and opinions expressed by the authors in this report do not necessarily reflect the views of the WBEA or the WBEA membership.

Stable Hg isotope ratios show an increase in $\Delta^{199}Hg$ and $\Delta^{201}Hg$ within 25 km of the AOSR, and we speculate that this is due to a change in the proportion of the various ligands to which Hg is bonded in the lichens, and a resulting change in the isotope fractionation during partial photochemical reduction and loss of Hg from lichen surfaces.

16.1 INTRODUCTION

Mercury (Hg) is a highly toxic and globally distributed atmospheric trace pollutant. It is released to the atmosphere from a range of industrial activities primarily as gaseous elemental mercury (Hg(0)) and reactive gaseous mercury (Hg(II)). Hg(0) has a long atmospheric residence time (~ 1 year) and can travel long distances before being oxidized to Hg(II) and deposited (Schroeder and Munthe, 1998). Hg(II) has a much shorter residence time than Hg(0) and is rapidly deposited locally in precipitation and dry deposition downwind of atmospheric point sources (Keeler et al., 2006; Sherman et al., 2012). Once Hg(II) is deposited to aquatic and terrestrial ecosystems, it can be methylated by sulfate- and iron-reducing bacteria producing highly toxic and bioaccumulative methylmercury. Biomagnification of atmospherically deposited Hg can result in the acquisition of high levels of methylmercury in high trophic position fish in remote lakes. Consumption of fish is a major human exposure pathway for methylmercury, especially for indigenous peoples (Chan and Receveur, 2000).

Hg bonds strongly to marine and terrestrial organic matter and is often associated with organic-rich geological deposits. The most important anthropogenic source of Hg to the atmosphere is through the burning of coal (Pirrone et al., 2010). However, natural gas processing (Spiric and Mashyanov, 2000) and oil refining (Wilhelm and Bloom, 2000) can also produce significant emissions of Hg. Although detailed studies have not been conducted previously, several studies have inferred that Athabasca oil sands region (AOSR) processing may result in enhanced atmospheric Hg emissions (Kelly et al., 2010; Timoney and Lee, 2009; Wilhelm and Bloom, 2000) and that Hg is methylated and bioaccumulated resulting in elevated levels in fish in the region (Timoney and Lee, 2009). A study by Kelly et al. (2010) also reported that Hg, along with other trace metals, was elevated in snow near the AOSR developments.

There is a great need for monitoring of Hg deposition spatially near potential atmospheric point sources. Monitoring is particularly difficult in remote areas such as the AOSR where atmospheric concentrations and deposition rates are difficult to measure due to lack of road access and electrical power needed to run atmospheric monitoring sites. The industrial development of the AOSR north of Fort McMurray, Alberta, Canada, has resulted in increases in some industrial atmospheric emissions in the region (Chapter 12). The present study was designed to use a single species of naturally occurring tree lichens to evaluate if Hg deposition in the region close to the AOSR was enhanced relative to background values further from the AOSR developments.

Many different species of epiphytic tree lichen have been used previously as passive collectors of a variety of atmospheric pollutants (e.g., Purvis and Halls, 1996), including Hg (e.g., Bargagli and Barghigiani, 1991). Physiological differences between lichen species result in differing metal accumulation properties, and so we focused our study on a single species of epiphytic lichen, *Hypogymnia physodes*, that is widely distributed in the AOSR and that has been used extensively in previous studies of Hg deposition in other geographic regions (e.g., Horvat et al., 2000; Makholm and Bennett, 1998). Hg was measured in the same lichen samples used for major and trace elements analyses (Chapter 14) and Pb isotope studies (Chapter 15). A gridded circular sampling pattern was used to allow the use of Hg and other major and trace elements in receptor modeling and impact assessment (Chapter 18).

Some previous studies have analyzed variations in the Hg content of naturally growing *H. physodes* near atmospheric pollution sources (Makholm and Bennett, 1998 and references therein; Horvat et al., 2000; Sensen and Richardson, 2002) and others have transplanted *H. physodes* with low natural Hg concentrations to locations near atmospheric point sources and monitored changes in Hg concentration with time (Makholm and Bennett, 1998 and references therein; Bialonska and Dayan, 2005; Horvat et al., 2000; Williamson et al., 2008). Additionally, studies have been carried out on the organic compounds produced by *H. physodes* and their effect on metal binding in this species of lichen (Bialonska and Dayan, 2005; Hauck, 2008; Hauck and Huneck, 2007). Collectively, these studies have shown that Hg concentrations in lichen increase near atmospheric point sources for Hg, and that *H. physodes*, as well as other epiphytic lichens, can provide a useful semiquantitative proxy for atmospheric Hg deposition.

In addition to the many studies that have measured Hg concentrations in epiphytic lichens, two previous studies explored variations in the stable isotopic composition of Hg in epiphytic lichens; but in species other than *H. physodes*.. Carignan et al. (2009) and Estrade et al. (2010) measured Hg concentrations and isotopic compositions of four tree lichen species (*Usnea, Ramalinia, Evernia,* and *Bryoria*) from remote areas in Canada and Switzerland and rural and urban areas in France. They did not find a correlation between Hg concentration and that of other heavy metals (Cd and Pb), but they did document mass-independent isotope anomalies in the odd mass Hg isotopes for all lichens sampled. Furthermore, these studies observed decreased mass-independent Hg isotope anomalies near industrial atmospheric sources of Hg and suggested that this was the result of additions of local anthropogenic Hg.

In this study, we measured total Hg concentrations ($n = 121$) and stable isotopic compositions ($n = 38$) in *H. physodes* collected in 2008 from trees at proximal and remote sites in a nested circular grid pattern distributed over 0.5–150 km from the AOSR industrial complex north of Fort McMurray, Alberta, Canada (Chapter 14). Spatial trends in Hg concentration and isotopic composition are used to evaluate the extent to which the AOSR developments have affected atmospheric Hg deposition. Hg concentrations and isotopic compositions are compared to values from the literature for measurements made in remote areas and near known point sources of Hg. Hg isotopic compositions are also compared to values measured from a variety of geological materials from the AOSR developments and nearby areas.

16.2　METHODS

16.2.1　Sample Collection and Preparation

A detailed field-sampling program was implemented during August–September 2008. Most sites were accessed by helicopter and others were visited by ground. At each site, composite samples of the lichen *H. physodes* were collected from the branches of six trees within jack pine (*Pinus banksiana Lamb.*) tree stands. Samples were stored in polyester sample bags and air dried in a laboratory at the University of Minnesota. Using Teflon-coated tweezers, tissue samples were cleaned by removing all foreign materials (bark and other debris) and were then stored frozen until analysis. GPS location data for all sample collection sites were used to determine the distance from a midpoint located at the center of the AORS industrial operations (Chapter 14). In 2009, bulk geological samples of oil sands, bitumen, processed tailing sand, overburden, and mine road material were acquired from active mining and energy-processing operations in the AOSR. Bulk samples of other undeveloped oil sand deposits exposed in outcrop in the Athabasca river valley north of the AOSR industrial developments were also obtained. All samples were shipped to the University of Michigan where they were freeze-dried and ground to a powder. Lichen samples were ground in a ball mill (Spex 8000M) using a zirconium grinding vial and balls that were thoroughly cleaned between each sample. Geological materials were ground by hand in an agate mortar and pestle that was thoroughly cleaned between each sample. An ~ 50 mg aliquot of each powdered sample was used for Hg concentration analysis and a 500 mg to 4 g aliquot of a subset of the samples was used for Hg stable isotope analysis.

16.2.2　Hg Concentration Analysis

Hg concentrations were measured using quality control methods that included analysis of analytical standards, procedural blanks, and replicate samples. Lichens and geologic samples were analyzed for total Hg concentration by combustion at $850\,^{\circ}\text{C}$ and quantification by cold-vapor atomic absorption spectrometry (MA-2000, Nippon Instruments). Replicate analyses agreed within $\pm 7\%$ and the reporting limit was 0.5 ng Hg. The IRMM powdered lichen reference material BCR482 was run with each batch of eight samples and we obtained an average Hg concentration of 454 ± 8 ng/g (1 SD, $n = 19$), which is within 5% of the certified value of 480 ± 20 ng/g.

16.2.3   Sample Preparation for Isotope Analysis

Hg was separated for isotopic analysis from lichens and geologic samples by combustion at 750 °C, thermal decomposition at 1000 °C, and inline trapping in a 1% $KMnO_4$ solution. To further purify Hg in the $KMnO_4$ solutions, the solutions were reduced with $SnCl_2$, purged with Hg-free air, and retrapped in another 1% $KMnO_4$ solution. The Hg concentration of each of the $KMnO_4$ solutions was measured to determine recoveries and to allow matching of standard and sample concentrations for isotope analysis. Details of the method are given in Biswas et al. (2008) and Demers et al. (2012). Mean recovery following combustion for IRMM lichen reference material BCR482 was 94% ($\pm$4%, 1 SD, $n=5$), and mean recovery following both combustion and secondary purge and trap of BCR482 was 93% ($\pm$3%, 1 SD, $n=5$).

16.2.4   Mass Spectrometry

Hg in $KMnO_4$ solutions separated from lichens and bulk geologic samples was analyzed for isotopic composition with a multiple collector inductively coupled plasma mass spectrometer (Nu Instruments) using continuous flow cold-vapor generation with $SnCl_2$ reduction (Blum and Bergquist, 2007; Lauretta et al., 2001). Instrumental mass bias was corrected using an internal Tl standard (NIST SRM 997, $^{205}Tl/^{203}Tl = 2.38714$) and sample-standard bracketing with the NIST SRM 3133 Hg standard. Analyses of samples and standards were run at 5 ppb Hg. On-peak zero corrections were applied to all masses and isobaric interference from ^{204}Pb was monitored using mass 206, but was always negligible.

We report isotopic compositions as permil (‰) deviations from the average of NIST SRM 3133 bracketing standards using delta notation following Eq. (16.1):

$$\delta^{xxx}Hg(‰) = \left[\left[\left(^{xxx}Hg/^{198}Hg\right)_{unknown} \Big/ \left(^{xxx}Hg/^{198}Hg\right)_{NIST\,SRM\,3133}\right] - 1\right] \times 1000,$$

$$(16.1)$$

where xxx is the mass of each Hg isotope between ^{199}Hg and ^{204}Hg. We use $\delta^{202}Hg$ to describe isotopic differences due to mass-dependent fractionation (MDF). Mass-independent fractionation (MIF) is defined as the deviation of the isotope ratios involving odd mass isotopes from the theoretically expected values determined by the kinetic MDF law based on the measured $\delta^{202}Hg$ (Blum and Bergquist, 2007). MIF is reported in "capital delta" notation ($\Delta^{xxx}Hg$) in units of permil (‰) and is well approximated for small ranges in delta values ($\leq$10‰) by Eq. (16.2):

$$\Delta^{xxx}\mathrm{Hg}(\text{‰}) = \delta^{xxx}\mathrm{Hg} - \left(\delta^{202}\mathrm{Hg} \times \beta\right), \qquad (16.2)$$

where xxx is the mass of each Hg isotope 199 and 201 and β is a constant (0.252 and 0.752, respectively; Blum and Bergquist, 2007).

16.2.5 Analytical Uncertainty

The uncertainty of Hg isotope measurements was characterized using a secondary standard solution that is widely distributed (UM-Almaden) and an international lichen reference material (BCR482). The UM-Almaden standard solution is useful for interlaboratory comparison and we obtained the following values for 10 replicate analyses carried out during the analyses of samples for this study: $\delta^{202}\mathrm{Hg} = -0.59 \pm 0.03$‰ (2 SD), $\Delta^{199}\mathrm{Hg} = -0.01 \pm 0.04$‰ (2 SD), and $\Delta^{201}\mathrm{Hg} = -0.04 \pm 0.03$‰ (2 SD). The lichen IRMM BCR482 is useful for assessment of the uncertainty of complete Hg separation and analyses and we obtained the following values for six replicate analyses carried out during the analyses of samples for this study: $\delta^{202}\mathrm{Hg} = -1.67 \pm 0.04$‰ (2 SD), $\Delta^{199}\mathrm{Hg} = -0.67 \pm 0.04$‰ (2 SD), and $\Delta^{201}\mathrm{Hg} = -0.68 \pm 0.03$‰ (2 SD).

16.3 RESULTS AND DISCUSSION

16.3.1 Hg Concentrations of *H. physodes*

Total Hg concentrations for the 121 *H. physodes* samples analyzed ranged from 71 to 268 ng/g with a mean value of 144 ± 33 ng/g (1 SD) (Figure 16.1). These values are in the range typical for remote areas unaffected by point source Hg emissions. Makholm and Bennett (1998) reported background Hg concentrations for *H. physodes* in northern Wisconsin of 110–155 ng/g and Bennett and Wetmore (1997) reported background values of 103–143 ng/g for northern Minnesota. Makholm and Bennett (1998 and references within) compiled background Hg concentration values for European studies as of 1998 and reported that most values (nine separate studies) ranged from 200 to 400 ng/g, whereas two studies in Scandinavia reported background values of 100 and 160 ng/g, respectively. Since 1998 two additional studies measured background Hg values in *H. physodes*. Horvat et al. (2000) measured a mean background value of 110 ± 10 ng/g at a remote location in Slovenia, and Sensen and Richardson (2002) measured a mean background Hg value of 148 ± 46 ng/g at a remote location in New Brunswick, Canada. Altogether, comparison of the values reported here for Hg in *H. physodes* in the AOSR are consistent with background values measured in a large number of studies of Hg in *H. physodes* in both North America and Europe. This suggests that any Hg deposition associated with the AOSR development has not significantly elevated the Hg concentrations in lichens.

Previous studies have shown that Hg concentrations in natural *H. physodes* collected near atmospheric Hg point sources, and Hg in *H. physodes*

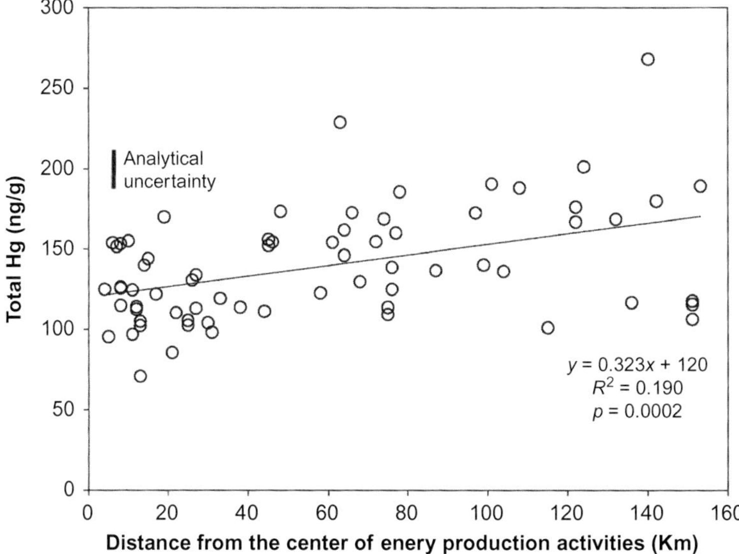

FIGURE 16.1   The concentration of Hg in lichens (in ng/g) versus the distance of the sample collection point from the center of the energy production activities (in Km).

transplanted to the vicinity of Hg point sources, have lichen tissue concentrations much higher than those that we measured in the AOSR. Makholm and Bennett (1998) transplanted *H. physodes* downwind of a chlor-alkali plant and after a year of exposure concentrations ranged up to 4418 ng/g at a distance of 250 m from the plant and up to 403 ng/g at a distance of 1250 m. Makholm and Bennett (1998 and references therein) also compiled the results of earlier studies of Hg in *H. physodes* near point sources and found that in each of seven studies Hg concentrations were highly elevated near the source with values ranging from 770 to 36,000 ng/g. In a more recent study, Horvat et al. (2000) transplanted *H. physodes* to locations around a natural gas treatment plant in Molve, Croatia, and locations around a Hg mining area in Idrija, Slovenia, where measurements of the concentration of ambient total gaseous mercury (TGM) in air were also made. They found a strong correlation between the logarithm of the total Hg concentration in *H. physodes* and the logarithm of the TGM concentration in air. Hg concentrations in *H. physodes* were 110 ng/g at the time of transplant and ranged from 540 to 4430 ng/g after 6 months in the mining area. In another recent study, Sensen and Richardson (2002) analyzed Hg in naturally growing *H. physodes* on spruce trees along transects near a Hg cell chlor-alkali plant in New Brunswick, Canada. They documented systematic increases in Hg concentration from a background value of 148 ng/g at a distance of 8 km from the plant to maximum values of 790 and 980 ng/g at a distance of 125 m along two different transects. In sum, these studies show that within about 1–8 km of atmospheric

Hg point sources the Hg concentrations of *H. physodes* are generally elevated far above background values. This reinforces our conclusion that the AOSR development is not a significant atmospheric Hg source that is detectable using lichen analysis and certainly does not generate atmospheric Hg levels near the magnitude of the chlor-alkali plants, smelters, gas-processing plants, or Hg mining areas that have been studied previously.

16.3.2 Hg Concentrations of Other Species of Epiphytic Lichens

The literature is rich with investigations that have used epiphytic lichens other than *H. physodes* to study Hg levels in the atmosphere. Because of physiological differences between species, it is not advisable to directly compare metal concentrations between species. Nevertheless, these studies do confirm that for most species a concentration range of 100–200 ng/g total Hg is a common background concentration that is often observed in remote areas. Large increases in Hg concentrations are often observed within about 8 km of large atmospheric Hg point sources, and lichen Hg concentrations generally correlate with TGM concentrations in air (Bargagli and Barghigiani, 1991; Grangeon et al., 2012; Guevara et al., 2004; Tretiach et al., 2011). TGM concentrations measured by Environment Canada at a Wood Buffalo Environmental Association (WBEA) air monitoring station in Fort McMurray between October 2010 and May 2011 averaged 1.41 ± 0.15 ng/m^3, similar to other regions in Canada (Chapter 4). Using four different lichen species (but not including *H. physodes*), Landers et al. (1995) observed increased Hg concentrations in Alaska with closer proximity to the marine coastline. Carignan and Sonke (2010) measured Hg in three lichen species (but again not in *H. physodes*) in Canada (mostly in Quebec) and found that Hg concentrations were well correlated with Br concentrations and that both elements increased in concentration with decreasing distance to Hudson Bay. These authors attributed the spatial trends to a marine source of Br to the atmosphere, which is known to oxidize background Hg(0) in the atmosphere, which then led to increased deposition of Hg(II). The Carignan and Sonke (2010) study documents lichen Hg concentrations as high as 2060 ng/g in the lichen species *Bryoria*, 20 km from Hudson Bay but far from anthropogenic point sources. Hg concentrations dropped below 250 ng/g at distances greater than 800 km from Hudson Bay. The AOSR is about 800 km from both the Pacific Ocean and Hudson Bay and appears to be unaffected by this Hg deposition mechanism.

16.3.3 Spatial Variation in Hg Concentrations of *H. physodes*

In order to explore whether the observed variability in the Hg concentration of *H. physodes* in the vicinity of the AOSR is affected by the mining and energy production activities, we plotted the Hg concentrations of lichens as a function of the distance from the center of the production area (Figure 16.1). Regression of the data reveals a small (slope $=0.33$ ng/g/km) but significant ($p=0.0002$) decrease in the Hg concentration of lichens as the active mining and upgrading facilities are approached. The regression line drops from a Hg concentration of 171 ng/g at 155 km to 122 ng/g at 5 km. This behavior is the opposite of what has been documented in the literature near atmospheric point sources of Hg and is (as far as we are aware) the first documentation of decreasing Hg concentration in lichen as an industrial development is approached. We do note, however, that previous studies documenting increases in the Hg concentration of *H. physodes* in the vicinity of atmospheric Hg sources, found elevated Hg in lichens only at distances of 8 km or less from the sources. The sampling distances in our study were generally much farther from the potential source than previous studies, and only eight of the 121 *H. physodes* samples were from 8 km or less distance from the center of the AOSR mining and energy production areas. However, even these samples within close proximity of the AOSR industrial development did not have elevated Hg concentrations.

Although a decrease in Hg in epiphytic lichens near industrial facilities has not been previously reported, a study comparing naturally growing and transplanted *H. physodes* within a kilometer of a chlor-alkali plant in Norway may provide some insight into this phenomenon. Steinnes and Krog (1977) found that Hg concentrations in naturally occurring *H. physodes* reached a maximum of 950 ng/g in close proximity to a chlor-alkali plant. However, when they transplanted *H. physodes* from a remote site to the industrial site they observed an increase in Hg content with time and found Hg concentrations in *H. physodes* as high as 6000–8000 ng/g after 16 weeks of exposure. The authors suggested that the lower uptake of Hg by the naturally occurring lichens compared to the transplants may have been related to SO_2 exposure and resulting loss of vigor in the *H. physodes* (Steinnes and Krog, 1977). In the vicinity of the AOSR, the sulfur content of *H. physodes* is higher due to SO_2 deposition (Chapters 14 and 18), and we suggest that this may have resulted in decreased Hg accumulation as suggested in the study of Steinnes and Krog (1977). Graney et al. (Chapter 15) observed a subtle decrease in Zn and Mn concentrations toward the center of the AOSR region in the same lichen samples we analyzed. They suggested that this trend may be related to canopy interactions or physiological processes in the lichens.

16.3.4   Hg Isotopic Composition of *H. physodes*

A subset of 38 *H. physodes* samples that were analyzed for Hg concentration was also analyzed for their stable Hg isotopic composition. Variations in the isotopic composition related to MDF of Hg are reported as δ^{202}Hg and samples ranged in values from -2.66‰ to -1.41‰ with an average value of -1.87 ± 0.22‰ (1 SD) (Figure 16.2). MIF is reported here as Δ^{199}Hg (but

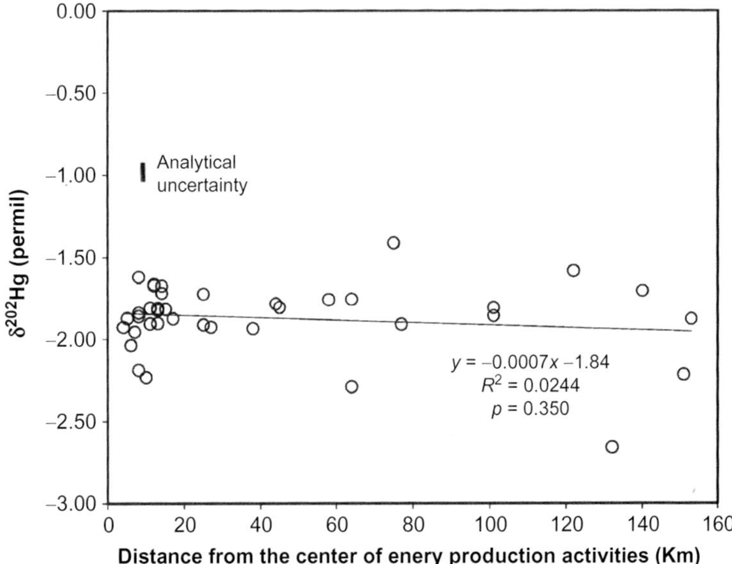

FIGURE 16.2 The δ^{202}Hg value of Hg in lichens (in permil) versus the distance of the sample collection point from the center of the energy production activities (in Km).

note that Δ^{201}Hg shows the same behavior) and samples ranged in values from $-0.55‰$ to $+0.33‰$ with an average value of $-0.24 \pm 0.17‰$ (1 SD) (Figure 16.3). The ratio of Δ^{199}Hg/Δ^{201}Hg is 0.997 (Figure 16.4), suggesting that the MIF results from photochemical reduction of Hg(II) to Hg(0) involving short-lived radical pair reactions and is caused by the magnetic isotope effect (Bergquist and Blum, 2007; Blum, 2011).

Two other investigations of Hg isotopic compositions of lichens have been published previously, but there have been no previous Hg isotopic studies of the lichen species *H. physodes*. Carignan et al. (2009) and Estrade et al. (2010) measured Hg isotope values in remote areas in Canada and Switzerland and both urban and industrial areas in France in many different lichen species. These lichens ranged widely in δ^{202}Hg values from $1.5‰$ down to $-2.2‰$. The lichens analyzed in these previous studies also displayed negative values for Δ^{199}Hg (and Δ^{201}Hg), with values ranging from $-0.2‰$ to $-0.9‰$. Lichens from remote areas showed the lowest Δ^{199}Hg values (Carignan et al., 2009) and values tended to become less negative in urban and industrial areas (Estrade et al., 2010). At that time these authors interpreted the negative Δ^{199}Hg values to be indicative of regional atmospheric Hg isotopic values and the increase in Δ^{199}Hg near industrial areas as the impact of adding industrial point sources of Hg that had near-zero Δ^{199}Hg values.

More recently, several studies of atmospheric Hg isotopes have shown that Hg (II) deposited from the atmosphere in precipitation has positive Δ^{199}Hg (and Δ^{201}Hg) values, and that gaseous Hg(0) in the atmosphere has negative Δ^{199}Hg

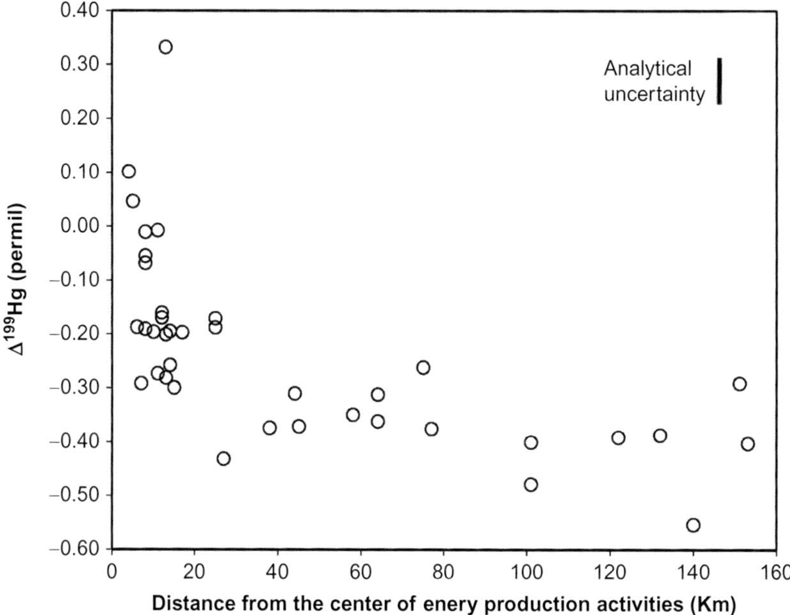

FIGURE 16.3 The Δ^{199}Hg value of Hg in lichens (in permil) versus the distance of the sample collection point from the center of the energy production activities (in Km).

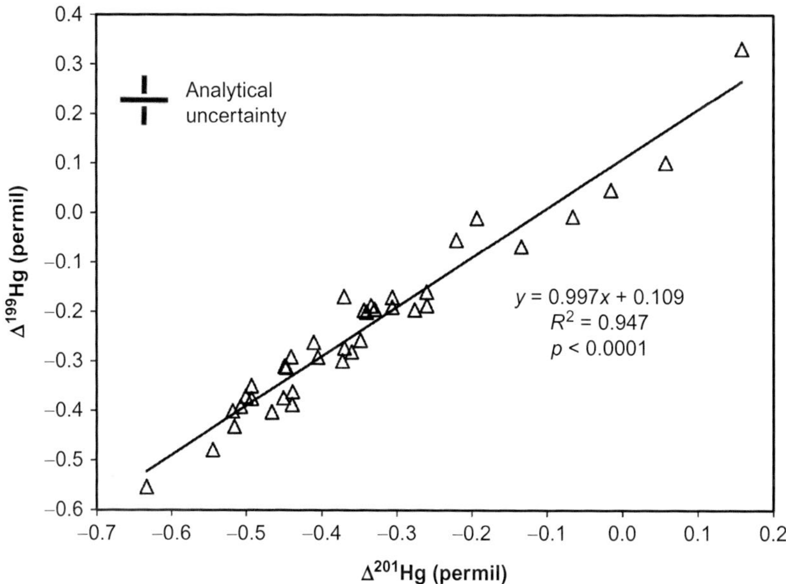

FIGURE 16.4 The Δ^{199}Hg value of Hg in lichens (in permil) versus the Δ^{201}Hg value of Hg in lichens (in permil).

(and $\Delta^{201}Hg$) values. Furthermore, these studies have demonstrated that there are fractionation mechanisms that can cause Hg with more positive $\Delta^{199}Hg$ to be released from surfaces following photochemical reduction of Hg(II) to Hg(0) (Demers et al., 2012; Gratz et al., 2010; Sherman et al., 2010, 2012). In a recent investigation of Hg isotope systematics in a forested ecosystem in northern Wisconsin (USA), Demers et al. (2012) found that gaseous elemental Hg(0) with $\Delta^{199}Hg$ of $-0.2‰$ is oxidized and deposited to tree leaf surfaces and that subsequent photochemical reduction and loss of a portion of this Hg results in increasingly negative values for $\Delta^{199}Hg$ in leaves, averaging about $-0.3‰$. This same Hg isotope fractionation process may occur on lichen surfaces as well as on leaves. We suggest that Hg(0) is deposited to lichen surfaces, becoming oxidized to Hg(II) as it binds with ligands in organic acids. Photochemical reactions then cause partial reduction and loss of some Hg(II) as Hg(0) with more positive $\Delta^{199}Hg$, and the Hg(II) that remains in the lichen acquires more negative $\Delta^{199}Hg$ values.

16.3.5 Spatial Variation in Hg Isotopic Composition of *H. physodes*

The $\delta^{202}Hg$ values of the lichens from this study are plotted versus distance from the active mining and energy production activities on Figure 16.2. $\delta^{202}Hg$ values do not vary significantly with distance ($p=0.350$) from the active energy production areas and are similar to values for foliage from trees in northern Wisconsin (USA) measured by Demers et al. (2012). Values are unaffected by proximity to the AOSR developments, as would be expected, because Hg concentrations show no evidence for addition of Hg from this source.

In contrast, $\Delta^{199}Hg$ (and $\Delta^{201}Hg$) values shift dramatically with distance from the AOSR development (Figure 16.3). At distances of >25 km, the average $\Delta^{199}Hg$ value is $-0.38 \pm 0.07‰$, whereas at <25 km, the average $\Delta^{199}Hg$ value is $-0.13 \pm 0.15‰$ and reaches a maximum value of $+0.33‰$. The dramatic increase in $\Delta^{199}Hg$ values combined with the decrease in Hg concentrations of *H. physodes* near the AOSR development is somewhat puzzling. It appears that either (i) the AOSR development is emitting Hg with very high $\Delta^{199}Hg$ values, but accumulation of Hg in the lichens near the AOSR is suppressed to the point that Hg concentrations actually appear to decrease near the active energy production areas or (ii) there are physiological effects in the lichens near the active energy production areas due to deposition of other pollutants (in particular SO_2) that both suppresses Hg accumulation and alters the MIF of Hg on lichen surfaces resulting in an increase in $\Delta^{199}Hg$ values. Each of these alternate explanations is discussed in more detail below.

16.3.6 Hg Isotopic Composition of Oil Sands Materials

To explore the possibility that the AOSR development is emitting Hg with highly elevated $\Delta^{199}Hg$ values and that this is responsible for the high

Δ^{199}Hg values in lichens in close proximity to the AOSR, we analyzed bulk samples of geological materials from the AOSR (Chapter 18). Oil sand, bitumen, processed tailings sand, overburden, and road materials from mines between Ft McMurray and Ft McKay were analyzed. Weathered oil sand that naturally outcrops along the Athabasca River north of Fort McKay (and has not been mined) was also analyzed. The concentration of Hg in the oil sands ranged from 4.1 to 38.8 ng/g, processed sands from tailings ranged from 1.9 to 10.6 ng/g, and the bitumen, overburden, and road material were 14.3, 13.1, and 32.8 ng/g, respectively. In comparison, Hg concentrations are 3.5 ng/g in mean volume weighted US crude oil and 100 ng/g in mean volume weighted US coal (Toole-O'Neil et al., 1999; Wilhelm et al., 2007). The mean background Hg concentration for lake sediments in Alberta was recently reported as 36 ng/g (Nasr et al., 2011) and indicates that the oil sands and associated materials do not have Hg concentrations that are far above background sediment concentrations.

Δ^{199}Hg and δ^{202}Hg values of the geologic materials analyzed are plotted along with values for the lichens on Figure 16.5. Oil sand, bitumen, processed sand, and overburden from mines between Fort McMurray and Fort McKay fall in a narrow range of Hg isotope values that overlap with values for *H. physodes* collected >25 km from the mines, but have lower Δ^{199}Hg and

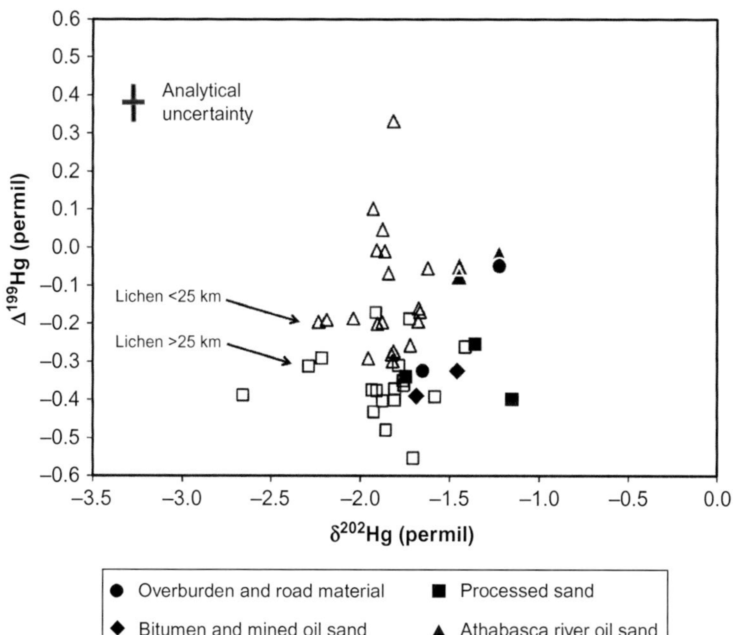

FIGURE 16.5   The Δ^{199}Hg value of Hg (in permil) versus the δ^{202}Hg value of Hg (in permil) in lichens as well as in a variety of geological materials from the AOSR region.

higher δ^{202}Hg than *H. physodes* collected <25 km from the mines. A single sample of road material from the mines has Δ^{199}Hg ~0.3‰ higher than the other mine samples and could possibly represent addition of Hg from oil, brake pads, or diesel fuel. The three samples of oil sand from the Athabasca River north of Fort McKay also have Δ^{199}Hg values about ~0.3‰ higher than the mine samples. These samples have Δ^{199}Hg closer to the values of the *H. physodes* <25 km from the mines, and δ^{202}Hg values about ~0.5‰ higher, but these deposits are not currently mined or processed.

Potential Hg sources from the mining operations are similar to the background isotopic values in lichens >25 km from the mining areas. Therefore, addition of this potential source of Hg would not shift the lichen Hg isotopic values appreciably. Even if all of the Hg in the lichens near the AOSR had the maximum observed Δ^{199}Hg from the undeveloped AOSR, this could only explain the lichens with intermediate values and could not explain the highest lichen Δ^{199}Hg values. Therefore, we can rule out addition of oil sand-derived Hg as the sole cause for the increase in Δ^{199}Hg values of *H. physodes* within 25 km of the AOSR area.

16.4 PROPOSED MECHANISM TO EXPLAIN HG ISOTOPIC VARIABILITY

We have argued above that the addition of Hg from the active mining and energy production activities does not adequately explain the changes in Hg isotopic composition observed in lichens within 25 km of the AOSR. Furthermore, the observed lower Hg concentrations in lichens near the AOSR are not consistent with addition of Hg from a point source to the regional background. We suggest that the increase in Δ^{199}Hg values of *H. physodes* within 25 km of the AOSR is caused by a change in the processes by which *H. physodes* binds Hg (II) species onto ligands present in the lichens. Following deposition of Hg to the lichens, there appears to be a change in the reaction pathway and degree of photochemical reduction of Hg (II) from the lichen surfaces. This reduction in the loss of Hg(0) from the lichen surfaces appears to result in a change in the lichen Δ^{199}Hg values.

Experimental studies of Hg isotope fractionation have shown that large magnitude MIF (>0.2‰) is generally the result of photochemical reactions involving short-lived radical pairs (Blum, 2011). The ratio of Δ^{199}Hg to Δ^{201}Hg, and the observation of either the preferential loss or gain of odd isotopes of Hg in volatilized Hg(0) during reduction, has been shown to be related to the ligands to which Hg is bonded (Zheng and Hintelmann, 2010). Regression of Δ^{199}Hg versus Δ^{201}Hg for lichens from this study yield a slope of 0.997 ($r^2=0.947$; Figure 16.4) and values are negative (meaning odd isotopes were preferentially lost). Hg (II) is known to be photochemically reduced to Hg(0) and emitted from leaf surfaces (Mowat et al., 2011). Zheng and Hintelmann (2010) showed that when Hg (II) is bonded to a sulfur-containing ligand, photochemical reduction and loss of Hg(0) from aqueous solutions result in residual Hg(II) with negative Δ^{199}Hg and Δ^{201}Hg. When the binding ligand is sulfurless, residual Hg(II) is observed to have positive Δ^{199}Hg and Δ^{201}Hg.

Lichens are known to produce a wide range of organic ligands, both containing and not containing sulfur, that play an important role in metal binding. Both intracellular binding of metals to sulfur-containing peptides (metallothioneins) and extracellular binding to sulfur-less organic acids are known to occur in lichens (Sarret et al., 1998). Although experiments exploring Hg isotope fractionation in the presence of these ligands have not yet been performed, we would expect that both positive and negative MIF could be produced by photochemical reduction of Hg(II) complexed with these various compounds. It seems likely that the net isotopic composition of lichen is the result of simultaneous negative and positive MIF of Hg associated with different bonding environments in the lichen. Given this assumption, we suggest that *H. physodes* collected >25 km from the active mining and energy production facilities have negative Δ^{199}Hg and Δ^{201}Hg that resulted from the dominance of photochemical reduction of Hg(II) from sulfur-bearing ligands compared to sulfur-less ligands. The increase in Δ^{199}Hg and Δ^{201}Hg that is observed closer to the active mining and energy production facilities could then be interpreted as a shift in the balance toward reduction of Hg(II) from sulfur-less ligands.

A study by Bialonska and Dayan (2005) measured changes in the production of organic acids by samples of *H. physodes* when they were transplanted to four different sites where they were exposed to industrial emissions of metals and SO_2. Site-specific changes in production of various organic acids were observed, with some increasing and others decreasing with exposure. For example, an increase in the content of physodic acid was observed in all transplanted lichens and the authors suggested a possible role for this compound in defense against stress caused by exposure to pollutants (Bialonska and Dayan, 2005). Although highly speculative, we suggest that exposure of *H. physodes* in the AOSR near active mining and energy production facilities to SO_2 and other metals could have changed the balance of the various organic acids produced in the lichens. Increased complexation of Hg(II) to these sulfur-less compounds, with resulting enhanced photochemical reduction of Hg and loss back to the atmosphere, is a possible explanation for both the increase in Δ^{199}Hg and Δ^{201}Hg and decrease in total Hg concentration observed in the lichens <25 km from the mining and energy production facilities.

16.5   CONCLUSIONS

A single species of epiphytic tree lichen (*H. physodes*) was sampled in 2008 at remote sites in a nested, circular grid pattern 0.5–150 km from the center of industrial development in the AOSR north of Fort McMurray, Alberta,

Canada. Total Hg concentrations in epiphytic lichens have been used in previous studies as a measure of spatial variations in atmospheric Hg deposition and to delineate the near field influence of atmospheric point sources of Hg. The stable Hg isotopic composition has also recently been used as a monitor of additions of local anthropogenic Hg to regional background deposition. Hg concentrations in *H. physodes* in the AOSR ranged from 71 to 268 ng/g, which is similar to that measured by others for remote sites in other regions. Based on concentration alone, there is no evidence for anomalously high atmospheric Hg deposition near the AOSR mining and energy production facilities. In fact, small variations with distance show that Hg concentrations decrease with proximity to the facilities. This conclusion is in contrast with the finding of Kelly et al. (2010) who observed increased Hg concentrations in snow near the AOSR energy development facilities.

The stable Hg isotope composition of the lichens does, however, change systematically with distance from the potential industrial sources. Most notably, the lichens display varying levels of MIF with both Δ^{199}Hg and Δ^{201}Hg ranging from near 0.0‰ close to the oil sands developments and systematically falling to -0.4‰ about 25 km away. The Δ^{199}Hg/Δ^{201}Hg slope is close to one, suggesting that the MIF results from photochemical reduction of Hg(II) to Hg(0). MDF does not change systematically with distance from the oil sands developments, and δ^{202}Hg averages -1.9‰. The spatial trends show that the energy production activities influence the Δ^{199}Hg and Δ^{201}Hg values without significantly affecting the δ^{202}Hg values and have the effect of lowering the Hg concentrations of the lichens. Measurements of the Hg isotopic composition of the oils sands themselves show that they cannot explain the variation in Hg isotopic composition of *H. physodes* near the energy production facilities. It is possible that other atmospheric pollutants, such as SO_2 (which is known to affect the vitality of lichens), are influencing Hg retention in the lichens. *H. physodes* is also known to respond to SO_2 by changing its production of organic acids. This may affect the proportions of different ligands to which Hg is bonded, thus influencing the photochemical reduction of Hg on lichen surfaces, which we suggest is the cause of the observed MIF.

ACKNOWLEDGMENTS

We wish to thank Shanti Berryman and Justin Straker for their efforts in sample collection and cleaning. Sarah North and Elaina Shope provided assistance in the laboratory. This work was funded by the WBEA.

REFERENCES

Bargagli, R., Barghigiani, C., 1991. Lichen biomonitoring of mercury emission and deposition in mining, geothermal and volcanic areas of Italy. Environ. Monit. Assess. 16 (3), 265–275.

Bennett, J.P., Wetmore, C.M., 1997. Chemical element concentrations in four lichens on a transect entering Voyageurs-National Park. Envir. Exper. Bot. 37 (2–3), 173–185.

Bergquist, B.A., Blum, J.D., 2007. Mass-dependent and -independent fractionation of Hg isotopes by photoreduction in aquatic systems. Science 318 (5849), 417–420.

Bialonska, D., Dayan, F.E., 2005. Chemistry of the lichen *Hypogymnia physodes* transplanted to an industrial region. J. Chem. Ecol. 31 (12), 2975–2991.

Biswas, A., Blum, J.D., Bergquist, B.A., Keeler, G.J., Xie, Z.Q., 2008. Natural mercury isotope variation in coal deposits and organic soils. Environ. Sci. Technol. 42 (22), 8303–8309.

Blum, J.D., 2011. Applications of stable mercury isotopes to biogeochemistry. In: Baskaran, M. (Ed.), Handbook of Environmental Isotope Geochemistry. Springer, Berlin, pp. 229–246.

Blum, J.D., Bergquist, B.A., 2007. Reporting of variations in the natural isotopic composition of mercury. Anal. Bioanal. Chem. 388 (2), 353–359.

Carignan, J., Sonke, J., 2010. The effect of atmospheric mercury depletion events on the net deposition flux around Hudson Bay, Canada. Atmos. Environ. 44 (35), 4372–4379.

Carignan, J., Estrade, N., Sonke, J.E., Donard, O.F.X., 2009. Odd isotope deficits in atmospheric Hg measured in lichens. Environ. Sci. Technol. 43 (15), 5660–5664.

Chan, H.M., Receveur, O., 2000. Mercury in the traditional diet of indigenous peoples in Canada. Environ. Pollut. 110 (1), 1–2.

Demers, J.D., Blum, J.D., Zak, D.R., 2012. Mercury isotopes in a forested ecosystem: new insights into biogeochemical cycling and the global mercury cycle. Global Biogeochem. Cycles (in review).

Estrade, N., Carignan, J., Donard, O.F.X., 2010. Isotope tracing of atmospheric mercury sources in an urban area of Northeastern France. Environ. Sci. Technol. 44 (16), 6062–6067.

Grangeon, S., Guedron, S., Asta, J., Sarret, G., Charlet, L., 2012. Lichen and soil as indicators of an atmospheric mercury contamination in the vicinity of a chlor-alkali plant (Grenoble, France). Ecol. Indic. 13 (1), 178–183.

Gratz, L.E., Keeler, G.J., Blum, J.D., Sherman, L.S., 2010. Isotopic composition and fractionation of mercury in Great Lakes precipitation and ambient air. Environ. Sci. Technol. 44 (20), 7764–7770.

Guevara, S.R., Bubach, D., Arribere, M., 2004. Mercury in lichens of Nahuel Huapi National Park, Patagonia, Argentina. J. Radioanal. Nucl. Chem. 261 (3), 679–687.

Hauck, M., 2008. Metal homeostasis in *Hypogymnia physodes* is controlled by lichen substances. Environ. Pollut. 153 (2), 304–308.

Hauck, M., Huneck, S., 2007. Lichen substances affect metal adsorption in *Hypogymnia physodes*. J. Chem. Ecol. 33 (1), 219–223.

Horvat, M., Jeran, Z., Spiric, Z., Jacimovic, R., Miklavcic, V., 2000. Mercury and other elements in lichens near the INA Naftaplin gas treatment plant, Molve, Croatia. J. Environ. Monit. 2 (2), 139–144.

Keeler, G.J., Landis, M.S., Norris, G.A., Christianson, E.M., Dvonch, J.T., 2006. Sources of mercury wet deposition in Eastern Ohio, USA. Environ. Sci. Technol. 40 (19), 5874–5881.

Kelly, E.N., Schindler, D.W., Hodson, P.V., Jeffrey, W.S., Roseanna, R., Nielsena, C.C., 2010. Oil sands development contributes elements toxic at low concentrations to the Athabasca River and its tributaries. Proc. Natl. Acad. Sci. U.S.A. 107 (37), 16178–16183.

Landers, D.H., Ford, J., Gubala, C., Monetti, M., Lasorsa, B.K., Martinson, J.J., 1995. Mercury in vegetation and lake-sediments from the Arctic. Water Air Soil Pollut. 80 (1–4), 591–601.

Lauretta, D.S., Klaue, B., Blum, J.D., Buseck, P.R., 2001. Mercury abundances and isotopic compositions in the Murchison (CM) and Allende (CV) carbonaceous chondrites. Geochim. Cosmochim. Acta 65 (16), 2807–2818.

Makholm, M.M., Bennett, J.P., 1998. Mercury accumulation in transplanted *Hypogymnia physodes* lichens downwind of Wisconsin chlor-alkali plant. Water Air Soil Pollut. 102 (3–4), 427–436.

Mowat, L.D., St Louis, V.L., Graydon, J.A., Lehnherr, I., 2011. Influence of forest canopies on the deposition of methylmercury to boreal ecosystem watersheds. Environ. Sci. Technol. 45 (12), 5178–5185.

Nasr, M., Ogilvie, J., Castonguay, M., Rencz, A., Arp, P.A., 2011. Total Hg concentrations in stream and lake sediments: discerning geospatial patterns and controls across Canada. Appl. Geochem. 26 (11), 1818–1831.

Pirrone, N., Cinnirella, S., Feng, X., Finkelman, R.B., Friedli, H.R., Leaner, J., Mason, R., Mukherjee, A.B., Stracher, G.B., Streets, D.G., Telmer, K., 2010. Global mercury emissions to the atmosphere from anthropogenic and natural sources. Atmos. Chem. Phys. 10 (13), 5951–5964.

Purvis, O.W., Halls, C., 1996. A review of lichens in metal-enriched environments. Lichenologist 28, 571–601.

Sarret, G., Manceau, A., Cuny, D., Van Haluwyn, D.C., Deruelle, S., Hazemann, J.L., Soldo, Y., Eybert-Berard, L., Eybert-Berard, J.J., 1998. Mechanisms of lichen resistance to metallic pollution. Environ. Sci. Technol. 32 (21), 3325–3330.

Schroeder, W.H., Munthe, J., 1998. Atmospheric mercury—an overview. Atmos. Environ. 32 (5), 809–822.

Sensen, M., Richardson, D.H.S., 2002. Mercury levels in lichens from different host trees around a chlor-alkali plant in New Brunswick, Canada. Sci. Total Environ. 293 (1–3), 31–45.

Sherman, L.S., Blum, J.D., Johnson, K.P., Keeler, G.J., James, A., Barres, J.A., Douglas, T.A., 2010. Mass-independent fractionation of mercury isotopes in Arctic snow driven by sunlight. Nat. Geosci. 3 (3), 173–177.

Sherman, L.S., Blum, J.D., Keeler, G.J., Demers, J.D., Dvonch, J.T., 2012. Investigation of local mercury deposition from a coal-fired power plant using mercury isotopes. Environ. Sci. Technol. 46 (1), 382–390.

Spiric, Z., Mashyanov, N.R., 2000. Mercury measurements in ambient air near natural gas processing facilities. Fresenius J. Anal. Chem. 366 (5), 429–432.

Steinnes, E., Krog, H., 1977. Mercury, arsenic and selenium fallout from an industrial complex studied by means of lichen transplants. Oikos 28 (2–3), 160–164.

Timoney, K.P., Lee, P., 2009. Does the Alberta Tar Sands Industry Pollute? The Scientific Evidence. Open Conserv. Biol. J. 3, 65–81.

Toole-O'Neil, B., Tewalt, S.J., Finkelman, R.B., Akers, D.J., 1999. Mercury concentration in coal—unraveling the puzzle. Fuel 78 (1), 47–54.

Tretiach, M., Fabio Candotto Carniel, F., Loppi, S., Alberto Carniel, A., Bortolussi, A., Mazzilis, D., Del Bianc, C., 2011. Lichen transplants as a suitable tool to identify mercury pollution from waste incinerators: a case study from NE Italy. Environ. Monit. Assess. 175 (1–4), 589–600.

Wilhelm, S.M., Bloom, N., 2000. Mercury in petroleum. Fuel Process. Technol. 63 (1), 1–27.

Wilhelm, S.M., Liang, L., Cussen, D., Kirchgessner, D.A., 2007. Mercury in crude oil processed in the United States (2004). Environ. Sci. Technol. 41 (13), 4509–4514.

Williamson, B.J., Purvis, O.W., Mikhailova, I.N., Spiro, B., Udachin, V., 2008. The lichen transplant methodology in the source apportionment of metal deposition around a copper smelter in the former mining town of Karabash, Russia. Environ. Monit. Assess. 141 (1–3), 227–236.

Zheng, W., Hintelmann, H., 2010. Isotope fractionation of mercury during its photochemical reduction by low-molecular-weight organic compounds. J. Phys. Chem. A 114 (12), 4246–4253.

Chapter 17

Measurement of Polynuclear Aromatic Hydrocarbons (PAHs) in Epiphytic Lichens for Receptor Modeling in the Athabasca Oil Sands Region (AOSR): A Pilot Study

W.B. Studabaker*[1], S. Krupa[†], R.K.M. Jayanty* and J.H. Raymer*

*RTI International, Post Office Box 12194, Research Triangle Park, Durham, North Carolina, USA

[†]Plant Pathology, University of Minnesota-Twin Cities, St. Paul, Minnesota, USA

[1]Corresponding author: e-mail: wstudabaker@rti.org

ABSTRACT

The use of measurements of polycyclic aromatic hydrocarbons (PAHs) in lichens was evaluated in a pilot study as a tool for receptor modeling of air pollution from mining and oil extraction activities in the Athabasca Oil Sands Region (AOSR). Lichen samples (*Hypogymnia physodes*) were collected at 20 locations within 150 km of the mining and oil extraction center located near Fort McKay. Samples were cleaned, homogenized using a cryogenic impactor, and extracted with cyclohexane. Extracts were cleaned up using Florisil solid-phase extraction and analyzed by gas chromatography with mass selective detection (GC/MS). Total PAHs (20 compounds) ranged from 52 to 350 µg/kg, comparable to values reported from other studies of PAHs in lichens. Analysis of air samples collected using polyurethane foam or dichotomous air sampler filters showed that PAHs were dominated by naphthalene and 3-ring congeners, whereas lichen samples from comparable locations yielded significant contributions from 4-, 5-, and 6-ring PAHs. The concentration of total PAHs in lichen samples and the relative contribution of higher ring number

Developments in Environmental Science, Vol. 11. http://dx.doi.org/10.1016/B978-0-08-099443-7.00017-X

PAHs both decreased with increasing distance from the mining and oil extraction centers, consistent with less efficient vapor-phase transport of the less-volatile higher ring number PAHs. PAH congener profiles for samples varied depending on distance from the mines. Principal components analysis incorporating analytical data from a variety of matrices indicated multiple factors contribute to PAH content in a given sample. Measurements of metals in the 20 lichen samples showed significant correlations ($r^2 > 0.8$, $p < 0.05$) between crustal element levels and total PAHs. Taken together, the results suggest that PAH concentrations and profiles in lichens depend on influences from multiple sources and transport mechanisms. A more fully validated approach is needed to develop an understanding of sampling and analytical variability and bias, and a review of recent research in this area indicates that this is a significant gap in many lichen-PAH biomonitoring studies.

17.1 INTRODUCTION

Polycyclic (or polynuclear) aromatic hydrocarbons, PAHs, are pollutants that derive from combustion, industrial, or natural sources. PAHs in air pollution are commonly associated with particulate matter (PM). However, PAHs occur in both PM and gas phases depending on their molecular weight and chemical structure. Human exposure to PM is linked to adverse cardiovascular (Gill et al., 2011; Korashy and El-Kadi, 2006; Sun et al., 2010b) and respiratory (Bosetti et al., 2007; Di Giampaolo et al., 2011; Sacks et al., 2011) health effects, although specific associations of PAHs to those outcomes have yet to be made (Korashy and El-Kadi, 2006). Nevertheless, certain PAHs (notably benzo[*a*]pyrene) are well established as human carcinogens (Bostrom et al., 2002). Air pollution by PM and PAHs in Alberta and the Canadian Arctic has been implicated in human health outcomes, including asthma (Villeneuve et al., 2007), migraines and headaches (Szyszkowicz et al., 2009), and stroke (Johnson et al., 2010), as well as to environmental effects, such as elevated hormone levels in tree swallows (Gentes et al., 2007), increased PAH levels in aquatic insects (Wayland et al., 2008), and increased mortality in minnow embryos (Kelly et al., 2010). The generation of PM and PAHs from industrial activities associated with the development of the Athabasca oil sands has raised concern over potential impacts on both human and ecological health.

The sources of air pollution may be local, regional, or global, and the types and amounts of pollutants found in any location are dependent on atmospheric transport mechanisms on those scales. Both source and receptor models are used in relating emissions of pollutants to measurements at a particular site. Source (or dispersion) models begin with the identification of a specific pollutant source and employ meteorological data to project where, and in what amounts, the pollutant may be taken up by a receptor. Receptor models employ multiple measurements, at multiple locations, to assign contributions of pollutants from multiple sources; in that way, receptor data can be used to validate source models. The distribution of PAH compounds that may typically be found in samples can vary significantly and is influenced by the nature of the source (Baumard et al., 1998; Dahle et al., 2003; Yunker and Macdonald, 2003), as well as the distance from the source, because low-molecular weight PAHs are transported over greater distances in the vapor phase than high-molecular weight PAHs, which are transported in association with PM (Cautreels and Vancauwenberghe, 1978). Receptor modeling has become an important tool in our understanding of pollution by PAHs.

The most sophisticated receptor models are able to take advantage of the existence of large pollution-monitoring networks. Frequently, however, such networks are not located in areas of interest, and it may be either economically or logistically unfeasible to deploy monitoring stations to obtain the required data. For low-volatility or nonvolatile analytes, plant samples are frequently used as passive media for air sampling. The use of biomonitoring for pollution by inorganic species and trace elements is well established (Conti and Cecchetti, 2001; Wolterbeek, 2002). More recently, a variety of plant species, including lichens, mosses, and pine trees (using needles) have been employed for monitoring of semivolatile or persistent organic pollutants, including PAHs (Augusto et al., 2011; Foan et al., 2010; Lehndorff and Schwark, 2009a,b, 2010; Migaszewski et al., 2009; Ratola et al., 2009, 2010, 2011; Rodriguez et al., 2010; Sun et al., 2010a; Usenko et al., 2010; Wang et al., 2009). The use of biomonitoring approaches is well suited for monitoring PAHS in the largely undeveloped region surrounding the oil sands development.

17.1.1   Lichens as Bioaccumulators of PAHs

Epiphytic lichens are an important marker of ecosystem health (Conti and Cecchetti, 2001; Estrabou et al., 2011). Moreover, they are well-established bioaccumulators of trace metals (Wolterbeek, 2002). Two approaches have typically been taken in using lichens in monitoring studies: mapping all lichen species over a region and correlating variation or changes in species composition with pollutant levels; or transplanting lichen species from remote (low contamination) regions to locations where higher pollutant levels are found or expected, and evaluating changes in lichen health as a function of exposure to air pollution. Many different inorganic markers of exposure have been used and include metals, in particular, heavy metals, sulfur, nitrogen and phosphorus, ozone, halides, and radionuclides. On the other hand, studies describing the incidence or impacts of pollution by organic compounds using biomonitors are rare, and the authors specifically pointed to a paucity of studies examining uptake of polychlorodibenzodioxins and polychlorodibenzofurans (Conti and Cecchetti, 2001). Although, as a result, there is little information regarding mechanisms for uptake of organic chemicals by lichens, more is known about uptake by other plant species. The cuticle, the waxy outer coating of leaves and needles, plays a central role in mobilizing organic chemicals from deposited atmospheric PM; once mobilized, nonpolar organic compounds (such as PAHs) move by simple diffusion. The chemistry of the cuticle is highly species dependent and varies considerably according to the development state of the leaf and the plant (Schwarz and Jones, 1997; Simonich and Hites, 1995) which greatly influence interactions between leaf cuticles and the atmosphere (see Percy et al., 1993). Lichens do not have cuticles; instead, they secrete an array of secondary metabolites called lichen acids onto their surfaces. The makeup of lichen acids is diverse across lichen species (Oksanen, 2006), and while little is known about their involvement in the uptake of organic pollutants by lichens, it is not unreasonable to expect a role comparable to that played by the cuticle of the higher plants.

Investigations of lichens as bioaccumulators of PAHs date back nearly 30 years (Carlberg et al., 1983; Thomas et al., 1984), and the past decade has seen a progressive increase in the use of epiphytic lichens for PAH monitoring, apportionment, and modeling (summarized in Table 17.1). Study designs specifically targeting PAHs have fallen broadly into two categories, although a given study may incorporate elements of both. At one extreme, lichens constitute only one among a number of environmental sample types, such as plant materials (mosses, bark, pine needles), soil, sediment, or snow. Analytes include PAHs as well as other organic and inorganic chemical markers of atmospheric pollution. These studies, for the large part, function as surveys of ecological markers of pollution for a specific locale. At the other extreme, lichens (from one or more species) and PAHs (with frequent inclusion of metals also associated with pollution by PM) are the exclusive focus. Data from these studies describe spatial or temporal variability in PAH pollution that can be associated with local variation in industrial or transportation activity, and data are often analyzed using multivariate approaches.

The first reported measurements of PAHs in lichens were part of biomarker survey-type studies in Norway and in Sweden; in each study, a single sample was analyzed for PAHs as well as chlorinated pesticides, PCBs, alkanes, phthalates, and/or metals (Carlberg et al., 1983; Thomas et al., 1984). In both cases, measurements of reported individual PAHs were low (<10 ng/g, except for fluoranthene ca. 60 ng/g in Thomas et al., 1984), despite the samples having been collected in a location or region with significant industrial activity or elevated levels of air pollution. In each of these studies, limitations in the analytical methodology may have contributed to the measurement of low values. In contrast, a later study in Poland that included lichen samples in addition to mosses, bark, pine needles, and soils from up to three different locations, reported total PAHs of 1184–2253 ng/g (Migaszewski et al., 2002). Total PAH concentrations in lichens were much higher than those found in other biological matrices (next highest were mosses at ~600 ng/g), establishing lichens as efficient bioaccumulators. The distribution of PAH congeners, as measured by total PAHs by ring number, varied significantly by matrix. Humic soil fractions had PAH concentrations comparable to those in lichens but were

TABLE 17.1 Listing of Studies Investigating Lichens as Bioaccumulators of PAHs

Citation	Specie(s)	Study location	#Samples	Time points	Sample type	Other analytes	Analytical method	$\sum$PAH range (ng/g lichen)	Features or findings
Carlberg et al. (1983)	*Hypogymnia physodes*	Norway	1	1	Endogenous	Organochlorines, alkanes, phthalates	Cyclohexane-isopropanol, sulfuric acid cleanup, GC–MS	20–40 (estimate)	Mainly 3- and 4-ring PAHs, attributed to regional background
Thomas et al. (1984)	*Cladonia rangiferina*	Sweden	1	1	Endogenous	Organochlorines	Acetone/Soxhlet, alumina/silica cleanup, thin layer chromatography/UV	ca. 74	Investigators also analyzed leaf litter and several species of mosses and evergreens, highest $\sum$PAH in litter
Migaszewski et al. (2002)	*Hypogymnia physodes*	Poland	5	1	Endogenous	Phenols, Cd, Cu, Hg, Pb, S, Zn	CH_2Cl_2/Soxhlet, no cleanup, GC–MS	1184–2253	Investigators analyzed various receptors (soil, moss, pines); lichens had the highest concentrations. Collection efficiency perhaps superior to bark but may have longer integration period compared to mosses or pine needles

Continued

TABLE 17.1 Listing of Studies Investigating Lichens as Bioaccumulators of PAHs—Cont'd

Citation	Specie(s)	Study location	#Samples	Time points	Sample type	Other analytes	Analytical method	$\sum$PAH range (ng/g lichen)	Features or findings
Naeth and Wilkinson (2008)	Three spp.	Northwest Territories, Canada	24	1	Endogenous	Phthalates; 33 trace elements; sulfate, nitrate, ammonium	Soxhlet/gel permeation/GC–MS	<10	All samples below detection limits for all analytes
Usenko et al. (2010)	Not identified	West USA	14	1	Endogenous	None	CH_2Cl_2/accelerated solvent extraction, gel permeation/ silica cleanup, GC–MS	Lipid adjusted data only	Total PAHs in snowpack, sediment, and lichens used to identify aluminum smelter as a significant local source
Owczarek et al. (2001)	*Physcia adscendens*	Rieti, Italy	6	1	Endogenous	Al, As, Cd, Co, Cr, Mn, Ni, Pb, Zn	CH_2Cl_2/ultrasound, no cleanup, GC–MS	217–1879	PAH and metals correlate with traffic volume but not each other
Guidotti et al. (2003)	*Pseudevernia furfuracea*	Rieti, Italy	41	5	Transplanted	None	Cyclohexane/ ultrasound, silica gel cleanup, GC–MS	36–375	$\sum$PAH correlates with traffic volume and changes over time are readily measurable
Guidotti et al. (2009)	*Pseudevernia furfuracea*	Viterbo, Italy	6	2	Transplanted	Cd, Cr, Cu, Ni, Pb, Zn	Cyclohexane/ ultrasound, silica	33–395	$\sum$PAH correlates with traffic volume (metals less so) and

							gel cleanup, GC–MS		changes over time are readily measurable
Domeno et al. (2006)	*Xanthoria parietina*	Zaragoza, Spain	1	1	Endogenous	None	CH_2Cl_2/dynamic sonication (DSASE), alumina cleanup, GC–MS	ca. 330	Extraction method development
Blasco et al. (2006)	*Parmelia sulcata*	Pyrenees, Spain	6	1	Endogenous	None	CH_2Cl_2/dynamic sonication (DSASE), alumina cleanup, GC–MS	910–1920	Colocated active sampling, ratios of PAH compounds consistent with traffic sources
Blasco et al. (2007)	*Parmelia sulcata*	Pyrenees, Spain	8	1	Endogenous	None	CH_2Cl_2/DSASE, aminopropyl SPE cleanup, GC–MS	352–1654	SPE method development
Blasco et al. (2008)	*Evernia prunastri*	Pyrenees, Spain	10	1	Endogenous	None	CH_2Cl_2/DSASE, aminopropyl SPE cleanup, GC–MS	696–6240	Analyzed samples in triplicate, 95% confidence interval ranged from 1% to 18%. Seven other species monitored as sentinels of ecosystem health, suggest referencing of Index of Air Purity to PAHs

Continued

TABLE 17.1 Listing of Studies Investigating Lichens as Bioaccumulators of PAHs—Cont'd

Citation	Specie(s)	Study location	#Samples	Time points	Sample type	Other analytes	Analytical method	$\sum$PAH range (ng/g lichen)	Features or findings
Blasco et al. (2011)	Six spp.	Pyrenees, Spain	156	1	Endogenous	None	CH_2Cl_2/DSASE, aminopropyl SPE cleanup, GC–MS	238–6240	Significant interspecies variation in PAH uptake, dependence of total PAH on proximity to traffic, ratio of PAH compounds consistent with traffic sources
Shukla et al. (2010)	Seven spp.	Himalayas region, India	11	1	Endogenous	None	CH_2Cl_2/Soxhlet, silica cleanup, HPLC–UV	3380–25010	Predominantly naphthalene and 3-ring PAHs, mainly vehicular source with some petrogenic source; very high concentrations of acenaphthylene relative to all other studies
Shukla and Upreti (2009)	*Phaeophyscia hispidula*	Himalayas region, India	9	1	Endogenous	None	CH_2Cl_2/Soxhlet, silica cleanup, HPLC–UV	683–33720	Very high concentrations of acenaphthylene relative to all other studies

Satya et al. (2012)	*Rinodina sophodes*	Kanpur City, India	18	1	Endogenous	None	CH_2Cl_2/Soxhlet, silica cleanup, HPLC–UV	189–494	Collected triplicate samples from each of six sites; RSD for $\sum$PAH ranged from 8% to 21%. Found mainly 3-ring PAHS and pyrene, attributed to vehicular sources
Augusto et al. (2009), Augusto et al. (2010)	*Parmotrema hypoleucinum, Xanthoria parietina*	Sines, Portugal	34	1	Endogenous	Si, Ni, Mn, Ca, Cr, Al, K, Fe, Co, Ti, Zn, and Mg; S and N	Acetonitrile/Soxhlet, Florisil cleanup, HPLC–UV	127–599	Collected from agricultural, industrial, forest areas. Highest $\sum$PAH in industrialized urban areas; multivariate analysis (PCA) showed differing PAH analyte profiles depending on sampling location. $\sum$PAH higher in lichen than in soil or pine needles from same locations
Fernandez et al. (2011)	*Pyxine coralligera*	Caracas, Venezuela	11	1	Endogenous	None	Cyclohexane-CH_2Cl_2/ultrasound, silica cleanup, HPLC–UV/ fluorescence	240–9080	PAH analyte ratios reflected vehicular sources except near refinery

　　　　　　　　　　　　　　　　　　　　　Alberta Oil Sands

dominated by 5- and 6-ring PAHs (50% or more of the total PAHs). PAHs with four rings constituted the largest fraction in lichens, whereas 3-ring congeners were the largest fraction in mosses; however, the relative amounts of 5- and 6-ring congeners were higher in mosses. Other matrices exhibited much lower total uptake of PAHs, and particularly much lower uptake of the 5- and 6-ring congeners. These differences likely reflect relative differences in deposition rates of PM, which contributes the higher ring congeners, on the different substrates. They also reflect differences in vapor-phase uptake and retention of the lower ring number congeners. The study results highlight the importance of fully characterizing the deposition and uptake characteristics of a proposed biomonitor and suggest a need for longitudinal validation against a standardized sampling method.

Two recent reports have described studies of comparable scope and similar design in North American environments. Three species of lichen were collected along with underlying soil samples from sites around a diamond mine in Northwest Territories, Canada (Naeth and Wilkinson, 2008). Samples from eight of the sites were analyzed and none of them, from any of the three species, yielded any PAH congener above the limit of detection. Although the sampling was conducted in a locale with some industrial activity, this is not completely surprising, as background levels of PAHs in both the vapor and particulate phase in the Canadian arctic are historically <1 ng/m^3 (Hung et al., 2005). In a study of pollution in U.S. National Parks, lichens from nine species were collected along with sediment cores and snowpack from eight different lake catchments and were used to identify an aluminum smelter as a significant local emissions source of PAHs (Usenko et al., 2010). PAH concentration data from this study cannot be directly compared with other studies as the authors adjusted all concentrations (reported as total PAH only) to μg/g lipid to reflect interspecies difference in uptake of PAHs. This is a useful approach for normalizing concentrations of PAHs absorbed from the vapor phase, and the detection primarily of 3- and 4-ring PAHs in the study lichen samples, indicating primarily vapor-phase uptake, supports its use. However, the use of lipid-normalized measurements of higher ring number PAHs associated with PM is problematic, as it is not obvious what role lipids (as opposed to, e.g., lichen acids) may play in the accumulation of PM. PAH congener ratios, commonly used for source apportionment (Kavouras et al., 2001), are unaffected by this correction.

17.1.2　Factors Affecting Bioaccumulation of PAHs by Lichens

While lichens can be effective PAH accumulators, uptake of PAHs by lichens may depend on a variety of factors beyond source and transport considerations, including growth rates (which may range from tens of millimeters per year to a few millimeters per century), local environmental conditions, and species morphology. Longitudinal exposure studies have shown that uptake of PAHs by a lichen species can take place at measurable levels over a period of months (Guidotti et al., 2003, 2009). Samples of *Pseudevernia furfuracea* were collected from a remote location, composited, and analyzed to provide a background concentration, then redivided and suspended in nylon nets at locations of varying traffic density in the cities of Rieti and Viterbo, Italy. Portions of the lichen were collected at intervals of 3–6 months for a study duration of 20 (Rieti) or 3 (Viterbo) months. PAH concentrations, measured in all samples at the conclusion of the sampling period, showed continuous increase from the baseline PAH_{TOTAL} of 33–36 µg/kg, with maximum observed PAH_{TOTAL} of 350–400 µg/kg in both studies. In Rieti, individual and total PAHs increased during each time period, and while the increases were variable, no seasonal bias was evident. In both cities, the primary contributors to PAH_{TOTAL} were the 3- and 4-ring species phenanthrene, anthracene, fluoranthene, and pyrene, and the amounts of each increased over the exposure period, commensurate with subjective assessments of local motor vehicle activity (described as none, light, moderate, or heavy traffic).

The longitudinal data create a context for interpreting data for a previous study in Rieti, during which endogenous *Physcia adscendens* was collected from six sites, again designated as low-, medium-, or high-traffic volume (Owczarek et al., 2001). Total PAHs from the six samples ranged from 217 (from a location in a park) to 1879 ng/g (at a location described as "medium"-traffic density, but where there was a concurrent community event described as a "sample fair," which was implied by the authors to have been associated with increased emissions). The authors suggested that the high levels in the latter sample ($\sim 2\times$ the next closest sample) may have reflected the ability of *P. adscendens* to accumulate PAHs quickly over a short period of time. However, given the gradual accumulation of PAHs seen in the subsequent longitudinal Rieti study (~ 5–10 ng/g/month), even in high-traffic areas, such an increase would seem improbably attributed only to vehicular traffic. In fact, the high contribution of 5- and 6-ring PAHs to total PAHs from the "sample fair" sample, relative to four of the other locations, would suggest looking for alternate explanations. Although differences in PAH accumulation rates between the two lichen species could be responsible (Blasco et al., 2011), the data point to the hazards of using measurements taken at a single point in time and highlight the need for longitudinal studies of PAH uptake specific to the lichen species being investigated.

17.1.3   Use of Lichens in Source Apportionment of PAHs

Three studies in India have explored the use of measurements of PAHs in lichens for performing source apportionment using this approach (Satya et al., 2012; Shukla and Upreti, 2009; Shukla et al., 2010). The investigators relied on congener ratios to differentiate the origins of PAHs according to petrogenic sources (uncombusted fossil fuel), and combustion from a variety of sources, as well as to distinguish between differing combustion sources (Baumard et al., 1998; Dahle et al., 2003; Pandey et al., 1999; Yunker and Macdonald, 2003). PAHs were determined by HPLC–UV$_{254}$ in all three studies. In two studies, in the vicinity of Dehradun, Himalayas, similar (and very high, relative to all other studies) total PAH measurements were obtained (680–33,000 and 3400–25,000 ng/g, respectively). Examination of reported concentrations of individual PAHs revealed that particularly in samples with high concentrations, two or three congeners were responsible for 90% or more of the total PAHs present; furthermore, 5- and 6-ring PAHs are reported to be absent. Acenaphthylene was a dominant congener in many of the samples of the Dehradun studies. Data from other studies have shown that in a sample that has a comparatively high total PAH concentration, the distribution of PAHs is not so skewed, and includes 5- and 6-ring compounds (Augusto et al., 2010; Blasco et al., 2008; Migaszewski et al., 2002). The Kanpur City study in India included collection of triplicate samples from each field location. They reported coefficients of variation for all detected analytes at each site, and values often exceed 100%. While the confidence value for $n=3$ is low, this is to date the only published evidence for sampling variability in measurements of PAHs in lichens, and points to a need to include sampling replicates in future studies. The authors concluded that although PAHs arose from multiple sources, vehicular traffic was the primary contributor.

Similar conclusions were drawn from measurements of PAHs in *Pyxine coralligera* samples collected in Caracas, Venezuela (Fernandez et al., 2011). Samples were collected in triplicate (although data reflecting sampling variability were not reported) from 11 locations, described as very low, low, low-medium, medium-high, and high pollution. Complete data were reported for 10 of the sites (all but the "very low" control site), as well as colocated PM$_{10}$ sampler data. PAHs were determined by HPLC–UV/fluorescence without confirmatory analysis. Total PAHs ranged from 240 to 9080 ng/g, with a median of 4620 ng/g. There was no pattern of agreement among any of total PAHs, the descriptive pollution level, and PM$_{10}$ values, outside of the "low" pollution sample being distinct from the others. Principal component analysis (PCA) showed clustering of PAHs into three groups: 2- and 3-ring PAHs (naphthalene, acenaphthylene, acenaphthene, fluorene, and anthracene), 4-ring (fluoranthene, chrysene, and pyrene), and 5-ring + phenanthrene. Phenanthrene concentrations were much lower than for other 2- and 3-ring compounds. Based on clustering patterns and PAH ratios, the authors found that for most sampling locations, vehicular traffic was the primary contributor to the PAHs, except for two sites near a petroleum refining center, where the primary contributor was petrogenic.

As was the case in the Rieti studies, specific local variables in both the Indian and Venezuelan studies might argue for alternatives to the primary explanation for the PAH distributions as being dependent on local traffic. These might include, for example, selective uptake (as suggested by the authors of the Caracas study), slow lichen growth rates resulting in a very long integration period for volatile PAHs, or local alternative fuels yielding unusual PAH distributions, such as the high acenaphthylene concentrations in samples from Dehradun not observed elsewhere.

A clearer picture emerges from studies that incorporate additional metrics into their design. Near Sines, Portugal, 34 samples of *Parmotrema hypoleucinum* were collected from a 20 km × 30 km, industrialized area that included a coal-fired power plant and an oil refinery (Augusto et al., 2009, 2010). The authors selected diverse sites, analyzed pine needles and soil samples, and included trace elements in their analysis. Concentrations of PAHs in lichens, measured by HPLC–UV$_{254}$-fluorescence, are described in Table 17.3. Lichen and soil concentrations were highly correlated, with $R = 0.95$ ($p < 0.05$) for total PAHs and $R > 0.8$ for many of the individual PAHs. Concentrations in lichens were approximately twice as high as in pine needles. PAH profiles were compared by summing PAH species by the number of rings. Statistically significant differences between lichens and the two other ecological monitors were found only for naphthalene in soil, and 3-ring PAHs in pine needles; given the use of 3-ring PAH data for source apportionment, this points to the importance of interpreting the data in the context of the biological matrix involved. PCA showed associations of 3-ring PAHs with forested environments, 4-ring PAHs with urban environments, and 5- and 6-ring PAHs with urban environments. Significant ($p < 0.05$) correlations were found for measurements of the elements Cr, Al, and Fe and 5-ring PAHs, and between Cr and Zn and 6-ring PAHs, and a negative correlation was found for phenanthrene and lead. During the study, two species of lichens (*P. hypoleucinum* and *Xanthoria parietina*) were collected from sites with colocated high-volume air samplers, and both species showed a bias toward accumulation of 2- and 3-ring PAHs from the vapor phase and against accumulation of 5- and 6-ring PAHs via deposition of PM. The authors concluded that lichens are a superior bioaccumulator to soil or pine needles and could be used to distinguish between environments with differing local source contributions, and suggested that lichen biomonitoring data should be translatable to other types of monitoring data. In this study, the use of colocated samplers as a reference sampling methodology is an essential element in validating the performance of the biomonitor.

A series of studies in Spain, including analytical method development, have culminated in the largest effort to date using lichens as PAH monitors (Blasco et al., 2006, 2007, 2008, 2011; Domeno et al., 2006). Sampling for these studies was conducted near major roadways on both sides of the border between France and Spain, in the Pyrenees mountains. Samples were analyzed using a fully characterized analytical method employing sonication, SPE, and GC–MS. A pilot study (Blasco et al., 2006) showed that selected PAH ratios were similar for *Parmelia sulcata* collected outside the Somport tunnel and air samples collected outside the tunnel and were characteristic for vehicular traffic as a source. A second study found an inverse correlation ($R = -0.394$, $p = 0.046$) between combustion-associated PAHs in *Evernia prunastri*, collected near a highway south of the Somport tunnel, with a quantitative measure of lichen biodiversity (Blasco et al., 2008). These results prompted a study encompassing six lichen species (*P. sulcata*, *E. prunastri*, *Ramalina farinacea*, *P. furfuracea*, *Usnea* sp., and *Lobaria pulmonaria*), sampling locations from diverse sites throughout the valleys north and south of the Somport tunnel, and replicate analysis of pooled samples (Blasco et al., 2011). Summary data across all species are shown in Table 17.4. Total PAH concentrations and distributions by ring number varied by lichen species, and although some of the variability can be ascribed to the absence of some species from many of the sampling sites, the differences point to the potential importance of building models based on the behavior of a specific species. For example, while PAH concentrations decreased with increasing distance from the roadway across species, the important phenanthrene/anthracene ratio varied significantly, so that use of a single species without characterization of uptake properties could lead to a biased source attribution. Using results from discriminant analysis, the authors related interspecies differences in PAH profiles to morphological and water sorption characteristics of the species, so that fruticose lichens dependent on wet deposition tend to accumulate more 3-ring PAHs while high-surface area foliose lichens tend to accumulate more higher ring number PAHs.

17.1.4 Use of Lichens as Bioaccumulators in the AOSR

During the summer of 2008, the Wood Buffalo Environmental Association (WBEA) collected samples of the epiphytic lichen, *Hypogymnia physodes*, from sites within ca. 150 km of the center of the oil sands development near Ft. McKay, Alberta, for the purpose of determining total sulfur and nitrogen accumulation in plant tissue. This study reports the results of the analysis of those samples for PAHs, and compares the data to prior studies, as well as to data from analysis of filter samples collected using conventional polyurethane foam (PUF) and dichotomous samplers, and to refining process samples from the oil sands development. Analytical data are further correlated with trace metals analyses performed on the same samples (Chapters 14 and 18).

17.2 METHODS

17.2.1 Sample Collection

The 20 lichen sample sites used in this pilot study were selected from a total of 300+ sites to provide a range of atmospheric PAH exposures. Specifically, locations were selected both north and south of the existing oil sands operations within the Athabasca River Valley that have described microclimatic and topographic conditions (Figure 17.1). Where possible, colocated lichen

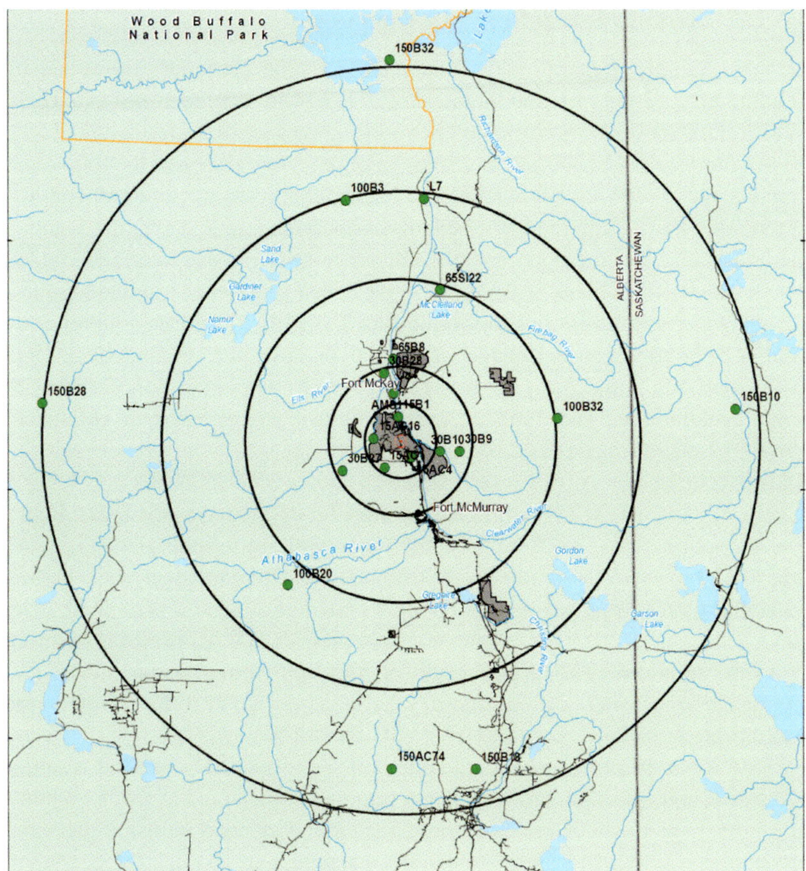

FIGURE 17.1   Lichen sampling locations.

and air quality monitoring sites were also selected. More distant locations, outside the potential direct industrial sources of influence, were identified to provide suitable background information on PAH concentrations in the region using sulfur and nitrogen concentrations (high or low) in the lichen tissue, as an additional criterion (Berryman et al., 2010).

A more detailed field-sampling program was implemented during August–September 2008. Most sites were accessed by helicopter and others were visited by ground. At each site, compound samples of the lichen *H. physodes* were collected from the branches of six trees interior in the stand and the samples were frozen. Using Teflon-coated tweezers, tissue samples were cleaned by removing all foreign materials (e.g., bark, etc.) before subsequent chemical analyses. GPS location data for all sites were used to determine distance from a midpoint located in the area of the existing facilities.

17.2.2 Sample Extraction and Cleanup

Lichens: Sample preparation was adapted from the procedure described by Guidotti et al. (2003). Lichen samples (ca. 0.5 g) were homogenized in a SPEX freezer mill (10–20 s). A portion of the resulting powder (0.2 g) was spiked with 20 µL internal standard solution (250 ng/mL, i.e., 5 ng, of each of the following: naphthalene-d_8, acenaphthylene-d_8, phenanthrene-d_{10}, fluoranthene-d_{10}, pyrene-d_{10}, benzo[*a*]pyrene-d_{12}, benzo[*g,h,i*]perylene-d_{12}). A reagent blank, a method blank, and method control were processed with the single analytical batch. The method blank was a composite lichen sample; the method control sample was a composite lichen sample fortified with 20 µL of a solution containing 250 ng/mL (5 ng) of each target analyte. After the spiking solvent had evaporated, the sample was extracted two times by sonicating for 30 min with 3 mL cyclohexane. The extracts were combined and reduced in volume to about 0.2 mL using a TurboVap LV, with decane as a keeper solvent. The concentrated extracts were purified using automated solid-phase extraction (SPE) on a Gilson GX-271, with a custom rack for use with glass SPE cartridges. The 1-g silica cartridges were preconditioned with 4 mL methylene chloride, then 4 mL hexane. The sample was then applied and the cartridges were washed with a total of 2.7 mL hexane. PAHs were eluted with 2.5 mL 40% methylene chloride–60% hexane. The eluate was spiked with 200 µL recovery internal standard solution (250 ng/mL each of the following: acenaphthene-d_{10}, anthracene-d_{10}, benz[*a*]anthracene-d_{12}, perylene-d_{12}). The final solution was concentrated to 0.2 mL for analysis by GC–MS.

Source materials: These included raw oil sands materials, process residues and waste, and environmental materials. Samples were stored at room temperature prior to analysis. Samples were extracted and cleaned up using the same procedures as the lichen samples, using 1 g sample and 10 mL cyclohexane in each extraction step. Extracts that were heavily contaminated with petroleum materials were subsampled and diluted 10× (20 µL into 200 µL) before cleanup.

17.2.3 GC–MS Analysis

Samples were analyzed on an Agilent 6890/5973 GC–MS. Samples (1 µL) were injected using an autosampler via a splitless injector at 250 °C onto an Agilent HP-5MS column (30 m × 0.25 mm × 0.2 µm) with helium carrier at 1 mL/min constant flow. The oven temperature was programmed with the initial temperature set at 70 °C for 4 min, then ramped at 7 °C/min to 300 °C, with a final hold time of 23.14 min (total run time, 60 min). The interface temperature was 290 °C, the source temperature 230 °C, and the quadrupole temperature 150 °C. The 5973 MSD was operated in electron ionization mode at 70 eV with specific ion monitoring. Acquisition parameters are provided in Table 17.2. The instrument was calibrated using six solvent-based standards ranging from 5 to 250 ng/mL. Calibration curves were a linear fit without

TABLE 17.2 GC–MS Acquisition Parameters

	Retention time	Primary ion	Secondary ion
Naphthalene-d_8	15.12	136	108
Naphthalene	15.19	128	102
Acenaphthylene-d_8	20.58	160	80
Acenaphthylene	20.64	152	150
Acenaphthene-d_{10}	21.14	164	160
Acenaphthene	21.25	154	76
Fluorene	23.04	166	165
Phenanthrene-d_{10}	26.26	188	94
Phenanthrene	26.34	178	89
Anthracene-d_{10}	26.46	188	94
Anthracene	26.52	178	89
Fluoranthene-d_{10}	30.37	212	106
Fluoranthene	30.43	202	101
Pyrene-d_{10}	31.13	212	106
Pyrene	31.2	202	101
Benzo[c]phenanthrene	34.58	228	113
Benz[a]anthracene-d_{12}	35.22	240	120
Benz[a]anthracene	35.29	228	113
Chrysene	35.42	228	113
Benzo[b,j,k]fluoranthene	38.91	252	126
7,12-Dimethylbenz[a]anthracene	39.01	252	126
Benzo[e]pyrene	39.95	252	126
Benzo[a]pyrene-d_{12}	40.06	264	132
Benzo[a]pyrene	40.16	252	126
Perylene-d_{12}	40.39	264	132
3-Methylcholanthrene	41.67	268	252
Indeno[1,2,3-cd]pyrene	45.54	276	138
Dibenz[a,h]anthracene	45.72	278	139

Continued

TABLE 17.2 GC–MS Acquisition Parameters—Cont'd

	Retention time	Primary ion	Secondary ion
Benzo[g,h,i]perylene-d_{12}	46.92	288	144
Benzo[g,h,i]perylene	47.09	276	138
Dibenzo[ah,ai,al]pyrene	56.19	302	150

weighting, and correlation coefficients for all analytes were greater than 0.999. Quantification based on the secondary ion that was required to be $\pm20\%$ from the primary ion for reporting.

17.2.4 Quality Control

Each matrix type was analyzed as a separate batch on the GC–MS. Solvent blanks and calibration check standards (at two concentrations) were analyzed at the beginning and end of each run, and after every 10 samples. Solvent blanks were required to be below the instrument detection limit. Check standards were required to be 85–115% of the expected value. Method control recoveries were expected to be 70–130% after correcting for the method blank. Samples exceeding the upper limit of the calibration curve were diluted and spiked with additional internal standard, then reanalyzed.

17.2.5 Air Samples

Time-integrated ambient air samples were collected by WBEA using URG (Chapel Hill, NC) $1'' \times 3''$ polyurethane foam (PUF) samplers and 47-mm glass fiber filters (Chapters 4 and 9). Sample analysis was provided by Airzone One Ltd (Mississauga, Ontario, Canada). Briefly, samples were extracted using accelerated solvent extraction (hexane/acetone) and analyzed by GC–MS. Quantification was accomplished using a six-point standard curve with $R^2 > 0.99$. All results were blank corrected.

17.2.6 Statistical Analysis

PCA was used to classify lichen samples based on their PAH profile. The eigen functions obtained in that way were used to generate PC values for other sample types and observe how other sample types cluster compared to the lichens. Analyses were performed using JMP version 5.0.1 for Windows (SAS Institute, Cary, NC).

17.3   RESULTS

17.3.1   Method Development

The GC–MS method was adapted from U.S. EPA method TO-13A, "Determination of Polycyclic Aromatic Hydrocarbons (PAHs) in Ambient Air Using Gas Chromatography/Mass Spectrometry (GC/MS)." In our method, the standard 16-component method was enhanced with methylated and 6-ring PAHs included in the mixture referred to as the Quebec Ministry of the Environment Mix. The sample preparation method was adapted from Guidotti et al. (2003). The sample size used by those authors was substantial (2 g), but they did not minimize their extract volume for analysis by GC–MS. In adapting their method, we reduced the sample size by a factor of 10 in order to minimize sample consumption; to compensate, we concentrated the final extract to 200 μL and scaled down all processing steps by a factor of 10.

In the published method, samples were ground manually prior to extraction. For *H. physodes*, both manual grinding and ball milling at room temperature gave an incompletely homogenized sample. However, milling at liquid nitrogen temperature (0.5 g sample, magnetic impactor, 20 s) afforded a uniform powder (Figure 17.2). This was important because the greater surface area increases solvent contact with the sample, affording a more complete recovery of the target analytes, and because the more uniform preparation of the sample should give greater reproducibility.

The published cleanup method, using silica gel chromatography, was implemented using silica gel SPE on an automated processor. Automated sample cleanup provides advantages over manual techniques by providing increased precision, reducing opportunities for operator error, and increasing throughput. Method performance was assessed by analysis of seven replicate composite lichen samples spiked with 25 ng/g each analyte. A single blank was used to correct analyte recoveries from each sample. Instrument and final method performance data, with comparison to published values, are presented in Table 17.3.

The recovery of chrysene (57%) was below the target range (70–130%) and was possibly related to the blank subtraction. Dibenzopyrene recovery (34%) was also low; in this case, the cleanup method is suspect, as dibenzopyrene is the most strongly retained analyte on the SPE column. Methylcholanthrene recovery is slightly high (139%). The high concentration of phenanthrene in the composite matrix resulted in a high measured standard deviation (SD) and an elevated coefficient of variation after blanks subtraction; the high SD translated to an estimated method detection limit (MDL, calculated as $3.2 \times SD$) that is likely much higher than the true value. Method performance was otherwise acceptable.

The MDLs were somewhat higher ($2–4\times$) in some cases than the published values but were low enough relative to published sample analyte concentrations to be acceptable; we anticipated detectable concentrations in all samples such that meaningful comparisons can be made. Because of the small

Ball mill, 10 min

Freezer mill (liquid nitrogen), 20 s

FIGURE 17.2   Comparison of techniques for milling *H. physodes*.

number of filter and processed oil sands samples analyzed, no additional method characterization was performed for those matrices.

17.3.2   Sample Analyses

Results for analysis of lichen samples are presented as summary statistics in Table 17.4 and compared to published data from several other recent studies (Augusto et al., 2010; Blasco et al., 2008, 2011). For the most part, phenanthrene was the most abundant PAH from among the analytes for which data were provided, with substantial contributions from naphthalene, fluoranthene, benz[*a*]anthracene, and pyrene. Ranges of total PAHs were roughly comparable (52–350 ng/g vs. 95–874 ng/g).

TABLE 17.3 Analytical Method Performance

	Mean (ng/g)	Recovery (%)	Standard deviation	CV (%)	Method detection limit (ng/g)
Naphthalene	21.7	87	0.62	2.9	2.7
Acenaphthylene	28.4	114	0.51	1.8	3.5
Acenaphthene	29.2	117	0.65	2.2	3.6
Fluorene	21.2	85	0.69	3.3	2.6
Phenanthrene	24.9	100	2.94	11.8	3.1
Anthracene	23.9	96	1.19	5.0	3.0
Fluoranthene	25.1	101	0.80	3.2	3.1
Pyrene	22.3	89	0.49	2.2	2.8
Benzo[c]phenanthrene	20.7	83	0.53	2.6	2.6
Benz[a]anthracene	18.4	74	0.65	3.5	2.3
Chrysene	14.4	57	0.77	5.4	1.8
Benzo[b,j,k]fluoranthene[a]	88.3	118	4.32	4.9	3.6
7,12-Dimethylbenz(a)anthracene	28.2	113	0.85	3.0	3.5
Benzo[e]pyrene	28.2	113	1.12	4.0	3.5
Benzo[a]pyrene	26.3	105	1.45	5.5	3.3
3-Methylcholanthrene	34.8	139	2.14	6.1	4.3
Indeno[1,2,3-cd]pyrene	26.2	105	0.57	2.2	3.3
Dibenz[a,h]anthracene	25.1	100	0.29	1.2	3.1
Benzo[g,h,i]perylene	26.2	105	0.39	1.5	3.3
Dibenzo[ah,ai,al]pyrene[a]	25.5	34	0.52	2.0	1.1

Lichen samples spiked at 25 ng/g.
[a]*Three compounds each at 25 ng/g elute as single peak.*

TABLE 17.4 Summary Statistics for PAHs in Lichens by Analyte and by Study

	This study (N=20)			Augusto et al. (2010) (N=34)			Blasco et al. (2011) (N=156)		
	Min	*Med*	*Max*	*Min*	*Med*	*Max*	*Min*	*Med*	*Max*
Naphthalene	14.3	18.0	28.7	7.7	16.3	156	21	94	474
Acenaphthylene	nd	nd	nd	d	d	d	7	15	85
Acenaphthene	nd	nd	nd	d	1.2	7	9	60	210
Fluorene	nd	d	d	2.8	4.5	19	6	27	509
Phenanthrene	13.7	22.1	71.6	27.7	50.8	202	26	133	2059
Anthracene	nd	nd	13.4	d	1.8	12.3	4	16	140
Fluoranthene	11.7	14.3	31.1	17.6	37.3	174	7	74	1170
Pyrene	d	14.7	33.8	d	32	209	5	37	276
Benzo[c]phenanthrene	nd	nd	d						
Benz[a]anthracene	nd	d	24.3	2.9	8.5	54	9	57	344
Chrysene	nd	15.0	36.2	3.9	10.3	30.3	4	23	119
Benzo[b,j,k]fluoranthene	nd	nd	31.1	2.8	8.1	61	11	47	289
7,12-Dimethylbenz[a]anthracene	nd	nd	d						
Benzo[e]pyrene	nd	nd	29.3						
Benzo[a]pyrene	nd	d	30.0	d	1.6	26	7	14	92
3-Methylcholanthrene	nd	nd	nd						
Indeno[1,2,3-cd]pyrene	nd	nd	nd	nd	d	21.7			
Dibenz[a,h]anthracene	nd	nd	nd				10	17	123
Benzo[g,h,i]perylene	nd	nd	19.3	d	3.1	44.4	8	19	153
Dibenzo[ah,ai,al]pyrene	nd	nd	nd	d	d	4.5			
Total PAH	52.2	124	352	95.5	196	874	238	726	6240

d, Detected, below quantitation limit; nd, no peak detected.

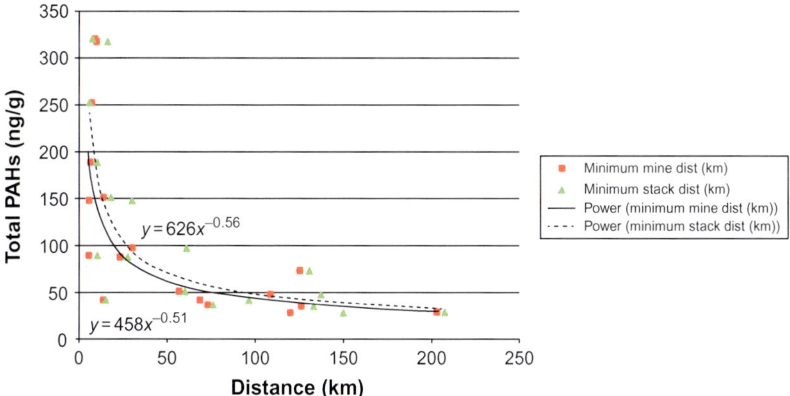

FIGURE 17.3   Total PAH concentration in study samples as a function of their distance from the source.

The geographical coordinates for each sample were used to calculate the distance from the sampling location to the nearest mine and to the nearest upgrading operation. Total PAHs are plotted versus each distance in Figure 17.3. In both cases, PAHs decline approximately with the inverse square root of the distance. Mine operations encompass three emissions sources, from vehicular operations, from refinery operations, and from mine dust, which has native bituminous mineral associated with it. The more rapid drop-off in PAHs with mine distance compared to stack distance suggests that the mine contribution may be predominantly from large particle size fugitive dust that settles out of the atmosphere quickly, whereas stack and vehicular emissions involve more vapor-phase PAHs and smaller PM that is transported farther than dust. PAHs in samples from the most remote locations were near the limits of detection, consistent previous findings (Naeth and Wilkinson, 2008).

PAH congener distributions were used to compare lichen-PAH data to WBEA ambient air PAH levels. Direct quantitative comparison of lichens PAH data with conventional data (air samples collected using a filter or PUF sampler) is not meaningful, due to the inherent differences in volume and integration time between passive collectors (lichens) and active samplers. However, a comparison of the PAH "fingerprint" for each matrix can be informative. Figure 17.4 compares the PAH profiles for the 20 lichens and for 14 PUF samples collected during the summer of 2009 from WBEA air-monitoring stations approximately 30 km from the mines. The profiles were

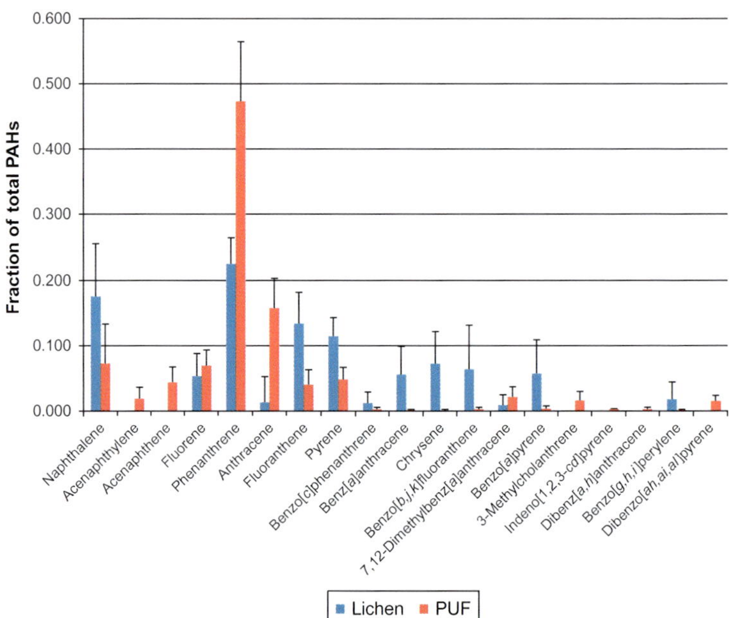

FIGURE 17.4 PAH distribution (relative contribution of individual PAHs to total PAHs) from lichens, PUF samplers.

generated by determining the ratio of each congener to the total PAHs for each sample, the averaging the ratio for each analyte across all samples. While the PUF samples were dominated by phenanthrene and anthracene, which, when combined, account for over 60% of the total PAH mass in the filter samples, the lichen samples yielded more heterogeneous mixtures, with a significant contribution from 4- and 5-ring PAHs to the total. Notably, anthracene was undetected in all but four lichen samples, suggesting a petroleum-based origin, whereas in the PUF samples the phenanthrene/anthracene ratio was <10, suggesting a combustion origin (Baumard et al., 1998). Larger ring PAHs are mostly associated with PM, whereas 2- and 3-ring PAHs are transported in the gas phase (Bostrom et al., 2002). These data suggest that either lichen may be preferentially accumulating PM-associated PAHs relative to gas-phase PAHs or that the lichens on average are exposed to more PM than was present at the WBEA air-monitoring stations.

Potential contributions of oil sands-derived PAHs to the PAHs measured in media for measuring atmospheric deposition could be assessed in the same way. PAH concentrations in oil sands-derived samples varied widely. Samples 510-ARA1 and 515-ARA2 contained <20 ng/g total PAHs each, while the other samples each yielded measurements in the range of 1000–3000 ng/g

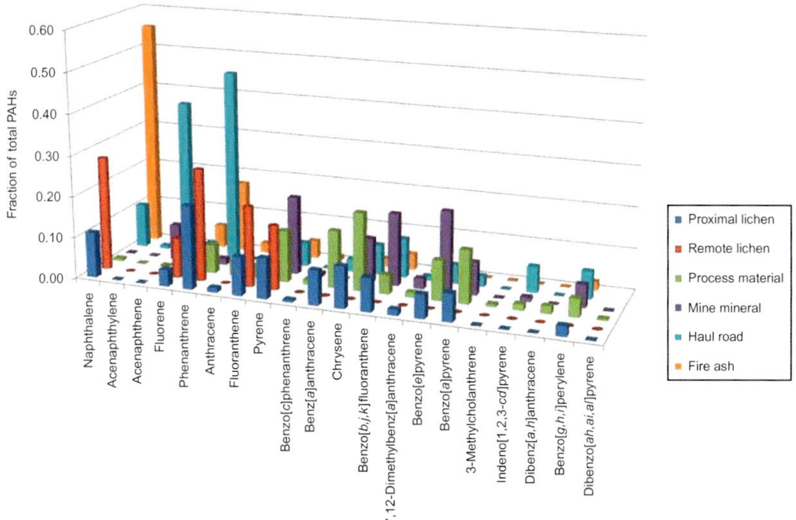

FIGURE 17.5   PAH distribution from lichen, source materials. n: Proximal lichens (4), remote lichens (4), refinery process material (2), mine mineral (2), haul road (2), and forest fire ash (2).

total PAHs. Although the PAH profiles of these samples are distinct from those observed for lichens, PUF samplers, and dichotomous samplers, there are still similarities between mine mineral samples and some lichen samples. Figure 17.5 compares mean fractional composition of each analyte for four samples collected within 20 km of the mines, four samples collected 150 km or more from the mines, and three mine mineral samples. The remote lichens yielded only low ring number PAHS, mine mineral contained mostly high ring number PAHs, and the lichen samples collected near the mine reflected contributions from both types.

Of the 20 samples analyzed during this study, we obtained metals analytical data for 16 (Chapter 14). For those samples, we summed PAH concentrations sample and determined correlation coefficients for total PAHs versus each metal (Table 17.5). Significant correlations were obtained for a number of elements. The most prominent among these ($|r| > 0.8$, $p < 0.05$) were found for the early lanthanide metals cerium, neodymium, samarium, and praseodymium, as well as the Group 6b metals, molybdenum, and tungsten. The correlation with these crustal metals, viewed in context of the earlier observation that PAHs drop off sharply as the distance from the mines increase, support the notion that a significant portion of the PAHs that are measured originate within the mines and, are transported with the mine dust, and derive from the endogenous bitumen.

TABLE 17.5 Correlation of Total PAH Concentrations to Metals Concentrations in 16 Study Samples

Element	R	ρ
Al	0.64	0.01
As	0.54	0.03
Ba	0.17	0.52
Be	0.54	0.03
Bi	0.36	0.17
Ca	0.39	0.14
Cd	−0.32	0.22
Ce	0.86	<0.0001
Co	0.13	0.64
Cr	0.47	0.07
Cs	0.55	0.03
Cu	0.59	0.02
Fe	0.77	0.00
K	0.31	0.24
La	0.52	0.04
Li	0.70	0.00
Mg	0.16	0.54
Mn	−0.15	0.57
Mo	0.87	<.0001
Na	0.53	0.03
Nb	0.68	0.01
Nd	0.84	<0.0001
Ni		0.01
P	−0.07	0.81
Pb	0.34	0.20
Pd	0.62	0.01
Pr	0.84	<0.0001
Pt	0.35	0.20
Rb	0.41	0.11
Sb	0.00	0.99

TABLE 17.5 Correlation of Total PAH Concentrations to Metals Concentrations in 16 Study Samples—Cont'd

Element	R	ρ
Se	0.79	0.00
Si	0.50	0.05
Sm	0.84	0.0001
Sn	0.70	0.00
Sr	−0.06	0.84
Ta	0.12	0.66
Th	0.57	0.02
Ti	0.49	0.05
Tl	0.68	0.00
U	0.53	0.04
V	0.75	0.00
W	0.86	<0.0001
Zn	−0.17	0.52

17.4   DISCUSSION

17.4.1   Analytical Methodology

While biomonitoring of a variety of pollutants using lichens has a long history, measurement of PAH levels in lichens for apportionment and modeling is a recent development. Samples have been collected from disparate geological locations, climates, and environmental conditions; and, a variety of species have been collected and analyzed, with different analytical procedures employed. Most of the studies have been, like the present work, of limited scope, with a sample total numbering 20 or fewer. As discussed elsewhere in this volume (Chapter 14), sampling strategy will impact application and power of data from a study to address hypotheses. Given all these considerations, one must be careful in ascribing significance in comparing results between or among studies in other than the most general terms.

Analytical protocols are not uniform among these studies but most include three distinct steps: solvent extraction, cleanup of the extracts, and quantitative determination. Methylene chloride is the most frequently used extraction solvent; cyclohexane has also been used alone or as a cosolvent with methylene chloride. Although methylene chloride is a stronger solvent, it also extracts far more plant material that can potentially interfere with analysis.

Published data for recovery of target analytes from spiked matrix using cyclohexane and sonication were acceptable (Guidotti et al., 2003), but to maximize recovery, we added a freezer-milling step to generate a finely divided powder and ensure maximum solvent contact with the matrix. Extraction techniques have included sonication and dynamic sonication, Soxhlet, and accelerated solvent extraction; based on method development data provided by the various authors, all approaches appear to be effective.

Extracts, even using cyclohexane, contain substantial amounts of plant material that interfere with GC–MS analysis, and cleanup of the extracts by a column chromatographic method or SPE was performed almost universally, using gel permeation or sorbents that included Florisil, alumina, silica, and aminopropyl-modified silica. A cautionary note regarding SPE is that most commercial cartridges are manufactured using polyethylene or polypropylene barrels and frits, and contact with the sample provides an opportunity for sample contamination by residues in the plastic as well as for nonspecific binding of nonpolar analytes. For that reason, we eliminated all plastics from the analytical procedure, using glass SPE cartridges with glass fiber frits.

GC–MS is the analytical method of choice for unambiguous identification and quantification of PAHs, but high-performance liquid chromatography with ultraviolet and/or fluorescence detection (HPLC–UV/fluorescence) has been widely used in the past (e.g., EPA SW-846 Method 8310) and is still used in some laboratories as an economic alternative. The method does not provide confirmation of analyte identity, however, and the EPA has suggested confirmatory analysis can be performed using of a column with a cyanopropyl stationary phase (http://www.epa.gov/osw/hazard/testmethods/faq/faqs_org.htm).

Lichens, and, presumably, other samples of biological origin, produce a wide variety of polycyclic metabolites (Muller, 2001) with the potential to interfere with either UV or fluorescence signals from PAHs, and while many of these will be separated from the sample during the cleanup step, it is reasonable to expect that others will be passed into the final analytical aliquot. Data obtained using HPLC–UV/MS should be supported by confirmatory analysis of at least some samples to help rule out false positives.

17.4.2   Implications of the Current Study for Receptor Modeling

The size of our study area, 300 km in diameter, is such that transport mechanisms become a critical factor in interpreting data. In addition, natural petrogenic mineral dusts containing PAHs are released not only as a result of mining activities but also occur naturally at the surface in some areas. The frequent occurrence of wildfires in the region provides a further complication. Consistent with the findings of Blasco et al. (2008, 2011) for other species, we found *H. physodes* to be an excellent receptor for 5- and 6-ring PAHs, which, while normally associated with combustion processes, also happen to be significant constituents of the native petroleum minerals.

Untangling these influences will require a study of a larger scope than our pilot study; however, even with a small number of samples, some clues emerge. Applying PCA to our 20 samples and 13 analytes that were detected in at least one sample, we determined three principal components that contributed 86% of the total variability among the samples. At the same time, the eigen function and eigenvector values were applied to air and source samples. The results are plotted for the first two principal components (PC1 and PC2) in Figure 17.6A and for PC2 and PC3 in Figure 17.6B. Eigenvectors for the individual PAHs are listed in Table 17.6. The eigenvectors for PC1, which accounts for 64% of the variability are negative up to pyrene, and are positive for benzanthracene and higher ring PAHs.

In the context of the decline of total PAHs with distance shown in Figure 17.3, and the high correlation with crustal elements in Table 17.5, we interpret this component to represent distance from the mines. Notably, as seen in both Figure 17.6A and B, four of the lichen samples at a distance >100 km, as well as two of the "intermediate" distance samples (50–100 km), are tightly clustered and have large negative values for PC1. The other "intermediate" distance samples are unclustered, and the "proximal" samples are clustered but strung out (and in three dimensions form a spiral around the PC2 axis) with mainly positive values for PC1.

Overlaying data from other matrices provides insights to sources and processes that might affect t uptake. For examples, air samples, which do not have a sampling bias for PM-associated PAHs that contribute to PC1, and forest fire ash samples, which have significant amounts of 3- and 4-ring PAHs as well as higher ring number PAHs, are clustered in the first plot, but dramatically separated by PC3 in the second plot. Mine mineral samples and refining process materials all have large values for PC1, consistent with the notion of these as important source materials contributing to the PAH burden in locally collected lichens. However, haul road samples had lower values for PC1, despite the presence of significant amounts of mine mineral in these samples. The difference is in the contribution of the high relative concentrations of acenaphthene and phenanthrene in these samples, likely a result of the heavy traffic by mining vehicles. There is a suggestion in the data that both haul road and lichen data are representing a combination of inputs from bituminous and vehicular sources.

17.5   CONCLUSIONS

Measured concentrations and relative abundance of PAHs in lichens from the AOSR are consistent with those obtained by other investigators in other areas. Total PAH concentrations were found to correlate to metals in crustal minerals and to decrease with increasing distance from the center of the study area, suggesting that similar transport mechanisms are involved. PCA of the distribution of PAH analytes in lichens and source materials is consistent with the mines as a primary source of PAHs in samples collected near the mines.

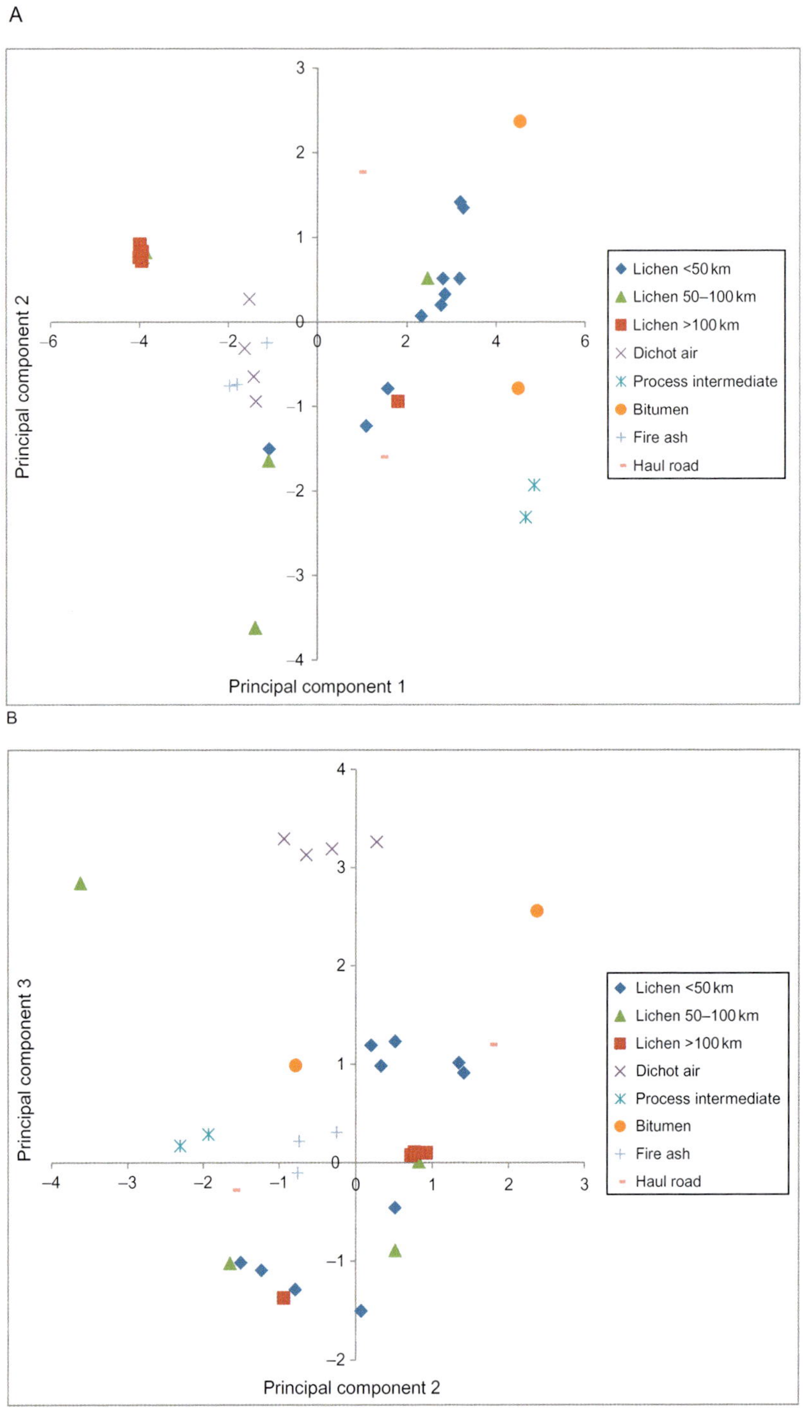

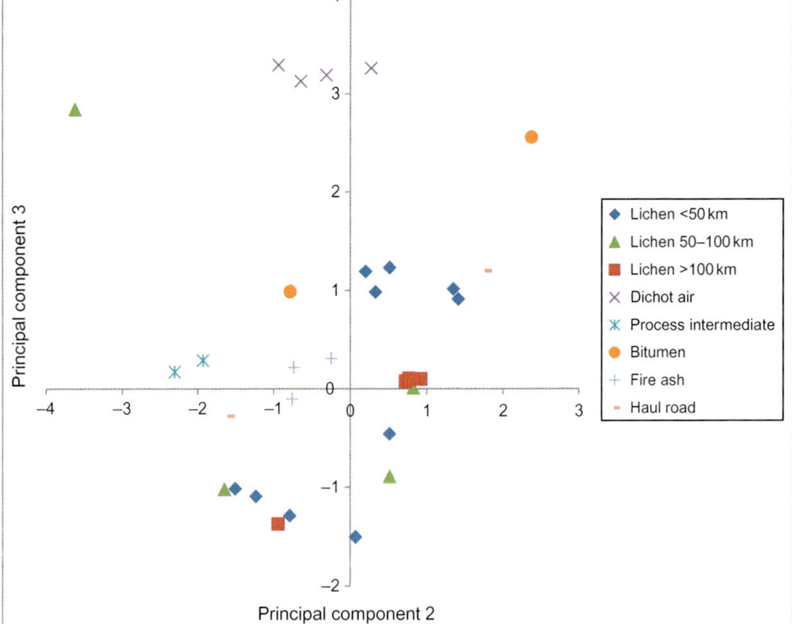

FIGURE 17.6 Principal components of lichen samples and application to other sample types.

TABLE 17.6 Eigenvectors from PCA (Principal Component Analysis) of Lichen Samples

	PC1	PC2	PC3
Naphthalene	−0.327	0.022	−0.066
Fluorene	−0.313	−0.046	0.082
Phenanthrene	−0.287	0.250	0.020
Anthracene	−0.006	−0.521	0.612
Fluoranthene	−0.319	0.045	−0.126
Pyrene	−0.314	0.174	0.086
Benzo[c]phenanthrene	0.224	0.291	−0.030
Benz[a]anthracene	0.232	−0.240	−0.403
Chrysene	0.244	−0.492	0.019
Benzo[b,j,k]fluoranthene	0.280	0.156	−0.173
7,12-Dimethylbenz[a]anthracene	0.199	0.312	0.489
Benzo[e]pyrene	0.281	0.096	0.028
Benzo[a]pyrene	0.304	0.092	−0.163
Benzo[g,h,i]perylene	0.244	0.329	0.361

More than a third of the total variance in sample composition is unexplained by distance variability, and transport mechanisms and other source compositions need to be more clearly elucidated for a better explanatory model. Furthermore, studies by other groups indicate a significant potential for biomonitor-dependent bias, and a combination of transport, uptake, and retention factors may be introducing a bias to the sampling process.

It is vitally important for future studies to characterize the biomonitor population and sampling variability against standardized sampling methods so that corrections for sampling bias can be built into modeling approaches.

ACKNOWLEDGMENTS

We thank WBEA and Phil Fellin at Airzone One for providing ambient air PAH data. Shanti Berryman and Justin Straker collected and cleaned lichens, and Todd Ennis, Keith Briggs, and Jocelin Deese-Spruill assisted with laboratory analysis. The work described herein was supported by the Wood Buffalo Environment Association, Fort McMurray, Alberta, Canada. We are grateful.

REFERENCES

Augusto, S., Maguas, C., Matos, J., Pereira, M.J., Soares, A., Branquinho, C., 2009. Spatial modeling of PAHs in lichens for fingerprinting of multisource atmospheric pollution. Environ. Sci. Technol. 43, 7762–7769.

Augusto, S., Maguas, C., Matos, J., Pereira, M.J., Branquinho, C., 2010. Lichens as an integrating tool for monitoring PAH atmospheric deposition: a comparison with soil, air and pine needles. Environ. Pollut. 158, 483–489.

Augusto, S., Gonzalez, C., Vieira, R., Maguas, C., Branquinho, C., 2011. Evaluating sources of PAHs in urban streams based on land use and biomonitors. Environ. Sci. Technol. 45, 3731–3738.

Baumard, P., Budzinski, H., Garrigues, P., 1998. Polycyclic aromatic hydrocarbons in sediments and mussels of the western Mediterranean sea. Environ. Toxicol. Chem. 17, 765–776.

Berryman, S., Straker, J., Krupa, S., Davies, M., Ver Hoef, J., Brenner, G., 2010. Mapping the Characteristics of Air Pollutant Deposition Patterns in the Athabasca Oil Sands Region Using Epiphytic Lichens as Bio-Indicators. Wood Buffalo Environmental Association (WBEA), Terrestrial Environmental Effects Monitoring (TEEM) Committee, Fort McMurray, AB, Canada.www.wbea.org/.

Blasco, M., Domeno, C., Nerin, C., 2006. Use of lichens as pollution biomonitors in remote areas: comparison of PAHs extracted from lichens and atmospheric particles sampled in and around the Somport tunnel (Pyrenees). Environ. Sci. Technol. 40, 6384–6391.

Blasco, M., Domeno, C., Bentayeb, K., Nerin, C., 2007. Solid-phase extraction clean-up procedure for the analysis of PAHs in lichens. Int. J. Environ. Anal. Chem. 87, 833–846.

Blasco, M., Domeno, C., Nerin, C., 2008. Lichens biomonitoring as feasible methodology to assess air pollution in natural ecosystems: combined study of quantitative PAHs analyses and lichen biodiversity in the Pyrenees Mountains. Anal. Bioanal. Chem. 391, 759–771.

Blasco, M., Domeno, C., Lopez, P., Nerin, C., 2011. Behaviour of different lichen species as biomonitors of air pollution by PAHs in natural ecosystems. J. Environ. Monit. 13, 2588–2596.

Bosetti, C., Boffetta, P., La Vecchia, C., 2007. Occupational exposures to polycyclic aromatic hydrocarbons, and respiratory and urinary tract cancers: a quantitative review to 2005. Ann. Oncol. 18, 431–446.

Bostrom, C.E., Gerde, P., Hanberg, A., Jernstrom, B., Johansson, C., Kyrklund, T., Rannug, A., Tornqvist, M., Victorin, K., Westerholm, R., 2002. Cancer risk assessment, indicators, and guidelines for polycyclic aromatic hydrocarbons in the ambient air. Environ. Health Perspect. 110, 451–488.

Carlberg, G.E., Ofstad, E.B., Drangsholt, H., Steinnes, E., 1983. Atmospheric deposition of organic micropollutants in Norway studied by means of moss and lichen analysis. Chemosphere 12, 341–356.

Cautreels, W., Vancauwenberghe, K., 1978. Experiments on distribution of organic pollutants between airborne particulate matter and corresponding gas-phase. Atmos. Environ. 12, 1133–1141.

Conti, M.E., Cecchetti, G., 2001. Biological monitoring: lichens as bioindicators of air pollution assessment—a review. Environ. Pollut. 114, 471–492.

Dahle, S., Savinov, V.M., Matishov, G.G., Evenset, A., Naes, K., 2003. Polycyclic aromatic hydrocarbons (PAHs) in bottom sediments of the Kara Sea shelf, Gulf of Ob and Yenisei Bay. Sci. Total Environ. 306, 57–71.

Di Giampaolo, L., Quecchia, C., Schiavone, C., Cavallucci, E., Renzetti, A., Braga, M., Di Gioacchino, M., 2011. Environmental pollution and asthma. Int. J. Immunopathol. Pharmacol. 24, 31S–38S.

Domeno, C., Blasco, M., Sanchez, C., Nerin, C., 2006. A fast extraction technique for extracting polycyclic aromatic hydrocarbons (PAHs) from lichens samples used as biomonitors of air pollution: dynamic sonication versus other methods. Anal. Chim. Acta 569, 103–112.

Estrabou, C., Filippini, E., Soria, J.P., Schelotto, G., Rodriguez, J.M., 2011. Air quality monitoring system using lichens as bioindicators in Central Argentina. Environ. Monit. Assess. 182, 375–383.

Fernandez, R., Galarraga, F., Benzo, Z., Marquez, G., Fernandez, A.J., Requiz, M.G., Hernandez, J., 2011. Lichens as biomonitors for the determination of polycyclic aromatic hydrocarbons (PAHs) in Caracas Valley, Venezuela. Int. J. Environ. Anal. Chem. 91, 230–240.

Foan, L., Sablayrolles, C., Elustondo, D., Lasheras, E., Gonzalez, L., Ederra, A., Simon, V., Santamaria, J.M., 2010. Reconstructing historical trends of polycyclic aromatic hydrocarbon deposition in a remote area of Spain using herbarium moss material. Atmos. Environ. 44, 3207–3214.

Gentes, M.L., McNabb, A., Waldner, C., Smits, J.E.G., 2007. Increased thyroid hormone levels in tree swallows (*Tachycineta bicolor*) on reclaimed wetlands of the Athabasca oil sands. Arch. Environ. Contam. Toxicol. 53, 287–292.

Gill, E.A., Curl, C.L., Adar, S.D., Allen, R.W., Auchincloss, A.H., O'Neill, M.S., Park, S.K., Van Hee, V.C., Diez Roux, A.V., Kaufman, J.D., 2011. Air pollution and cardiovascular disease in the Multi-Ethnic Study of atherosclerosis. Prog. Cardiovasc. Dis. 53, 353–360.

Guidotti, M., Stella, D., Owczarek, M., De Marco, A., De Simone, C., 2003. Lichens as polycyclic aromatic hydrocarbon bioaccumulators used in atmospheric pollution studies. J. Chromatogr. A 985, 185–190.

Guidotti, M., Stella, D., Dominici, C., Blasi, G., Owczarek, M., Vitali, M., Protano, C., 2009. Monitoring of traffic-related pollution in a Province of Central Italy with transplanted lichen Pseudovernia furfuracea. Bull. Environ. Contam. Toxicol. 83, 852–858.

Hung, H., Blanchard, P., Halsall, C.J., Bidleman, T.F., Stern, G.A., Fellin, P., Muir, D.C.G., Barrie, L.A., Jantunen, L.M., Helm, P.A., Ma, J., Konoplev, A., 2005. Temporal and spatial variabilities of atmospheric polychlorinated biphenyls (PCBs), organochlorine (OC) pesticides and polycyclic aromatic hydrocarbons (PAHs) in the Canadian Arctic: results from a decade of monitoring. Sci. Total Environ. 342, 119–144.

Johnson, J.Y., Rowe, B.H., Villeneuve, P.J., 2010. Ecological analysis of long-term exposure to ambient air pollution and the incidence of stroke in Edmonton, Alberta, Canada. Stroke 41, 1319–1325.

Kavouras, I.G., Koutrakis, P., Tsapakis, M., Lagoudaki, E., Stephanou, E.G., Von Baer, D., Oyola, P., 2001. Source apportionment of urban particulate aliphatic and polynuclear aromatic hydrocarbons (PAHs) using multivariate methods. Environ. Sci. Technol. 35, 2288–2294.

Kelly, E.N., Schindler, D.W., Hodson, P.V., Short, J.W., Radmanovich, R., Nielsen, C.C., 2010. Oil sands development contributes elements toxic at low concentrations to the Athabasca River and its tributaries. Proc. Natl. Acad. Sci. USA 107, 16178–16183.

Korashy, H.M., El-Kadi, A.O., 2006. The role of aryl hydrocarbon receptor in the pathogenesis of cardiovascular diseases. Drug Metab. Rev. 38, 411–450.

Lehndorff, E., Schwark, L., 2009a. Biomonitoring airborne parent and alkylated three-ring PAHs in the Greater Cologne Conurbation I: temporal accumulation patterns. Environ. Pollut. 157, 1323–1331.

Lehndorff, E., Schwark, L., 2009b. Biomonitoring airborne parent and alkylated three-ring PAHs in the Greater Cologne Conurbation II: regional distribution patterns. Environ. Pollut. 157, 1706–1713.

Lehndorff, E., Schwark, L., 2010. Biomonitoring of air quality in the Cologne Conurbation using pine needles as a passive sampler—Part III: major and trace elements. Atmos. Environ. 44, 2822–2829.

Migaszewski, Z.M., Galuszka, A., Paslawski, P., 2002. Polynuclear aromatic hydrocarbons, phenols, and trace metals in selected soil profiles and plant bioindicators in the Holy Cross Mountains, south-central Poland. Environ. Int. 28, 303–313.

Migaszewski, Z.M., Galuszka, A., Crock, J.G., Lamothe, P.J., Dolegowska, S., 2009. Interspecies and interregional comparisons of the chemistry of PAHs and trace elements in mosses *Hylocomium splendens* (Hedw.) BSG and *Pleurozium schreberi* (Brid.) Mitt. from Poland and Alaska. Atmos. Environ. 43, 1464–1473.

Muller, K., 2001. Pharmaceutically relevant metabolites from lichens. Appl. Microbiol. Biotechnol. 56, 9–16.

Naeth, M.A., Wilkinson, S.R., 2008. Lichens as biomonitors of air quality around a diamond mine, northwest territories. Can. J. Environ. Qual. 37, 1675–1684.

Oksanen, I., 2006. Ecological and biotechnological aspects of lichens. Appl. Microbiol. Biotechnol. 73, 723–734.

Owczarek, M., Guidotti, M., Blasi, G., De Simone, C., De Marco, A., Spadoni, M., 2001. Traffic pollution monitoring using lichens as bioaccumulators of heavy metals and polycyclic aromatic hydrocarbons. Fresen. Environ. Bull. 10, 42–45.

Pandey, P.K., Patel, K.S., Lenicek, J., 1999. Polycyclic aromatic hydrocarbons: need for assessment of health risks in India? Study of an urban-industrial location in India. Environ. Monit. Assess. 59, 287–319.

Percy, K.E., Cape, J.N., Jagels, R., Simpson, C.M. (Eds.), 1993. Air Pollutants and the Leaf Cuticle. In: NATO ASI Series, vol. 36. Springer-Verlag, Heidelberg 391 pp.

Ratola, N., Lacorte, S., Barcelo, D., Alves, A., 2009. Microwave-assisted extraction and ultrasonic extraction to determine polycyclic aromatic hydrocarbons in needles and bark of *Pinus pinaster* Ait. and *Pinus pinea* L. by GC-MS. Talanta 77, 1120–1128.

Ratola, N., Amigo, J.M., Alves, A., 2010. Levels and sources of PAHs in selected sites from Portugal: biomonitoring with *Pinus pinea* and *Pinus pinaster* needles. Arch. Environ. Contam. Toxicol. 58, 631–647.

Ratola, N., Alves, A., Psillakis, E., 2011. Biomonitoring of polycyclic aromatic hydrocarbons contamination in the island of Crete using pine needles. Water Air Soil Pollut. 215, 189–203.

Rodriguez, J.H., Pignata, M.L., Fangmeier, A., Klumpp, A., 2010. Accumulation of polycyclic aromatic hydrocarbons and trace elements in the bioindicator plants *Tillandsia capillaris* and *Lolium multiflorum* exposed at PM10 monitoring stations in Stuttgart (Germany). Chemosphere 80, 208–215.

Sacks, J.D., Stanek, L.W., Luben, T.J., Johns, D.O., Buckley, B.J., Brown, J.S., Ross, M., 2011. Particulate matter-induced health effects: who is susceptible? Environ. Health Perspect. 119, 446–454.

Satya, Upreti, D.K., Patel, D.K., 2012. *Rinodina sophodes* (Ach.) Massal.: a bioaccumulator of polycyclic aromatic hydrocarbons (PAHs) in Kanpur City, India. Environ. Monit. Assess. 184, 229–238. http://dx.doi.org/10.1007/s10661-011-1962-5.

Schwarz, O.J., Jones, L.W., 1997. Bioaccumulation of xenobiotic chemicals by terrestrial plants. In: Corsuch, J.W., Hughes, J.S. (Eds.), Plants for Environmental Studies. CRC Press, Boca Raton, FL.

Shukla, V., Upreti, D.K., 2009. Polycyclic aromatic hydrocarbon (PAH) accumulation in lichen, Phaeophyscia hispidula of Dehra Dun City, Garhwal Himalayas. Environ. Monit. Assess. 149, 1–7.

Shukla, V., Upreti, D.K., Patel, D.K., Tripathi, R., 2010. Accumulation of polycyclic aromatic hydrocarbons in some lichens of Garhwal Himalayas, India. Int. J. Environ. Waste Manage. 5, 104–113.

Simonich, S.L., Hites, R.A., 1995. Organic pollutant accumulation in vegetation. Environ. Sci. Technol. 29, 2905–2914.

Sun, F.F., Wen, D.Z., Kuang, Y.W., Li, J.O., Li, J.L., Zuo, W.D., 2010a. Concentrations of heavy metals and polycyclic aromatic hydrocarbons in needles of Masson pine (*Pinus massoniana* L.) growing nearby different industrial sources. J. Environ. Sci. (China) 22, 1006–1013.

Sun, Q., Hong, X., Wold, L.E., 2010b. Cardiovascular effects of ambient particulate air pollution exposure. Circulation 121, 2755–2765.

Szyszkowicz, M., Stieb, D.M., Rowe, B.H., 2009. Air pollution and daily ED visits for migraine and headache in Edmonton, Canada. Am. J. Emerg. Med. 27, 391–396.

Thomas, W., Ruhling, A., Simon, H., 1984. Accumulation of airborne pollutants (PAH, chlorinated hydrocarbons, heavy metals) in various plant species and humus. Environ. Pollut. Ser. A. Ecol. Biol. 36, 295–310.

Usenko, S., Smonich, S.L.M., Hageman, K.J., Schrlau, J.E., Geiser, L., Campbell, D.H., Appleby, P.G., Landers, D.H., 2010. Sources and deposition of polycyclic aromatic hydrocarbons to Western US National Parks. Environ. Sci. Technol. 44, 4512–4518.

Villeneuve, P.J., Chen, L., Rowe, B.H., Coates, F., 2007. Outdoor air pollution and emergency department visits for asthma among children and adults: a case-crossover study in northern Alberta, Canada. Environ. Health 6, 40.

Wang, Z., Chen, J.W., Yang, P., Tian, F.L., Qiao, X.L., Bian, H.T., Ge, L.K., 2009. Distribution of PAHs in pine (*Pinus thunbergii*) needles and soils correlates with their gas-particle partitioning. Environ. Sci. Technol. 43, 1336–1341.

Wayland, M., Headley, J.V., Peru, K.M., Crosley, R., Brownlee, B.G., 2008. Levels of polycyclic aromatic hydrocarbons and dibenzothiophenes in wetland sediments and aquatic insects in the oil sands area of Northeastern Alberta, Canada. Environ. Monit. Assess. 136, 167–182.

Wolterbeek, B., 2002. Biomonitoring of trace element air pollution: principles, possibilities and perspectives. Environ. Pollut. 120, 11–21.

Yunker, M.B., Macdonald, R.W., 2003. Alkane and PAH depositional history, sources and fluxes in sediments from the Fraser River Basin and Strait of Georgia, Canada. Org. Geochem. 34, 1429–1454.

Chapter 18

Receptor Modeling of Epiphytic Lichens to Elucidate the Sources and Spatial Distribution of Inorganic Air Pollution in the Athabasca Oil Sands Region

M.S. Landis[*,1], J.P. Pancras[†], J.R. Graney[‡], R.K. Stevens[§], K.E. Percy[¶] and S. Krupa[‖]

[*]*US EPA, Office of Research and Development, Research Triangle Park, North Carolina, USA*

[†]*Allion Science and Technology, Research Triangle Park, North Carolina, USA*

[‡]*Geological Sciences and Environmental Studies, Binghamton University, Binghamton, New York, USA*

[§]*Cary, North Carolina, USA - Formerly with U.S. EPA Office of Research and Development, Research Triangle Park, North Carolina, USA*

[¶]*Wood Buffalo Environmental Association, Fort McMurray, Alberta, Canada*

[‖]*Plant Pathology, University of Minnesota-Twin Cities, St. Paul, Minnesota, USA*

[1]*Corresponding author: e-mail: landis.matthew@epa.gov*

ABSTRACT

The contribution of inorganic air pollutant emissions to atmospheric deposition in the Athabasca Oil Sands Region (AOSR) of Alberta, Canada, was investigated in the surrounding boreal forests, using a common epiphytic lichen bioindicator species (*Hypogymnia physodes*) and applying multiple receptor models. Source materials from anthropogenic and natural emitters of air pollution in the AOSR were obtained and chemically characterized to aid in the assessment. The lichens selected for analysis were collected in 2008 using a stratified, nested grid approach radiating away from the central area of oil sands production at 121 sampling sites extending as far as 150 km.

Disclaimer: The EPA through its Office of Research and Development collaborated in this research. It has been subjected to EPA Agency review and approved for publication.

The content and opinions expressed by the authors do not necessarily reflect the views of the EPA, WBEA, or the WBEA membership.

Source and lichen samples were extracted and analyzed for 43 elements using dynamic reaction cell quadrupole inductively coupled plasma mass spectroscopy (DRC-ICPMS). Source apportionment of the lichen tissue analytical results was conducted using principal component analysis (PCA), chemical mass balance (CMB), positive matrix factorization (PMF), and Unmix models.

Initial Varimax-rotated PCA screening analysis indicated that there were five principal components that could explain 89% of the variance contained in the lichen data set, with the majority of the variance lumped into a fugitive dust factor. This fugitive dust source could be separated into tailing sand, haul road, and overburden components using CMB on lichen samples collected near the mining and oil processing facilities. However, the CMB model performance was limited by the similarity of sources and the lack of total nitrogen measurements in the emission source profiles. The PMF and Unmix models were found to perform best with this unique AOSR lichen data set, providing very similar results at near-source as well as remote lichen collection sites. The PMF results showed that sources significantly contributing to concentrations of elements in the lichen tissue include combustion processes ($\sim$23%), tailing sand ($\sim$19%), haul roads and limestone ($\sim$15%), oil sand and processed materials ($\sim$15%), and a general anthropogenic urban source ($\sim$15%).

The spatial patterns of CMB, PMF, and Unmix receptor models estimated that source impacts on the *H. physodes* tissue elemental concentrations from the oil sand processing and fugitive dust sources had a significant association with the distance from the primary oil sands surface mining operations and related production facilities. The spatial extent of the fugitive dust impact was limited to a radius of approximately 20 km around the major mining and oil production facilities, which is indicative of ground-level coarse particulate fugitive emissions from these sources. The impact of the general urban source was found to be enhanced in the southern portion of the sampling domain in the vicinity of the Fort McMurray urban area. The receptor model results showed lower Mn concentrations in lichen tissues near oil sands production operations suggesting a biogeochemical response. Overall, the largest impact on elemental concentrations of *H. physodes* tissue in the AOSR was related to fugitive dust, suggesting that implementation of a fugitive dust abatement strategy could minimize the near-field atmospheric deposition of any future mining-related production activities.

18.1 INTRODUCTION

The Athabasca Oil Sands Region (AOSR) in northern Alberta, Canada, contains recoverable petroleum reserves estimated to be in excess of 170 billion barrels consisting mostly of bitumen (Attanasi and Meyer, 2010). These proven reserves rank the AOSR third in the world, behind only Saudi Arabia and Venezuela. Oil production in the AOSR has been steadily increasing over the last decade from 0.6 million barrels per day in 2000, to 1.6 million barrels per day in 2011. Production is expected to be in excess of 3.5 million barrels per day by 2025 (Chapter 2). Synthetic crude oil production from bitumen in the AOSR is accomplished using a combination of surface mining and *in situ* production. Of the proven reserves, it is estimated that 20% of the bitumen will ultimately be recovered through surface mining and 80% from *in situ* production techniques (Government of Alberta, 2008). The type and magnitude of inorganic air pollutants emitted from these two extraction techniques are unique. Quantifying their relative contribution to observed ambient concentrations and atmospheric deposition is critical to be able to mitigate and manage local environmental impacts as production levels are increased.

Surface mining in the AOSR results in large land disturbances (ca. 715 km^2; Chapter 2) and is similar to coal, copper, and other traditional mining operations. Currently, the soil and glacial till overlying the deposits (overburden) is removed and the exposed oil sands are excavated and transported for processing using large-scale shovel and truck hauling operations. Atmospheric pollution from shovel and truck fleet operations consists mainly of fugitive particulate matter (PM) emissions (wind-blown dust) and diesel engine combustion exhaust. Bitumen is separated from sand and clay components and recovered using a warm water separation technique. The water, sand, and clay waste stream is pumped to large tailings ponds where the water is removed and recycled; and the sand and clay are consolidated and used for mine reclamation activities. After the oil sands are removed, the mines reach the underlying limestone bedrock. Limestone in the area is quarried, crushed, and used for development of haul roads and other construction activities. Overburden stored for future mine reclamation, haul roads, and tailings ponds are all potential sources of fugitive wind-blown dust.

In situ production refers to the extraction of bitumen from oil sand deposits that are present at depths that make it uneconomical to access using traditional surface mining (currently about 75 m), and using techniques to separate the bitumen in place. Steam-assisted gravity drainage (SAGD) is currently the primary *in situ* stimulation technique employed. SAGD involves the drilling of two parallel wells, one above the other. Steam is injected into the upper well to thermally separate the bitumen from the host material. The reduced viscosity of the heated bitumen causes it to drain down into the second underlying well where it is pumped to the surface and recovered. The natural gas- and syngas-fired boilers used to generate the steam are sources of atmospheric pollutant emissions (e.g., NO, NO$_x$).

The bitumen recovered from the AOSR is considered a "sour" extra heavy crude oil (Attanasi and Meyer, 2010). It has a high specific gravity (>10°API), is extremely viscous at ambient temperatures, and contains elevated concentrations of sulfur (>0.5%) and some metals (e.g., nickel, vanadium). The bitumen is upgraded to synthetic crude oil by thermal/catalytic cracking to break down large long-chain molecules and removal of excess sulfur (hydro-desulfurization) to facilitate the production of valuable light (e.g., gasoline) and medium (e.g., diesel) distillate fuels. Some facilities upgrade the bitumen to synthetic crude on site in the AOSR, while others dilute the bitumen with naptha and transport it to refineries via pipeline to other parts of Canada or the United States. Upgrading, refining, and power generation are significant sources of atmospheric NO, NO$_x$, PM, and SO$_2$ emissions. In addition to the anthropogenic sources of atmospheric emissions from the petroleum industry in the AOSR, there are significant light-duty mobile source emissions, commercial boilers, and residential heating sources as well as natural pollutant emitters such as forest fires.

The AOSR is located in a remote boreal forest ecosystem. Other than Fort McMurray, much of the region has no ready access by land transportation and is not serviced by commercial electric power infrastructure. Current active ambient air monitoring is concentrated in the relatively narrow, north/south transportation corridor where the main oil sands operations in the northern portion of the AOSR are situated (Chapter 4). Therefore, the Wood Buffalo Environmental Association (WBEA)–Terrestrial Environmental Effects Monitoring (TEEM) program used the epiphytic lichen, *Hypogymnia physodes,* growing predominantly on jack pine (*Pinus banksiana*) and black spruce (*Picea mariana*), as a bioindicator of the atmospheric deposition and accumulation of air pollutants for on-going terrestrial impact assessment. *H. physodes* was selected as the bioindicator species of choice because it is an epiphytic lichen that extracts all its nutrients from the air, has a high tolerance for SO_2, is prevalent in all areas of the AOSR, and is commonly used in air quality monitoring (Garty, 2001; Jeran et al., 2002). Our investigation focused on total sulfur (S), total nitrogen (N) (Berryman et al., 2010), 43 metals, stable isotopes of lead (Pb) (Chapter 15), mercury (Hg) (Chapter 16), and poly-aromatic hydrocarbons (Chapter 17). Initially, spatial maps of S and N accumulation in the lichen were developed for locations up to 150 km from the center of the oil sands production emission source area, with sampling at sites distributed as a nested grid. Based on the S and N distributions, 43 metals and the Pb and Hg isotopes were quantified at a subset of the lichen sampling locations and the contributions of specific emission types were investigated (Chapters 15 and 16).

Deterministic, or atmospheric dispersion models, are routinely used by environmental managers and government regulators as a tool to estimate the transport, transformation, and deposition of atmospheric pollutants (Chapter 12). The ability of these models (e.g., CALPUFF, CMAQ, ISC3, AERMOD) to reliably simulate the fate of emitted pollutants on the spatial scales of interest in the AOSR is highly dependent on the (i) quality of emission inventory data, (ii) completeness of the chemical kinetics module, (iii) accuracy and resolution of the underlying gridded meteorological fields, (iv) topography, and (v) proper parameterization of gas/particle interaction and wet and dry deposition phenomena. In practice, it is extremely difficult to accurately model air pollution in remote areas such as the AOSR where nonpoint mobile sources, fugitive sources, batch processes, and forest fires are significant emission sources, and where few local meteorological measurements are available for four-dimensional data assimilation to "nudge" the underlying meteorological drivers.

Receptor models provide another approach to understanding the impacts of air pollution sources as the model results are based on measurement data at receptor, or sampling, locations. Receptor models quantify the impact of air emission sources retrospectively by using advanced mathematical methods on a matrix of elements or compounds in atmospheric samples, or bioindicators, as tracers for the presence of materials from specific sources (Gordon, 1985; Hopke, 1985, 2009). The goal of receptor modeling is to apportion the sources into specific identifiable categories (e.g., combustion, refining, motor vehicles, incineration, metals smelting, etc.) and quantify their relative importance. Receptor models can also be used to constrain the uncertainty in deterministic modeling estimates and help identify sources that may not be accurately represented in emission inventories.

The main objectives of this study were to (i) identify the major sources of air pollution in the AOSR, (ii) collect and analyze samples to develop chemical source profiles (fingerprints), (iii) conduct a quantitative source apportionment analysis to determine the major sources impacting the atmospheric deposition and accumulation of potentially phytotoxic levels of S and N in the tissue of *H. physodes*, and (iv) provide supportive data for Forest Health Monitoring (Chapters 9 and 13).

18.2   METHODS

18.2.1   Lichen Sampling and Analysis

A discussion on the selection of *H. physodes* as a species for study can be found in Chapter 15. A complete description of the collection, selection strategy, and analysis methods for the *H. physodes* samples is contained in Chapter 14. Briefly, in 2008, WBEA–TEEM funded the collection of lichen samples from 369 sampling locations using a stratified nested grid approach, with higher density sampling at the center of the grid in close proximity to the main oil sands production sites (Figure 14.1 in Chapter 14). All samples were analyzed for total S and total N at the University of Minnesota Research Analytical Laboratory (Berryman et al., 2010). A subset of samples from 121 of the sites were selected for microwave-assisted acid extraction and analysis for 43 elements using DRC-ICPMS (Chapter 14).

18.2.2   Source Sampling and Analysis

Bulk material samples representing various steps in the oil sands production cycle and other background materials were collected by WBEA for our analysis including overburden, raw oil sand, aged oil sand, limestone, materials used to construct haul roads, bitumen, fluid coke, petroleum coke, vacuum tower bottoms, tailings sand, and ash from forest fires. All bulk samples were extracted and analyzed by Atmospheric Research & Analysis (ARA; Cary, NC) using the same microwave-assisted acid extraction and DRC-ICPMS analysis methods used on the lichen samples (Chapter 14). In addition, diluted PM source sample emissions were collected on filters from the main stacks of an AOSR upgrading facility by Desert Research Institute (DRI) (Chapter 8) and the exhaust of heavy-duty hauling trucks by DRI (Chapter 7); also, ambient $PM_{2.5}$ on filters collected by WBEA at Fort McKay during heavy smoke-impacted days following forest fire ($PM_{2.5} > 420\ \mu g\ m^{-3}$) were extracted and analyzed by ARA using DRC-ICPMS.

18.2.3   Theory and Concepts of Source Apportionment and Receptor Models

According to Hopke (2009), source apportionment is the estimation of the contributions to the pollutant concentrations resulting from emissions from multiple natural and anthropogenic sources. Forensic data analysis tools called *receptor models* (mathematical and/or statistical) are applied to extract information on the sources of air pollutants from the measured constituent concentrations at a receptor location. Unlike deterministic dispersion air quality models, receptor models generally do not use pollutant emissions, meteorological data, and chemical transformation mechanisms to estimate the contribution of sources to receptor concentrations. Instead, receptor models use mathematically detectable characteristics (chemical and physical) of gasses and particles measured at a monitoring or receptor site to both identify and quantify source contributions to receptor concentrations. These models are therefore a natural complement to deterministic air quality models. The United States Environmental Protection Agency (EPA) Office of Research and Development has developed several integrated receptor modeling software tools such as Chemical Mass Balance (CMB), Unmix, and Positive Matrix Factorization (PMF). Each of the EPA-implemented receptor model programs have a graphical user interface, data screening and analysis tools, and data visualization capabilities. EPA has made all of these models available to the public for use by students, researchers, industry, and government regulators (http://www.epa.gov/scram001/receptorindex.htm).

Typically, receptor models use repeated measurements of the chemical composition data for airborne PM samples collected at a monitoring site (spatially fixed, temporally resolved). In such cases, the outcome is the identification of the pollution source types and estimates of the contribution of each source type to the observed concentrations (Table 18.1). *In lieu* of a lack of sufficient data on the chemical composition of PM in AOSR, we used data from the epiphytic lichen, *H. physodes,* as an accumulator or bioindicator of various elements through atmospheric wet and dry deposition (Kuik et al., 1993; Sloof, 1995). It is believed that *H. physodes* samples represent 3–5 years of accumulated atmospheric deposition in their tissue (Berryman et al., 2010; Chapter 12). We developed our AOSR epiphytic lichen concentration data matrix based on samples collected at 121 locations in 2008 (spatially resolved, temporally fixed).

18.2.3.1   Principal Component Analysis

Principal component analysis (PCA) is often used as a preliminary data reduction technique to identify a small number of factors that explain the majority of the variance observed in a much larger number of measured variables. According to Jolliffe (2002), PCA is probably the oldest (first introduced in 1901) and best known technique of multivariate analysis. Like many multivariate methods, it was not widely used until the advent of computers, but it is

TABLE 18.1 Recent Examples of PMF and Unmix Receptor Modeling Studies

Date	Authors	Study location
2005	Begum et al.	Washington, DC
2011	Cao et al.	Xi'an, PRC
2011	Chan et al.	Brisbane, Australia
2009	Cheng et al.	Hong Kong, PRC
2011	Gu et al.	Augsburg, Germany
2003	Hopke et al.	Baltimore, MD
2003	Hsu et al.	Chicago, IL
2006	Kim and Hopke	Great Smoky Mountains, NC-TN
2004	Kim et al.	Atlanta, GA
2003	Lewis et al.	Phoenix, AZ
2003	Maykut et al.	Seattle, WA
2004	Olson et al.	World Trade Center, New York, NY
2006	Pancras et al.	Tampa, FL
2011	Pancras et al.	Tampa, FL
2009	Sosa et al.	Mexico City, MX
2004	Wang et al.	Beijing, PRC
2006	Zhao and Hopke	Mammoth Cave National Park, KY
2004	Zhao et al.	Houston, TX
2004a	Zhou et al.	Pittsburgh, PA
2004b	Zhou et al.	Rural New York, NY

now available in virtually every statistical computer software package. The central idea of PCA is to reduce the dimensionality of a data set in which there are a large number of interrelated variables, while retaining as much as possible of the variation present in the data set. This reduction is achieved by transforming the data to a new set of variables, the principal components, that are minimally correlated.

Principal components may be viewed as the eigenvectors of a positive semidefinite symmetric matrix. The eigenvectors are "characteristic" vectors of a matrix. They are unique in that they remain directionally invariant under linear transformation by its parent matrix. Thus, the definition and computation of principal components are straightforward and have a wide variety of applications (e.g., Pratt and Krupa, 1985; Voukantsis et al., 2011). The general form for the Equation (18.1) to compute scores on the first (main) component extracted (created) in a PCA:

$$C1 = b11(X1) + b12(X2) + \cdots + b1p(Xp) \qquad (18.1)$$

where $C1$, the subject's score on principal component 1 (the first component extracted); $b1p$, the regression coefficient (or weight) for observed variable p, as used in creating principal component 1; and Xp, the subject's score on observed variable p.

In previous air pollution studies, the principal components have been found to represent sources such as soil, motor vehicles, iron and steel production, metal smelting, coal combustion, incineration, and oil combustion (Gaarenstroom et al., 1977; Hopke et al., 1976). In many cases, interpretations of the principal components have been difficult because most of the variability of the data was loaded onto a single component (Thurston and Spengler, 1985). This is not surprising, as PCA is designed to incorporate the maximal amount of variance into the first factor (Hopke, 1985). In this study, Varimax orthogonal rotation was performed in a manner described by Harmon (1976) to make physical interpretation of the principal components easier (Thurston, 1981). Only rotated principal components with eigenvalues >1 were retained for consideration (Hopke, 1983).

Thurston and Spengler (1985) introduced an absolute principle component (APC) scores calculation scheme utilizing an arbitrary zero-concentration sample wherein all elemental concentrations are zero. Regressing mass concentration data on the APC scores gave estimates of the coefficients which convert the APC score into pollutant source mass contribution for each sample. An attractive feature of this modeling framework is that no prior knowledge of the number or chemical composition of possible sources is required. However, some of the major chemical characteristics of the emission source must be present to correctly attribute the principal component (PC) to a particular source type.

18.2.3.2 Chemical Mass Balance

EPA-implemented CMB version 8.2 was used for this analysis (U.S. EPA, 2004a). In receptor modeling, a mass balance equation can be written to account for "m" chemical species in the "n" samples as contributions from "p" independent sources (Equation 18.2).

$$x_{ij} = \sum_{k=1}^{k=p} g_{ik} f_{kj} + e_{ij} \qquad (18.2)$$

where x_{ij}, the measured concentration of the jth species in the ith sample; f_{kj}, the concentration of the jth species in material emitted by source p; g_{ik}, the contribution of the pth source to the ith sample; and e_{ij}, the portion of the measurement that cannot be fitted by the model.

If the number and nature of the sources in the region are known (e.g., p and f_{pj}s), then the only unknown is the mass contribution of each source to each sample, g_{ip} (Miller et al., 1972; Winchester and Nifong, 1971). The problem is typically solved using an effective-variance least-squares approach (Cooper et al., 1984) that is generally referred to as the CMB model (Watson et al., 1990; U.S. EPA, 2004a,b). In an ideal case, location-specific source profiles are generated using the same extraction and analytical techniques as the receptor samples. Typically, this type of source characterization is not feasible, and profiles from a source library are utilized such as those available in the EPA SPECIATE version 4.3, profile repository (www.epa.gov/ttn/CHIEF/software/speciate). In this case, AOSR-specific source profiles were generated (see Section 18.3.1).

Three statistical measurements are commonly used to evaluate the CMB model's ability to match the calculated species concentrations and the receptor data (U.S. EPA, 2004b): r^2 values, χ^2 values, and the percent of total mass explained by the fit. An r^2 value is the fraction of the variance in the measured concentrations explained by the variance in the calculated species concentrations. It is determined by linear regression of calculated versus model-measured values for the fitting species. Ranges are from 0 to 1, with values >0.8, indicating that the measured concentrations are well explained by the source contribution estimates. The χ^2 value is the weighted sum of squares of the differences between the measured and calculated element concentrations. Ideally, there should be no difference, resulting in χ^2 of 0. A large χ^2 (>4.0) means that one or more of the calculated species concentrations significantly differs from the measured concentrations. The values for these statistics exceed their targets when (i) contributing sources have been omitted from the CMB calculation, (ii) one or more source profiles have been selected that do not represent the contributing source types, (iii) uncertainty estimates of receptor or source profile data are underestimated, and/or (iv) there are errors or inconsistencies between analytical measurements used for source and receptor data. Percent mass explained is the ratio of the difference between the sum of the model-calculated source contribution estimates and the measured mass concentrations. Ratios should equal 100%, but values between 80% and 120% are acceptable. In our CMB application, the total variable (PM mass in lichen) is not measurable. Also, receptor concentrations are normalized to lichen mass. As a result, the CMB calculation estimates potential source contributions ("g" matrix in Equation 18.2) in the form of the total lichen mass concentration attributable to sources.

CMB is most useful for primary emissions where the chemical characteristics of the particles are sufficient to characterize their apportionment. The inclusion of profiles for secondary particles are difficult as they represent the product of atmospheric transformations of gaseous emissions into particles and are generally treated as specific chemical species (e.g., sulfate, nitrate, ammonium or ammonium sulfate and ammonium nitrate). Unlike the multivariate receptor models like PMF and Unmix, CMB can be used to determine contributions with a single sample.

18.2.3.3 Positive Matrix Factorization

The EPA-implemented PMF version 4.2 was used for this analysis (U.S. EPA, 2011). PMF is a constrained eigenvector, implicit least-squares analysis aimed at minimizing the sum of squared residuals for the model. Paatero and Tapper (2003) showed that in a PCA analysis, there is scaling of the data by column or by row and that scaling will lead to distortions in the analysis. They further showed that the optimum method for scaling uncertainty in the data matrix would be to scale each data point individually. In this way, the more precise data will have more influence on the solution than points that have higher uncertainties. However, point-by-point scaling results in a scaled data matrix that cannot be reproduced by a conventional factor analysis based on the singular value decomposition.

PMF allows each data point to be individually weighted. This feature allows the modeler to adjust the influence of each data point, depending on the confidence in the measurement. For example, data below detection limit can be retained for use in the model, with the associated uncertainty adjusted, so these data points have less influence on the solution than measurements above the detection limit. A speciated data set can be viewed as a data matrix X of i by j dimensions, in which i is the number of samples and j is the chemical species measured. Thus, PMF uses an explicit least-squares method that minimizes the object function Q with respect to g (mass) and f (species profile) based on the uncertainties u (Equation 18.3), while constraining the results so that no sample can have a significant negative source contribution.

$$Q = \sum_{i=1}^{n} \sum_{j=1}^{m} \left[\frac{x_{ij} - \sum_{k=1}^{p} g_{ik} f_{kj}}{u_{ij}} \right]^2 \tag{18.3}$$

Initially, a unique algorithm (PMF2, Paatero, 1997) was used for solving the factor analysis equation. For small- and medium-sized problems, this algorithm was found to be more efficient than alternate least squares (ALS) methods (Hopke et al., 1998). Subsequently, a different approach that provides a flexible modeling system was developed for solving the various PMF factor analyses least-squares problems (Paatero, 1999). This approach, called the multilinear engine (ME), has been applied to environmental problems that involve the solution of more complex models (Begum et al., 2005; Chueinta et al., 2004; Hopke et al., 2003; Paatero and Tapper, 2003; Zhao et al., 2004).

Block bootstrap is the widely used method to estimate variability or modeling uncertainty in a PMF solution (U.S. EPA, 2011). The block bootstrap method captures effects from random errors in the solution, and also partially accounts for errors from computational rotational ambiguity. EPA–PMF performs bootstrapping by randomly selecting blocks of samples, and creating new input data for the selected sample, with the same dimensions as the original dataset. PMF is then run on the newly created dataset, and each factor from the bootstrap run is mapped to the base run factor by comparing the contributions of each factor. The newly created bootstrap factor is assigned to the base factor with which the bootstrap factor has the highest uncentered correlation above a user-specified threshold. If no base factors have a correlation above the threshold for a given bootstrap factor, that factor is considered "unmapped." If more than one bootstrap factor from the same run is correlated with the same base factor, they will all be mapped to that base factor. This process is repeated for as many bootstrap runs as the user specifies. A solution is considered valid when the occurrence of unmapped factors is $<10\%$ of the total bootstrap runs. EPA–PMF reports variability in factor strengths as various (5, 25, 50, 75, and 95) percentiles of factor strengths.

PMF2 was used to analyze data sets of major ion compositions of daily precipitation samples collected at a number of sites in Finland (Juntto and Paatero, 1994) and bulk precipitation (Anttila et al., 1995) to obtain information on the sources of those ions. Polissar et al. (1996) applied PMF2 data from seven Alaska National Park sites to resolve the major source contributions quantitatively.

Lee et al. (1999) applied PMF to urban aerosol compositions in Hong Kong. They were able to identify up to nine sources that provided a good apportionment of the airborne PM. Similarly, Huang et al. (1999) analyzed elemental composition of PM at Narragansett, RI, using both PMF and conventional PCA analysis. They were able to resolve more components, with PMF using physically realistic compositions. Thus, the approach does have some inherent advantages particularly due to its ability to individually weight each data point. PMF is somewhat more complex and harder to use, but it provides improved resolution of sources and better quantification of those sources than PCA (Huang et al., 1999).

Chueinta et al. (2000) introduced a directional source contribution analogous to a wind "rose" to help provide information on the direction of the source relative to the receptor site. Ramadan et al. (2000) applied PMF to a set of daily data from Phoenix, AZ. In this analysis, separate profiles were resolved for diesel and spark-ignition vehicles. Analogously, Lewis et al. (2003) analyzed the same data using Unmix and found similar results for sources that contribute the largest amounts to the ambient mass concentrations.

Chemical composition of $PM_{2.5}$ samples collected from 1988 to 1995 at Underhill, Vermont were analyzed by Polissar et al. (2001a). Sources representing wood burning, coal and oil combustion, photochemical sulfate production, metal production plus municipal waste incineration, and the emissions from motor vehicles were identified. In addition, emissions from smelting of nonferrous metal ores and arsenic, as well as soil particles and particles with high concentrations of Na, were identified by PMF.

18.2.3.4 Unmix

The EPA-implemented Unmix version 6.0 was used for this analysis (U.S. EPA, 2007). Unmix is a constrained multivariate receptor model which seeks to solve a general mixture problem where the data are assumed to be a linear combination of an unknown number of sources of unknown composition which contribute an unknown amount to each sample (Henry, 1997). Like PMF, Unmix also assumes that the compositions and contributions of the sources are all nonnegative. Unfortunately, it has been shown that nonnegativity conditions alone are not sufficient to give a unique solution and more constraints are needed (Henry, 1987). To mitigate this constraint, Unmix assumes that for each source there are at least a few samples that contain little or no contribution from that source. This has been found to be a reasonable assumption as, in an ambient monitoring example, the wind could be blowing away from the source, or for a lichen biomonitoring example, the receptor location may be too far away from the source to make a significant impact. Using only the concentration data for a given selection of species, Unmix estimates the number of sources, source compositions, and source contributions to each sample. It should be noted that, unlike PMF, Unmix does not allow for downweighting using data uncertainty values.

Unmix is also based on an eigenvalue analysis. The model uses a transformation method based on the self-modeling curve resolution (SMCR) technique. The SMCR technique identifies the feasible region of the real solution with explicit physical constraints, for example, source compositions must be nonnegative. Explicit physical conditions form linear inequality constraints in the space spanned by the eigenvectors, and these constraints form the feasible region in eigenvectors' space.

The Unmix model users' manual (U.S. EPA, 2007) has a good description of how SMCR identifies specific source impacts by using "edges." Briefly, if the data consists of many observations of M species, then the data can be plotted in an M-dimensional data space where the coordinates of a data point are the observed concentrations of the species during a sampling period. If there are N sources, the data space can be reduced to an $(N-1)$-dimensional space. Edges are drawn using the assumption that for each source there are some data points where the contribution of the source is not present or is small compared to the other sources. These are called edge points and Unmix works by finding these points and fitting a hyperplane through them; this hyperplane is called an edge (if $N=3$, the hyperplane is a line). By definition, each edge defines the points where a single source is not contributing. If there are N sources, then the intersection of $(N-1)$ of these hyperplanes defines a point that has only one source contributing. Thus, this point gives the source composition. In this way, the composition of the N sources is found, and from this, the source contributions are calculated so as to give a best fit to the data.

As an example, the Unmix model was applied to PM composition data from Phoenix (Lewis et al., 2003). The analysis generated source profiles and overall average percentage source contribution estimates for five source categories: gasoline engines ($33\pm4\%$), diesel engines ($16\pm2\%$), secondary sulfate ($19\pm2\%$), crustal/soil ($22\pm2\%$), and biomass burning ($10\pm2\%$). One of the unique aspects of this study was the ability to separate motor vehicle contributions into separate diesel and gasoline sources. Diesel emissions were identified by high elemental carbon relative to the organic carbon whereas gasoline vehicles had a profile with more organic than elemental carbon. In addition, a substantial difference was found in the contribution of diesel emissions between weekend and weekday samples.

The Unmix's use of hyperplane edges was found to be particularly useful when modeling high time resolution (30 min) $PM_{2.5}$ measurements in Tampa, FL (Pancras et al., 2011). Multiple sources such as residual oil combustion, lead smelting, coal combustion, biomass burning, marine aerosol, general industrial, and a Cd-rich source were clearly identified.

18.3   RESULTS AND DISCUSSION

18.3.1   AOSR Source Characterization

The sources of inorganic atmospheric emissions in the AOSR are dominated by the mining, processing, and upgrading of oil sand. While there are numerous sources of emissions, most are different mixtures of similar components. The raw material that drives the oil production activities in the AOSR is oil sand, made up primarily of sand, clay, bitumen, and water. The mining and processing of the oil sand aims to separate the bitumen (produced material) from the sand, clay, and water (tailings). The bitumen is upgraded and refined creating targeted products such as synthetic crude, diesel fuel, and gasoline; and byproducts such as petroleum coke and elemental sulfur (e.g., used in agriculture). Petroleum coke is burned to produce electrical power and steam. Diesel is used to fuel mine fleets, heavy haul trucks, and buses. Limestone and overburden are used to construct haul roads. Tailing sand is stored in large ponds for use in mine pit reclamation. Superimposed over the oil sands mining and processing emissions are regional contributions from forest fires, a common occurrence in the AOSR. This reality makes source apportionment modeling a challenge in the AOSR.

The analytical results from the bulk material and stack test filters showed extensive overlap or colinearity among the samples. We ultimately found it helpful to consolidate similar source categories for developing the emission profiles for CMB, and interpreting the PCA, PMF, and Unmix receptor modeling results. Table 18.2 summarizes the composited source samples, their sample types, and sampling locations. Table 18.3 presents the analytical emission profiles (mean $\pm$ standard deviation) for the consolidated sources used in CMB model runs.

TABLE 18.2 Summary of AOSR Composite Source Samples used in CMB Analysis

No	Sample name	Type	Sampling description	Comments
1	Haul road dust	Composite	Average of two grab samples from different haul roads	Haul roads are constructed with mined materials (limestone, overburden, low grade oil sand). Expected as fugitive dust source due to mining-related traffic on these roads.
2	Overburden	Grab sample	Sampled from an overburden pile	Expected to be airborne under windy conditions. Profile represents soil and glacial till. Does not contain limestone component.
3	Processed materials	Composite	Bitumen, fluid coke, vacuum tower bottoms, and petroleum coke	Petroleum coke stored in large stockpiles and expected to be airborne under windy conditions.
4	Tailing sand	Composite	Average of three tailing pond sand samples from different locations	Expected to be airborne under windy conditions.
5	Fleet emissions	Composite	Average of 14 samples from three facilities	Filters collected by DRI (Chapter 7).
6	Upgrader stack	Composite	Average of 12 stack samples from facility A main upgrader stack	Filters collected by DRI (Chapter 8).
7	Forest fire	Composite	Average of four WBEA Fort McKay site dichot filter samples impacted by forest fire	Filters collected by WBEA. Mean $PM_{2.5} = 474\ \mu g\ m^{-3}$.

18.3.2 Modeling Information

Elemental concentrations measured in the *H. physodes* samples that exhibited a signal-to-noise ratio > 2 (2σ above MDL; Chapter 14) were chosen for inclusion in PCA, CMB, PMF, and Unmix modeling. For PCA, CMB, and PMF runs, a total of 28 species were retained (Al, As, Ba, Ca, Ce, Cr, Cu, Fe, K, La, Li, Mg,

TABLE 18.3 WBEA Source Profile Composition Table

Element	Forest fire PM			Haul road dust			Overburden			Tailing sand		
	Avg. $\mu g\,g^{-1}$		SD	Avg. $\mu g\,g^{-1}$		SD	Avg. $\mu g\,g^{-1}$		SD	Avg. $\mu g\,g^{-1}$		SD
(a)												
Li	0.33	±	0.05	11.02	±	6.79	21.92	±	2.18	2.65	±	0.84
Be				0.51	±	0.26	0.38	±	0.26	0.06	±	0.03
Na	3.38	±	1.48	852.03	±	2.46	164.28	±	29.74	175.59	±	134.33
Mg	27.9	±	19.0	6303.0	±	3119.3	1004.4	±	130.6	129.5	±	91.7
Al	78.1	±	51.9	16,797.8	±	9362.7	15,391.5	±	2228.5	3512.3	±	1937.1
Si	315.2	±	56.3	14,753.2	±	8645.5	13,262.2	±	1933.1	51,701.4[a]	±	15,510.4[a]
P	129.3	±	19.9	276.6	±	167.5	70.9	±	17.1	43.1	±	39.7
K	2084.9	±	285.2	3529.7	±	732.0	2494.2	±	202.9	1735.7	±	1079.4
Ca				27,874.8	±	23,726.7	1303.4	±	331.6	355.5	±	300.5
S	4099.7	±	1138.1	2464.8	±	258.3	1623.8	±	164.4	85.0	±	70.2
Ti	1.55	±	0.82	122.50	±	12.40	72.97	±	25.79	41.33	±	41.53
V	0.27	±	0.09	28.25	±	23.92	19.26	±	2.27	2.38	±	1.46
Cr	3.95	±	1.64	14.55	±	11.47	11.38	±	2.05	1.49	±	0.71
Mn	12.0	±	6.6	226.9	±	38.9	103.7	±	7.9	54.7	±	37.5
Fe	485.4	±	259.1	13,663.2	±	1851.7	4929.8	±	323.5	912.7	±	865.0

Continued

TABLE 18.3 WBEA Source Profile Composition Table—Cont'd

Element	Forest fire PM		Haul road dust		Overburden		Tailing sand	
	Avg. $\mu g\,g^{-1}$	SD	Avg. $\mu g\,g^{-1}$	SD	Avg. $\mu g\,g^{-1}$	SD	Avg. $\mu g\,g^{-1}$	SD
(a)								
Co	0.07 ±	0.05	4.79 ±	2.94	4.03 ±	0.64	0.49 ±	0.45
Ni	0.55 ±	0.44	11.85 ±	5.62	8.48 ±	0.95	0.78 ±	0.77
Cu	2.48 ±	0.41	7.51 ±	4.96	3.46 ±	1.06	0.47 ±	0.20
Zn	159.91 ±	30.53	31.22 ±	17.54	18.28 ±	4.81	8.31 ±	5.44
Se	1.74 ±	0.26	6.94 ±	3.36	5.30 ±	0.90	1.23 ±	0.37
Rb	5.00 ±	1.00	19.22 ±	11.39	15.05 ±	1.54	5.51 ±	2.90
Sr	0.85 ±	0.39	65.31 ±	7.07	24.56 ±	2.41	18.92 ±	8.22
As	2.70 ±	0.33	3.73 ±	1.59	1.10 ±	0.53		
Nb			0.27 ±	0.07	0.23 ±	0.17	0.11 ±	0.13
Mo	0.23 ±	0.19	0.33 ±	0.03	0.47 ±	0.11	0.04 ±	0.02
Pd	0.02 ±	0.01	0.23 ±	0.16	0.27 ±	0.10	0.09 ±	0.08
Cd	8.82 ±	5.84	0.05 ±	0.02	0.04 ±	0.01		
Sn	0.87 ±	0.53	0.22 ±	0.09	0.29 ±	0.35		
Sb	0.04 ±	0.02	0.13 ±	0.05	0.04 ±	0.03	0.01 ±	0.01

Element											
Cs	0.09	±	0.02	1.14	±	1.11	0.99	±	0.11	0.14	±	0.09
Ba				111.96	±	1.13	50.79	±	6.00	60.16	±	33.07
La	0.08	±	0.03	8.62	±	2.98	7.65	±	1.07	3.04	±	0.91
Ce	0.15	±	0.07	18.00	±	7.23	18.01	±	2.00	6.36	±	1.59
Pr				2.15	±	0.91	1.97	±	0.36	0.70	±	0.18
Nd	0.05	±	0.02	8.71	±	4.00	7.62	±	0.95	2.52	±	0.63
Sm	0.01	±	0.00	1.64	±	0.88	1.43	±	0.20	0.44	±	0.11
Ta							0.02	±	0.04			
W				0.04	±	0.05	0.03	±	0.07			
Pt				0.01	±	0.01	0.01	±	0.04			
Tl	0.15	±	0.02	0.15	±	0.06	0.09	±	0.02	0.03	±	0.02
Pb	4.04	±	1.82	5.14	±	2.45	4.03	±	0.39	2.05	±	0.57
Bi	0.07	±	0.02	0.04	±	0.04	0.03	±	0.01	0.01	±	0.00
Th	0.03	±	0.02	2.28	±	1.58	2.15	±	0.36	0.57	±	0.23
U	0.00	±	0.00	0.49	±	0.33	0.42	±	0.06	0.09	±	0.04

Continued

Element	Processed materials			Heavy hauler fleet			Main upgrader stack		
	Avg. $\mu g\,g^{-1}$		*SD*	*Avg.* $\mu g\,g^{-1}$		*SD*	*Avg.* $\mu g\,g^{-1}$		*SD*
(b)									
Li	0.97	±	0.16				3.12	±	0.78
Be	0.03	±	0.01				0.09	±	0.03
Na	16.45	±	2.12	104.19	±	266.36	56.09	±	17.58
Mg	40.5	±	3.3	62.2	±	106.0	90.4	±	31.6
Al	518.1	±	112.9	194.2	±	1826.2	809.7	±	281.7
Si	754.5	±	161.6	594.2	±	1286.0	12,691.0	±	3623.1
P	10.9	±	2.1	5114.6	±	1923.6	61.7	±	31.2
K	58.3	±	11.2				117.9	±	36.0
Ca				9914.1	±	4115.4	411.1	±	120.0
SO	754.5	±	161.6	2157.9	±	1897.4	14,8443.8	±	20,627.0
Ti	28.54	±	7.72	20.92	±	32.75	168.15	±	43.85
V	21.85	±	13.30				101.55	±	30.40
Cr	0.88	±	0.23	10.65	±	33.69	3.11	±	4.52
Mn	8.7	±	4.2	3.3	±	7.5	50.7	±	19.9

Fe	386.3	±	117.0	148.4	±	369.8	1792.2	±	773.3
Co	0.73	±	0.23				2.68	±	0.91
Ni	5.73	±	2.03	5.57	±	20.74	41.82	±	15.56
Cu	1.26	±	0.15	138.06	±	257.18	9.01	±	11.31
Zn	1.93	±	0.40	5147.34	±	1300.00	25.70	±	22.54
Se	0.61	±	0.15	0.29	±	1.41	12.30	±	2.59
Rb	0.23	±	0.03				0.91	±	0.27
Sr	1.64	±	0.35	5.14	±	3.19	8.91	±	3.19
As							3.70	±	0.63
Nb	0.06	±	0.03				0.38	±	0.12
Mo	3.29	±	1.01	3.86	±	2.41	7.61	±	2.51
Pd	0.02	±	0.01	0.14	±	0.66			
Cd	0.01	±	0.00				0.12	±	0.05
Sn				1.33	±	11.25	0.53	±	0.41
Sb	0.02	±	0.00	0.84	±	9.05	0.27	±	0.21
Cs	0.02	±	0.00				0.06	±	0.02
Ba	2.23	±	0.72				6.68	±	3.59
La	0.60	±	0.13	0.06	±	2.24	2.55	±	0.75

Continued

Element	Processed materials			Heavy hauler fleet			Main upgrader stack		
	Avg. $\mu g\,g^{-1}$		SD	Avg. $\mu g\,g^{-1}$		SD	Avg. $\mu g\,g^{-1}$		SD
(b)									
Ce	1.30	±	0.27	0.07	±	4.94	4.60	±	1.37
Pr	0.15	±	0.03	0.05	±	0.67	0.51	±	0.15
Nd	0.58	±	0.13	0.07	±	2.40	1.92	±	0.57
Sm	0.11	±	0.03	0.01	±	0.37	0.35	±	0.10
Ta									
W	0.04	±	0.02				0.12	±	0.08
Pt							0.01	±	0.00
Tl	0.01	±	0.00				0.06	±	0.02
Pb	1.11	±	0.30	1.45	±	5.02	2.97	±	1.25
Bi	0.02	±	0.01				0.32	±	0.13
Th	0.17	±	0.04				0.60	±	0.17
U	0.04	±	0.01	0.03	±	0.04	0.14	±	0.04

Note: Concentration values < the reported MDL (Chapter 14) were deleted.
[a]Residue in the extraction vessels suggests the digestion of the tailings sand was incomplete. Based on the elemental composition expected for sandstone as listed in Faure (1991), it is likely that the amount of HF used during the digestion procedure was insufficient to dissolve all of the SiO_2 in the tailings sand. The Si/Al ratio of sandstone listed in Faure (1991) was 14.72; the value of Si reported for the tailings sand reflects this Si/Al ratio.

Mn, Mo, N, Na, Nd, Ni, P, Pb, S, Se, Si, Sm, Sr, Ti, V, Zn). For Unmix runs, Ba and Ca were dropped as better model fit statistics were observed in the absence of those two species. Two samples exhibited several outlier concentration points, and were therefore excluded from the data modeling. The following results are based on the remaining 119 samples. Variation in elemental concentrations of a lichen specimen may arise due to its age, chronic exposure and the corresponding tissue gain or loss, and their governing genetic and morphological variations. For these reasons, a total of 10 field duplicate samples were also collected and analyzed. A mean relative percent deviation for every element from the field duplicate results was then calculated and used as sampling precision in Equation (18.4) to estimate the measurement uncertainty in elemental concentrations needed in PMF and CMB.

$$\text{Uncertainty} = \sqrt{(\text{MDL})^2 \times (\text{sampling precision})^2} \qquad (18.4)$$

18.3.3   PCA: Multilinear Regression

A PCA analysis of the lichen-speciation data yielded five factors with eigenvalues >1.0 after Varimax rotation. The overall model explained 89% of the total variance. Communalities of all elements were over 80% with the exception of Cu, Pb, Ca, and Zn, whose communalities were $>65\%$. The rotated component matrix (factor loading) is presented in Table 18.4.

The first factor component (FC #1) accounted for 58% of the total variance and showed high loadings of fossil fuel marker elements (V, Ni, Mo, As, Se) and crustal elements (Li, Na, Al, Si, Ti, Fe, Mo, La, Ce, Nd, Sm). Given the AOSR source composition data presented in Table 18.3, this factor very likely represents a composite of all coarse PM sources such as oil sand, process material, and fugitive emissions. The FC #2 accounted for 14% of the total variance with significant loadings for Mg, P, K, Ca, Sr, and Ba. These elements are consistent with composition of the limestone bedrock material mined in the AOSR region. FC #3 explained 7% of the total variance and has a high loading of S and N. Oxides of sulfur and nitrogen can be either primary (stack) or secondary products of high temperature combustion processes. Therefore, this factor likely represents emissions from stack (stationary) and fleet vehicles (mobile). FC #4 presented high loadings for Zn, Ba, and Cu, and accounted for 6% of the total variances. Identification of this source is difficult as Zn, Ba, and Cu may be attributed to motor vehicle brake/tire wear, combustion of synthetic lubricants, or general anthropogenic activities. The last factor showed a strong negative loading for Mn. Graney et al. (Chapter 15) observed Mn depletion in *H. physodes* near the active mining and bitumen upgrading facilities, perhaps caused by biological inhibition of uptake or losses from tissue degradation. Nevertheless, it accounted for only 4% of the total variance, and therefore FC #5 is neglected in further analysis.

TABLE 18.4 Results of Varimax Rotated PCA Factor Loadings

Element	FC #1	FC #2	FC #3	FC #4	FC #5
S	0.50	0.00	0.74	−0.07	0.16
N	0.59	0.17	0.67	0.13	0.07
Li	0.93	0.25	0.20	0.01	0.06
Na	0.88	0.25	0.16	0.17	0.08
Mg	0.57	0.72	0.04	0.04	−0.08
Al	0.95	0.22	0.16	0.01	0.07
Si	0.91	0.26	0.12	0.13	0.09
P	−0.18	0.82	0.55	−0.07	−0.04
K	0.27	0.82	0.51	−0.13	0.02
Ca	0.49	0.65	−0.11	0.23	0.01
Ti	0.96	0.20	0.12	0.07	0.10
V	0.90	0.08	0.21	0.06	0.15
Cr	0.94	0.20	0.13	0.08	0.10
Mn	−0.23	−0.08	−0.11	0.11	−0.95
Fe	0.96	0.19	0.05	−0.01	0.05
Ni	0.90	0.23	0.16	0.06	0.10
Cu	0.69	0.17	0.42	0.26	−0.04
Zn	−0.19	0.31	0.05	0.80	−0.15
Se	0.96	0.17	0.13	−0.01	0.08
Sr	0.27	0.80	−0.05	0.22	0.19
As	0.93	0.25	0.15	0.06	0.03
Mo	0.88	0.08	0.30	0.10	0.12
Ba	0.27	0.72	−0.01	0.44	0.06
La	0.97	0.16	0.11	0.06	0.08
Ce	0.97	0.16	0.12	0.03	0.08
Nd	0.97	0.16	0.09	0.02	0.07
Sm	0.97	0.16	0.08	0.01	0.07
Pb	0.55	0.04	−0.06	0.63	0.02

TABLE 18.5 PCA-MLR Contribution of Key Elements to Identify Sources

| | Contribution from each source category (%) | | | | |
Element	Oil sand and fugitive dust	Haul road and limestone	Combustion processes	General anthropogenic	Unexplained (%)
N	40	13	46	0	1
S	28	0	40	−10	43
Na	59	17	9	10	6
Mg	24	31	0	0	45
Al	81	19	12	0	−12
Si	65	18	8	9	0
P	−19	80	45	−6	0
K	10	30	14	−6	53
Ca	38	50	−11	22	0
Ti	78	15	8	5	−6
V	98	8	24	0	−30
Cr	86	18	12	6	−22
Fe	60	11	0	0	30
Ni	70	18	12	0	1
Cu	42	10	24	12	11
Zn	−10	11	0	40	59
Sr	22	63	−9	24	0
Mo	64	6	24	6	0
La	86	13	14	4	−17
Ce	85	13	10	0	−8
Pb	34	0	0	35	31

Percent contribution of every element in a factor was determined by running a multilinear regression (MLR) model on each measured variable as dependent and all four absolute factor scores as independent variables (Thurston and Spengler, 1985). Table 18.5 summarizes the apportionment of measured concentrations by the PCA-MLR method. Over 65% of the measured concentrations of

Al, Ce, La, Mo, Ni, Si, Ti, and V were found to contribute to FC #1, which may be related to oil sand mining and processing activities. Elevated Ca, P, and Sr contributions confirm FC #2 as limestone, while the 40–45% of the measured N and S in FC #3 suggest that this factor is combustion related. Zn and Pb are the dominant contributing elements to the factor identified as general anthropogenic. A significant fraction of the measured Ba, Mg, S, Pb, K, and Zn concentrations were not explained by the PCA-MLR model.

18.3.4 Chemical Mass Balance

The selection of appropriate source profiles is a challenge when utilizing CMB. In this case, we used all the individual source sample profiles collected in the AOSR in the initial CMB model runs. Many of the local sources were observed to be not estimable by CMB due to excessive colinearity between the source profiles such as haul road dust emissions, limestone bedrock, tailing sand, oil sand, and overburden samples. Crustal, limestone, and oil component signatures (e.g., Ni, V) were present in all of these source materials (because bitumen extraction from oil sand is not 100% quantitative). The general CMB model (and other receptor models) assumes that (i) composition of source emissions are constant over the ambient and source sampling period, (ii) chemical species do not react with each other, (iii) chemical species add linearly, (iv) all major contributing sources are identified and characterized, (v) number of sources are less than the number of chemical constituent measured, (vi) source profiles are linearly independent, and (vii) measurement error is available, and it is random, uncorrelated, and normally distributed. However, studies show that deviations from one or more of the previously mentioned assumptions can still yield acceptable apportionment results. Nonetheless, "nearly collinear" sources affect CMB apportionment and often lead to unacceptable solutions. Chemically similar sources without unique marker species to distinguish between them are termed collinear sources. If two or more sources exhibit similar composition profiles, negative contributions are outputted by CMB. Such situations can be mitigated by variable selection, (e.g., eliminating one or more analytical species or entire sources that are nearly collinear). However, care must be taken not to eliminate a known source to improve the numerical performance of a receptor model.

Mined oil sand (raw material) is physically and chemically dissociated into bitumen (target product), by-products (petroleum coke) and residual materials (tailings). But the individual sources can still retain chemical similarities. To illustrate this point, the linear combination of tailing sand + processed material (y axis) is plotted against raw oil sand (x axis) composition in Figure 18.1. Therefore, either oil sand or processed material and tailing sand can be included in the model, but not all three together. Limestone source material was also not included in the CMB run as this crushed bedrock construction mineral was found to be colinear with the haul road dust source profile. Upon closer examination, it was clear that the haul road dust profile was dominated by limestone

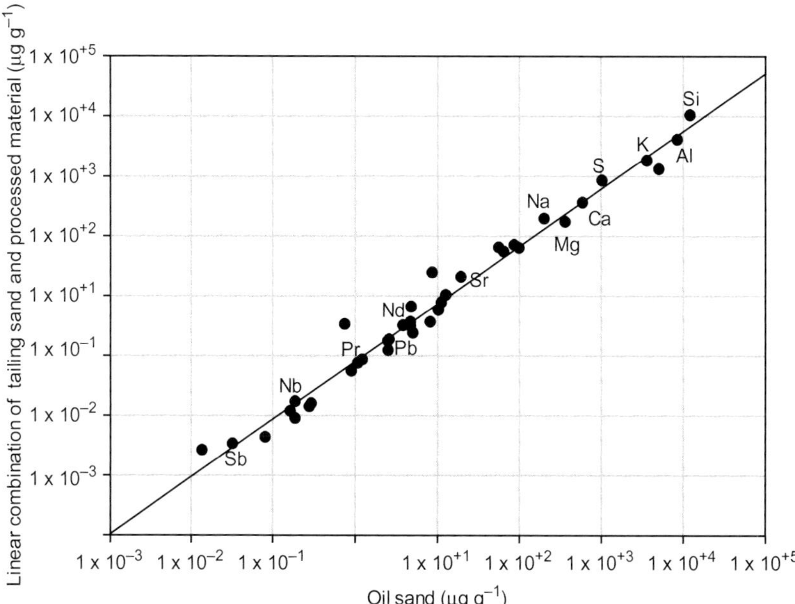

FIGURE 18.1   Relationship between profile concentrations of oil sand and a linear combination of processed material and tailing sand.

mineral. This finding was not surprising, because these temporary roads are constructed primarily of mined limestone minerals, overburden, and low grade oil sand. Large variability in emission signatures from the main upgrader stack and heavy-duty hauler fleet source profiles were other major areas of concern as species with large uncertainties are likely to be noninfluential in the CMB apportionment (U.S. EPA, 2004b).

In order to overcome these obstacles, similar source materials were combined into composite source profiles (Table 18.2) and the CMB model rerun with these carefully chosen seven local source profiles such as haul road dust, processed materials, tailing sand, fleet vehicles, main upgrader stack, forest fire/wood smoke, and overburden. The fit statistics ($r^2 > 0.8$ and $\chi^2 > 2$) were excellent for samples collected near the mining location and worse for the distal samples. For receptors located within a 20-km radius ($n = 28$), $72 \pm 23\%$ of the lichen mass was explained by these six sources. Median PM contributions of haul road dust, processed materials, tailing sand, overburden, forest fires, fleet vehicles, and main upgrader stack to the near-field lichens were estimated to be 242 ± 78, 190 ± 116, 178 ± 100, 87 ± 57, 45 ± 30, 6 ± 2, and 1 ± 0 mg g^{-1} of lichen mass, respectively.

Figure 18.2 presents individual sample contribution as a function of distance (km) from the center of the surface mining oil production activities. The strong influence of fugitive dust from the oil sand mining and processing operations on the near-field (<20 km) lichen samples is clear. The relative magnitude of the fugitive dust sources was found to be haul road $>$ tailing

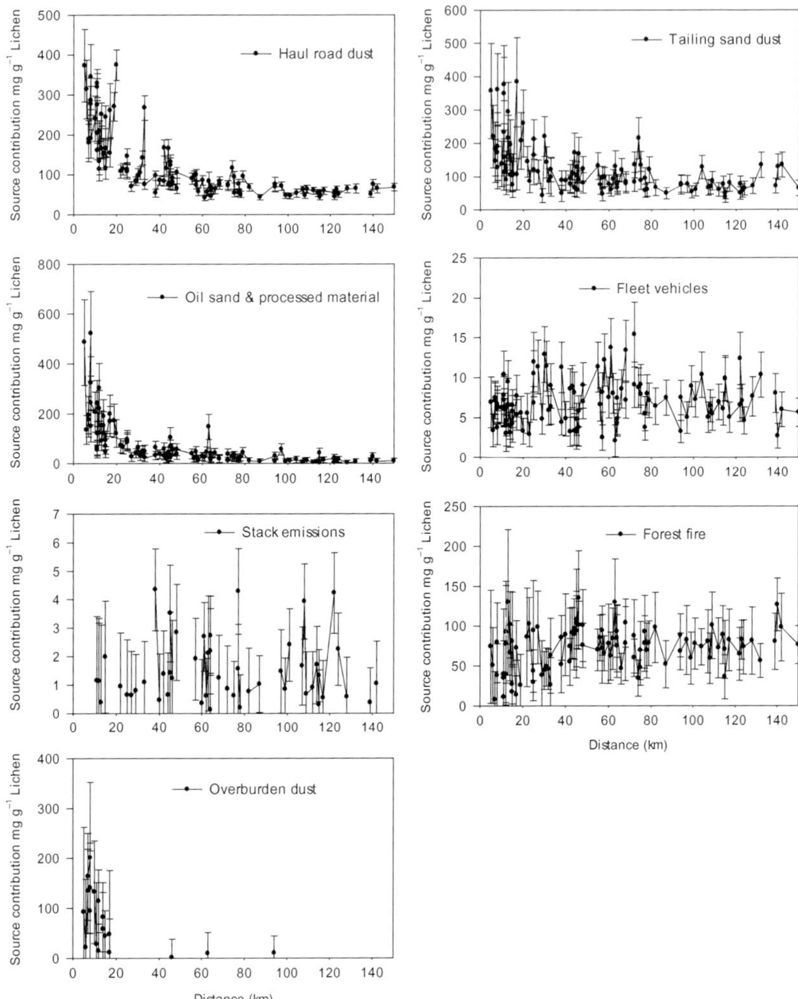

FIGURE 18.2 CMB source contribution estimates as a function of distance from the midpoint of oil sand mining and upgrading operations.

sand > overburden. Distal samples (>20 km) were possibly underestimated because of underrepresentation of contributing sources in the CMB model itself. Edgerton et al. (Chapter 14) documented that the lichen tissue concentrations collected in distal site locations were found to be lower in element concentrations than those from other areas in North America. It has been observed that underrepresenting the number of sources had little effect on the calculated source contribution estimates (SCEs) if the dominant species of the missing sources were excluded from the solution (U.S. EPA, 2004b). As the objective of this study was to evaluate air pollution from the AOSR region, no further attempts were made to explain all of the measured concentrations in distal receptor samples.

18.3.5   PMF and Unmix Modeling

18.3.5.1   Description of Factors

A six-factor solution was found to be optimal by both the PMF and Unmix models. Figure 18.3 presents and compares the source profiles generated by

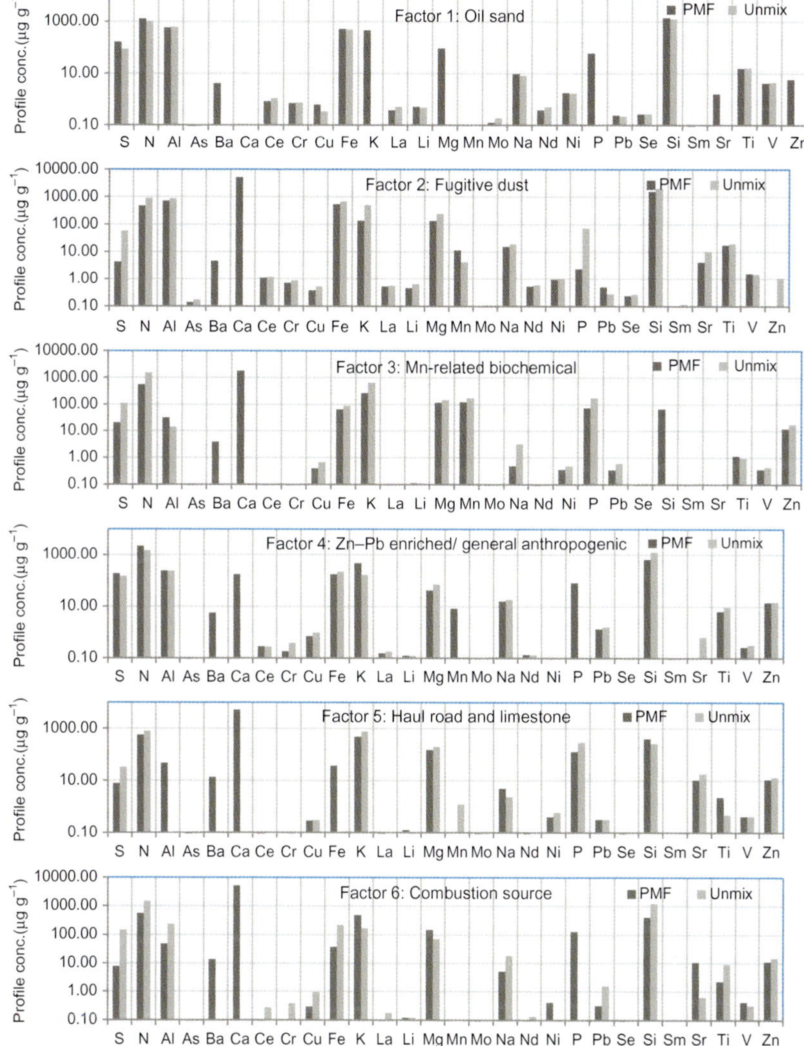

FIGURE 18.3   PMF and Unmix factor profiles of the lichen samples collected in the AOSR.

TABLE 18.6 Linear Regression Analysis Between PMF and Unmix (y/x)
Factor Contribution Estimates

Factors	Slope (PMF/Unmix)	r^2	Intercept
Oil sand	0.95	0.91	0.05
Fugitive dust	1.23	0.92	0.23
Haul road and limestone	0.98	0.54	0.02
Combustion	0.66	0.55	0.36
Mn-related biochemical	1.07	0.75	0.07
General anthropogenic	0.85	0.58	0.17

the models. In general, all factors showed good agreement between the two modeling approaches used, except the factor attributed to combustion sources. The block bootstrap method was used to evaluate modeling uncertainty in both PMF and Unmix solutions. There were no rejected (uncorrelated) factors from either model runs. Factor contributions were paired, and linear regression analysis was performed between the pairs of Unmix and PMF factor contribution estimates. As shown in Table 18.6, all of the six factors, interpreted as sources in the following section, showed good agreement between the two modeling results ($r^2 > 0.5$, slope > 0.6). The following source identifications are for PMF and Unmix indentified factors:

Oil sand and processed material: High loadings of V (59%), Ni (46%), Mo (51%), La (34%), Ce (34%), and Sm (35%), with La/Ce and V/Ni ratios close to source material values identifies this factor as the oil component in the oil sand+processed material signature. Modeled V/Ni and La/Ce ratios are 2.40 and 0.48, respectively. The composite oil sand source profile contains V/Ni of 1.95 and La/Ce of 0.42. This factor does not include a significant Ca loading, which is also a characteristic of oil sand source profile. A source contribution estimate (SCE) map (Figure 18.4) depicts an area of high source impact at the very center of the oil sand mining and processing activities. This type of clear near-field enhancement is consistent with ground-level emission of coarse particle fugitive dust. Coarse PM is produced mainly by mechanical forces such as crushing and abrasion, and therefore, consists primarily of finely divided minerals such as oxides of aluminum, silicon, iron, calcium, and potassium. Coarse particles of soil or dust mainly result from entrainment by wind or from other mechanical action. As the size range of these particles is quite large, their corresponding deposition velocities by sedimentation are relatively high. Therefore, particle retention time in the atmosphere and transport scales are generally short, resulting in enhancement of near-field deposition gradients (Davidson and Wu, 2002; Landis and Keeler, 2002).

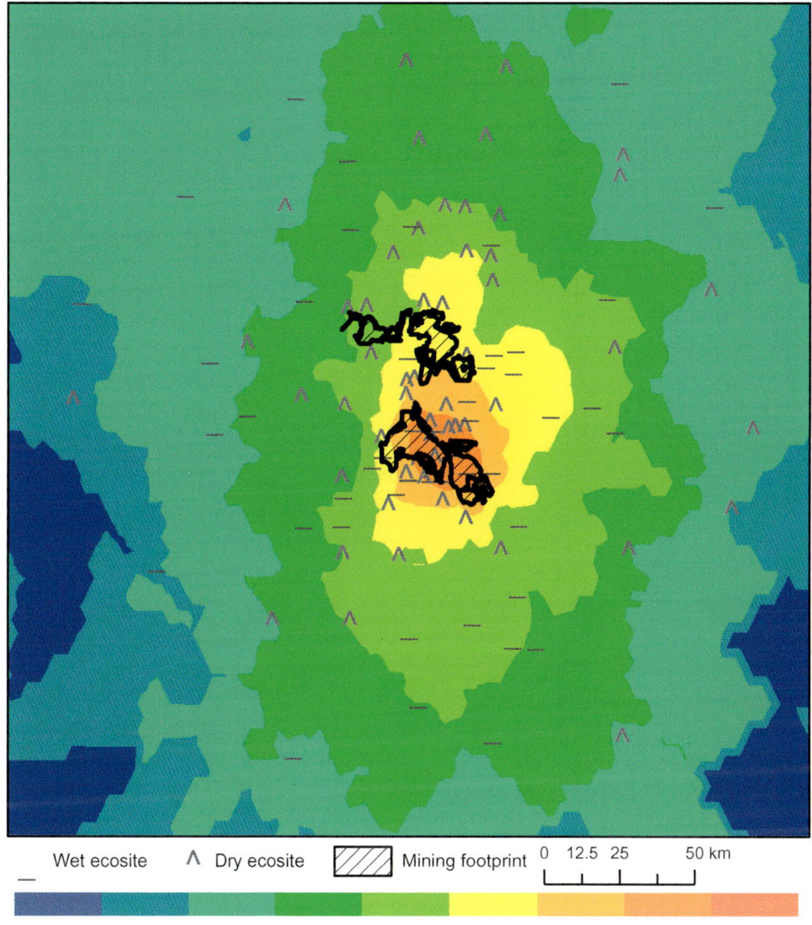

FIGURE 18.4   Modeled source contribution estimate of oil sand factor.

Fugitive dust from tailings sand: This factor comprises elements such as Al, Si, Ti, Ca, Ba, La, and Sm with a large relative occurrence. The composite tailing sand source sample shows a close resemblance with this factor. As tailing sand is processed and has had the bitumen removed, it is physically (smaller aggregate particle size) and chemically (lower concentration of the oil tracer species such as Ni and V) different from the mined raw oil sand particles. The SCE map (Figure 18.5) clearly supports that this factor is local to the central oil sand mining and production areas, but with a slightly more easterly extent and more widely distributed in space as compared to the oil sand factor. This factor is therefore assigned to represent local fugitive sand resulting from the exposed tailings ponds and various ground-based hauling activities.

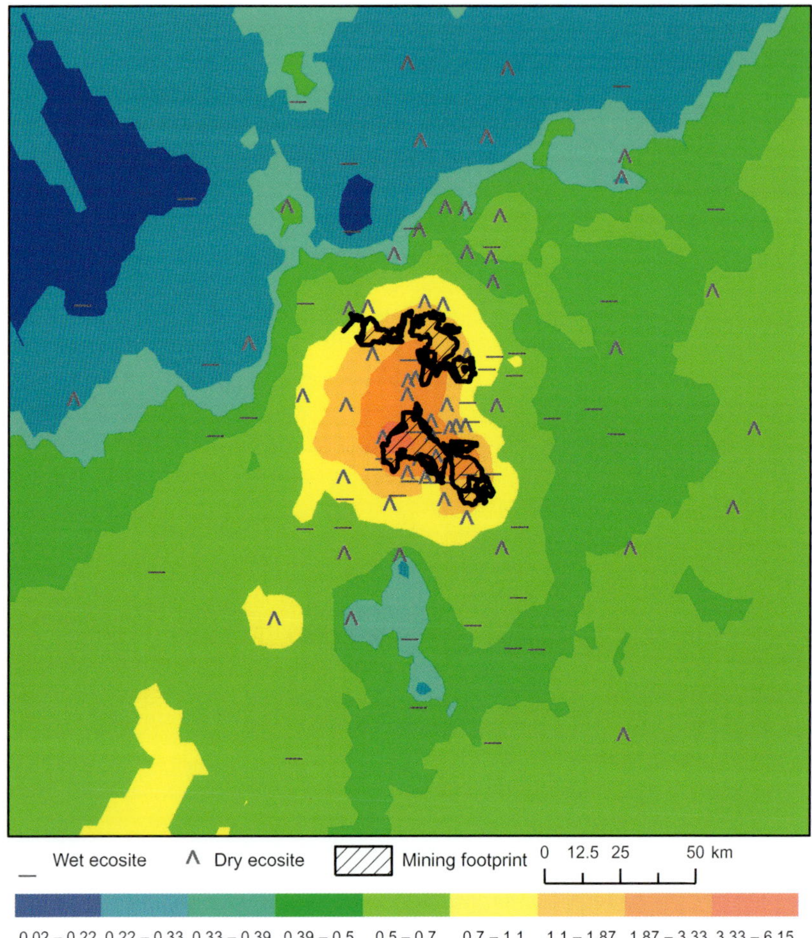

FIGURE 18.5   Modeled source contribution estimate of fugitive dust factor.

Haul road and limestone mixture: Elevated levels of Ca, Mg, Sr, and Ba characterize this factor. The ratio of Ca/Sr and Ca/Mg are very close to the limestone source material collected in the active mining areas. The limestone bedrock mined in this region underlying the oil sands is used to construct temporary roads for truck-hauling operations. Spatial contribution estimates presented in Figure 18.6 matches our expectation with high loading estimates near the active mining areas in 2008 and haul road activity.

Combustion source emissions: S, N, P, K, and Cu contributed 47%, 39%, 52%, 42%, and 26% of their respective modeled concentrations to this factor. Oxides of nitrogen and sulfur are primarily combustion-related emissions such as from SAGD boilers and upgrader main stacks (N, S), and fleet vehicle (P, Ca, Cu) emissions. The spatial distribution of contributions (Figure 18.7)

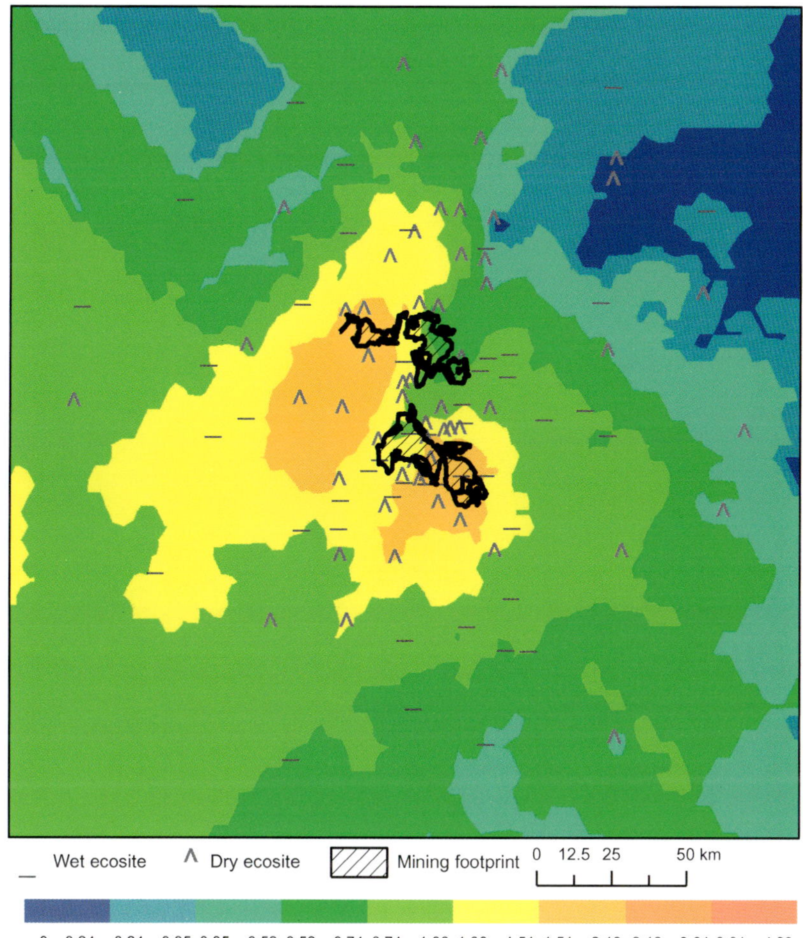

FIGURE 18.6   Modeled source contribution estimate of haul road and limestone factor.

for this factor shows the area of highest impact was farther away from the oil production facilities, which is consistent with an elevated stack emitting a plume with thermal buoyancy. The impact of ecosite classification of the lichen collection sites (Chapter 15) showed significant differences for this factor. Mean source contribution estimate from dry (1.5) sites were significantly higher than wet (0.6) sites. High loadings for P and K and larger contribution estimates in dry ecosite locations may signify that this factor also includes contributions from forest fire emissions.

Mn/biochemical: PMF and Unmix models attribute 74% and 82% of the measured Mn concentrations to this factor, respectively. The spatial map of this factor contribution (Figure 18.8) in some ways resembles surface topography and also clearly shows that Mn is depleted in close proximity to the

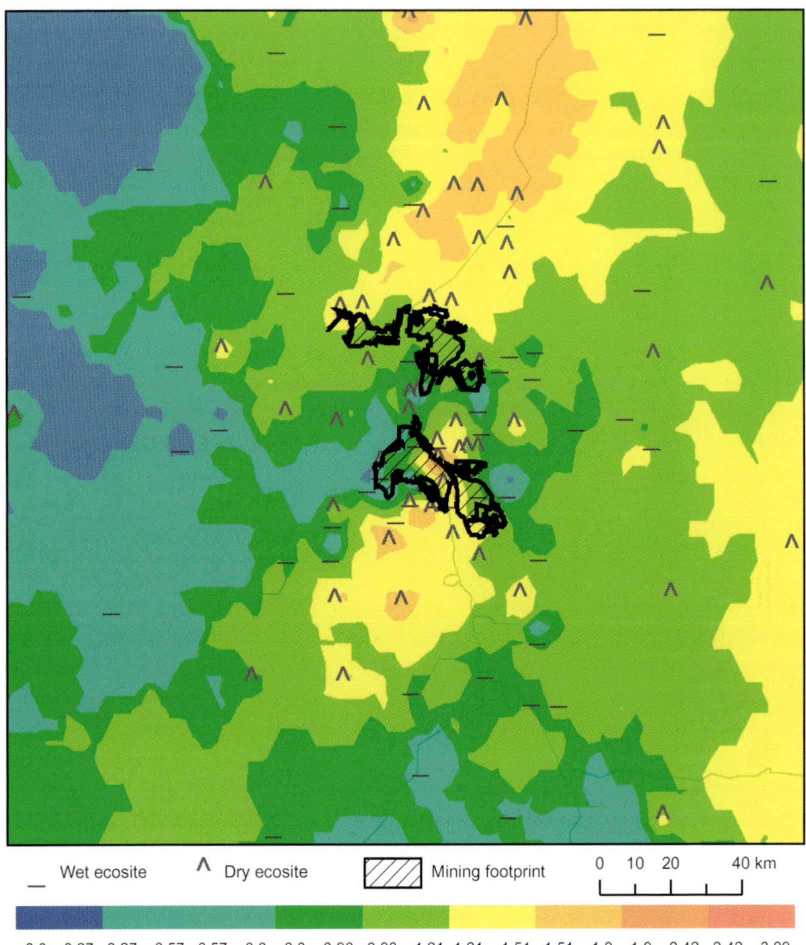

FIGURE 18.7 Modeled source contribution estimate of combustion source factor.

main oil sand mining and production areas. Larger source contributions are observed at higher elevation sites, and minimal contributions are seen in samples collected in lower elevation areas. There is also a significant difference between mean source contributions between wet (1.2) and dry (0.8) ecosite classification. Therefore, this factor is thought to represent a biochemical response from the *H. physodes* mobilizing these elements in response to their metabolic needs, and to the impact of near-field deposition of other chemical species from the oil production activities. Previous investigators have documented morphological responses (e.g., less diversity, smaller size) in lichen colonies in response to proximity to air pollution sources (Berryman et al., 2010), but the observations presented here of a potential Mn inhibition/suppression response in *H. physodes* tissue may be unique.

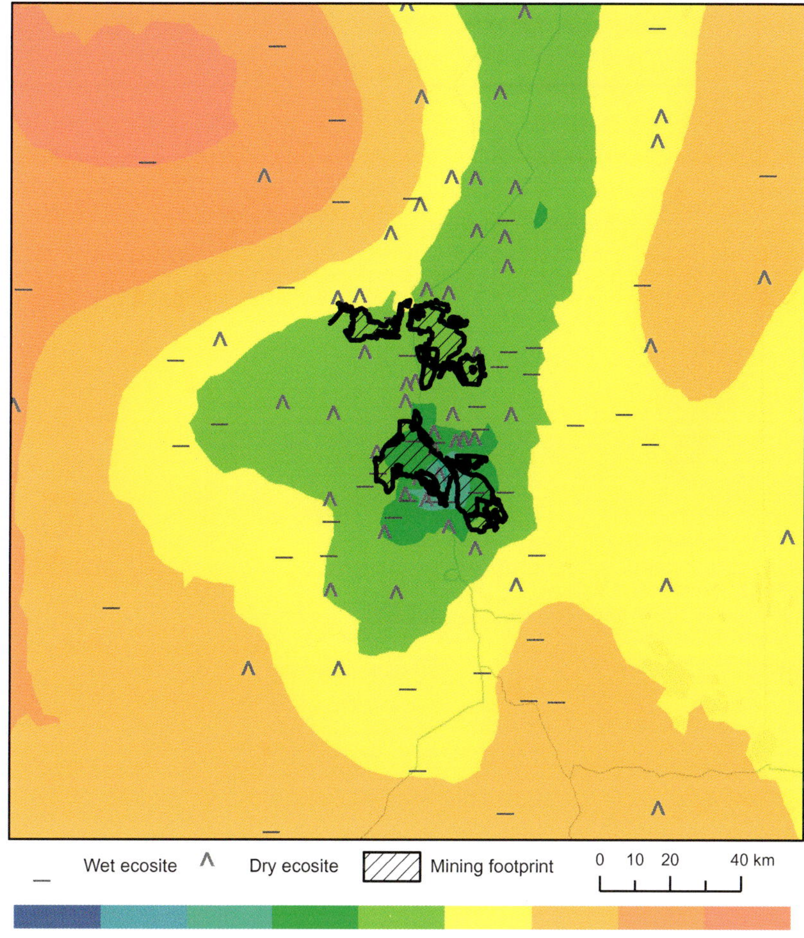

FIGURE 18.8   Modeled source contribution estimate of Mn-related biochemical factor.

General anthropogenic: This factor contains significant loadings of Zn and Pb, which are the typical tracer elements of general anthropogenic pollution. The source contribution estimate plot (Figure 18.9) shows larger contribution to this factor from the south in the vicinity of Fort McMurray, where urban activities are expected.

18.3.5.2 Apportionment

Percent contributions of total sulfur, total nitrogen, soil, and other trace metal oxides to the Unmix/PMF-identified factors are presented in Table 18.7. Confidence intervals for total sulfate and nitrate apportionment in absolute lichen mass (mg g^{-1}) is also included in the table. Soil contribution was calculated

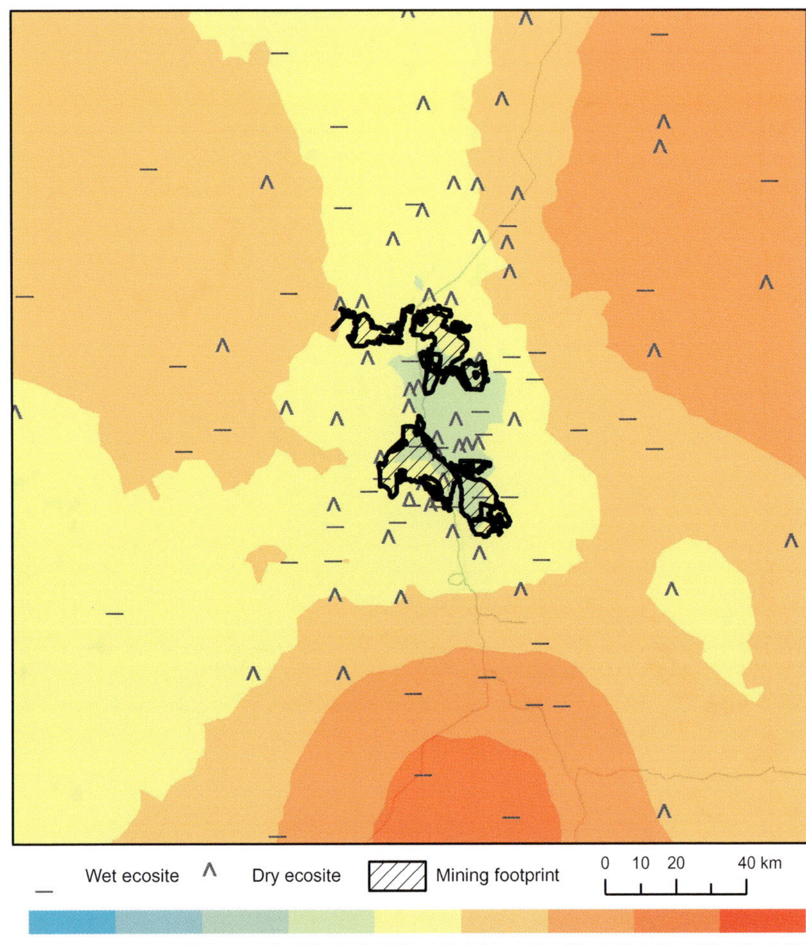

FIGURE 18.9 Modeled source contribution estimate of Zn-Pb/general anthropogenic factor.

as the sum of oxides of Si, Ca, Fe, and Ti. Sum of other atmospheric metal oxides were also calculated as described by Landis et al. (2001). Both models explained >97% of the measured total sulfur and total nitrogen concentrations. Metal oxide contributions estimated by the two models differ. This was most likely due to Ca not being included in the Unmix model (see Section 18.3.2). While Unmix overpredicted other metal oxides, PMF did not explain 8% and 6% of the soil and other metal oxide contributions, respectively.

Nearly 40% of the measured sulfate and nitrate concentrations were explained by combustion sources that include fleet vehicles, stack, and forest fire emissions. General anthropogenic background emerges as the next

TABLE 18.7 Percent Contributions of Total Sulfur, Total Nitrogen, Soil, and Other Metal Oxide Sources from Factors Identified by Unmix and PMF to the Measured Concentration in the Lichen Samples

Components	Oil sand		Tailing sand		Mn/biochemistry		Zn-Pb/general anthropogenic		Haul road dust/ limestone		Combustion processes		Unexplained	
	PMF	*Unmix*	*PMF*	*Unmix*	*PMF*	*Unmix*	*PMF*	*Unmix*	*PMF*	*Unmix*	*PMF*	*Unmix*	*PMF*	*Unmix*
Total sulfur[a] (%)	22.5 (0.2–0.6)	12.3 (0.1–0.5)	0.6 (0.0–0.5)	8 (0.0–0.4)	2.7 (0.0–0.4)	15.7 (0.2–0.5)	24.9 (0.0–0.6)	20.5 (0.2–0.6)	1.1 (0.0–0.4)	4.6 (0.0–0.2)	45.7 (0.8–1.4)	40.8 (0.7–1.3)	2.5	0.7
Total nitrogen[a] (%)	15 (2.4–7.2)	11.9 (0–7.7)	5.4 (0.7–6.9)	10.6 (1.1–7.9)	6.5 (1.4–6.9)	18.3 (3.7–8.7)	25.7 (3.0–11.2)	17.9 (2.4–10.5)	6.5 (2.0–8.6)	9.5 (0.6–5.7)	38 (12.3–18.6)	34.6 (10.5–19.9)	2.8	0.1
Soil-related (%)	14.4	12.8	34.7	43.9	7.3	8.9	7	16.4	21.9	25.3	6.6	1.8	8	−1.6
Metal oxides (%)	11.4	0.1	3	13.7	10.5	24.7	12.6	3.7	15.1	29.2	41.5	45.6	5.8	−10.2

[a] *5th and 95th percentile concentrations from the block bootstrap error estimation are given in mg g^{-1} lichen mass.*

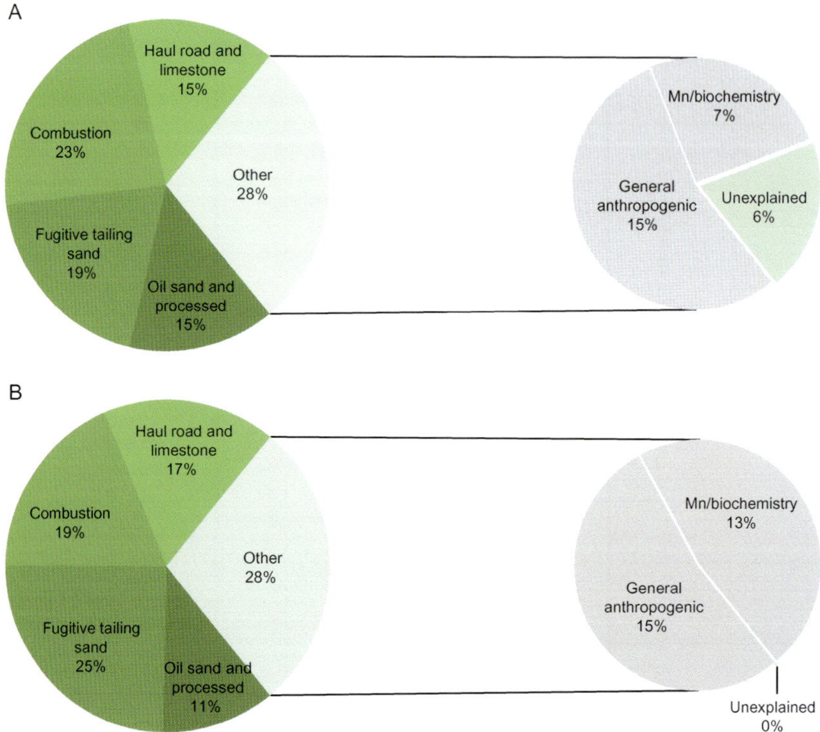

FIGURE 18.10 (A) Estimated percent source contributions by PMF model. (B) Estimated percent source contributions by Unmix model.

significant source for sulfate and nitrate. As expected, soil-related contributions are significant from tailing sand fugitive dust and haul road fugitive dust factors. Of all the sources identified, oil sand and processed materials, tailing sand fugitive dust, haul road fugitive dust, and combustion emissions appear to be originating from the AOSR oil sand mining and production operations. Together, these sources explain 72% of the measured element concentrations (as oxides) found in *H. physodes* tissue samples (Figure 18.10).

18.4 CONCLUSIONS

Overall, the concentration of elements observed in *H. physodes* tissue samples in the boreal forests in the AOSR were consistent with those reported in other areas of Canada, the United States, and in other areas in the northern hemisphere (Chapter 14). However, near-field samples collected within 20 km of the main surface mining and oil sand production/upgrading operations had significantly higher concentrations of both crustal (La, Ce, Nd, Ti, Fe, Ca, Sr) and anthropogenic elements (Ni, V, Sn, Mo, Cr, Cu, Sb). The anthropogenic and natural sources of air pollution in the AOSR including oil sands mining and processing activities and forest fires were identified, sampled, and chemically characterized. The relative contributions of the different inorganic air pollutant source types in the AOSR on the epiphytic lichen *H. physodes* tissue concentrations observed in the surrounding boreal forests were investigated using PCA, CMB, PMF, and Unmix receptor models.

Initial PCA screening analysis indicated that there were five principal components that could explain 89% of the variance contained in the lichen data set. The use of the CMB model was hindered by source collinearity issues, but the model was able to successfully apportion near-field sampling locations (<20 km of mining and upgrading facilities). CMB determined that six of the seven composited source profiles significantly contributed to the near-field lichen tissue concentrations. The PMF and Unmix multivariate receptor models provided very consistent results and indicated that there were six significant source factors. Five of the sources impacting the lichen tissue concentrations were primarily anthropogenic, including (i) oil sand and processed material, (ii) tailing sand fugitive dust, (iii) combustion processes, (iv) limestone and haul road fugitive dust, and (v) a general urban source. The remaining significant source was an Mn-dominated biogeochemistry factor.

The spatial patterns of CMB, PMF, and Unmix receptor model-estimated source impacts on the *H. physodes* tissue concentrations from the oil sand/produced material and fugitive dust sources were significantly correlated to the distance from the primary oil sand surface mining operations and related production facilities. The spatial extent of the impact was limited to an approximately 20-km radius around the major mining and oil production facilities, which is clearly indicative of ground-level coarse particulate fugitive emissions from these sources. The relative impact of the general urban background source was found to be enhanced in the lichens in proximity to the Fort McMurray urban area. The receptor models also show an Mn-related biogeochemical response factor that is a combination of ecological factors (wet vs. dry ecosite) as well as an Mn response to near-field oil sands production operations.

Overall, the largest impact on elemental concentrations of *H. physodes* tissue in the AOSR was related to fugitive dust, suggesting that implementation of a fugitive dust abatement strategy could minimize the near-field impact of future mining-related production activities. Over the next decade, as oil production increases in the AOSR, (i) new surface mining operations will expand the footprint of land disturbance, (ii) *in situ* techniques will represent a larger percentage of overall bitumen extraction volume, (iii) new production and upgrading technologies will improve extraction efficiencies while reducing energy demand, (iv) new techniques for treating tailings will emerge, and (v) mine remediation activities will accelerate. How these changes impact atmospheric deposition in the surrounding boreal forests remains to be seen. It is recommended that the combination of epiphytic lichen biomonitoring and the application of receptor models continue to be used to inform residents and stakeholders in the AOSR on the potential impact of bitumen production on their communities and natural forest resources.

ACKNOWLEDGMENTS

We thank Shanti Berryman and Justin Straker (Stantec) for their efforts in lichen sample collection and cleaning, Joel Blum (University of Michigan) for lichen sample grinding, and Mike Fort and Eric Edgerton (ARA) for lichen sample extraction and DRC-ICPMS analysis. This work was funded by WBEA.

REFERENCES

Anttila, P., Paatero, P., Tapper, U., Järvinen, O., 1995. Application of positive matrix factorization to source apportionment: results of a study of bulk deposition chemistry in Finland. Atmos. Environ. 29, 1705–1718.

Attanasi, E.D., Meyer, R.F., 2010. Natural bitumen and extra-heavy oil, Survey of Energy Resources, 22nd ed. World Energy Council, London, UK, ISBN 0-946121-26-5, pp. 123–140.

Begum, B., Hopke, P.K., Zhao, W., 2005. Source identification of fine particles in Washington DC by expanded factor analysis modeling. Environ. Sci. Technol. 55, 227–240.

Berryman, S., Straker, J., Krupa, S., Davies, M., Ver Hoef, J., Brenner, G., 2010. Mapping the characteristics of air pollutant deposition patterns in the Athabasca Oil Sands Region using epiphytic lichens as bioindicators. Interim Report Submitted to the Wood Buffalo Environmental Association, Fort McMurray, AB, Canada.

Cao, J., Li, H., Chow, J.C., Watson, J.G., Lee, S., Rong, B., Dong, J.G., Ho, K.F., 2011. Chemical composition of indoor and outdoor atmospheric particles at Emperor Qin's Terra-cotta Museum, Xi'an, China. Aerosol & Air Qual. Res. 11, 70–79.

Chan, Y.C., Hawas, O., Hawker, D., Vowles, P., Cohen, D.D., Stelcer, E., Simpson, R., Golding, G., Christensen, E., 2011. Using multiple type composition data and wind data in PMF analysis to apportion and locate sources of air pollutants. Atmos. Environ. 45, 439–449.

Cheng, Y., Lee, S.C., Cao, J.J., Ho, K.F., Chow, J.C., Watson, J.G., 2009. Elemental composition of airborne aerosols at a traffic site and a suburban site in Hong Kong. Int. J. Environ. Pollut. 36, 166–179.

Chueinta, W., Hopke, P.K., Paatero, P., 2000. Investigation of sources of atmospheric aerosol urban and suburban residential areas in Thailand by positive matrix factorization. Atmos. Environ. 34, 3319–3329.

Chueinta, W., Hopke, P.K., Paatero, P., 2004. A multi-linear model for spatial pattern analysis of the measurement of haze and visual effects project. Environ. Sci. Technol. 38, 544–554.

Cooper, J.A., Watson, J.G., Huntzicker, J.J., 1984. The effective variance weighting for least squares calculations applied to the mass balance receptor model. Atmos. Environ. 18, 1347–1355.

Davidson, C.I., Wu, Y.L., 2002. Dry deposition of particle and vapors. In: Lindberg, S.E., Page, A.L., Norton, S.A. (Eds.), Acidic Precipitation. 3: Sources, Deposition and Canopy Interactions. Springer-Verlag, New York.

Faure, G., 1991. Principles and Applications of Inorganic Geochemistry. Macmillan Publishing Company, New York.

Gaarenstroom, P.D., Perone, S., Moyers, J.P., 1977. Application of pattern recognition and factor analysis for characterization of atmospheric particulate composition in southwest desert atmosphere. Environ. Sci. Technol. 11, 795–800.

Garty, J., 2001. Biomonitoring atmospheric heavy metals with lichens: theory and application. Crit. Rev. Plant Sci. 20 (4), 309–371.

Gordon, G.E., 1985. Receptor models. Environ. Sci. Technol. 22, 1132–1142.

Government of Alberta, 2008. Alberta's oil sands: opportunity, balance. 978-0-7785-7348-7 Edmonton, AB, Canada.

Gu, J.W., Pitz, M., Schnelle-Kreis, J., Diemer, J., Reller, A., Zimmermann, R., Soentgen, J., Stoelzel, M., Wichmann, H.E., Peters, A., Cyrys, J., 2011. Source apportionment of ambient particles: comparison of positive matrix factorization analysis applied to particle size distribution and chemical composition data. Atmos. Environ. 45, 1849–1857.

Harmon, H.H., 1976. Modern Factor Analysis, third ed. rev. University of Chicago Press, Chicago, IL.

Henry, R.C., 1987. Current factor analysis models are ill-posed. Atmos. Environ. 21, 1815–1820.

Henry, R.C., 1997. History and fundamentals of multivariate air quality receptor models. Chemom. Intell. Lab. Syst. 37, 525–530.

Hopke, P.K., 1983. An Introduction to multivariate analysis of environmental data. In: Natusch, D.F.S., Hopke, P.K. (Eds.), Analytical Aspects of Environmental Chemistry. Wiley, New York, pp. 219–261.

Hopke, P.K., 1985. Receptor modeling in environmental chemistry. J.W. Wiley & Sons, Hoboken, NJ.

Hopke, P.K., 2009. Theory and application of atmospheric source apportionment. In: Legge, A.H. (Ed.), Air Quality and Ecological Impacts. Elsevier, Amsterdam, The Netherlands, pp. 1–33.

Hopke, P.K., Gladney, G., Gordon, W., Zoller, W., Jones, A., 1976. The use of multivariate analysis to identify sources of selected elements in Boston urban aerosol. Atmos. Environ. 10, 1015–1025.

Hopke, P.K., Paatero, P., Jia, H., Ross, R.T., Harshman, R.A., 1998. Three-way (PARAFAC) factor analysis: examination and comparison of alternative computational methods as applied to ill-conditioned data. Chemom. Intell. Lab. Syst. 43, 25–42.

Hopke, P.K., Ramadan, Z., Paatero, P., Norris, G., Landis, M., Williams, R., Lewis, C.W., 2003. Receptor modeling of ambient and personal exposure samples: (1998). Baltimore particulate matter epidemiology-exposure study. Atmos. Environ. 37, 3289–3302.

Hsu, Y.-K., Holsen, T.M., Hopke, P.K., 2003. Comparison of hybrid receptor models to locate PCB sources in Chicago. Atmos. Environ. 37, 545–562.

Huang, S., Rahn, K.A., Arimoto, R., 1999. Testing and optimizing two factor-analysis techniques on aerosol at Narragansett, Rhode Island. Atmos. Environ. 33, 2169–2185.

Jeran, Z., Jacimovic, R., Batic, F., Mavsar, R., 2002. Lichens as integrating air pollution monitors. Environ. Pollut. 120, 107–113.

Jolliffe, I.T., 2002. Principal component analysis. Springer, Berlin.

Juntto, S., Paatero, P., 1994. Analysis of daily precipitation data by positive matrix factorization. Environmetrics 5, 127–144.

Kim, E., Hopke, P.K., 2006. Characterization of fine particle sources in the Great Smoky Mountains Area. Sci. Total Environ. 368, 781–794.

Kim, E., Hopke, P.K., Edgerton, E.S., 2004. Improving source identification of Atlanta aerosol using temperature resolved carbon fractions in positive matrix factorization. Atmos. Environ. 38, 3349–3362.

Kuik, P., Sloof, J.E., Wolterbeek, H.Th., 1993. Application of Monte Carlo-assisted factor analysis to large sets of environmental pollution data. Atmos. Environ. 27, 1975–1983.

Landis, M.S., Keeler, G.J., 2002. Atmospheric mercury deposition to Lake Michigan during the Lake Michigan Mass Balance Study. Environ. Sci. Technol. 36, 4518–4524.

Landis, M.S., Norris, G.A., Williams, R.W., Weinstein, J.P., 2001. Personal exposures to $PM_{2.5}$ mass and trace elements in Baltimore, MD, USA. Atmos. Environ. 35, 6511–6524.

Lee, E., Chan, C.K., Paatero, P., 1999. Application of positive matrix factorization in source apportionment of particulate pollutants in Hong Kong. Atmos. Environ. 33, 3201–3212.

Lewis, C.W., Norris, G.A., Henry, R.C., Conner, T.L., 2003. Source apportionment of Phoenix $PM_{2.5}$ aerosol with the UNMIX receptor model. J. Air Waste Manage. Assoc. 53, 325–338.

Maykut, N.N., Lewtas, J., Kim, E., Larson, T.V., 2003. Source apportionment of $PM_{2.5}$ at an urban IMPROVE site in Seattle, WA. Environ. Sci. Technol. 37, 5135–5142.

Miller, M.S., Friedlander, S.K., Hidy, G.M., 1972. A chemical element balance for the Pasadena aerosol. J. Colloid Interface Sci. 39, 65–176.

Olson, D.A., Norris, G.A., Landis, M.S., Vette, A.F., 2004. Chemical characterization of ambient particulate matter near the World Trade Center: elemental carbon, organic carbon, and mass reconstruction. Environ. Sci. Technol. 38, 4465–4473.

Paatero, P., 1997. Least squares formulation of robust, non-negative factor analysis. Chemom. Intell. Lab. Syst. 37, 23–35.

Paatero, P., 1999. The multilinear engine – a table-driven least squares program for solving multilinear problems, including the n-way parallel factor analysis model. J. Comput. Graph. Stat. 8, 854–888.

Paatero, P., Tapper, U., 2003. Analysis of different modes of factor analysis as least squares fit problems. Chemom. Intell. Lab. Syst. 18, 183–194.

Pancras, J.P., Ondov, J.M., Poor, N., Landis, M.S., Stevens, R.K., 2006. Identification of sources and estimation of emission profiles from highly time-resolved pollutant measurements in Tampa, FL. Atmos. Environ. 40, 467–481.

Pancras, J.P., Vedantham, R., Landis, M.S., Norris, G.A., Ondov, J.M., 2011. Application of EPA UNMIX and non-parametric wind regression on high time resolution trace elements and speciated mercury in Tampa, Florida Aerosol. Environ, Sci, Technol. 45, 3511–3518.

Polissar, A.V., Hopke, P.K., Malm, W.C., Sisler, J.F., 1996. The ratio of aerosol optical absorption coefficients to sulfur concentrations, as an indicator of smoke from forest fires when sampling in polar regions. Atmos. Environ. 30, 1147–1157.

Polissar, A.V., Hopke, P.K., Poirot, R.L., 2001. Atmospheric aerosol over Vermont: chemical composition and sources. Environ. Sci. Technol. 35, 4604–4621.

Pratt, G.C., Krupa, S.V., 1985. Aerosol chemistry in Minnesota and Wisconsin and its relation to rainfall chemistry. Atmos. Environ. 19, 961–971.

Ramadan, Z., Song, X.-H., Hopke, P.K., 2000. Identification of sources of Phoenix aerosol by positive matrix factorization. J. Air Waste Manage. Assoc. 50, 1308–1320.

Sloof, J.E., 1995. Pattern recognition in lichens for source apportionment. Atmos. Environ. 29, 333–343.

Sosa, E.R., Bravo, A.H., Mugica, A.V., Sanchez, A.P., Bueno, L.E., Krupa, S., 2009. Levels and source apportionment of volatile organic compounds in southwestern area of Mexico City. Environ. Pollut. 157, 1038–1044.

Thurston, G.D., 1981. Discussion of "Multivariate Analysis of Particulate Sulfate and Other Air Quality Variables by Principal Components-Part I. Annual Data from Los Angles and New York" In: Henry, R.C., Hidy, G.M. (Eds.), Atmos. Environ. 15, 424–425.

Thurston, G.D., Spengler, J.D., 1985. A quantitative assessment of source contributions to inhalable particulate matter pollution in Metropolitan Boston. Atmos. Environ. 19, 19–25.

U.S. Environmental Protection Agency, 2004a. EPA-CMB8.2 Users Manual. Report No. EPA-452/R-04-011. December 2004. Office of Air Quality Planning & Standards, Research Triangle Park, NC, USA. www.epa.gov/ttn/SCRAM/receptor_cmb.htm, last accessed on June 21, 2012.

U.S. Environmental Protection Agency, 2004b. Protocol for applying and validating the CMB model for $PM_{2.5}$ and VOC. Report No. EPA-451/R-04-001. December 2004. Office of Air Quality Planning & Standards, Research Triangle Park, NC, USA

U.S. Environmental Protection Agency, 2007. EPA Unmix 6.0 Fundamentals & User Guide. Report No. EPA/600/R-07/089. June 2007. Office of Research and Development, Washington, DC, USA. www.epa.gov/heasd/products/unmix/unmix.html. June 21, 2012.

U.S. Environmental Protection Agency, 2011. EPA Positive Matrix Factorization (PMF) 4.2 Fundamentals & User Guide. Report No. EPA-600/R-11/117. Nov 2011. Office of Research and Development, Washington, DC, USA. www.epa.gov/heasd/products/pmf/pmf.html.

Voukantsis, D., Karatzas, K., Kukkonen, J., Räsänen, T., Karppinen, A., Kolehmainen, M., 2011. Inter-comparison of air quality data using principal component analysis, and forecasting of PM10 and PM2.5 concentrations using artificial neural networks, in Thessaloniki and Helsinki. Sci. Total Environ. 409, 1266–1276.

Wang, Y.Q., Zhang, X.Y., Arimoto, R., Cao, J.J., Shen, Z.X., 2004. The transport pathways and sources of PM_{10} pollution in Beijing during spring 2001, 2002 and 2003. Geophys. Res. Lett. 31, Art. No. L14110. 4PP, doi:10.1029/2004GL019732.

Watson, J.G., Robinson, N.F., Chow, J.C., Henry, R.C., Kim, B.M., Pace, T.G., Meyer, E.L., Nguyen, Q., 1990. The USEPA/DRI chemical mass balance receptor model, CMB 7.0. Env. Soft. 5, 38–49.

Winchester, J.W., Nifong, G.D., 1971. Water pollution in Lake Michigan by trace elements from pollution aerosol fallout. Water Air Soil Pollut. 1, 50–64.

Zhao, W., Hopke, P.K., 2006. Source identification for fine aerosols in Mammoth Cave National Park. Atmos. Res. 80, 309–322.

Zhao, W., Hopke, P.K., Karl, T., 2004. Source identification of volatile organic compounds in Houston. Environ. Sci. Technol. 38, 1338–1347.

Zhou, L., Kim, E., Hopke, P.K., Stanier, C., Pandis, S., 2004a. Advanced factor analysis on Pittsburgh particle size distribution data. Aerosol Sci. Technol. 38, 118–132.

Zhou, L., Hopke, P.K., Liu, W., 2004b. Comparison of two trajectory-based models for locating particle sources for two rural New York sites. Atmos. Environ. 38, 1955–1963.

Chapter 19

Concluding Remarks

K.E. Percy[*,1] and S. Krupa[†]
[*]*Wood Buffalo Environmental Association, Fort McMurray, Alberta, Canada*
[†]*Plant Pathology, University of Minnesota-Twin Cities, St. Paul, Minnesota, USA*
[1]*Corresponding author: e-mail: kpercy@wbea.org*

ABSTRACT

The Athabasca Oil Sands Region (AOSR) in northeastern Alberta is the largest of Canada's oil sand deposits. Oil production in the AOSR reached 1.6 million barrels per day in 2011. Atmospheric emissions of sulfur, nitrogen, trace metals and organic compounds, and their possible effects on air quality and the terrestrial environment in AOSR require rigorous monitoring. The Wood Buffalo Environmental Association (WBEA) has been monitoring air quality and the terrestrial environment in the 68,454 km^2 airshed since 1997. In 2008, WBEA adopted a source to sink approach yielding science-based, and practical monitoring of the AOSR environment. Measuring and monitoring at key points along the air pollutant pathway, four years of measurements enabled by a three-fold increase in funding have provided scientific data on emissions, transport, air quality, deposition, as well as chemical and biological measurements in receptor ecosystems. Emission source chemical fingerprints including stable isotopes and PAHs have been successfully used in receptor modelling to apportion elemental concentrations measured in terrestrial receptors back to source type. The results of this *practical scientific monitoring and measurement* have clearly demonstrated the efficacy of a combination of vegetation receptor bio-monitoring and the application of atmospheric receptor models, including "real-world" source characterization, to fully integrate air and terrestrial systems in time and space in the changing landscape. This information will be used by decision makers on air shed management, contribute new knowledge to support environmental impact assessments, inform stakeholders and the public on air quality. The Wood Buffalo Environmental Association's scientifically-enhanced air and terrestrial monitoring programs will also be essential in meeting government regulatory requirements and expectations under cumulative effects management policies.

19.1 INTRODUCTION

An inexorable global increase by some 46% in energy demand is expected over the next 20 years (Chapter 1). The Canadian Oil Sands represents the third largest oil reserve in the world, after Saudi Arabia and Venezuela.

Among the three deposits in Alberta, Canada, the Athabasca Oil Sands Region (AOSR) in northeastern Alberta is the largest (Figure 19.1).

Oil sands are a mixture of sand, water, clay, and bitumen. Bitumen is heavy oil that is too thick to flow or to be pumped without being diluted or heated. The resource has been known since the eighteenth century when early fur traders arrived and found aboriginal people using bitumen to seal their canoes (Chapter 2). Technologies used to produce oil from the oil sands have evolved considerably since first commercial development in 1967. Industry continues to find ways to make the process more efficient and improve environmental performance (Chapter 3).

Oil production in the AOSR has been steadily increasing over the past decade, from 0.6 million barrels per day in 2000 to 1.6 million barrels per day in 2011. Production is expected to be in excess of 3.5 million barrels per day by 2025. Synthetic crude oil production from bitumen in the AOSR is accomplished using a combination of surface mining and drilling (*in situ*). Of the proved reserves, it is estimated that 20% of the bitumen will ultimately be recovered through surface mining and 80% from *in situ* production techniques (Chapter 2). The type and magnitude of inorganic air pollutants emitted from these two extraction techniques are unique. Quantifying their relative contributions to observed ambient concentrations and atmospheric deposition are critical to be able to mitigate and manage any local environmental impacts as production levels are increased.

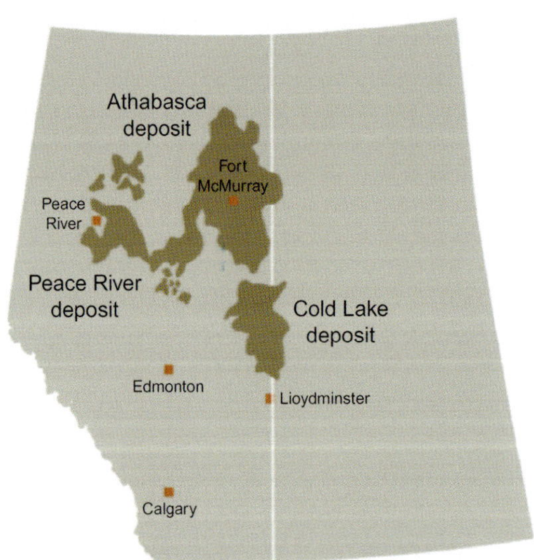

FIGURE 19.1   Location and extent of Canadian oil sands deposits.

Atmospheric emissions of sulfur, nitrogen, trace metals, and organic compounds, and their potential effects on the terrestrial environment, including soil acidification, and boreal forest health in AOSR are all issues requiring close attention. Thus, the Wood Buffalo Environmental Association (WBEA) (www.wbea.org) has been monitoring air quality and the terrestrial environment since 1997. The airshed that WBEA is responsible for monitoring is one of eight in Alberta and is the largest at some 68,454 km^2.

19.2   SUMMARY OF BOOK CONTENT

Currently, there are 15 WBEA air quality monitoring stations measuring air quality using continuous and time-integrated techniques (Chapter 4). WBEA also operates 22 other locations in the boreal forest accessible only by a helicopter that are used for passive monthly sampling of five gaseous air pollutants (http://www.wbea.org/monitoring-stations-aamp-data). These data are quality controlled and assured as per standard operating protocols of WBEA, and the combined guidelines of Alberta Environment, Environment Canada, and the U.S. EPA.

In 2011, maximum 1-h SO_2 concentrations ranged from 56 to 122 ppb at compliance/attribution stations and 12 to 83 ppb at community stations. There were no exceedances of the Alberta Ambient Air Quality Objectives (AAAQO) or primary U.S. EPA National Ambient Air Quality Standard (NAAQS) for SO_2 in 2011. Maximum 1-h NO_2 concentrations ranged from 52 to 154 ppb at compliance/attribution stations and 42 to 66 ppb at community stations. There were no exceedances of the 1-h AAAQO for NO_2 (159 ppb/1 h) or the primary NAAQS. Maximum 1-h O_3 concentrations measured at community stations ranged from 77 to 89 ppb. In 2011, there were 15 exceedances of the 82 ppb/1 h AAAQO for O_3, all in the period of intense forest fire activity. In 2011, ambient fine particulate matter ($PM_{2.5}$) levels were highly influenced by the heavy particulate loading from forest fire smoke, resulting in periods of extremely reduced visibility. Maximum 1-h $PM_{2.5}$ concentrations ranged from 406 to 451 $\mu g\ m^{-3}$. There were 97 exceedances (smoke from the forest fires) of the 24-h AAAQO for $PM_{2.5}$ of 30 $\mu g\ m^{-3}$. Beta attenuation technology being evaluated in 2011 for deployment by WBEA measured short-term concentrations as high as 900 $\mu g\ m^{-3}$. Ninety-four exceedances (97% of total) of $PM_{2.5}$ occurred during the fire event. Maximum 1-h H_2S/TRS (total reduced sulfur) concentrations ranged from 6 to 98 ppb at compliance/attribution stations and from 3 to 7 ppb at community stations. Maximum 24-h concentrations ranged from 1 to 23 ppb at compliance/attribution stations and from 1 to 3 ppb at community related stations. In 2011, there were 23 exceedances of the H_2S/TRS 1-h AAAQO. However, this was 74% less than in 2010 and 89% less than in 2009. There were no exceedances in 2011 at community stations.

Time-integrated air quality measurements included canister sampling for 60 VOC (volatile organic compounds) and 20 RSC (reduced sulfur compounds) at nine air-monitoring stations. Most frequently measured VOCs were benzene, toluene, acetone, butane, isopentane, isobutane, *m,p*-xylene, 2-methylpentane, hexane, pentane, and ethylbenzene. Mean ambient 24-h sampling concentrations for these VOCs ranged from 3.37 (acetone) to 0.68 ppb (benzene), to 0.20 ppb (*m,p*-xylene). The maximum 24-h benzene concentration recorded in 2011 was 8.69 ppb. There is no 24-h AAAQO for benzene. The 24-h average AAAQO for toluene and xylene are 106 and 161 ppb, respectively.

Most frequently measured RSC were carbonyl sulfide, carbon disulfide, H_2S, and dimethyl disulfide. Twenty-three polycyclic aromatic hydrocarbons (PAHs) species were routinely measured in ambient air at WBEA community stations. Phenanthrene (4.35 ng m^{-3}), fluorene (1.04 ng m^{-3}), fluoranthene (0.84 ng m^{-3}), pyrene (0.74 ng m^{-3}), acenapthylene (0.46 ng m^{-3}), anthracene (0.42 ng m^{-3}), and acenapthene (0.25 ng m^{-3}) had the highest mean concentrations. Mean (± 1 SD) benzo(*a*)pyrene concentration in 2011 was 0.06 ± 0.05 ng m^{-3} (Chapter 4).

In the context of environmental accumulation, PAHs were also measured in an epiphytic lichen species (*Hypogymniaphysodes*), a bioindicator, at 20 locations within the AOSR representing major air pollutant dispersion and deposition patterns (Chapter 17). PAHs are of concern due to their direct health effects and their effects through the food chain. Total PAHs (20 compounds) ranged from 52 to 350 µg/kg, comparable to values reported from other studies of PAHs in lichens. Naphthalene, phenanthrene, and fluoranthene were the key species measured. Principal component analysis incorporating analytical data from a variety of matrices indicated multiple factors contributed to PAH content.

Measurements of metals in the 20 lichen samples showed significant correlations ($r^2 > 0.8$, $p < 0.05$) between crustal element levels and total PAHs. That is very important finding and confirms other studies reported in this volume that mining, fugitive dust, and coarse particle deposition govern the receptors within ~ 20 km from the source complex, after which deposition patterns are significantly less. Additional data on PAHs could be combined into receptor modeling, so that their spatial profiles and sources can be well defined.

With regard to other potential air quality issues, between October 21, 2010 and May 31, 2011, Environment Canada operated a gaseous mercury analyzer in at the WBEA Patricia McInnes (AMS 6) air monitoring station in Fort McMurray (Chapter 4). Air quality observations in 2011 were dominated by a massive forest fire complex north of Fort McMurray that consumed some 700,000 ha and burned to within a few kilometers of the community of Fort McKay. Excluding data collected during the forest fire period, average ambient total gaseous mercury (TGM) concentration was 1.40 ± 0.15 ng m^{-3}. This value is similar to average TGM concentrations measured elsewhere in Canada.

Concerning two other important issues, frequency distribution of data presented on O_3 and $PM_{2.5}$ was influenced by occasional higher than the frequently occurring normal values as indicated by published literature (see the Special Issue of the journal Atmospheric Environment 2003, on NARSTO). Air quality health index (AQHI) values calculated for the four reporting WBEA air monitoring stations indicated that air quality, according to the Environment/Health Canada index and metric, posed a low risk to health between 96% and 99.3% of the time in 2011. AQHI values presenting a high health risk (0.9–1.3% of data available, 8760 h per station) all occurred during the forest fire smoke episodes.

While exceedance of the odor perception-based AAAQO for H_2S/TRS (10 ppb 1 h; 3 ppb 24 h) has decreased significantly in recent years, odors remain a concern in some WBEA-monitored communities. Therefore, odor-related compounds in the air are being characterized by WBEA using state-of-the art technology, pneumatic focusing gas chromatography, and mass spectrometry (MS) (Chapter 6). In 2011, RSC were present at concentrations below 100 parts-per-trillion during odor episodes at the community station where measurements are conducted. Confirmed compounds included carbonyl sulfide, carbon disulfide, 2-methylthiophene, 3-methylthiophene, 2-ethylthiophene, 2,5-dimethylthiophene, and 2,4-dimethylthiophene.

Ammonia (NH_3) in the air represents a key mechanism for the formation of regional scale $PM_{2.5}$ sulfate and nitrate aerosols that can result in loss of visibility due to their deliquescence, their ability for causing the refraction of sunlight, and also their ability to form clouds and subsequent precipitation. Therefore, ammonia concentrations were measured at two locations within AOSR. The 98th percentile concentrations at both sites were below detection indicating very infrequent measurements of NH_3 concentrations in the ambient air. There was no exceedance of the 2000 ppb (2.0 ppm), 1 h NH_3 AAAQO (Chapter 4).

In order to examine regional scale issues, deterministic or atmospheric dispersion models are routinely used by environmental managers and government regulators, as a tool to estimate the transport, transformation, and deposition of atmospheric pollutants. The ability of these models (e.g., CALPUFF, CMAQ, ISC3, AERMOD) to reliably simulate the fate of emitted pollutants on the spatial scales of interest in the AOSR is highly dependent on the (1) quality of emission inventory data, (2) completeness of the chemical kinetics module, (3) accuracy and resolution of the underlying gridded meteorological fields, (4) topography, and (5) proper parameterization of gas/particle interaction and wet and dry deposition phenomena. Davies (Chapter 12) indicated that, in practice, it is extremely difficult to accurately model air pollution in remote areas such as the AOSR where nonpoint mobile sources, fugitive sources, batch processes, and forest fires are significant emission sources and also where few local meteorological measurements are available for four-dimensional data assimilation in order to "nudge" the underlying meteorological drivers.

In a WBEA case study (Chapter 12), CALPUFF was applied to investigate the accuracy of model predictions. The sulfur and nitrogen contents in lichen tissues collected at 359 sites located up to 150 km from the primary emission source region were compared to sulfur and nitrogen deposition predictions. Both the model predictions and the lichen measurements indicate that the main air quality footprint is within 20 km of the main emission sources. The primary influence of the SO_2 sources tended to be within a nominal 20 km radius where the deposition was greater than 5 kg S/ha/yr. The influence of the sources decreased from 5 kg/ha/yr at 20 km to about 3 kg S/ha/yr at 50 km. At a distance of 100 km, the deposition converged to the background value of about 2.2 kg S/ha/yr.

There was a clear trend for decreasing deposition with increasing distance from the main NO_X emission sources (Chapter 12). The primary influence of the NO_X sources tended to be within a nominal 20 km radius where the deposition was greater than 4 kg N/ha/yr. The influence of the sources decreased from 4 kg/ha/yr at 20 km to about 2 kg N/ha/yr at 50 km. At a distance of 100 km, the deposition converged to the background value of about 1.8 kg N/ha/yr. As with the sulfur content, the relative scatter of the nitrogen content measured in the lichens at 359 sites was greater than that associated with the predicted nitrogen deposition. The scatter represented contributions from other smaller emitting sources, upwind/downwind influences, meteorological influences, and terrain influences. Additional influences were due to one or more of the following: site elevation, terrain slope, terrain aspect, vegetation type, canopy edge location, canopy closure, and natural variability.

Total N deposition to a boreal forest can have either, no measurable effect on ecosystem processes, a stimulatory and/or an inhibitory effect as determined under the toxicological theory of *hormesis* (Jäger and Krupa, 2009). Nitrogen deposited by wet/dry mechanisms can (1) alter understory species composition favoring nitrophilous species, (2) act as a fertilizer to speedup growth, possibly followed in extreme cases by ecosystem collapse from general nutrient imbalances (e.g., plant C/N ratio), (3) or act in soil acidification, cause nutrient imbalances (i.e., N saturation/leaching), possibly leading to eutrophication of adjacent aquatic systems. The aerial extents where the N deposition is predicted to exceed the indicated deposition loadings are as follows: 256 km^2 at 10 kg N/ha/yr, 433 km^2 at 8 kg N/ha/yr, and 1348 km^2 at 5kgN/ha/yr. These predicted values are based on the application of the CALPUFF model as determined by Davies (Figure 12.8, Chapter 12). Building upon Aherne (2008, 2011), Clair et al. (2011) mapped the 5th percentile of empirical critical loads of nutrient nitrogen for terrestrial (forest) ecosystems. The authors showed that the boreal forest in the mid-northern AOSR has nitrogen critical loads in the 400–700 eq/ha/yr (ca. 4–10 kg N/yr as $NH_4 + NO_3$) range. Forests south of the mid-AOSR and upper ca. 20% of forests in northern Saskatchewan have critical loads in the 700–1000 eq/ha/yr (ca. >10 kg N/ha/yr) range.

In comparison to air quality models, *receptor models* provide another approach to understanding the impacts of air pollution sources since the model results are based on measurement data at receptor or sampling locations. Receptor models quantify the impact of air emission sources retrospectively by using advanced mathematical methods on a matrix of elements or compounds in atmospheric samples, or bioindicators, as tracers for the presence of materials from specific sources (Gordon, 1985; Hopke, 1985, 2009). The goal of receptor modeling is to apportion the sources into specific identifiable categories (e.g., combustion, refining, motor vehicles, incineration, metals smelting) and quantify their relative importance. Receptor models can also be used to constrain the uncertainty in deterministic modeling estimates and help identify sources that may not be accurately represented in emission inventories.

The main objectives (Chapter 13) of the overall effort were to (1) identify the major source types of air pollution in the AOSR, (2) collect and analyze air samples to develop chemical source profiles (finger prints), (3) conduct a quantitative source apportionment analysis to determine the major source types impacting the atmospheric deposition and accumulation of potentially phytotoxic levels of S and N in the terrestrial vegetation, and (4) provide supportive data for Forest Health Monitoring (Chapter 9).

Because of logistics considerations (lack of ready access by roads or electrical power), an epiphytic lichen *Hypogymniaphysodes* growing on jack pine (Pinusbanksiana Lamb.) and black spruce (*Piceamariana* Mill. B.S.P.) was used as a bioindicator (accumulator of sulfur, nitrogen, trace metals, and organic compounds) of relative air quality. *Hypogymniaphysodes* is prevalent throughout AOSR and is the most studied epiphytic lichen—air quality bioindicator in the northern hemisphere (Berryman et al., 2010). Guderian (1977) observed that both *Hypogymniaphysodes* and needles of higher plant hosts such as pine and spruce accumulate sulfur under similar exposure conditions. This is an important finding since overload of sulfur in the plant foliage can result in reduced carbon assimilation, growth, and productivity (Legge and Krupa, 2002). Thus, bioindicators and their elemental accumulation can serve as sentinels of ecosystem function and health.

Lichen samples collected during 2008 using a nested grid, radiating up to 150 km from the AOSR industrial complex, were analyzed for total sulfur (S) and nitrogen (N) at 359 sites. The results showed that there were locations where lichen S and N levels or their ratios were high compared to others. Guderian (1977) found in controlled SO_2 exposures that the accumulation of S in conifer needles was governed by the pollutant concentration while the accumulation in *Hypogymnia* was influenced by the exposure duration (chronic exposure).

Nevertheless, there is a need to provide a mechanistic reasoning for the observed results of the spatial patterns of high levels of S and N in the lichens. Therefore, using the 2008 database, a subset of some 121 appropriate locations were selected, and the corresponding samples were analyzed for detailed elemental (43 metals) concentrations using dynamic reaction cell—inductively coupled plasma (ICP)—MS. Of all the metal data, 28 element concentrations were used in the modeling, because their signal-to-noise ratio was >2. These data were then used to develop spatially resolved emission source apportionment through the use of the state-of-the-art receptor modeling (Chapter 18). This approach is expected to be very relevant to any future emission control strategies.

Because of the importance of the issue, several different approaches were used in receptor modeling (see Chapter 18): (1) principal component analysis (PCA), (2) PCA coupled with multilinear regression, (3) chemical mass balance, (4) positive matrix factorization (PMF), and (5) Unmix. Each of the methods has advantages and limitations. Importantly, the results from those multiple approaches agree, providing much credence to the results of the present study.

Some 89% of the variability in the overall data was explained by the results. The PMF (one most used) based on its detailed results showed that sources significantly contributing to concentrations of elements in the lichen tissue included combustion processes ($\sim$23%), tailing sand ($\sim$19%), haul roads and limestone ($\sim$15%), oil sand and processed materials ($\sim$15%), and a general anthropogenic urban source ($\sim$15%). It is important to note that much of the measurable atmospheric deposition is occurring within the first $\sim$20 km of the emission sources (coarse particles). These results have been further supported by the studies on stable lead and mercury isotopes in *Hypogymnia* in AOSR (Chapters 15 and 16).

The WBEA results showed that Pb stable isotope ratios were a better predictor of the extent of the source impacts than the element concentrations because the Pb isotope ratios were not affected by either the metabolic processing of elements by the lichens or by moisture-related controls on atmospheric deposition processes at the collection sites (Chapter 15). In contrast, concentrations of Hg in the lichen were comparable to background values measured in previous studies from remote areas and were far below values observed near significant atmospheric industrial sources of Hg. Spatial patterns provide no evidence for a significant atmospheric point source of Hg from the oil sands developments and Hg accumulation actually decreases in lichens within 25 km of the AOSR development (Chapter 16).

A key question that naturally follows is what do the reported air quality characterizations mean with regard to effects on terrestrial ecosystems? Ecosystem changes are reflected by modifications of energy (carbon) and water (hydrology) flow within (Karnosky et al., 2003; Kickert et al., 1999). Accumulation of S, N, and other chemical constituents in the soil or in vegetation does not necessarily mean adverse effects on the entire ecosystem (Bell and Treshow, 2002). Elemental concentrations in a given receptor will have to exceed its normal range of requirement of essential elements for maintenance and growth, such as S, N, and many trace metals and its capacity to tolerate any excess before negative effects can be measured. The WBEA Acid Deposition Monitoring Program (ADMP) was established at 13 forested sites in 1996–1997, with intensive measurements made in 1998 and 2004. By 2007, there was evidence of increased elemental concentrations in plant foliage with increasing predicted deposition levels. However, there was no evidence of a negative effect on nutrient cycling processes or on forest productivity.

WBEA subsequently enhanced its ADMP with a more pollution-pattern-oriented and ecosystem-based *Forest Health Monitoring* (FHM) Program (Chapter 9). As part of FHM, based on quantitative vegetation composition-cover-relative frequency-stand structure, soil conditions, bedrock and surficial parent material types, mesoclimatic conditions, and inferred microclimatic conditions, 12 new ecologically analogous sites (Chapter 10) are now integrated within the FHM network. At 23 forested sites, a range of above- and below-ground indicators were measured in 2011 for the purposes of detection, evaluation, and cause-effect linkage. Future studies at the sites should involve N species, so that one can better explain any changes in the needle biomass. More importantly, changes in the needle biomass will need to be interpreted within the context of any changes in tree growth and productivity. New solar-powered forest tower infrastructure is being deployed to eight forest health sites to provide continuous meteorological and soil measurements in the boreal forest in order to account for the influence of interannual changes in climate (Chapter 9).

In the past, with isolated sources in Alberta, stable sulfur isotopes (e.g., ^{34}S) could be used as a tracer of air emissions and its effects on the associated forests (see, e.g., Krupa and Legge, 1998). So, such an approach was explored in AOSR. Here, both stable isotopes of sulfur (^{34}S) and nitrogen (^{15}N) were examined in samples of $PM_{2.5}$, bulk and throughfall, soil solution, the lichen *Everniamesomorpha*, and jack pine (*Pinusbanksiana*) needles. $\delta^{15}N$ and $\delta^{34}S$ values of conifer needles from samples taken within the forest stands did not reveal clear evidence of industrial impact (Chapter 11). Individual sources with the AOSR industrial complex did not have distinct S and N source signatures or fingerprints of their relative stable isotope ratios. Therefore, it is recognized that any future studies should consider the use of stable isotopes of one or two trace metals identified as source signatures in the receptor modeling studies (Chapter 18).

19.3   SYMPOSIUM PANEL DISCUSSION

19.3.1   The Panel

As outlined in the Introduction (Percy, Preface, this volume), the WBEA (www.wbea.org) hosted the International Symposium *Alberta Oil Sands: Energy, Industry and the Environment* on May 23, 2011 in Fort McMurray, Alberta, Canada. The content for Chapters 1–5, 7, 8, and 12–18 in this volume of the same title was drawn from presentations made at the Symposium. The content for Chapters 6 and 9–11 in this volume is drawn from associated projects also funded by WBEA as part of its strategic science enhancement begun in 2008.

Immediately following the Symposium scientific presentations, a panel was convened to discuss, in general terms, key messages coming from the scientific presentations just completed. Four invited panel members were drawn from academia, industry, and an environmental nongovernmental organization. The panel chair introduced the session and emphasized the necessity for strong science in support of policy decisions being made by regulators in the AOSR. The chair also highlighted the role that WBEA has played through its science enhancement in addressing the requirement for good science. Also stated was the important role that the multistakeholder process has in the air quality management model adopted by Alberta under the Clean Air Strategic Alliance (CASA, 2012) during the 1990s. In this process, representatives from government, industry, and nongovernment organizations use consensus to actively participate in a comprehensive air quality management system. Airsheds are defined on the basis of emissions sources and volumes, dispersion characteristics, impacts, and administrative characteristics. WBEA is one of eight current Alberta airshed zone partners in CASA. This model is unique to Canada at the scale in which it operates.

The four panel members were then introduced by the chair and invited to speak to the presentations made. The first panelist reiterated the chair's remarks and emphasized the absolute requirement for "good policy" to be based on "good science." As a former senior policy maker in the United States, he complimented the multistakeholder, consensus-based model being used in the AOSR, and exemplified in WBEA. The panelist described how other jurisdictions have been struggling to seek ways to inject better science into policy, in part, achieved through greater "scientific literacy." Following on, one point of emphasis made was that modeling without measurement is inappropriate. The panelist concluded by stating that what was manifest at the Symposium was a "remarkable" concentration of government, industry, and nongovernmental partners coming together on monitoring.

The second panelist supported previous comments made. He stated that the Symposium had demonstrated a successful model of how to move from compliance monitoring to embrace a broader array of scientific monitoring and measurement projects that could serve to bridge the gap between measurement and modeling through knowledge transfer. The panelist stated one challenge for the AOSR is to move toward *expected case* modeling scenarios, as opposed to the current *worst case* approach. He then stated the case for applying more the scientific risk analysis approach to risk management in the AOSR. A better balancing between the proportions contributed by actual drivers of the cumulative risk was called for.

The third panelist echoed previous positive comments made to WBEA and Symposium organizers on stakeholder interaction and the scope of work presented.

As a leading academic with manifest influence through science into regulatory policy, he spoke on personal international experience in "leapfrogging opportunities for air quality improvement" through the introduction of new science into the regulatory process. The panelist further stated that what was evident in the CASA–WBEA model might be applied in other countries, such as India and China. In conclusion, he reiterated the prerequisite for in place, on the ground air quality and receptor measurement as the preferred method to link air quality with effects.

The fourth panelist remarked upon the good science, and its wide scope as presented at the Symposium. He supported its utility in making informed policy decisions, but questioned how the science was to be implemented going forward in the AOSR. The panelist firmly supported its use in cumulative effects management and environmental impact assessment processes. It was important to him that science drive environmental management. The panelist expressed concern that the scope and scale of environmental monitoring was behind the pace of industrial development, and that the predicted 3.5 times (from 1995) production levels might be attained before the stated target date of 2025. He concluded by stating that the "cutting edge" monitoring and measurements presented at the Symposium should have been in place many years ago.

19.3.2   Discussion

Prior to inviting questions for the panelist from Symposium attendees, the Panel chair stated that it was ". . . essential for all stakeholders to be part of the solution" by getting out ahead of the issues as much as possible. Among questions raised or statements made by symposium attendees were those related to the context surrounding the content presented and the opinions voiced by the panelists.

The requirement for science informing good policy and regulation was again strongly emphasized. From a First Nations (native people) perspective, it was stated that "western" science was only one component of public science policy. Aboriginal peoples have a different context for what science can tell us. The Aboriginal people believe that one does not put things into the ground; what is left (i.e., sulfur recovered prior to emission to air) when industry leaves and others remain? Context was highlighted as being very important. A comment was made from a community standpoint, that the AAAQO might not be fully protective.

A panelist remarked that industry does vigorously engage with First Nations and stated that the process was one of the better ones in Canada. The application of sulfur capture technology was said to be part of the balance between applied science and risk assessment, often requiring huge investment in pollution abatement. Regulators (government) are in fact provided appropriate information, and it is they who ultimately make the risk assessments.

Another panelist commented on the perceived conflict between "providing information and transparency." Using the example of the shale gas fracturing debate, he remarked that the cost of not making data on chemical processes public has greatly exceeded the cost of doing so at an early stage. Not working with the public was seen to result in huge costs in the end.

In the case of the AOSR, industrial procedures were stated as having undergone a long, slow evolution. The standard set now is that of open sharing of technology between oil sands operators. Much information is also shared with multistakeholder associations like WBEA and researchers to enable the work, as was manifest in a number of presentations at the Symposium. The big challenge was stated as tailings management, although considerable technological process has recently been made.

The subject of accounting for the total "cost" of effects due to oil sands operations was raised. It was recommended that industry do so through internalization of total costs. A panelist answered by stating the difficulty of coming up with the true valuation. How does one value the costs of poor visibility when it occurs? Another panelist mentioned the *Stern Report on the Economics of Climate Change* (http://www.hm-treasury.gov.uk/stern_review_report.htm) that has developed a costing for dealing with the case of carbon dioxide emissions. While Alberta benefits from the oil sands, it also bears some of the biggest costs. Another panelist thought that this question was beyond the scope of the Symposium.

19.4 FUTURE PERSPECTIVES

Since 2003, annual global oil and condensate production has ranged between 72 and 74 Mb/d. Global energy demand is expected to grow by 39% by 2030. By 2030, BP (www.BP.com) predicts that coal, oil, and gas together will contribute to meeting most of this demand. The Canadian Oil Sands deposits lay under 142,220 km^2 of Boreal Forest, and they comprise the third largest oil reserve in the world, with 170 Bb recoverable using today's technology. To date, some 751 km^2 of the boreal forest overlying the Oil Sands have been disturbed by operations (Chapter 2). Contributing some 1.5 Mb/d in 2012, by 2030 this is expected to have increased by 2 Mb/d. This level of production will by then comprise 12% of the global production increase expected to have occurred by 2030.

Large-scale industrial development activity inevitably brings the challenge of managing it sustainably. In recent years, the development in the AOSR has come under intense public scrutiny. The results we have summarized in this final chapter in "Alberta Oil Sands: Energy, Industry and the Environment" come from WBEA *practical science* projects conducted as part of the WBEA strategic science enhancement begun in 2008. Measuring and monitoring the environment at key points along the air pollutant pathway, these chapters provide original scientific data on emissions, transport, air quality, deposition, and source contributions to terrestrial receptors. This information will be used by decision makers on airshed management, contribute new knowledge to support environmental impact assessments, inform stakeholders and the public on air quality, provide guidance on what has been learned since 2008.

What is still needed, then, to achieve a more holistic evaluation of the role of oil sands development on the air and terrestrial environments? To an as yet undetermined extent, the enhanced level of research and measurements proposed in the February 2, 2012 Joint Canada/Alberta Implementation Plan for Oil Sands Monitoring (www.ec.gc.ca; www.environment.alberta.ca) will address some of this need. Nevertheless, what has unquestionably emerged from WBEA's 4 years of practical science is the absolute prerequisite for an on-the-ground, long-term integrated suite of measurements that by design link the air and terrestrial systems. These systems must then be linked with the aquatic systems.

It is certainly intuitive that concentrations of trace elements we have measured in terrestrial receptors in the AOSR reflected proximity to oil sands mining and processing operations, influenced to an extent by eco-site variability at the lichen collection sites. Elemental groupings have then been linked back to source origin. AOSR anthropogenic contributions have been delineated from AOSR "natural" (e.g., exposed sand/bitumen, forest fires) or urban contributions. Concentration ranges for sulfur, nitrogen, trace metals, and PAHs in lichens within the AOSR have been shown to be consistent with those obtained by other investigators in other areas. Together with deposition tracer studies using stable isotopes and regional dispersion modeling for sulfur and nitrogen (dry + wet), a zone of enrichment above background deposition exists extending out some 20–25 km from the main oil sands mining and processing operations.

Previous investigators have focused on the nearer-to-source, Athabasca River Valley-influenced transport of emissions from fixed sources such as stacks. This first scientific investigation integrating air and land systems through source–receptor modeling has unequivocally demonstrated through convergence of five models, each holding quite different assumptions, that the largest impact on elemental concentrations in the ubiquitous terrestrial receptor *Hypogymniaphysodes* in the AOSR is, in fact, related to the resuspension into the atmosphere of coarse particulate, or dust. This conclusion suggests that implementation of a fugitive dust abatement strategy could minimize the near-field impact of future mining-related production activities.

Going forward, it is recommended that in order to continue to inform residents, stakeholders, and regulators in the AOSR on the potential impact of bitumen production on their communities and natural forest resources, a continued commitment be made, as demonstrated since 2008 in WBEA's considerable science enhancement. This commitment should adhere to the key factors identified from retrospective analysis of monitoring systems that have been shown to be responsible for the successful implementation of monitoring over the long term (Percy, 2002; Percy and Ferretti, 2004). Such *practical scientific monitoring and measurement* must to be supported by (1) continuity in funding, (2) an international, multidisciplinary team of distinguished scientists, (3) a robust and responsive air quality monitoring program that has an increased emphasis on speciation and quantification of particulate matter,

(4) a fully developed, infrastructure-equipped detection, evaluation, and reporting network of plots to assess the state of forest health, and (5) the combination of vegetation receptor biomonitoring and the application of atmospheric receptor models including source characterization to fully integrate air and terrestrial systems in time and space in the changing landscape.

REFERENCES

Aherne, J., 2008. Calculating critical loads of acid deposition for forest soils in Alberta: Final report: critical load, exceedance and limitations. Report to the Canadian Council of Ministers of the Environment (CCME). 14 pp.

Aherne, J., 2011. Uncertainty in critical load exceedance [UNCLE]: critical loads uncertainty and risk analysis for Canadian forest ecosystems. Report to Canadian Council of Ministers of the Environment (CCME), 19 pp.

Bell, J.N.B., Treshow, M. (Eds.), 2002. Air Pollution and Plant Life. second ed. John Wiley & Sons Ltd, Chichester, UK.

Berryman, S., Straker, J., Krupa, S., Davies, M., VerHoef, J., Brenner, G., 2010. Mapping the characteristics of air pollutant deposition patterns in the Athabasca Oil Sands Region using epiphytic lichens as bio-indicators. Interim Report Submitted to the Wood Buffalo Environmental Association, Fort McMurray, AB, Canada.

Clair, T.A., Dennis, I., Jeffries, D., Aherne, J., Wolniewicz, 2011. In: Aherne, J., Dennis, D., Jeffries, D., Clair T. (Eds.), Environment Canada Submission From The National Focus Centre to the Co-ordination Centre for Effects Under the UN ECE Convention on LRTAP, pp. 83–86.

Clean Air Strategic Alliance, 2012. Airshed zones. http://www.casahome.org/Partners/Airshed-Zones.aspx. Accessed August 31

Gordon, G.E., 1985. Receptor models. Environ. Sci. Technol. 22, 1132–1142.

Guderian, R., 1977. Air Pollution: Phyto-Toxicity of Acid Gases and Its Significance in Air Pollution Control. Springer-Verlag, Berlin, Germany 122 pp + color plates.

Hopke, P.K., 1985. Receptor Modeling in Environmental Chemistry. John Wiley & Sons, Hoboken, NJ, USA.

Hopke, P.K., 2009. Theory and application of atmospheric source apportionment. In: Legge, A.H. (Ed.), Air Quality and Ecological Impacts. Elsevier, Amsterdam, The Netherlands.

Jäger, H.-J., Krupa, S.V., 2009. Hormesis—its relevance in phytotoxicology. In: Legge, A.H. (Ed.), Air Quality and Ecological Impacts: Relating Sources to Effects. Elsevier, Amsterdam, The Netherlands, pp. 137–152.

Karnosky, D.F., Percy, K.E., Chappelka, A.H., Simpson, C., Pikkarainen, J. (Eds.), 2003. Air Pollution, Global Change and Forests in the New Millenium. Elsevier Science, Amsterdam, The Netherlands.

Kickert, R.N., Tonella, G., Simonov, A., Krupa, S.V., 1999. Predictive modeling of effects under global change. Environ. Pollut. 100, 87–132.

Krupa, S.V., Legge, A.H., 1998. Sulfur dioxide, particulate sulfur and their impacts on a boreal forest ecosystem. In: Ambasht, R.S. (Ed.), Modern Trends in Ecology and Environment. Backhuys Publishers, Leiden, The Netherlands, pp. 285–306.

Legge, A.H., Krupa, S.V., 2002. Effects of sulfur dioxide. In: Bell, J.N.B., Treshow, M. (Eds.), Air Pollution and Plant Life. second ed. John Wiley & Sons Ltd., Chichester, UK, pp. 135–162.

Percy, K.E., 2002. Is air pollution an important factor in international forest health? In: Szaro, R.C., Bytnerowicz, A., Oszlanyi, J. (Eds.), Effects of Air Pollution on Forest Health and Biodiversity in Forests of the Carpathian Mountains. IOS Press, Amsterdam, The Netherlands, pp. 23–42.

Percy, K.E., Ferretti, M., 2004. Air pollution and forest health: toward new monitoring concepts. Environ. Pollut. 130 (1), 113–126.

Index

Note: Page numbers followed by "*f*" indicate figures, and "*t*" indicate tables.